MW01715539

Free! PDF version of this costbook!

Just call us at **1.888.264.2665** to confirm your purchase* and we will send you your digital costbook. You can search for individual cost items and have your cost data available everywhere you go.

Free 30-Day trial! Simpl•Est Unit Price

The online resource that makes it simple

With **Simpl•Est Unit Price** you get a digital copy of the Costbook and you can use it to create a complete, detailed estimate just by clicking on the individual items. Everything is automatically saved for you in the Cloud and you can download them to Excel and adjust quantities, add comments, modify labor rates — update your entire estimate, all with a click of a mouse.

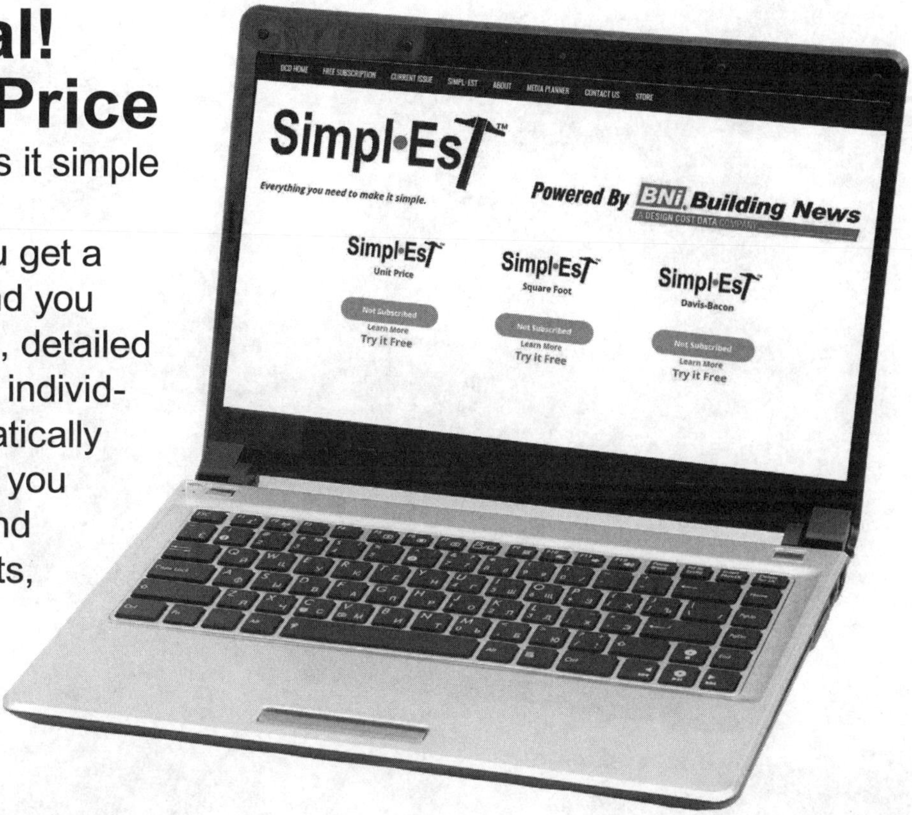

*You can start your 30-day free trial of **Simpl•Est Unit Price** by going to www.dcd.com/simpl_est (free registration on dcd.com is required).*

SPECIAL BONUS for buyers of this costbook

What's more if you call us* to confirm your purchase of this costbook, we will extend your access to **Simpl•Est Unit Price** for a full year (a $154.95 value!), plus we will give you unlimited free access to ***Simpl•Est Davis-Bacon***. Updated weekly, ***Simpl•Est Davis-Bacon*** gives you the latest prevailing wage rates for commercial construction in over 300 regions of the country.

*Call us at **1.888.264.2665** *(M-F between 9 am and 7 pm Eastern time)*

31ST EDITION

BNi Building News
A DESIGN COST DATA COMPANY
DATA YOU CAN TRUST

GENERAL CONSTRUCTION
2021 COSTBOOK

CSI MasterFormat®

BNi Building News
A DESIGN COST DATA COMPANY
DATA YOU CAN TRUST

EDITOR-IN-CHIEF

William D. Mahoney, P.E.

TECHNICAL SERVICES

Tony De Augustine
Joan Hamilton
Eric Mahoney, AIA
Ana Varela

GRAPHIC DESIGN

Robert O. Wright Jr.

BNi Publications, Inc.

VISTA
990 PARK CENTER DRIVE, SUITE E
VISTA, CA 92081

1-888-BNI-BOOK (1-888-264-2665)
www.bnibooks.com

ISBN 978-1-58855-193-1

Copyright © 2020 by BNi Publications, Inc. All rights reserved. Printed in the United States of America. Except as permitted under the United States Copyright Act of 1976, no part of this publication may be reproduced or distributed in any form or by any means, or stored in a data base or retrieval system, without the prior written permission of the publisher.

While diligent effort is made to provide reliable, accurate and up-to-date information, neither BNi Publications Inc., nor its authors or editors, can place a guarantee on the correctness of the data or information contained in this book. BNi Publications Inc., and its authors and editors, do hereby disclaim any responsibility or liability in connection with the use of this book or of any data or other information contained therein.

Table of Contents

Preface

For over 74 years, BNi Building News has been dedicated to providing construction professionals with timely and reliable information. Based on this experience, our staff has researched and compiled thousands of up-to-the-minute costs for the **BNi Costbooks**. This book is an essential reference for contractors, engineers, architects, facility managers — any construction professional who must provide an estimate for any type of building project.

Whether working up a preliminary estimate or submitting a formal bid, the costs listed here can be quickly and easily tailored to your needs. All costs are based on prevailing labor rates. Overhead and profit should be included in all costs. Man-hours are also provided.

All data is categorized according to the CSI division format. This industry standard provides an all-inclusive checklist to ensure that no element of a project is overlooked. In addition, to make specific items even easier to locate, there is a complete alphabetical index.

The "Features of this Book" section presents a clear overview of the many features of this book. Included is an explanation of the data, sample page layout and discussion of how to best use the information in the book.

Of course, all buildings and construction projects are unique. The information provided in this book is based on averages from well-managed projects with good labor productivity under normal working conditions (eight hours a day). Other circumstances affecting costs, such as overtime, unusual working conditions, savings from buying bulk quantities for large projects, and unusual or hidden costs, must be factored in as they arise.

The data provided in this book is for estimating purposes only. Check all applicable federal, state and local codes and regulations for local requirements.

Format

All data is categorized according to the ***CSI MASTERFORMAT***. This industry standard provides an all-inclusive checklist to ensure that no element of a project is overlooked.

DIVISION 00 ...PROCUREMENT & CONTRACTING REQUIREMENTS

00 10 00 SOLICITATION
00 20 00 INSTRUCTIONS FOR PROCUREMENT
00 30 00 AVAILABLE INFORMATION
00 40 00 PROCUREMENT FORMS AND SUPPLEMENTS
00 50 00 CONTRACTING FORMS AND SUPPLEMENTS
00 60 00 PROJECT FORMS
00 70 00 CONDITIONS OF THE CONTRACT
00 80 00 Reserved
00 90 00 REVISIONS, CLARIFICATIONS, AND MODIFICATIONS

DIVISION 01 .... GENERAL REQUIREMENTS

01 10 00 SUMMARY
01 20 00 PRICE AND PAYMENT PROCEDURES
01 30 00 ADMINISTRATIVE REQUIREMENTS
01 40 00 QUALITY REQUIREMENTS
01 50 00 TEMPORARY FACILITIES AND CONTROLS
01 60 00 PRODUCT REQUIREMENTS
01 70 00 EXECUTION AND CLOSEOUT REQUIREMENTS
01 80 00 PERFORMANCE REQUIREMENTS
01 90 00 LIFE CYCLE ACTIVITIES

DIVISION 02 .... EXISTING CONDITIONS

02 30 00 SUBSURFACE INVESTIGATION
02 40 00 DEMOLITION AND STRUCTURE MOVING
02 50 00 SITE REMEDIATION
02 60 00 CONTAMINATED SITE MATERIAL REMOVAL
02 70 00 WATER REMEDIATION
02 80 00 FACILITY REMEDIATION

DIVISION 03 .... CONCRETE

03 10 00 CONCRETE FORMING AND ACCESSORIES
03 20 00 CONCRETE REINFORCING
03 30 00 CAST-IN-PLACE CONCRETE
03 40 00 PRECAST CONCRETE
03 50 00 CAST DECKS AND UNDERLAYMENT
03 60 00 GROUTING
03 70 00 MASS CONCRETE
03 80 00 CONCRETE CUTTING AND BORING

DIVISION 04 .... MASONRY

04 20 00 UNIT MASONRY
04 30 00 Reserved
04 40 00 STONE ASSEMBLIES
04 50 00 REFRACTORY MASONRY
04 60 00 CORROSION-RESISTANT MASONRY
04 70 00 MANUFACTURED MASONRY
04 80 00 Reserved
04 90 00 Reserved

DIVISION 05.. METALS

05 10 00 STRUCTURAL METAL FRAMING
05 12 00 STRUCTURAL STEEL FRAMING
05 20 00 METAL JOISTS
05 30 00 METAL DECKING
05 40 00 COLD-FORMED METAL FRAMING
05 50 00 METAL FABRICATIONS
05 60 00 Reserved
05 70 00 DECORATIVE METAL
05 80 00 Reserved
05 90 00 Reserved

DIVISION 06........................... WOOD, PLASTICS, AND COMPOSITES

06 10 00 ROUGH CARPENTRY
06 30 00 Reserved
06 40 00 ARCHITECTURAL WOODWORK
06 50 00 STRUCTURAL PLASTICS
06 60 00 PLASTIC FABRICATIONS
06 70 00 STRUCTURAL COMPOSITES
06 80 00 COMPOSITE FABRICATIONS
06 90 00 Reserved

DIVISION 07....................... THERMAL AND MOISTURE PROTECTION

07 10 00 DAMPPROOFING AND WATERPROOFING
07 30 00 STEEP SLOPE ROOFING
07 40 00 ROOFING AND SIDING PANELS
07 50 00 MEMBRANE ROOFING
07 60 00 FLASHING AND SHEET METAL
07 70 00 ROOF AND WALL SPECIALTIES AND ACCESSORIES
07 80 00 FIRE AND SMOKE PROTECTION
07 90 00 JOINT PROTECTION

DIVISION 08.. OPENINGS

08 10 00 DOORS AND FRAMES
08 20 00 Reserved
08 30 00 SPECIALTY DOORS AND FRAMES
08 40 00 ENTRANCES, STOREFRONTS, AND CURTAIN WALLS
08 50 00 WINDOWS
08 60 00 ROOF WINDOWS AND SKYLIGHTS
08 70 00 HARDWARE
08 80 00 GLAZING
08 90 00 LOUVERS AND VENTS

DIVISION 09.. FINISHES

09 20 00 PLASTER AND GYPSUM BOARD
09 30 00 TILING
09 40 00 Reserved
09 50 00 CEILINGS
09 60 00 FLOORING
09 70 00 WALL FINISHES
09 80 00 ACOUSTIC TREATMENT
09 90 00 PAINTING AND COATING

Format *(Continued)*

DIVISION 10.. SPECIALTIES

10 20 00 INTERIOR SPECIALTIES
10 30 00 FIREPLACES AND STOVES
10 40 00 SAFETY SPECIALTIES
10 50 00 STORAGE SPECIALTIES
10 60 00 Reserved
10 70 00 EXTERIOR SPECIALTIES
10 80 00 OTHER SPECIALTIES

DIVISION 11.. EQUIPMENT

11 10 00 VEHICLE AND PEDESTRIAN EQUIPMENT
11 15 00 SECURITY, DETENTION AND BANKING EQUIPMENT
11 20 00 COMMERCIAL EQUIPMENT
11 30 00 RESIDENTIAL EQUIPMENT
11 40 00 FOODSERVICE EQUIPMENT
11 50 00 EDUCATIONAL AND SCIENTIFIC EQUIPMENT
11 60 00 ENTERTAINMENT EQUIPMENT
11 65 00 ATHLETIC AND RECREATIONAL EQUIPMENT
11 80 00 COLLECTION AND DISPOSAL EQUIPMENT
11 90 00 OTHER EQUIPMENT

DIVISION 12.. FURNISHINGS

12 10 00 ART
12 20 00 WINDOW TREATMENTS
12 30 00 CASEWORK
12 40 00 FURNISHINGS AND ACCESSORIES
12 50 00 FURNITURE
12 60 00 MULTIPLE SEATING
12 70 00 Reserved
12 80 00 Reserved
12 90 00 OTHER FURNISHINGS

DIVISION 13.. SPECIAL CONSTRUCTION

13 10 00 SPECIAL FACILITY COMPONENTS
13 20 00 SPECIAL PURPOSE ROOMS
13 30 00 SPECIAL STRUCTURES
13 40 00 INTEGRATED CONSTRUCTION
13 50 00 SPECIAL INSTRUMENTATION

DIVISION 14.. CONVEYING EQUIPMENT

14 10 00 DUMBWAITERS
14 30 00 ESCALATORS AND MOVING WALKS
14 40 00 LIFTS
14 70 00 TURNTABLES
14 80 00 SCAFFOLDING
14 90 00 OTHER CONVEYING EQUIPMENT

DIVISION 21.. FIRE SUPPRESSION

21 10 00 WATER-BASED FIRE-SUPPRESSION SYSTEMS
21 20 00 FIRE-EXTINGUISHING SYSTEMS
21 30 00 FIRE PUMPS
21 40 00 FIRE-SUPPRESSION WATER STORAGE
21 50 00 Reserved
21 60 00 Reserved
21 70 00 Reserved
21 80 00 Reserved
21 90 00 Reserved

DIVISION 22.. PLUMBING

22 10 00 PLUMBING PIPING AND PUMPS
22 30 00 PLUMBING EQUIPMENT
22 40 00 PLUMBING FIXTURES
22 50 00 POOL AND FOUNTAIN PLUMBING SYSTEMS
22 60 00 GAS AND VACUUM SYSTEMS FOR LABORATORY AND HEALTHCARE FACILITIES

DIVISION 23........ HEATING, VENTILATING, AND AIR-COND. (HVAC)

23 10 00 FACILITY FUEL SYSTEMS
23 20 00 HVAC PIPING AND PUMPS
23 30 00 HVAC AIR DISTRIBUTION
23 40 00 HVAC AIR CLEANING DEVICES
23 50 00 CENTRAL HEATING EQUIPMENT
23 60 00 CENTRAL COOLING EQUIPMENT
23 70 00 CENTRAL HVAC EQUIPMENT
23 80 00 DECENTRALIZED HVAC EQUIPMENT
23 90 00 Reserved

DIVISION 25.. INTEGRATED AUTOMATION

25 10 00 INTEGRATED AUTOMATION NETWORK EQUIPMENT
25 20 00 Reserved
25 30 00 INTEGRATED AUTOMATION INSTRUMENTATION AND TERMINAL DEVICES
25 40 00 Reserved
25 50 00 INTEGRATED AUTOMATION FACILITY CONTROLS
25 60 00 Reserved
25 70 00 Reserved
25 80 00 Reserved
25 90 00 INTEGRATED AUTOMATION CONTROL SEQUENCES

DIVISION 26.. ELECTRICAL

26 10 00 MEDIUM-VOLTAGE ELECTRICAL DISTRIBUTION
26 20 00 LOW-VOLTAGE ELECTRICAL TRANSMISSION
26 30 00 FACILITY ELEC. POWER GENERATING & STORING EQUIP
26 40 00 ELECTRICAL AND CATHODIC PROTECTION
26 50 00 LIGHTING
26 60 00 Reserved
26 70 00 Reserved
26 80 00 Reserved
26 90 00 Reserved

DIVISION 27.. COMMUNICATIONS

27 10 00 STRUCTURED CABLING
27 20 00 DATA COMMUNICATIONS
27 30 00 VOICE COMMUNICATIONS
27 40 00 AUDIO-VIDEO COMMUNICATIONS
27 50 00 DISTRIBUTED COMMUNICATIONS AND MONITORING SYSTEMS
27 60 00 Reserved
27 70 00 Reserved
27 80 00 Reserved
27 90 00 Reserved

DIVISION 28............................ ELECTRONIC SAFETY AND SECURITY

28 10 00 ELEC. ACCESS CONTROL & INTRUSION DETECTION
28 20 00 ELECTRONIC SURVEILLANCE
28 30 00 ELECTRONIC DETECTION AND ALARM
28 40 00 ELECTRONIC MONITORING AND CONTROL
28 50 00 Reserved
28 60 00 Reserved
28 70 00 Reserved
28 80 00 Reserved
28 90 00 Reserved

DIVISION 31.. EARTHWORK

31 10 00 SITE CLEARING
31 20 00 EARTH MOVING
31 30 00 EARTHWORK METHODS
31 40 00 SHORING AND UNDERPINNING
31 50 00 EXCAVATION SUPPORT AND PROTECTION
31 60 00 SPECIAL FOUNDATIONS AND LOADBEARING ELEMENTS
31 70 00 TUNNELING AND MINING
31 80 00 Reserved
31 90 00 Reserved

Format *(Continued)*

DIVISION 32 .... EXTERIOR IMPROVEMENTS

32 10 00 BASES, BALLASTS, AND PAVING
32 20 00 Reserved
32 30 00 SITE IMPROVEMENTS
32 40 00 Reserved
32 50 00 Reserved
32 60 00 Reserved
32 70 00 WETLANDS
32 80 00 IRRIGATION
32 90 00 PLANTING

DIVISION 33 .... UTILITIES

33 10 00 WATER UTILITIES
33 20 00 WELLS
33 30 00 SANITARY SEWERAGE UTILITIES
33 40 00 STORM DRAINAGE UTILITIES
33 50 00 FUEL DISTRIBUTION UTILITIES
33 60 00 HYDRONIC AND STEAM ENERGY UTILITIES
33 70 00 ELECTRICAL UTILITIES
33 80 00 COMMUNICATIONS UTILITIES
33 90 00 Reserved

DIVISION 34 .... TRANSPORTATION

34 10 00 GUIDEWAYS/RAILWAYS
34 20 00 TRACTION POWER
34 30 00 Reserved
34 40 00 TRANSPORTATION SIGNALING AND CONTROL EQUIP
34 50 00 TRANSPORTATION FARE COLLECTION
34 60 00 Reserved
34 70 00 TRANSPORTATION CONSTRUCTION AND EQUIPMENT
34 80 00 BRIDGES
34 90 00 Reserved

DIVISION 35 ....WATERWAY AND MARINE CONSTRUCTION

35 10 00 WATERWAY AND MARINE SIGNALING AND CONTROL EQUIP
35 20 00 WATERWAY AND MARINE CONSTRUCTION AND EQUIP
35 30 00 COASTAL CONSTRUCTION
35 40 00 WATERWAY CONSTRUCTION AND EQUIPMENT
35 50 00 MARINE CONSTRUCTION AND EQUIPMENT
35 60 00 Reserved
35 70 00 DAM CONSTRUCTION AND EQUIPMENT
35 80 00 Reserved
35 90 00 Reserved

DIVISION 40 .... PROCESS INTEGRATION

40 10 00 GAS AND VAPOR PROCESS PIPING
40 20 00 LIQUIDS PROCESS PIPING
40 30 00 SOLID AND MIXED MATERIALS PIPING AND CHUTES
40 40 00 PROCESS PIPING AND EQUIPMENT PROTECTION
40 50 00 Reserved
40 60 00 Reserved
40 70 00 Reserved
40 80 00 COMMISSIONING OF PROCESS SYSTEMS
40 90 00 INSTRUMENTATION & CONTROL FOR PROC. SYSTEMS

DIVISION 41 ...MATERIAL PROCESSING & HANDLING EQUIPMENT

41 10 00 BULK MATERIAL PROCESSING EQUIPMENT
41 20 00 PIECE MATERIAL HANDLING EQUIPMENT
41 30 00 MANUFACTURING EQUIPMENT
41 40 00 CONTAINER PROCESSING AND PACKAGING
41 50 00 MATERIAL STORAGE
41 60 00 MOBILE PLANT EQUIPMENT
41 70 00 Reserved
41 80 00 Reserved
41 90 00 Reserved

DIVISION 42....... PROCESS HEATING, COOLING, & DRYING EQUIP.

42 10 00 PROCESS HEATING EQUIPMENT
42 20 00 PROCESS COOLING EQUIPMENT
42 30 00 PROCESS DRYING EQUIPMENT
42 40 00 Reserved
42 50 00 Reserved
42 60 00 Reserved
42 70 00 Reserved
42 80 00 Reserved
42 90 00 Reserved

DIVISION 43....... PROCESS GAS & LIQUID HDLG, PURIFICATION, & STORAGE EQUIPMENT

43 10 00 GAS HANDLING EQUIPMENT
43 20 00 LIQUID HANDLING EQUIPMENT
43 30 00 GAS AND LIQUID PURIFICATION EQUIPMENT
43 40 00 GAS AND LIQUID STORAGE
43 50 00 Reserved
43 60 00 Reserved
43 70 00 Reserved
43 80 00 Reserved
43 90 00 Reserved

DIVISION 44................................ POLLUTION CONTROL EQUIPMENT

44 10 00 AIR POLLUTION CONTROL
44 20 00 NOISE POLLUTION CONTROL
44 40 00 WATER TREATMENT EQUIPMENT
44 50 00 SOLID WASTE CONTROL
44 60 00 Reserved
44 70 00 Reserved
44 80 00 Reserved
44 90 00 Reserved

DIVISION 45. INDUSTRY-SPECIFIC MANUFACTURING EQUIPMENT

45 60 00 Reserved
45 70 00 Reserved
45 80 00 Reserved
45 90 00 Reserved

DIVISION 48................................ELECTRICAL POWER GENERATION

48 10 00 ELECTRICAL POWER GENERATION EQUIPMENT
48 20 00 Reserved
48 30 00 Reserved
48 40 00 Reserved
48 50 00 Reserved
48 60 00 Reserved
48 70 00 ELECTRICAL POWER GENERATION TESTING
48 80 00 Reserved
48 90 00 Reserved

BNi Building News

A DESIGN COST DATA COMPANY

DATA YOU CAN TRUST

Features of this Book

Sample pages with graphic explanations are included before the Costbook pages. These explanations, along with the discussions below, will provide a good understanding of what is included in this book and how it can best be used in construction estimating.

Material Costs

The material costs used in this book represent national averages for prices that a contractor would expect to pay plus an allowance for freight (if applicable) and handling and storage. These costs reflect neither the lowest or highest prices, but rather a typical average cost over time. Periodic fluctuations in availability and in certain commodities (e.g. copper, conduit) can significantly affect local material pricing. In the final estimating and bidding stages of a project when the highest degree of accuracy is required, it is best to check local, current prices.

Labor Costs

Labor costs include the basic wage, plus commonly applicable taxes, insurance and markups for overhead and profit. The labor rates used here to develop the costs are typical average prevailing wage rates. Rates for different trades are used where appropriate for each type of work.

Fixed government rates and average allowances for taxes and insurance are included in the labor costs. These include employer-paid Social Security/Medicare taxes (FICA), Worker's Compensation insurance, state and federal unemployment taxes, and business insurance.

Please note, however, most of these items vary significantly from state to state and within states. For more specific data, local agencies and sources should be consulted.

Man-Hours

These productivities represent typical installation labor for thousands of construction items. The data takes into account all activities involved in normal construction under commonly experienced working conditions such as site movement, material handling, start-up, etc.

Equipment Costs

Costs for various types and pieces of equipment are included in Division 1 - General Requirements and can be included in an estimate when required either as a total "Equipment" category or with specific appropriate trades. Costs for equipment are included when appropriate in the installation costs in the Costbook pages.

Overhead and Profit

Included in the labor costs are allowances for overhead and profit for the contractor/employer whose workers are performing the specific tasks. No cost allowances or fees are included for management of subcontractors by the general contractor or construction manager. These costs, where appropriate, must be added to the costs as listed in the book.

The allowance for overhead is included to account for office overhead, the contractors' typical costs of doing business. These costs normally include in-house office staff salaries and benefits, office rent and operating expenses, professional fees, vehicle costs and other operating costs which are not directly applicable to specific jobs. It should be noted for this book that office overhead as included should be distinguished from project overhead, the General Requirements (Division 1) which are specific to particular projects. Project overhead should be included on an item by item basis for each job.

Depending on the trade, an allowance of 10-15 percent is incorporated into the labor/installation costs to account for typical profit of the installing contractor. See Division 1, General Requirements, for a more detailed review of typical profit allowances.

Features of this Book *(Continued)*

Adjustments to Costs

The costs as presented in this book attempt to represent national averages. Costs, however, vary among regions, states and even between adjacent localities.

In order to more closely approximate the probable costs for specific locations throughout the U.S., a table of Geographic Multipliers is provided. These adjustment factors are used to modify costs obtained from this book to help account for regional variations of construction costs. Whenever local current costs are known, whether material or equipment prices or labor rates, they should be used if more accuracy is required.

Editor's Note: This **Costbook** is intended to provide accurate, reliable, average costs and typical productivities for thousands of common construction components. The data is developed and compiled from various industry sources, including government, manufacturers, suppliers and working professionals. The intent of the information is to provide assistance and guidelines to construction professionals in estimating. The user should be aware that local conditions, material and labor availability and cost variations, economic considerations, weather, local codes and regulations, etc., all affect the actual cost of construction. These and other such factors must be considered and incorporated into any and all construction estimates.

Sample Costbook Page

In order to best use the information in this book, please review this sample page and read the "Features In This Book" section.

Division

GENERAL CONSTRUCTION COSTBOOK — *03 CONCRETE*

Broadscope Category (First 2 Digits)

PLACING CONCRETE

Description		Output		Unit Costs			
ID Code	Component Descriptions	Unit of Meas.	Manhr / Unit	Material Cost	Labor Cost	Equipment Cost	Total Cost
03 - 31006	**GRADE BEAM CONCRETE**						**03 - 31006**
0960	Grade beam						
0980	2500# or 3000# concrete						
1000	By chute	CY	0.267	130	17.00		150
1020	By crane	"	0.686	130	43.75	30.00	200
1040	By pump	"	0.600	130	38.25	26.25	200
1060	By hand buggy	"	0.800	130	51.00		180
1070	3500# or 4000# concrete						
1080	By chute	CY	0.267	140	17.00		160
1100	By crane	"	0.686	140	43.75	30.00	210
1120	By pump	"	0.600	140	38.25	26.25	200
1140	By hand buggy	"	0.800	140	51.00		190
1150	5000# concrete						
1160	By chute	CY	0.267	150	17.00		170
1180	By crane	"	0.686	150	43.75	30.00	220
1190	By pump	"	0.600	150	38.25	26.25	210
1200	By hand buggy	"	0.800	150	51.00		200
03 - 31007	**PILE CAP CONCRETE**						**03 - 31007**
0970	Pile cap						
0980	2500# or 3000 concrete						
1000	By chute	CY	0.267	140	17.00		160
1005	By crane	"	0.800	140	51.00	35.00	230
1010	By pump	"	0.686	140	43.75	30.00	210
1020	By hand buggy	"	0.800	140	51.00		190
1980	3500# or 4000# concrete						
2000	By chute	CY	0.267	140	17.00		160
2010	By crane	"	0.800	140	51.00	35.00	230
2020	By pump	"	0.686	140	43.75	30.00	210
2040	By hand buggy	"	0.800	140	51.00		190
3980	5000# concrete						
4010	By chute	CY	0.267	150	17.00		170
4015	By crane	"	0.800	150	51.00	35.00	240
4020	By pump	"	0.686	150	43.75	30.00	220
4030	By hand buggy	"	0.800	150	51.00		200

Mediumscope Category (5 Digits)

Detailed Descriptions

Complete descriptions of items may include information listed above a particular line. Review of the whole category is recommended for a complete description.

Labor Cost

Labor cost represents U.S. prevailing wages plus applicable fringes.

Material Cost

Material cost represents average contractor prices plus an allowance for freight, handling and storage.

Equipment Cost

This cost includes equipment costs only, the wages for the crew operating the equipment are included in the Labor column.

Total Cost

The total cost is the sum of material and installation costs. This total represents typical contractors' costs including overhead and profit, but does not include markups for the general contractor or construction management fees.

Unit of Measurement

Each item (and cost) is defined in terms of the common estimating unit. All costs are listed in dollars per unit.

BNi Building News

A DESIGN COST DATA COMPANY

DATA YOU CAN TRUST

DIVISION 01
GENERAL

REQUIREMENTS

	Description	Output		Unit Costs			
ID Code	Component Descriptions	Unit of Meas.	Manhr / Unit	Material Cost	Labor Cost	Equipment Cost	Total Cost
01 - 21001	**ALLOWANCES**						**01 - 21001**
0090	Overhead						
1000	$20,000 project						
1020	Minimum	PCT					15.00
1040	Average	"					20.00
1060	Maximum	"					40.00
1080	$100,000 project						
1100	Minimum	PCT					12.00
1120	Average	"					15.00
1140	Maximum	"					25.00
1160	$500,000 project						
1170	Minimum	PCT					10.00
1180	Average	"					12.00
1200	Maximum	"					20.00
1220	$1,000,000 project						
1240	Minimum	PCT					6.00
1260	Average	"					10.00
1280	Maximum	"					12.00
1380	$10,000,000 project						
1385	Minimum	PCT					1.50
1386	Average	"					5.00
1388	Maximum	"					8.00
1480	Profit						
1500	$20,000 project						
1520	Minimum	PCT					10.00
1540	Average	"					15.00
1560	Maximum	"					25.00
1580	$100,000 project						
1600	Minimum	PCT					10.00
1620	Average	"					12.00
1640	Maximum	"					20.00
1660	$500,000 project						
1680	Minimum	PCT					5.00
1700	Average	"					10.00
1720	Maximum	"					15.00
1740	$1,000,000 project						
1760	Minimum	PCT					3.00
1780	Average	"					8.00
1800	Maximum	"					15.00

© 2020 BNi Publications Inc. ALL RIGHTS RESERVED

REQUIREMENTS

	Description	Output		Unit Costs			
ID Code	Component Descriptions	Unit of Meas.	Manhr / Unit	Material Cost	Labor Cost	Equipment Cost	Total Cost
01 - 21001	**ALLOWANCES, Cont'd...**						**01 - 21001**
2000	Professional fees						
2100	Architectural						
2120	$100,000 project						
2140	Minimum	PCT					5.00
2160	Average	"					10.00
2180	Maximum	"					20.00
2200	$500,000 project						
2220	Minimum	PCT					5.00
2240	Average	"					8.00
2260	Maximum	"					12.00
2280	$1,000,000 project						
2300	Minimum	PCT					3.50
2320	Average	"					7.00
2340	Maximum	"					10.00
2360	Structural engineering						
2380	Minimum	PCT					2.00
2400	Average	"					3.00
2420	Maximum	"					5.00
2440	Mechanical engineering						
2460	Minimum	PCT					4.00
2480	Average	"					5.00
2500	Maximum	"					15.00
2520	Electrical engineering						
2540	Minimum	PCT					3.00
2560	Average	"					5.00
2580	Maximum	"					12.00
4080	Taxes						
5000	Sales tax						
5020	Minimum	PCT					4.00
5040	Average	"					5.00
5060	Maximum	"					10.00
5080	Unemployment						
5100	Minimum	PCT					3.00
5120	Average	"					6.50
5140	Maximum	"					8.00
5200	Social security (FICA)	"					7.85

© 2020 BNi Publications Inc. ALL RIGHTS RESERVED

PROJECT MANAGEMENT AND COORDINATION

Description		Output		Unit Costs			
ID Code	Component Descriptions	Unit of Meas.	Manhr / Unit	Material Cost	Labor Cost	Equipment Cost	Total Cost
01 - 31130	FIELD STAFF						01 - 31130
1000	Superintendent						
1020	Minimum	YEAR					102,115
1040	Average	"					127,662
1060	Maximum	"					153,348
1080	Field engineer						
1100	Minimum	YEAR					100,718
1120	Average	"					115,601
1140	Maximum	"					132,858
1160	Foreman						
1180	Minimum	YEAR					67,872
1200	Average	"					108,544
1220	Maximum	"					127,084
1240	Bookkeeper/timekeeper						
1260	Minimum	YEAR					39,261
1280	Average	"					51,270
1300	Maximum	"					66,333
1320	Watchman						
1340	Minimum	YEAR					29,253
1360	Average	"					39,132
1380	Maximum	"					49,397

CONSTRUCTION PROGRESS DOCUMENTATION

ID Code	Component Descriptions	Unit of Meas.	Manhr / Unit	Material Cost	Labor Cost	Equipment Cost	Total Cost
01 - 32130	SCHEDULING						01 - 32130
0090	Scheduling for						
1000	$100,000 project						
1020	Minimum	PCT					1.02
1040	Average	"					2.05
1060	Maximum	"					5.12
1080	$500,000 project						
1100	Minimum	PCT					0.51
1120	Average	"					1.02
1140	Maximum	"					2.05
1160	$1,000,000 project						
1180	Minimum	PCT					0.34
1200	Average	"					0.77
1220	Maximum	"					1.53
4000	Scheduling software						

© 2020 BNi Publications Inc. ALL RIGHTS RESERVED

CONSTRUCTION PROGRESS DOCUMENTATION

	Description	Output		Unit Costs			
ID Code	Component Descriptions	Unit of Meas.	Manhr / Unit	Material Cost	Labor Cost	Equipment Cost	Total Cost
01 - 32130	**SCHEDULING, Cont'd...**						**01 - 32130**
4020	Minimum	EA					690
4040	Average	"					3,970
4060	Maximum	"					79,410
01 - 32230	**SURVEYING**						**01 - 32230**
0080	Surveying						
1000	Small crew	DAY					1,010
1020	Average crew	"					1,520
1040	Large crew	"					2,000
2000	Lot lines and boundaries						
2020	Minimum	ACRE					730
2040	Average	"					1,520
2060	Maximum	"					2,480
01 - 32330	**JOB REQUIREMENTS**						**01 - 32330**
1000	Job photographs, small jobs						
1020	Minimum	EA					130
1040	Average	"					200
1060	Maximum	"					460
1080	Large projects						
1100	Minimum	EA					660
1120	Average	"					990
1140	Maximum	"					3,310

QUALITY CONTROL

01 - 45230	**TESTING**						**01 - 45230**
1080	Testing concrete, per test						
1100	Minimum	EA					23.50
1120	Average	"					39.25
1140	Maximum	"					78.00
1160	Soil, per test						
1180	Minimum	EA					48.00
1200	Average	"					120
1220	Maximum	"					320
1240	Welding, per test						
1260	Minimum	EA					23.50
1280	Average	"					39.25
1300	Maximum	"					160

© 2020 BNi Publications Inc. ALL RIGHTS RESERVED

CONSTRUCTION FACILITIES

ID Code	Description: Component Descriptions	Output: Unit of Meas.	Output: Manhr / Unit	Unit Costs: Material Cost	Unit Costs: Labor Cost	Unit Costs: Equipment Cost	Unit Costs: Total Cost
01 - 52190	**SANITARY FACILITIES**						**01 - 52190**
0010	Porta Potty						
0100	Rental per Day minimum	EA					70.00
0120	Rental per Day Maximum	"					180
0140	Rental per Month minimum	"					200
0160	Rental per Month Maximum	"					350
0180	Specialized, ADA						
0200	Rental per Day minimum	EA					120
0220	Rental per Day Max	"					180
0240	Rental per Month, minimum	"					350
0260	Rental per Month, Maximum	"					580
0280	Purchase prices, minimum	"					820

CONSTRUCTION AIDS

ID Code	Component Descriptions	Unit of Meas.	Manhr / Unit	Material Cost	Labor Cost	Equipment Cost	Total Cost
01 - 54001	**CONSTRUCTION AIDS**						**01 - 54001**
1000	Scaffolding/staging, rent per month						
1020	Measured by lineal feet of base						
1040	10' high	LF					14.50
1060	20' high	"					26.50
1080	30' high	"					37.00
1100	40' high	"					42.75
1120	50' high	"					51.00
1140	Measured by square foot of surface						
1160	Minimum	SF					0.64
1180	Average	"					1.11
1200	Maximum	"					1.99
1220	Safety nets, heavy duty, per job						
1240	Minimum	SF					0.43
1260	Average	"					0.52
1280	Maximum	"					1.14
1300	Tarpaulins, fabric, per job						
1320	Minimum	SF					0.30
1340	Average	"					0.51
1360	Maximum	"					1.32

© 2020 BNi Publications Inc. ALL RIGHTS RESERVED

CONSTRUCTION AIDS

	Description	Output		Unit Costs			
ID Code	Component Descriptions	Unit of Meas.	Manhr / Unit	Material Cost	Labor Cost	Equipment Cost	Total Cost
01 - 54008	**MOBILIZATION**						**01 - 54008**
1000	Equipment mobilization						
1020	Bulldozer						
1040	Minimum	EA					220
1060	Average	"					460
1080	Maximum	"					780
1100	Backhoe/front-end loader						
1120	Minimum	EA					130
1140	Average	"					230
1160	Maximum	"					510
1180	Crane, crawler type						
1200	Minimum	EA					2,440
1220	Average	"					6,000
1240	Maximum	"					12,880
1260	Truck crane						
1280	Minimum	EA					560
1300	Average	"					860
1320	Maximum	"					1,490
1340	Pile driving rig						
1360	Minimum	EA					11,100
1380	Average	"					22,210
1400	Maximum	"					39,970
01 - 54009	**EQUIPMENT**						**01 - 54009**
0080	Air compressor						
1000	60 cfm						
1020	By day	EA					110
1030	By week	"					320
1040	By month	"					980
1100	300 cfm						
1120	By day	EA					230
1130	By week	"					700
1140	By month	"					2,140
1200	600 cfm						
1210	By day	EA					620
1220	By week	"					1,860
1230	By month	"					5,630
1300	Air tools, per compressor, per day						
1310	Minimum	EA					44.50

© 2020 BNi Publications Inc. ALL RIGHTS RESERVED

CONSTRUCTION AIDS

Description		Output		Unit Costs			
ID Code	Component Descriptions	Unit of Meas.	Manhr / Unit	Material Cost	Labor Cost	Equipment Cost	Total Cost
01 - 54009	**EQUIPMENT, Cont'd...**						**01 - 54009**
1320	Average	EA					56.00
1330	Maximum	"					78.00
1400	Generators, 5 kw						
1410	By day	EA					110
1420	By week	"					330
1430	By month	"					1,020
1500	Heaters, salamander type, per week						
1510	Minimum	EA					130
1520	Average	"					190
1530	Maximum	"					400
1600	Pumps, submersible						
1605	50 gpm						
1610	By day	EA					89.00
1620	By week	"					270
1630	By month	"					800
1640	100 gpm						
1650	By day	EA					110
1660	By week	"					330
1670	By month	"					1,000
1675	500 gpm						
1680	By day	EA					180
1690	By week	"					530
1700	By month	"					1,600
1900	Diaphragm pump, by week						
1920	Minimum	EA					160
1930	Average	"					270
1940	Maximum	"					550
2000	Pickup truck						
2020	By day	EA					170
2030	By week	"					490
2040	By month	"					1,510
2080	Dump truck						
2100	6 c.y. truck						
2120	By day	EA					440
2130	By week	"					1,330
2140	By month	"					4,010
2160	10 c.y. truck						
2170	By day	EA					550

© 2020 BNi Publications Inc. ALL RIGHTS RESERVED

CONSTRUCTION AIDS

ID Code	Description Component Descriptions	Output Unit of Meas.	Manhr / Unit	Unit Costs Material Cost	Labor Cost	Equipment Cost	Total Cost
01 - 54009	**EQUIPMENT, Cont'd...**						**01 - 54009**
2180	By week	EA					1,670
2190	By month	"					5,010
2300	16 c.y. truck						
2310	By day	EA					890
2320	By week	"					2,670
2340	By month	"					8,020
2400	Backhoe, track mounted						
2420	1/2 c.y. capacity						
2430	By day	EA					910
2440	By week	"					2,780
2450	By month	"					8,240
2500	1 c.y. capacity						
2510	By day	EA					1,450
2520	By week	"					4,340
2530	By month	"					13,030
2550	2 c.y. capacity						
2560	By day	EA					2,450
2570	By week	"					7,350
2580	By month	"					22,050
2600	3 c.y. capacity						
2620	By day	EA					4,680
2640	By week	"					14,030
2680	By month	"					42,100
3000	Backhoe/loader, rubber tired						
3005	1/2 c.y. capacity						
3010	By day	EA					550
3020	By week	"					1,670
3030	By month	"					5,010
3035	3/4 c.y. capacity						
3040	By day	EA					670
3050	By week	"					2,000
3060	By month	"					6,010
3200	Bulldozer						
3205	75 hp						
3210	By day	EA					780
3220	By week	"					2,340
3230	By month	"					7,020
3280	200 hp						

© 2020 BNi Publications Inc. ALL RIGHTS RESERVED

CONSTRUCTION AIDS

	Description	Output		Unit Costs			
ID Code	Component Descriptions	Unit of Meas.	Manhr / Unit	Material Cost	Labor Cost	Equipment Cost	Total Cost
01 - 54009	EQUIPMENT, Cont'd...						01 - 54009
3300	By day	EA					2,230
3310	By week	"					6,680
3320	By month	"					20,050
3330	400 hp						
3340	By day	EA					3,310
3350	By week	"					9,920
3360	By month	"					29,780
4000	Cranes, crawler type						
4005	15 ton capacity						
4010	By day	EA					1,000
4020	By week	"					3,010
4030	By month	"					9,020
4035	25 ton capacity						
4040	By day	EA					1,220
4050	By week	"					3,670
4060	By month	"					11,020
4070	50 ton capacity						
4080	By day	EA					2,230
4090	By week	"					6,680
4100	By month	"					20,050
4110	100 ton capacity						
4120	By day	EA					3,340
4130	By week	"					10,020
4140	By month	"					30,170
4145	Truck mounted, hydraulic						
4150	15 ton capacity						
4160	By day	EA					940
4170	By week	"					2,840
4180	By month	"					8,190
5380	Loader, rubber tired						
5385	1 c.y. capacity						
5390	By day	EA					670
5400	By week	"					2,000
5410	By month	"					6,020
5430	2 c.y. capacity						
5440	By day	EA					1,000
5450	By week	"					3,900
5460	By month	"					11,690

© 2020 BNi Publications Inc. ALL RIGHTS RESERVED

CONSTRUCTION AIDS

ID Code	Description: Component Descriptions	Output: Unit of Meas.	Output: Manhr / Unit	Unit Costs: Material Cost	Unit Costs: Labor Cost	Unit Costs: Equipment Cost	Unit Costs: Total Cost
01 - 54009	**EQUIPMENT, Cont'd...**						**01 - 54009**
5470	3 c.y. capacity						
5480	By day	EA					1,780
5490	By week	"					5,350
5500	By month	"					16,040

TEMPORARY BARRIERS AND ENCLOSURES

ID Code	Component Descriptions	Unit of Meas.	Manhr / Unit	Material Cost	Labor Cost	Equipment Cost	Total Cost
01 - 56230	**TEMPORARY FACILITIES**						**01 - 56230**
1000	Barricades, temporary						
1010	Highway						
1020	Concrete	LF	0.080	14.75	5.10		19.75
1040	Wood	"	0.032	5.10	2.04		7.14
1060	Steel	"	0.027	5.29	1.70		6.99
1090	Pedestrian barricades						
1100	Plywood	SF	0.027	4.54	1.70		6.24
1120	Chain link fence	"	0.027	3.85	1.70		5.55
1130	Trailers, general office type, per month						
2020	Minimum	EA					250
2040	Average	"					410
2060	Maximum	"					820
2070	Crew change trailers, per month						
2100	Minimum	EA					150
2120	Average	"					160
2140	Maximum	"					250

PROJECT IDENTIFICATION

ID Code	Component Descriptions	Unit of Meas.	Manhr / Unit	Material Cost	Labor Cost	Equipment Cost	Total Cost
01 - 58130	**SIGNS**						**01 - 58130**
0080	Construction signs, temporary						
1000	Signs, 2' x 4'						
1020	Minimum	EA					41.75
1040	Average	"					100
1060	Maximum	"					350
1080	Signs, 4' x 8'						
1100	Minimum	EA					88.00
1120	Average	"					230
1140	Maximum	"					980
1160	Signs, 8' x 8'						

© 2020 BNi Publications Inc. ALL RIGHTS RESERVED

PROJECT IDENTIFICATION

ID Code	Description: Component Descriptions	Output: Unit of Meas.	Output: Manhr / Unit	Unit Costs: Material Cost	Unit Costs: Labor Cost	Unit Costs: Equipment Cost	Unit Costs: Total Cost
01 - 58130	**SIGNS, Cont'd...**						**01 - 58130**
1180	Minimum	EA					110
1200	Average	"					350
1220	Maximum	"					3,540

CLOSEOUT SUBMITTALS

ID Code	Component Descriptions	Unit of Meas.	Manhr / Unit	Material Cost	Labor Cost	Equipment Cost	Total Cost
01 - 78330	**BONDS**						**01 - 78330**
1000	Performance bonds						
1020	Minimum	PCT					0.62
1040	Average	"					1.93
1060	Maximum	"					3.07

QUICK COSTS

ID Code	Component Descriptions	Unit of Meas.	Manhr / Unit	Material Cost	Labor Cost	Equipment Cost	Total Cost
01 - 99800	**SQUARE FOOT COSTS**						**01 - 99800**
2000	Fire Suppression						
2010	60-Model National Average Building						
2030	Minimum	SF					1.46
2050	Average	"					3.35
2070	Maximum	"					4.84
2200	Plumbing						
2210	60-Model National Average Building						
2230	Minimum	SF					8.54
2250	Average	"					15.00
2270	Maximum	"					16.25
2500	Rough-in						
2510	Minimum	SF					1.55
2530	Average	"					2.70
2550	Maximum	"					3.93
2600	Fixtures						
2610	Minimum	SF					2.05
2630	Average	"					3.24
2650	Maximum	"					3.45
3000	HVAC						
3010	60-Model National Average Building						
3030	Minimum	SF					19.50
3050	Average	"					27.25
3070	Maximum	"					29.25

© 2020 BNi Publications Inc. ALL RIGHTS RESERVED

QUICK COSTS

ID Code	Description: Component Descriptions	Output: Unit of Meas.	Output: Manhr / Unit	Unit Costs: Material Cost	Unit Costs: Labor Cost	Unit Costs: Equipment Cost	Unit Costs: Total Cost
01 - 99800	**SQUARE FOOT COSTS, Cont'd...**						**01 - 99800**
3200	Controls						
3210	Minimum	SF					3.68
3230	Average	"					4.04
3250	Maximum	"					4.88
3400	Ductwork						
3410	Minimum	SF					2.09
3430	Average	"					4.18
3450	Maximum	"					4.25
4000	Electrical						
4010	60-Model National Average Building						
4030	Minimum	SF					24.25
4050	Average	"					33.50
4070	Maximum	"					50.00
4200	Rough-in						
4210	Minimum	SF					2.38
4230	Average	"					4.26
4250	Maximum	"					8.29
4400	Lighting						
4410	Minimum	SF					3.21
4430	Average	"					7.66
4450	Maximum	"					8.85
4500	Switchgear						
4510	Minimum	SF					1.67
4530	Average	"					2.51
4550	Maximum	"					3.22
5000	Communications						
5010	60-Model National Average Building						
5030	Minimum	SF					1.46
5050	Average	"					4.74
5070	Maximum	"					6.75
5500	Electronic Safety and Security						
5510	60-Model National Average Building						
5530	Minimum	SF					1.65
5550	Average	"					3.21
5570	Maximum	"					3.71

© 2020 BNi Publications Inc. ALL RIGHTS RESERVED

DIVISION 02
SITE CONSTRUCTION

SITE PREPARATION

	Description	Output		Unit Costs			
ID Code	Component Descriptions	Unit of Meas.	Manhr / Unit	Material Cost	Labor Cost	Equipment Cost	Total Cost
02 - 32130	**SOIL BORING**						**02 - 32130**
1000	Borings, uncased, stable earth						
1022	2-1/2" dia.						
1024	Minimum	LF	0.200		12.75	11.00	23.75
1026	Average	"	0.300		19.00	16.50	35.75
1028	Maximum	"	0.480		30.50	26.50	57.00
1042	4" dia.						
1044	Minimum	LF	0.218		13.75	12.00	26.00
1046	Average	"	0.343		21.75	19.00	40.75
1048	Maximum	"	0.600		38.00	33.25	71.00
1500	Cased, including samples						
1522	2-1/2" dia.						
1524	Minimum	LF	0.240		15.25	13.25	28.50
1526	Average	"	0.400		25.25	22.25	47.50
1528	Maximum	"	0.800		51.00	44.25	95.00
1542	4" dia.						
1544	Minimum	LF	0.480		30.50	26.50	57.00
1546	Average	"	0.686		43.50	38.00	81.00
1548	Maximum	"	0.960		61.00	53.00	110
2000	Drilling in rock						
2022	No sampling						
2024	Minimum	LF	0.436		27.75	24.25	52.00
2026	Average	"	0.632		40.00	35.00	75.00
2028	Maximum	"	0.857		54.00	47.50	100
2042	With casing and sampling						
2044	Minimum	LF	0.600		38.00	33.25	71.00
2046	Average	"	0.800		51.00	44.25	95.00
2048	Maximum	"	1.200		76.00	67.00	140
3000	Test pits						
3022	Light soil						
3024	Minimum	EA	3.000		190	170	360
3026	Average	"	4.000		250	220	480
3028	Maximum	"	8.000		510	440	950
3042	Heavy soil						
3044	Minimum	EA	4.800		300	270	570
3046	Average	"	6.000		380	330	710
3048	Maximum	"	12.000		760	670	1,430

© 2020 BNi Publications Inc. ALL RIGHTS RESERVED

SELECTIVE SITE DEMOLITION

ID Code	Description: Component Descriptions	Output: Unit of Meas.	Output: Manhr / Unit	Unit Costs: Material Cost	Unit Costs: Labor Cost	Unit Costs: Equipment Cost	Unit Costs: Total Cost
02 - 41131	**CATCH BASIN / MANHOLE DEMOLITION**						**02 - 41131**
0102	Abandon catch basin or manhole (fill with sand)						
0104	Minimum	EA	3.000		190	170	360
0106	Average	"	4.800		300	270	570
0108	Maximum	"	8.000		510	440	950
0202	Remove and reset frame and cover						
0204	Minimum	EA	1.600		100	89.00	190
0206	Average	"	2.400		150	130	290
0208	Maximum	"	4.000		250	220	480
0280	Remove catch basin, to 10' deep						
0302	Masonry						
0304	Minimum	EA	4.800		300	270	570
0306	Average	"	6.000		380	330	710
0308	Maximum	"	8.000		510	440	950
0402	Concrete						
0403	Minimum	EA	6.000		380	330	710
0404	Average	"	8.000		510	440	950
0406	Maximum	"	9.600		610	530	1,140
02 - 41132	**FENCE DEMOLITION**						**02 - 41132**
0060	Remove fencing						
0080	Chain link, 8' high						
0100	For disposal	LF	0.040		2.55		2.55
0200	For reuse	"	0.100		6.38		6.38
0980	Wood						
1000	4' high	SF	0.027		1.70		1.70
1020	6' high	"	0.032		2.04		2.04
1040	8' high	"	0.040		2.55		2.55
1960	Masonry						
1980	8" thick						
2000	4' high	SF	0.080		5.10		5.10
2020	6' high	"	0.100		6.38		6.38
2040	8' high	"	0.114		7.29		7.29
2050	12" thick						
2060	4' high	SF	0.133		8.51		8.51
2080	6' high	"	0.160		10.25		10.25
2100	8' high	"	0.200		12.75		12.75
2120	12' high	"	0.267		17.00		17.00

© 2020 BNi Publications Inc. ALL RIGHTS RESERVED

SELECTIVE SITE DEMOLITION

02 - 41133 CURB & GUTTER DEMOLITION 02 - 41133

ID Code	Component Descriptions	Unit of Meas.	Manhr / Unit	Material Cost	Labor Cost	Equipment Cost	Total Cost
1000	Curb removal						
1100	Concrete, unreinforced						
1150	Minimum	LF	0.048		3.04	2.66	5.70
1155	Average	"	0.060		3.80	3.32	7.12
1160	Maximum	"	0.075		4.75	4.15	8.90
1200	Reinforced						
1250	Minimum	LF	0.077		4.90	4.29	9.19
1300	Average	"	0.086		5.43	4.75	10.25
1350	Maximum	"	0.096		6.08	5.32	11.50
1450	Combination curb and 2' gutter						
1480	Unreinforced						
1500	Minimum	LF	0.063		4.00	3.50	7.50
1700	Average	"	0.083		5.24	4.58	9.82
1710	Maximum	"	0.120		7.60	6.65	14.25
1720	Reinforced						
1730	Minimum	LF	0.100		6.33	5.54	11.75
2000	Average	"	0.133		8.44	7.38	15.75
2010	Maximum	"	0.240		15.25	13.25	28.50
2020	Granite curb						
2030	Minimum	LF	0.069		4.34	3.80	8.14
2100	Average	"	0.080		5.06	4.43	9.50
2150	Maximum	"	0.092		5.84	5.11	11.00
2160	Asphalt curb						
2170	Minimum	LF	0.040		2.53	2.21	4.75
2180	Average	"	0.048		3.04	2.66	5.70
2190	Maximum	"	0.057		3.62	3.16	6.78

REMOVAL AND SALVAGE

02 - 42132 GUARDRAIL DEMOLITION 02 - 42132

ID Code	Component Descriptions	Unit of Meas.	Manhr / Unit	Material Cost	Labor Cost	Equipment Cost	Total Cost
0080	Remove standard guardrail						
1000	Steel						
1020	Minimum	LF	0.060		3.80	3.32	7.12
1030	Average	"	0.080		5.06	4.43	9.50
1040	Maximum	"	0.120		7.60	6.65	14.25
2000	Wood						
2020	Minimum	LF	0.052		3.30	2.89	6.19
2040	Average	"	0.062		3.89	3.41	7.30

© 2020 BNi Publications Inc. ALL RIGHTS RESERVED

REMOVAL AND SALVAGE

	Description	Output		Unit Costs			
ID Code	Component Descriptions	Unit of Meas.	Manhr / Unit	Material Cost	Labor Cost	Equipment Cost	Total Cost
02 - 42132	**GUARDRAIL DEMOLITION, Cont'd...**						**02 - 42132**
2060	Maximum	LF	0.100		6.33	5.54	11.75
02 - 43133	**HYDRANT DEMOLITION**						**02 - 43133**
1002	Remove fire hydrant						
1004	Minimum	EA	3.000		190	170	360
1006	Average	"	4.000		250	220	480
1008	Maximum	"	6.000		380	330	710
5002	Remove and reset fire hydrant						
5004	Minimum	EA	8.000		510	440	950
5006	Average	"	12.000		760	670	1,430
5008	Maximum	"	24.000		1,520	1,330	2,850

PAVEMENT AND SIDEWALK DEMOLITION

02 - 44134	**PAVEMENT AND SIDEWALK DEMOLITION**						**02 - 44134**
0090	Bituminous pavement, up to 3" thick						
0102	On streets						
0104	Minimum	SY	0.069		4.34	3.80	8.14
0106	Average	"	0.096		6.08	5.32	11.50
0108	Maximum	"	0.160		10.25	8.86	19.00
0202	On pipe trench						
0204	Minimum	SY	0.096		6.08	5.32	11.50
0206	Average	"	0.120		7.60	6.65	14.25
0208	Maximum	"	0.240		15.25	13.25	28.50
0300	Concrete pavement, 6" thick						
0402	No reinforcement						
0404	Minimum	SY	0.120		7.60	6.65	14.25
0406	Average	"	0.160		10.25	8.86	19.00
0408	Maximum	"	0.240		15.25	13.25	28.50
0452	With wire mesh						
0454	Minimum	SY	0.185		11.75	10.25	22.00
0456	Average	"	0.240		15.25	13.25	28.50
0458	Maximum	"	0.300		19.00	16.50	35.75
0552	With rebars						
0554	Minimum	SY	0.240		15.25	13.25	28.50
0556	Average	"	0.300		19.00	16.50	35.75
0558	Maximum	"	0.400		25.25	22.25	47.50
1380	9" thick						
1402	No reinforcement						

© 2020 BNi Publications Inc. ALL RIGHTS RESERVED

PAVEMENT AND SIDEWALK DEMOLITION

	Description	Output		Unit Costs			
ID Code	Component Descriptions	Unit of Meas.	Manhr / Unit	Material Cost	Labor Cost	Equipment Cost	Total Cost
02 - 44134	**PAVEMENT AND SIDEWALK DEMOLITION, Cont'd...**						**02 - 44134**
1404	Minimum	SY	0.160		10.25	8.86	19.00
1406	Average	"	0.200		12.75	11.00	23.75
1408	Maximum	"	0.240		15.25	13.25	28.50
1412	With wire mesh						
1414	Minimum	SY	0.253		16.00	14.00	30.00
1416	Average	"	0.300		19.00	16.50	35.75
1418	Maximum	"	0.369		23.50	20.50	43.75
1422	With rebars						
1424	Minimum	SY	0.320		20.25	17.75	38.00
1426	Average	"	0.400		25.25	22.25	47.50
1428	Maximum	"	0.533		33.75	29.50	63.00
1440	12" thick						
1452	No reinforcement						
1454	Minimum	SY	0.200		12.75	11.00	23.75
1456	Average	"	0.240		15.25	13.25	28.50
1458	Maximum	"	0.300		19.00	16.50	35.75
1462	With wire mesh						
1464	Minimum	SY	0.282		18.00	15.75	33.50
1466	Average	"	0.343		21.75	19.00	40.75
1468	Maximum	"	0.436		27.75	24.25	52.00
1472	With rebars						
1474	Minimum	SY	0.400		25.25	22.25	47.50
1476	Average	"	0.480		30.50	26.50	57.00
1478	Maximum	"	0.600		38.00	33.25	71.00
1502	Sidewalk, 4" thick, with disposal						
1504	Minimum	SY	0.057		3.62	3.16	6.78
1506	Average	"	0.080		5.06	4.43	9.50
1508	Maximum	"	0.114		7.24	6.33	13.50
1802	Removal of pavement markings by waterblasting						
1804	Minimum	SF	0.003		0.21		0.21
1806	Average	"	0.004		0.25		0.25
1808	Maximum	"	0.008		0.51		0.51

© 2020 BNi Publications Inc. ALL RIGHTS RESERVED

DRAINAGE AND PIPING DEMOLITION

ID Code	Description Component Descriptions	Output Unit of Meas.	Manhr / Unit	Unit Costs Material Cost	Labor Cost	Equipment Cost	Total Cost
02 - 45134	**DRAINAGE PIPING DEMOLITION**						**02 - 45134**
1000	Remove drainage pipe, not including excavation						
1020	12" dia.						
1030	Minimum	LF	0.080		5.06	4.43	9.50
1040	Average	"	0.100		6.33	5.54	11.75
1050	Maximum	"	0.126		8.00	7.00	15.00
1100	18" dia.						
1110	Minimum	LF	0.109		6.91	6.04	13.00
1120	Average	"	0.126		8.00	7.00	15.00
1130	Maximum	"	0.160		10.25	8.86	19.00
1200	24" dia.						
1210	Minimum	LF	0.133		8.44	7.38	15.75
1220	Average	"	0.160		10.25	8.86	19.00
1230	Maximum	"	0.200		12.75	11.00	23.75
1300	36" dia.						
1310	Minimum	LF	0.160		10.25	8.86	19.00
1320	Average	"	0.200		12.75	11.00	23.75
1330	Maximum	"	0.253		16.00	14.00	30.00

GAS PIPING DEMOLITION

ID Code	Component Descriptions	Unit of Meas.	Manhr / Unit	Material Cost	Labor Cost	Equipment Cost	Total Cost
02 - 46134	**GAS PIPING DEMOLITION**						**02 - 46134**
0980	Remove welded steel pipe, not including excavation						
1002	4" dia.						
1004	Minimum	LF	0.120		7.60	6.65	14.25
1006	Average	"	0.150		9.50	8.31	17.75
1008	Maximum	"	0.200		12.75	11.00	23.75
2002	5" dia.						
2004	Minimum	LF	0.200		12.75	11.00	23.75
2006	Average	"	0.240		15.25	13.25	28.50
2008	Maximum	"	0.300		19.00	16.50	35.75
2022	6" dia.						
2024	Minimum	LF	0.253		16.00	14.00	30.00
2026	Average	"	0.300		19.00	16.50	35.75
2028	Maximum	"	0.400		25.25	22.25	47.50
2032	8" dia.						
2034	Minimum	LF	0.369		23.50	20.50	43.75
2036	Average	"	0.480		30.50	26.50	57.00
2038	Maximum	"	0.632		40.00	35.00	75.00

© 2020 BNi Publications Inc. ALL RIGHTS RESERVED

GAS PIPING DEMOLITION

	Description	Output		Unit Costs			
ID Code	Component Descriptions	Unit of Meas.	Manhr / Unit	Material Cost	Labor Cost	Equipment Cost	Total Cost
02 - 46134	**GAS PIPING DEMOLITION, Cont'd...**						**02 - 46134**
2042	10" dia.						
2044	Minimum	LF	0.480		30.50	26.50	57.00
2046	Average	"	0.600		38.00	33.25	71.00
2048	Maximum	"	0.800		51.00	44.25	95.00

SANITARY PIPING DEMOLITION

ID Code	Component Descriptions	Unit of Meas.	Manhr / Unit	Material Cost	Labor Cost	Equipment Cost	Total Cost
02 - 47134	**SANITARY PIPING DEMOLITION**						**02 - 47134**
0980	Remove sewer pipe, not including excavation						
1002	4" dia.						
1004	Minimum	LF	0.067		4.22	3.69	7.91
1006	Average	"	0.096		6.08	5.32	11.50
1008	Maximum	"	0.160		10.25	8.86	19.00
1022	6" dia.						
1024	Minimum	LF	0.075		4.75	4.15	8.90
1026	Average	"	0.109		6.91	6.04	13.00
1028	Maximum	"	0.200		12.75	11.00	23.75
1042	8" dia.						
1044	Minimum	LF	0.080		5.06	4.43	9.50
1046	Average	"	0.120		7.60	6.65	14.25
1048	Maximum	"	0.240		15.25	13.25	28.50
1062	10" dia.						
1064	Minimum	LF	0.086		5.43	4.75	10.25
1066	Average	"	0.126		8.00	7.00	15.00
1068	Maximum	"	0.267		17.00	14.75	31.75
1082	12" dia.						
1084	Minimum	LF	0.092		5.84	5.11	11.00
1086	Average	"	0.133		8.44	7.38	15.75
1088	Maximum	"	0.300		19.00	16.50	35.75
1102	15" dia.						
1104	Minimum	LF	0.100		6.33	5.54	11.75
1106	Average	"	0.141		8.94	7.82	16.75
1108	Maximum	"	0.343		21.75	19.00	40.75
1122	18" dia.						
1124	Minimum	LF	0.109		6.91	6.04	13.00
1126	Average	"	0.160		10.25	8.86	19.00
1128	Maximum	"	0.400		25.25	22.25	47.50
1142	24" dia.						

© 2020 BNi Publications Inc. ALL RIGHTS RESERVED

SANITARY PIPING DEMOLITION

ID Code	Description: Component Descriptions	Output: Unit of Meas.	Output: Manhr / Unit	Unit Costs: Material Cost	Unit Costs: Labor Cost	Unit Costs: Equipment Cost	Unit Costs: Total Cost
02 - 47134	**SANITARY PIPING DEMOLITION, Cont'd...**						**02 - 47134**
1144	Minimum	LF	0.120		7.60	6.65	14.25
1146	Average	"	0.200		12.75	11.00	23.75
1148	Maximum	"	0.480		30.50	26.50	57.00
1162	30" dia.						
1164	Minimum	LF	0.133		8.44	7.38	15.75
1166	Average	"	0.240		15.25	13.25	28.50
1168	Maximum	"	0.600		38.00	33.25	71.00
1182	36" dia.						
1184	Minimum	LF	0.160		10.25	8.86	19.00
1186	Average	"	0.300		19.00	16.50	35.75
1188	Maximum	"	0.800		51.00	44.25	95.00

WATER PIPING DEMOLITION

ID Code	Component Descriptions	Unit of Meas.	Manhr / Unit	Material Cost	Labor Cost	Equipment Cost	Total Cost
02 - 48134	**WATER PIPING DEMOLITION**						**02 - 48134**
0980	Remove water pipe, not including excavation						
1002	4" dia.						
1004	Minimum	LF	0.096		6.08	5.32	11.50
1006	Average	"	0.109		6.91	6.04	13.00
1008	Maximum	"	0.126		8.00	7.00	15.00
1022	6" dia.						
1024	Minimum	LF	0.100		6.33	5.54	11.75
1026	Average	"	0.114		7.24	6.33	13.50
1028	Maximum	"	0.133		8.44	7.38	15.75
1042	8" dia.						
1044	Minimum	LF	0.109		6.91	6.04	13.00
1046	Average	"	0.126		8.00	7.00	15.00
1048	Maximum	"	0.150		9.50	8.31	17.75
1062	10" dia.						
1064	Minimum	LF	0.114		7.24	6.33	13.50
1066	Average	"	0.133		8.44	7.38	15.75
1068	Maximum	"	0.160		10.25	8.86	19.00
1082	12" dia.						
1084	Minimum	LF	0.120		7.60	6.65	14.25
1086	Average	"	0.141		8.94	7.82	16.75
1088	Maximum	"	0.171		10.75	9.50	20.25
1102	14" dia.						
1104	Minimum	LF	0.126		8.00	7.00	15.00

© 2020 BNi Publications Inc. ALL RIGHTS RESERVED

WATER PIPING DEMOLITION

ID Code	Component Descriptions	Unit of Meas.	Manhr / Unit	Material Cost	Labor Cost	Equipment Cost	Total Cost
02 - 48134	WATER PIPING DEMOLITION, Cont'd... 02 - 48134						
1106	Average	LF	0.150		9.50	8.31	17.75
1108	Maximum	"	0.185		11.75	10.25	22.00
1122	16" dia.						
1124	Minimum	LF	0.133		8.44	7.38	15.75
1126	Average	"	0.160		10.25	8.86	19.00
1128	Maximum	"	0.200		12.75	11.00	23.75
1142	18" dia.						
1144	Minimum	LF	0.141		8.94	7.82	16.75
1146	Average	"	0.171		10.75	9.50	20.25
1148	Maximum	"	0.218		13.75	12.00	26.00
1162	20" dia.						
1164	Minimum	LF	0.150		9.50	8.31	17.75
1166	Average	"	0.185		11.75	10.25	22.00
1168	Maximum	"	0.240		15.25	13.25	28.50
1180	Remove valves						
1200	6"	EA	1.200		76.00	67.00	140
1220	10"	"	1.333		84.00	74.00	160
1240	14"	"	1.500		95.00	83.00	180
1260	18"	"	2.000		130	110	240

EXTERIOR DEMOLITION

ID Code	Component Descriptions	Unit of Meas.	Manhr / Unit	Material Cost	Labor Cost	Equipment Cost	Total Cost
02 - 49135	SAW CUTTING PAVEMENT 02 - 49135						
0100	Pavement, bituminous						
0110	2" thick	LF	0.016		1.01	1.20	2.21
0120	3" thick	"	0.020		1.26	1.50	2.76
0130	4" thick	"	0.025		1.55	1.84	3.40
0140	5" thick	"	0.027		1.68	2.00	3.68
0150	6" thick	"	0.029		1.81	2.14	3.95
0200	Concrete pavement, with wire mesh						
0210	4" thick	LF	0.031		1.94	2.30	4.25
0212	5" thick	"	0.033		2.11	2.50	4.61
0215	6" thick	"	0.036		2.30	2.72	5.03
0220	8" thick	"	0.040		2.53	3.00	5.53
0250	10" thick	"	0.044		2.81	3.33	6.14
0300	Plain concrete, unreinforced						
0320	4" thick	LF	0.027		1.68	2.00	3.68
0340	5" thick	"	0.031		1.94	2.30	4.25

© 2020 BNi Publications Inc. ALL RIGHTS RESERVED

EXTERIOR DEMOLITION

	Description	Output		Unit Costs			
ID Code	Component Descriptions	Unit of Meas.	Manhr / Unit	Material Cost	Labor Cost	Equipment Cost	Total Cost
02 - 49135	**SAW CUTTING PAVEMENT, Cont'd...**						**02 - 49135**
0360	6" thick	LF	0.033		2.11	2.50	4.61
0380	8" thick	"	0.036		2.30	2.72	5.03
0390	10" thick	"	0.040		2.53	3.00	5.53
02 - 49298	**WALL, EXTERIOR, DEMOLITION**						**02 - 49298**
0980	Concrete wall						
0990	Light reinforcing						
1000	6" thick	SF	0.120		7.60	6.65	14.25
1020	8" thick	"	0.126		8.00	7.00	15.00
1040	10" thick	"	0.133		8.44	7.38	15.75
1060	12" thick	"	0.150		9.50	8.31	17.75
1180	Medium reinforcing						
1200	6" thick	SF	0.126		8.00	7.00	15.00
1220	8" thick	"	0.133		8.44	7.38	15.75
1240	10" thick	"	0.150		9.50	8.31	17.75
1260	12" thick	"	0.171		10.75	9.50	20.25
1380	Heavy reinforcing						
1400	6" thick	SF	0.141		8.94	7.82	16.75
1420	8" thick	"	0.150		9.50	8.31	17.75
1440	10" thick	"	0.171		10.75	9.50	20.25
1460	12" thick	"	0.200		12.75	11.00	23.75
1980	Masonry						
1990	No reinforcing						
2000	8" thick	SF	0.053		3.37	2.95	6.33
2020	12" thick	"	0.060		3.80	3.32	7.12
2040	16" thick	"	0.069		4.34	3.80	8.14
2050	Horizontal reinforcing						
2060	8" thick	SF	0.060		3.80	3.32	7.12
2080	12" thick	"	0.065		4.10	3.59	7.70
2100	16" thick	"	0.077		4.90	4.29	9.19
2110	Vertical reinforcing						
2120	8" thick	SF	0.077		4.90	4.29	9.19
2140	12" thick	"	0.089		5.63	4.92	10.50
2160	16" thick	"	0.109		6.91	6.04	13.00
5000	Remove concrete headwall						
5020	15" pipe	EA	1.714		110	95.00	200
5040	18" pipe	"	2.000		130	110	240
5060	24" pipe	"	2.182		140	120	260

© 2020 BNi Publications Inc. ALL RIGHTS RESERVED

EXTERIOR DEMOLITION

	Description	Output		Unit Costs			
ID Code	Component Descriptions	Unit of Meas.	Manhr / Unit	Material Cost	Labor Cost	Equipment Cost	Total Cost
02 - 49298	**WALL, EXTERIOR, DEMOLITION, Cont'd...**						**02 - 49298**
5080	30" pipe	EA	2.400		150	130	290
5100	36" pipe	"	2.667		170	150	320
5120	48" pipe	"	3.429		220	190	410
5140	60" pipe	"	4.800		300	270	570

DEMOLITION

02 - 51061	**COMPLETE BUILDING DEMOLITION**						**02 - 51061**
0100	Building, complete with disposal						
0200	Wood frame	CF	0.003		0.17	0.24	0.41
0300	Concrete	"	0.004		0.25	0.36	0.61
0400	Steel frame	"	0.005		0.34	0.48	0.82
02 - 51190	**SELECTIVE BUILDING DEMOLITION**						**02 - 51190**
1000	Partition removal						
1100	Concrete block partitions						
1120	4" thick	SF	0.040		2.55		2.55
1140	8" thick	"	0.053		3.40		3.40
1160	12" thick	"	0.073		4.64		4.64
1200	Brick masonry partitions						
1220	4" thick	SF	0.040		2.55		2.55
1240	8" thick	"	0.050		3.19		3.19
1260	12" thick	"	0.067		4.25		4.25
1280	16" thick	"	0.100		6.38		6.38
1380	Cast-in-place concrete partitions						
1400	Unreinforced						
1421	6" thick	SF	0.160		10.25	8.86	19.00
1423	8" thick	"	0.171		10.75	9.50	20.25
1425	10" thick	"	0.200		12.75	11.00	23.75
1427	12" thick	"	0.240		15.25	13.25	28.50
1440	Reinforced						
1441	6" thick	SF	0.185		11.75	10.25	22.00
1443	8" thick	"	0.240		15.25	13.25	28.50
1445	10" thick	"	0.267		17.00	14.75	31.75
1447	12" thick	"	0.320		20.25	17.75	38.00
1500	Terra cotta						
1520	To 6" thick	SF	0.040		2.55		2.55
1700	Stud partitions						
1720	Metal or wood, with drywall both sides	SF	0.040		2.55		2.55

© 2020 BNi Publications Inc. ALL RIGHTS RESERVED

DEMOLITION

	Description	Output		Unit Costs			
ID Code	Component Descriptions	Unit of Meas.	Manhr / Unit	Material Cost	Labor Cost	Equipment Cost	Total Cost
02 - 51190	**SELECTIVE BUILDING DEMOLITION, Cont'd...**						**02 - 51190**
1740	Metal studs, both sides, lath and plaster	SF	0.053		3.40		3.40
2000	Door and frame removal						
2020	Hollow metal in masonry wall						
2030	Single						
2040	2'6"x6'8"	EA	1.000		64.00		64.00
2060	3'x7'	"	1.333		85.00		85.00
2070	Double						
2080	3'x7'	EA	1.600		100		100
2085	4'x8'	"	1.600		100		100
2140	Wood in framed wall						
2150	Single						
2160	2'6"x6'8"	EA	0.571		36.50		36.50
2180	3'x6'8"	"	0.667		42.50		42.50
2190	Double						
2200	2'6"x6'8"	EA	0.800		51.00		51.00
2220	3'x6'8"	"	0.889		57.00		57.00
2240	Remove for re-use						
2260	Hollow metal	EA	2.000		130		130
2280	Wood	"	1.333		85.00		85.00
2300	Floor removal						
2340	Brick flooring	SF	0.032		2.04		2.04
2360	Ceramic or quarry tile	"	0.018		1.13		1.13
2380	Terrazzo	"	0.036		2.27		2.27
2400	Heavy wood	"	0.021		1.36		1.36
2420	Residential wood	"	0.023		1.45		1.45
2440	Resilient tile or linoleum	"	0.008		0.51		0.51
2500	Ceiling removal						
2520	Acoustical tile ceiling						
2540	Adhesive fastened	SF	0.008		0.51		0.51
2560	Furred and glued	"	0.007		0.42		0.42
2580	Suspended grid	"	0.005		0.31		0.31
2600	Drywall ceiling						
2620	Furred and nailed	SF	0.009		0.56		0.56
2640	Nailed to framing	"	0.008		0.51		0.51
2660	Plastered ceiling						
2680	Furred on framing	SF	0.020		1.27		1.27
2700	Suspended system	"	0.027		1.70		1.70
2800	Roofing removal						

© 2020 BNi Publications Inc. ALL RIGHTS RESERVED

DEMOLITION

	Description	Output		Unit Costs			
ID Code	Component Descriptions	Unit of Meas.	Manhr / Unit	Material Cost	Labor Cost	Equipment Cost	Total Cost
02 - 51190	**SELECTIVE BUILDING DEMOLITION, Cont'd...**						**02 - 51190**
2820	Steel frame						
2840	Corrugated metal roofing	SF	0.016		1.02		1.02
2860	Built-up roof on metal deck	"	0.027		1.70		1.70
2900	Wood frame						
2920	Built-up roof on wood deck	SF	0.025		1.57		1.57
2940	Roof shingles	"	0.013		0.85		0.85
2960	Roof tiles	"	0.027		1.70		1.70
8900	Concrete frame	CF	0.053		3.40		3.40
8920	Concrete plank	SF	0.040		2.55		2.55
8940	Built-up roof on concrete	"	0.023		1.45		1.45
9200	Cut-outs						
9230	Concrete, elevated slabs, mesh reinforcing						
9240	Under 5 cf	CF	0.800		51.00		51.00
9260	Over 5 cf	"	0.667		42.50		42.50
9270	Bar reinforcing						
9280	Under 5 cf	CF	1.333		85.00		85.00
9290	Over 5 cf	"	1.000		64.00		64.00
9300	Window removal						
9301	Metal windows, trim included						
9302	2'x3'	EA	0.800		51.00		51.00
9304	2'x4'	"	0.889		57.00		57.00
9306	2'x6'	"	1.000		64.00		64.00
9308	3'x4'	"	1.000		64.00		64.00
9310	3'x6'	"	1.143		73.00		73.00
9312	3'x8'	"	1.333		85.00		85.00
9314	4'x4'	"	1.333		85.00		85.00
9315	4'x6'	"	1.600		100		100
9316	4'x8'	"	2.000		130		130
9317	Wood windows, trim included						
9318	2'x3'	EA	0.444		28.25		28.25
9319	2'x4'	"	0.471		30.00		30.00
9320	2'x6'	"	0.500		32.00		32.00
9321	3'x4'	"	0.533		34.00		34.00
9322	3'x6'	"	0.571		36.50		36.50
9324	3'x8'	"	0.615		39.25		39.25
9325	6'x4'	"	0.667		42.50		42.50
9326	6'x6'	"	0.727		46.50		46.50
9327	6'x8'	"	0.800		51.00		51.00

© 2020 BNi Publications Inc. ALL RIGHTS RESERVED

DEMOLITION

ID Code	Description: Component Descriptions	Output: Unit of Meas.	Output: Manhr / Unit	Unit Costs: Material Cost	Unit Costs: Labor Cost	Unit Costs: Equipment Cost	Unit Costs: Total Cost
02 - 51190	**SELECTIVE BUILDING DEMOLITION, Cont'd...**						**02 - 51190**
9329	Walls, concrete, bar reinforcing						
9330	Small jobs	CF	0.533		34.00		34.00
9340	Large jobs	"	0.444		28.25		28.25
9360	Brick walls, not including toothing						
9390	4" thick	SF	0.040		2.55		2.55
9400	8" thick	"	0.050		3.19		3.19
9410	12" thick	"	0.067		4.25		4.25
9415	16" thick	"	0.100		6.38		6.38
9420	Concrete block walls, not including toothing						
9440	4" thick	SF	0.044		2.83		2.83
9450	6" thick	"	0.047		3.00		3.00
9460	8" thick	"	0.050		3.19		3.19
9465	10" thick	"	0.057		3.64		3.64
9470	12" thick	"	0.067		4.25		4.25
9500	Rubbish handling						
9519	Load in dumpster or truck						
9520	Minimum	CF	0.018		1.13		1.13
9540	Maximum	"	0.027		1.70		1.70
9550	For use of elevators, add						
9560	Minimum	CF	0.004		0.25		0.25
9570	Maximum	"	0.008		0.51		0.51
9600	Rubbish hauling						
9640	Hand loaded on trucks, 2 mile trip	CY	0.320		20.25	24.00	44.25
9660	Machine loaded on trucks, 2 mile trip	"	0.240		15.25	13.25	28.50

SITE REMEDIATION

ID Code	Component Descriptions	Unit of Meas.	Manhr / Unit	Material Cost	Labor Cost	Equipment Cost	Total Cost
02 - 65006	**UNDERGROUND STORAGE TANK REMOVAL**						**02 - 65006**
1980	Remove underground storage tank, and backfill						
2000	50 to 250 gals	EA	8.000		510	440	950
2050	600 gals	"	8.000		510	440	950
2060	1000 gals	"	12.000		760	670	1,430
2100	4000 gals	"	19.200		1,220	1,060	2,280
2120	5000 gals	"	19.200		1,220	1,060	2,280
2140	10,000 gals	"	32.000		2,030	1,770	3,800
2160	12,000 gals	"	40.000		2,530	2,220	4,750
2180	15,000 gals	"	48.000		3,040	2,660	5,700
2200	20,000 gals	"	60.000		3,800	3,330	7,130

© 2020 BNi Publications Inc. ALL RIGHTS RESERVED

SITE REMEDIATION

ID Code	Description: Component Descriptions	Output: Unit of Meas.	Output: Manhr / Unit	Unit Costs: Material Cost	Unit Costs: Labor Cost	Unit Costs: Equipment Cost	Unit Costs: Total Cost
02 - 65007	**SEPTIC TANK REMOVAL**						**02 - 65007**
0980	Remove septic tank						
1000	1000 gals	EA	2.000		130	110	240
1020	2000 gals	"	2.400		150	130	290
1040	5000 gals	"	3.000		190	170	360
1060	15,000 gals	"	24.000		1,520	1,330	2,850
1080	25,000 gals	"	32.000		2,030	1,770	3,800
2000	40,000 gals	"	48.000		3,040	2,660	5,700

HAZARDOUS WASTE

ID Code	Component Descriptions	Unit of Meas.	Manhr / Unit	Material Cost	Labor Cost	Equipment Cost	Total Cost
02 - 82001	**ASBESTOS REMOVAL**						**02 - 82001**
1000	Enclosure using wood studs & poly, install & remove	SF	0.020	530	1.27		530
1020	Trailer (change room)	DAY					120
1100	Disposal suits (4 suits per man day)	"					47.00
1120	Type C respirator mask, includes hose & filters, per	"					23.50
1130	Respirator mask & filter, light contamination	"					9.39
1980	Air monitoring test, 12 tests per day						
2000	Off job testing	DAY					1,230
2020	On the job testing	"					1,640
6000	Asbestos vacuum with attachments	EA					710
6500	Hydraspray piston pump	"					950
6600	Negative air pressure system	"					950
6800	Grade D breathing air equipment	"					2,140
6900	Glove bag, 44" x 60" x 6 mil plastic	"					6.90
7980	40 CY asbestos dumpster						
8000	Weekly rental	EA					800
8100	Pick up/delivery	"					360
8400	Asbestos dump fee	"					230
02 - 82002	**DUCT INSULATION REMOVAL**						**02 - 82002**
0080	Remove duct insulation, duct size						
1000	6" x 12"	LF	0.044	240	2.83		240
1020	x 18"	"	0.062	180	3.92		180
1040	x 24"	"	0.089	120	5.67		130
1060	8" x 12"	"	0.067	160	4.25		160
1080	x 18"	"	0.073	150	4.64		150
1100	x 24"	"	0.100	110	6.38		120
1120	12" x 12"	"	0.067	160	4.25		160
1140	x 18"	"	0.089	120	5.67		130

© 2020 BNi Publications Inc. ALL RIGHTS RESERVED

HAZARDOUS WASTE

ID Code	Description: Component Descriptions	Output: Unit of Meas.	Output: Manhr / Unit	Unit Costs: Material Cost	Unit Costs: Labor Cost	Unit Costs: Equipment Cost	Unit Costs: Total Cost
02 - 82002	DUCT INSULATION REMOVAL, Cont'd...						02 - 82002
1160	x 24"	LF	0.114	96.00	7.29		100
02 - 82003	PIPE INSULATION REMOVAL						02 - 82003
0060	Removal, asbestos insulation						
0080	2" thick, pipe						
1000	1" to 3" dia.	LF	0.067		4.25		4.25
1020	4" to 6" dia.	"	0.076		4.86		4.86
1030	3" thick						
1040	7" to 8" dia.	LF	0.080		5.10		5.10
1060	9" to 10" dia.	"	0.084		5.37		5.37
1070	11" to 12" dia.	"	0.089		5.67		5.67
1080	13" to 14" dia.	"	0.094		6.00		6.00
1090	15" to 18" dia.	"	0.100		6.38		6.38

© 2020 BNi Publications Inc. ALL RIGHTS RESERVED

DIVISION 03
CONCRETE

CONCRETE RESTORATION

	Description	Output		Unit Costs			
ID Code	Component Descriptions	Unit of Meas.	Manhr / Unit	Material Cost	Labor Cost	Equipment Cost	Total Cost
03 - 01301	**CONCRETE REPAIR**						**03 - 01301**
0090	Epoxy grout floor patch, 1/4" thick	SF	0.080	8.89	5.10		14.00
0100	Grout, epoxy, 2 component system	CF					430
0110	Epoxy sand	BAG					29.00
0120	Epoxy modifier	GAL					190
0140	Epoxy gel grout	SF	0.800	4.32	51.00		55.00
0150	Injection valve, 1 way, threaded plastic	EA	0.160	12.00	10.25		22.25
0155	Grout crack seal, 2 component	CF	0.800	1,000	51.00		1,050
0160	Grout, non-shrink	"	0.800	100	51.00		150
0165	Concrete, epoxy modified						
0170	Sand mix	CF	0.320	160	20.50		180
0180	Gravel mix	"	0.296	110	19.00		130
0190	Concrete repair						
0195	Soffit repair						
0200	16" wide	LF	0.160	5.07	10.25		15.25
0210	18" wide	"	0.167	5.39	10.75		16.25
0220	24" wide	"	0.178	6.44	11.25		17.75
0230	30" wide	"	0.190	7.25	12.25		19.50
0240	32" wide	"	0.200	7.73	12.75		20.50
0245	Edge repair						
0250	2" spall	LF	0.200	2.41	12.75		15.25
0260	3" spall	"	0.211	2.41	13.50		16.00
0270	4" spall	"	0.216	2.57	13.75		16.25
0280	6" spall	"	0.222	2.66	14.25		17.00
0290	8" spall	"	0.235	2.82	15.00		17.75
0300	9" spall	"	0.267	2.89	17.00		20.00
0330	Crack repair, 1/8" crack	"	0.080	4.75	5.10		9.85
5000	Reinforcing steel repair						
5005	1 bar, 4 ft						
5010	#4 bar	LF	0.100	0.74	8.32		9.06
5012	#5 bar	"	0.100	1.00	8.32		9.32
5014	#6 bar	"	0.107	1.22	8.88		10.00
5016	#8 bar	"	0.107	2.22	8.88		11.00
5020	#9 bar	"	0.114	2.83	9.51		12.25
5030	#11 bar	"	0.114	4.43	9.51		14.00
7010	Form fabric, nylon						
7020	18" diameter	LF					19.25
7030	20" diameter	"					19.50
7040	24" diameter	"					32.00

© 2020 BNi Publications Inc. ALL RIGHTS RESERVED

CONCRETE RESTORATION

	Description	Output		Unit Costs			
ID Code	Component Descriptions	Unit of Meas.	Manhr / Unit	Material Cost	Labor Cost	Equipment Cost	Total Cost
03 - 01301	**CONCRETE REPAIR, Cont'd...**						**03 - 01301**
7050	30" diameter	LF					32.75
7060	36" diameter	"					37.75
7100	Pile repairs						
7105	Polyethylene wrap						
7108	30 mil thick						
7110	60" wide	SF	0.267	20.75	17.00		37.75
7120	72" wide	"	0.320	22.75	20.50		43.25
7125	60 mil thick						
7130	60" wide	SF	0.267	24.50	17.00		41.50
7140	80" wide	"	0.364	28.50	23.25		52.00
8010	Pile spall, average repair 3'						
8020	18" x 18"	EA	0.667	64.00	42.50		110
8030	20" x 20"	"	0.800	85.00	51.00		140

CONCRETE FORMING & ACCESSORIES

03 - 10030	**FORMWORK ACCESSORIES**						**03 - 10030**
1000	Column clamps						
1010	Small, adjustable, 24"x24"	EA					87.00
1020	Medium 36"x36"	"					90.00
1030	Large 60"x60"	"					91.00
2000	Forming hangers						
2010	Iron 14 ga.	EA					2.81
2020	22 ga.	"					2.81
3000	Snap ties						
3010	Short-end with washers, 6" long	EA					1.73
3020	12" long	"					1.96
4000	18" long	"					2.34
4010	24" long	"					2.52
4020	Long-end with washers, 6' long	"					2.04
4030	12" long	"					2.26
4040	18" long	"					2.56
4050	24" long	"					2.87
5000	Stakes						
5010	Round, pre-drilled holes, 12" long	EA					6.14
5020	18" long	"					6.87
5030	24" long	"					8.90
5040	30" long	"					11.50

© 2020 BNi Publications Inc. ALL RIGHTS RESERVED

CONCRETE FORMING & ACCESSORIES

ID Code	Component Descriptions	Unit of Meas.	Manhr / Unit	Material Cost	Labor Cost	Equipment Cost	Total Cost
03 - 10030	**FORMWORK ACCESSORIES, Cont'd...**						**03 - 10030**
5050	36" long	EA					13.75
5060	48" long	"					18.50
5070	I beam type, 12" long	"					5.05
6000	18" long	"					5.75
6010	24" long	"					8.73
6020	30" long	"					10.25
6030	36" long	"					13.00
7000	48" long	"					16.25
7010	Taper ties						
7020	50K, 1-1/4" to 1", 35" long	EA					85.00
8000	45" long	"					140
8010	55" long	"					170
8020	Walers						
8030	5" deep, 4' long	EA					240
8040	8' long	"					310
8050	12' long	"					510
9000	16' long	"					660
9010	8" deep, 4' long	"					320
9020	8' long	"					580
9030	12' long	"					840
9040	16' long	"					1,330

FORMWORK

ID Code	Component Descriptions	Unit of Meas.	Manhr / Unit	Material Cost	Labor Cost	Equipment Cost	Total Cost
03 - 11130	**BEAM FORMWORK**						**03 - 11130**
1000	Beam forms, job built						
1020	Beam bottoms						
1040	1 use	SF	0.133	5.27	10.75		16.00
1060	2 uses	"	0.127	3.10	10.25		13.25
1080	3 uses	"	0.123	2.38	10.00		12.50
1100	4 uses	"	0.118	1.97	9.59		11.50
1120	5 uses	"	0.114	1.80	9.32		11.00
2000	Beam sides						
2020	1 use	SF	0.089	3.77	7.24		11.00
2040	2 uses	"	0.084	2.24	6.86		9.10
2060	3 uses	"	0.080	1.97	6.52		8.49
2080	4 uses	"	0.076	1.81	6.21		8.02
2100	5 uses	"	0.073	1.60	5.93		7.53

© 2020 BNi Publications Inc. ALL RIGHTS RESERVED

FORMWORK

ID Code	Description: Component Descriptions	Output: Unit of Meas.	Output: Manhr / Unit	Unit Costs: Material Cost	Unit Costs: Labor Cost	Unit Costs: Equipment Cost	Unit Costs: Total Cost
03 - 11131	**BOX CULVERT FORMWORK**						**03 - 11131**
1000	Box culverts, job built						
1010	6' x 6'						
1020	1 use	SF	0.080	3.84	6.52		10.25
1040	2 uses	"	0.076	2.09	6.21		8.30
1060	3 uses	"	0.073	1.74	5.93		7.67
1080	4 uses	"	0.070	1.47	5.67		7.14
1100	5 uses	"	0.067	1.28	5.43		6.71
1110	8' x 12'						
1120	1 use	SF	0.067	3.84	5.43		9.27
1130	2 uses	"	0.064	2.09	5.21		7.30
1150	3 uses	"	0.062	1.74	5.01		6.75
1170	4 uses	"	0.059	1.47	4.83		6.30
1200	5 uses	"	0.057	1.28	4.66		5.94
03 - 11132	**COLUMN FORMWORK**						**03 - 11132**
1000	Column, square forms, job built						
1020	8" x 8" columns						
1040	1 use	SF	0.160	4.44	13.00		17.50
1060	2 uses	"	0.154	2.39	12.50		15.00
1080	3 uses	"	0.148	2.02	12.00		14.00
1100	4 uses	"	0.143	1.84	11.75		13.50
1120	5 uses	"	0.138	1.57	11.25		12.75
1200	12" x 12" columns						
1220	1 use	SF	0.145	4.05	11.75		15.75
1240	2 uses	"	0.140	2.25	11.50		13.75
1260	3 uses	"	0.136	1.80	11.00		12.75
1280	4 uses	"	0.131	1.57	10.75		12.25
1290	5 uses	"	0.127	1.32	10.25		11.50
1300	16" x 16" columns						
1320	1 use	SF	0.133	3.86	10.75		14.50
1340	2 uses	"	0.129	2.03	10.50		12.50
1360	3 uses	"	0.125	1.63	10.25		12.00
1380	4 uses	"	0.121	1.49	9.88		11.25
1390	5 uses	"	0.118	1.22	9.59		10.75
1400	24" x 24" columns						
1420	1 use	SF	0.123	3.86	10.00		13.75
1440	2 uses	"	0.119	1.80	9.73		11.50
1460	3 uses	"	0.116	1.50	9.45		11.00

© 2020 BNi Publications Inc. ALL RIGHTS RESERVED

FORMWORK

	Description	Output		Unit Costs			
ID Code	Component Descriptions	Unit of Meas.	Manhr / Unit	Material Cost	Labor Cost	Equipment Cost	Total Cost
03 - 11132	**COLUMN FORMWORK, Cont'd...**						**03 - 11132**
1480	4 uses	SF	0.113	1.22	9.18		10.50
1490	5 uses	"	0.110	1.12	8.93		10.00
1500	36" x 36" columns						
1520	1 use	SF	0.114	3.89	9.32		13.25
1540	2 uses	"	0.111	1.82	9.06		11.00
1560	3 uses	"	0.108	1.50	8.81		10.25
1580	4 uses	"	0.105	1.29	8.58		9.87
1590	5 uses	"	0.103	1.20	8.36		9.56
2000	Round fiber forms, 1 use						
2040	10" dia.	LF	0.160	5.79	13.00		18.75
2060	12" dia.	"	0.163	7.13	13.25		20.50
2080	14" dia.	"	0.170	9.36	14.00		23.25
2100	16" dia.	"	0.178	12.25	14.50		26.75
2120	18" dia.	"	0.190	20.00	15.50		35.50
2140	24" dia.	"	0.205	24.50	16.75		41.25
2160	30" dia.	"	0.222	36.75	18.00		55.00
2180	36" dia.	"	0.242	45.75	19.75		66.00
2200	42" dia.	"	0.267	83.00	21.75		100
03 - 11133	**CURB FORMWORK**						**03 - 11133**
0980	Curb forms						
0990	Straight, 6" high						
1000	1 use	LF	0.080	2.64	6.52		9.16
1020	2 uses	"	0.076	1.58	6.21		7.79
1040	3 uses	"	0.073	1.19	5.93		7.12
1060	4 uses	"	0.070	1.06	5.67		6.73
1080	5 uses	"	0.067	0.97	5.43		6.40
1090	Curved, 6" high						
2000	1 use	LF	0.100	2.86	8.15		11.00
2020	2 uses	"	0.094	1.80	7.67		9.47
2040	3 uses	"	0.089	1.37	7.24		8.61
2060	4 uses	"	0.085	1.25	6.94		8.19
2080	5 uses	"	0.082	1.16	6.65		7.81

© 2020 BNi Publications Inc. ALL RIGHTS RESERVED

FORMWORK

	Description	Output		Unit Costs			
ID Code	Component Descriptions	Unit of Meas.	Manhr / Unit	Material Cost	Labor Cost	Equipment Cost	Total Cost
03 - 11134	**ELEVATED SLAB FORMWORK**						**03 - 11134**
0100	Elevated slab formwork						
1000	Slab, with drop panels						
1020	1 use	SF	0.064	4.75	5.21		9.96
1040	2 uses	"	0.062	2.75	5.01		7.76
1060	3 uses	"	0.059	2.13	4.83		6.96
1080	4 uses	"	0.057	1.89	4.66		6.55
1100	5 uses	"	0.055	1.70	4.49		6.19
2000	Floor slab, hung from steel beams						
2020	1 use	SF	0.062	3.82	5.01		8.83
2040	2 uses	"	0.059	2.09	4.83		6.92
2060	3 uses	"	0.057	1.91	4.66		6.57
2080	4 uses	"	0.055	1.64	4.49		6.13
2100	5 uses	"	0.053	1.40	4.34		5.74
3000	Floor slab, with pans or domes						
3020	1 use	SF	0.073	6.82	5.93		12.75
3040	2 uses	"	0.070	4.43	5.67		10.00
3060	3 uses	"	0.067	4.13	5.43		9.56
3080	4 uses	"	0.064	3.85	5.21		9.06
3100	5 uses	"	0.062	3.41	5.01		8.42
9030	Equipment curbs, 12" high						
9035	1 use	LF	0.080	3.51	6.52		10.00
9040	2 uses	"	0.076	2.26	6.21		8.47
9060	3 uses	"	0.073	1.96	5.93		7.89
9080	4 uses	"	0.070	1.77	5.67		7.44
9100	5 uses	"	0.067	1.53	5.43		6.96
03 - 11135	**EQUIPMENT PAD FORMWORK**						**03 - 11135**
1000	Equipment pad, job built						
1020	1 use	SF	0.100	4.61	8.15		12.75
1040	2 uses	"	0.094	2.77	7.67		10.50
1060	3 uses	"	0.089	2.22	7.24		9.46
1080	4 uses	"	0.084	1.73	6.86		8.59
1100	5 uses	"	0.080	1.37	6.52		7.89

© 2020 BNi Publications Inc. ALL RIGHTS RESERVED

FORMWORK

	Description	Output		Unit Costs			
ID Code	Component Descriptions	Unit of Meas.	Manhr / Unit	Material Cost	Labor Cost	Equipment Cost	Total Cost
03 - 11136	**FOOTING FORMWORK**						**03 - 11136**
2000	Wall footings, job built, continuous						
2040	1 use	SF	0.080	2.14	6.52		8.66
2050	2 uses	"	0.076	1.51	6.21		7.72
2060	3 uses	"	0.073	1.24	5.93		7.17
2080	4 uses	"	0.070	1.10	5.67		6.77
2090	5 uses	"	0.067	0.95	5.43		6.38
3000	Column footings, spread						
3020	1 use	SF	0.100	2.27	8.15		10.50
3040	2 uses	"	0.094	1.69	7.67		9.36
3060	3 uses	"	0.089	1.20	7.24		8.44
3080	4 uses	"	0.084	1.01	6.86		7.87
3100	5 uses	"	0.080	0.92	6.52		7.44
03 - 11137	**GRADE BEAM FORMWORK**						**03 - 11137**
1000	Grade beams, job built						
1020	1 use	SF	0.080	3.38	6.52		9.90
1040	2 uses	"	0.076	1.89	6.21		8.10
1060	3 uses	"	0.073	1.48	5.93		7.41
1080	4 uses	"	0.070	1.23	5.67		6.90
1100	5 uses	"	0.067	1.02	5.43		6.45
03 - 11138	**PILE CAP FORMWORK**						**03 - 11138**
1500	Pile cap forms, job built						
1510	Square						
1520	1 use	SF	0.100	3.83	8.15		12.00
1540	2 uses	"	0.094	2.21	7.67		9.88
1560	3 uses	"	0.089	1.75	7.24		8.99
1580	4 uses	"	0.084	1.55	6.86		8.41
1600	5 uses	"	0.080	1.28	6.52		7.80
2000	Triangular						
2020	1 use	SF	0.114	4.07	9.32		13.50
2040	2 uses	"	0.107	2.68	8.69		11.25
2060	3 uses	"	0.100	2.14	8.15		10.25
2080	4 uses	"	0.094	1.75	7.67		9.42
2100	5 uses	"	0.089	1.39	7.24		8.63

© 2020 BNi Publications Inc. ALL RIGHTS RESERVED

FORMWORK

ID Code	Description: Component Descriptions	Output: Unit of Meas.	Output: Manhr / Unit	Unit Costs: Material Cost	Unit Costs: Labor Cost	Unit Costs: Equipment Cost	Unit Costs: Total Cost
03 - 11139	**SLAB / MAT FORMWORK**						**03 - 11139**
3000	Mat foundations, job built						
3020	1 use	SF	0.100	3.35	8.15		11.50
3040	2 uses	"	0.094	1.94	7.67		9.61
3060	3 uses	"	0.089	1.42	7.24		8.66
3080	4 uses	"	0.084	1.20	6.86		8.06
3100	5 uses	"	0.080	0.97	6.52		7.49
3980	Edge forms						
3990	6" high						
4000	1 use	LF	0.073	3.38	5.93		9.31
4001	2 uses	"	0.070	1.94	5.67		7.61
4002	3 uses	"	0.067	1.43	5.43		6.86
4003	4 uses	"	0.064	1.21	5.21		6.42
4004	5 uses	"	0.062	0.97	5.01		5.98
4006	12" high						
4010	1 use	LF	0.080	3.18	6.52		9.70
4011	2 uses	"	0.076	1.81	6.21		8.02
4012	3 uses	"	0.073	1.32	5.93		7.25
4013	4 uses	"	0.070	1.10	5.67		6.77
4014	5 uses	"	0.067	0.89	5.43		6.32
5000	Formwork for openings						
5020	1 use	SF	0.160	4.56	13.00		17.50
5040	2 uses	"	0.145	2.63	11.75		14.50
5060	3 uses	"	0.133	2.20	10.75		13.00
5080	4 uses	"	0.123	1.69	10.00		11.75
5100	5 uses	"	0.114	1.42	9.32		10.75
03 - 11140	**STAIR FORMWORK**						**03 - 11140**
1000	Stairway forms, job built						
1020	1 use	SF	0.160	5.27	13.00		18.25
1030	2 uses	"	0.145	2.95	11.75		14.75
1040	3 uses	"	0.133	2.29	10.75		13.00
1050	4 uses	"	0.123	2.10	10.00		12.00
1060	5 uses	"	0.114	1.76	9.32		11.00
2000	Stairs, elevated						
2020	1 use	SF	0.160	6.36	13.00		19.25
2040	2 uses	"	0.133	3.38	10.75		14.25
2060	3 uses	"	0.114	2.95	9.32		12.25
2080	4 uses	"	0.107	2.54	8.69		11.25

© 2020 BNi Publications Inc. ALL RIGHTS RESERVED

FORMWORK

	Description	Output		Unit Costs			
ID Code	Component Descriptions	Unit of Meas.	Manhr / Unit	Material Cost	Labor Cost	Equipment Cost	Total Cost
03 - 11140	**STAIR FORMWORK, Cont'd...**						**03 - 11140**
2100	5 uses	SF	0.100	2.10	8.15		10.25
03 - 11141	**WALL FORMWORK**						**03 - 11141**
2980	Wall forms, exterior, job built						
3000	Up to 8' high wall						
3120	1 use	SF	0.080	3.61	6.52		10.25
3140	2 uses	"	0.076	1.99	6.21		8.20
3160	3 uses	"	0.073	1.75	5.93		7.68
3180	4 uses	"	0.070	1.51	5.67		7.18
3190	5 uses	"	0.067	1.32	5.43		6.75
3200	Over 8' high wall						
3220	1 use	SF	0.100	3.96	8.15		12.00
3230	2 uses	"	0.094	2.26	7.67		9.93
3240	3 uses	"	0.089	2.06	7.24		9.30
3280	4 uses	"	0.084	1.88	6.86		8.74
3290	5 uses	"	0.080	1.62	6.52		8.14
3300	Over 16' high wall						
3320	1 use	SF	0.114	4.16	9.32		13.50
3340	2 uses	"	0.107	2.49	8.69		11.25
3360	3 uses	"	0.100	2.26	8.15		10.50
3380	4 uses	"	0.094	2.06	7.67		9.73
3400	5 uses	"	0.089	1.88	7.24		9.12
4000	Radial wall forms						
4020	1 use	SF	0.123	3.88	10.00		14.00
4040	2 uses	"	0.114	2.32	9.32		11.75
4060	3 uses	"	0.107	2.15	8.69		10.75
4080	4 uses	"	0.100	1.95	8.15		10.00
4090	5 uses	"	0.094	1.75	7.67		9.42
4591	Retaining wall forms						
4592	1 use	SF	0.089	3.35	7.24		10.50
4593	2 uses	"	0.084	1.78	6.86		8.64
4594	3 uses	"	0.080	1.54	6.52		8.06
4595	4 uses	"	0.076	1.34	6.21		7.55
4596	5 uses	"	0.073	1.14	5.93		7.07
4600	Radial retaining wall forms						
4620	1 use	SF	0.133	3.55	10.75		14.25
4640	2 uses	"	0.123	2.18	10.00		12.25
4660	3 uses	"	0.114	1.87	9.32		11.25

© 2020 BNi Publications Inc. ALL RIGHTS RESERVED

FORMWORK

ID Code	Description: Component Descriptions	Output: Unit of Meas.	Output: Manhr / Unit	Unit Costs: Material Cost	Unit Costs: Labor Cost	Unit Costs: Equipment Cost	Unit Costs: Total Cost
03 - 11141	**WALL FORMWORK, Cont'd...**						**03 - 11141**
4680	4 uses	SF	0.107	1.77	8.69		10.50
4690	5 uses	"	0.100	1.52	8.15		9.67
5000	Column pier and pilaster						
5020	1 use	SF	0.160	3.96	13.00		17.00
5040	2 uses	"	0.145	2.34	11.75		14.00
5060	3 uses	"	0.133	2.18	10.75		13.00
5080	4 uses	"	0.123	1.99	10.00		12.00
5090	5 uses	"	0.114	1.78	9.32		11.00
6980	Interior wall forms						
7000	Up to 8' high						
7020	1 use	SF	0.073	3.61	5.93		9.54
7040	2 uses	"	0.070	2.01	5.67		7.68
7060	3 uses	"	0.067	1.78	5.43		7.21
7080	4 uses	"	0.064	1.52	5.21		6.73
7100	5 uses	"	0.062	1.28	5.01		6.29
7200	Over 8' high						
7220	1 use	SF	0.089	3.96	7.24		11.25
7240	2 uses	"	0.084	2.26	6.86		9.12
7260	3 uses	"	0.080	2.06	6.52		8.58
7280	4 uses	"	0.076	1.88	6.21		8.09
7290	5 uses	"	0.073	1.64	5.93		7.57
7300	Over 16' high						
7320	1 use	SF	0.100	4.15	8.15		12.25
7340	2 uses	"	0.094	2.49	7.67		10.25
7360	3 uses	"	0.089	2.26	7.24		9.50
7380	4 uses	"	0.084	2.06	6.86		8.92
7390	5 uses	"	0.080	1.88	6.52		8.40
7400	Radial wall forms						
7420	1 use	SF	0.107	3.88	8.69		12.50
7440	2 uses	"	0.100	2.32	8.15		10.50
7460	3 uses	"	0.094	2.15	7.67		9.82
7480	4 uses	"	0.089	1.95	7.24		9.19
7490	5 uses	"	0.084	1.75	6.86		8.61
7500	Curved wall forms, 24" sections						
7520	1 use	SF	0.160	3.71	13.00		16.75
7540	2 uses	"	0.145	2.21	11.75		14.00
7560	3 uses	"	0.133	2.05	10.75		12.75
7580	4 uses	"	0.123	1.87	10.00		11.75

© 2020 BNi Publications Inc. ALL RIGHTS RESERVED

FORMWORK

	Description	Output		Unit Costs			
ID Code	Component Descriptions	Unit of Meas.	Manhr / Unit	Material Cost	Labor Cost	Equipment Cost	Total Cost
03 - 11141	**WALL FORMWORK, Cont'd...**						**03 - 11141**
7590	5 uses	SF	0.114	1.67	9.32		11.00
9000	PVC form liner, per side, smooth finish						
9010	1 use	SF	0.067	9.32	5.43		14.75
9020	2 uses	"	0.064	5.12	5.21		10.25
9030	3 uses	"	0.062	4.33	5.01		9.34
9040	4 uses	"	0.057	3.34	4.66		8.00
9050	5 uses	"	0.053	2.66	4.34		7.00
03 - 11242	**MISCELLANEOUS FORMWORK**						**03 - 11242**
1200	Keyway forms (5 uses)						
1220	2 x 4	LF	0.040	0.30	3.26		3.56
1240	2 x 6	"	0.044	0.44	3.62		4.06
1500	Bulkheads						
1510	Walls, with keyways						
1515	2 piece	LF	0.073	5.09	5.93		11.00
1520	3 piece	"	0.080	6.44	6.52		13.00
1560	Elevated slab, with keyway						
1570	2 piece	LF	0.067	5.84	5.43		11.25
1580	3 piece	"	0.073	8.60	5.93		14.50
1600	Ground slab, with keyway						
1620	2 piece	LF	0.057	6.03	4.66		10.75
1640	3 piece	"	0.062	7.37	5.01		12.50
2000	Chamfer strips						
2020	Wood						
2040	1/2" wide	LF	0.018	0.29	1.44		1.73
2060	3/4" wide	"	0.018	0.37	1.44		1.81
2070	1" wide	"	0.018	0.50	1.44		1.94
2100	PVC						
2120	1/2" wide	LF	0.018	1.29	1.44		2.73
2140	3/4" wide	"	0.018	1.40	1.44		2.84
2160	1" wide	"	0.018	2.03	1.44		3.47
2170	Radius						
2180	1"	LF	0.019	1.51	1.55		3.06
2200	1-1/2"	"	0.019	2.73	1.55		4.28
3000	Reglets						
3020	Galvanized steel, 24 ga.	LF	0.032	1.96	2.60		4.56
5000	Metal formwork						
5020	Straight edge forms						

© 2020 BNi Publications Inc. ALL RIGHTS RESERVED

FORMWORK

ID Code	Description: Component Descriptions	Output: Unit of Meas.	Output: Manhr / Unit	Unit Costs: Material Cost	Unit Costs: Labor Cost	Unit Costs: Equipment Cost	Unit Costs: Total Cost
03 - 11242	MISCELLANEOUS FORMWORK, Cont'd...						03 - 11242
5040	4" high	LF	0.050	23.50	4.07		27.50
5060	6" high	"	0.053	25.75	4.34		30.00
5080	8" high	"	0.057	35.25	4.66		40.00
5100	12" high	"	0.062	41.00	5.01		46.00
5120	16" high	"	0.067	48.25	5.43		54.00
5300	Curb form, S-shape						
5310	12" x						
5320	1'-6"	LF	0.114	52.00	9.32		61.00
5340	2'	"	0.107	57.00	8.69		66.00
5360	2'-6"	"	0.100	62.00	8.15		70.00
5380	3'	"	0.089	67.00	7.24		74.00

ACCESSORIES

ID Code	Component Descriptions	Unit of Meas.	Manhr / Unit	Material Cost	Labor Cost	Equipment Cost	Total Cost
03 - 15001	CONCRETE ACCESSORIES						03 - 15001
1000	Expansion joint, poured						
1010	Asphalt						
1020	1/2" x 1"	LF	0.016	0.90	1.02		1.92
1040	1" x 2"	"	0.017	2.82	1.11		3.93
1060	Liquid neoprene, cold applied						
1080	1/2" x 1"	LF	0.016	3.52	1.04		4.56
1100	1" x 2"	"	0.018	14.50	1.13		15.75
1110	Polyurethane, 2 parts						
1120	1/2" x 1"	LF	0.027	3.37	1.70		5.07
1140	1" x 2"	"	0.029	13.25	1.85		15.00
1150	Rubberized asphalt, cold						
1160	1/2" x 1"	LF	0.016	0.86	1.02		1.88
1180	1" x 2"	"	0.017	2.59	1.11		3.70
1190	Hot, fuel resistant						
1200	1/2" x 1"	LF	0.016	1.56	1.02		2.58
1220	1" x 2"	"	0.017	7.53	1.11		8.64
1300	Expansion joint, premolded, in slabs						
1310	Asphalt						
1320	1/2" x 6"	LF	0.020	1.00	1.27		2.27
1340	1" x 12"	"	0.027	1.67	1.70		3.37
1350	Cork						
1360	1/2" x 6"	LF	0.020	1.99	1.27		3.26
1380	1" x 12"	"	0.027	7.56	1.70		9.26

© 2020 BNi Publications Inc. ALL RIGHTS RESERVED

ACCESSORIES

ID Code	Description: Component Descriptions	Output: Unit of Meas.	Output: Manhr / Unit	Unit Costs: Material Cost	Unit Costs: Labor Cost	Unit Costs: Equipment Cost	Unit Costs: Total Cost
03 - 15001	**CONCRETE ACCESSORIES, Cont'd...**						**03 - 15001**
1390	Neoprene sponge						
1400	1/2" x 6"	LF	0.020	2.93	1.27		4.20
1420	1" x 12"	"	0.027	10.75	1.70		12.50
1430	Polyethylene foam						
1440	1/2" x 6"	LF	0.020	1.13	1.27		2.40
1460	1" x 12"	"	0.027	5.21	1.70		6.91
1560	Polyurethane foam						
1580	1/2" x 6"	LF	0.020	1.49	1.27		2.76
1600	1" x 12"	"	0.027	3.29	1.70		4.99
1610	Polyvinyl chloride foam						
1620	1/2" x 6"	LF	0.020	3.19	1.27		4.46
1640	1" x 12"	"	0.027	6.89	1.70		8.59
1650	Rubber, gray sponge						
1660	1/2" x 6"	LF	0.020	4.97	1.27		6.24
1680	1" x 12"	"	0.027	21.50	1.70		23.25
1700	Asphalt felt control joints or bond breaker, screed joints						
1780	4" slab	LF	0.016	1.34	1.02		2.36
1800	6" slab	"	0.018	1.68	1.13		2.81
1820	8" slab	"	0.020	2.19	1.27		3.46
1840	10" slab	"	0.023	3.09	1.45		4.54
1900	Keyed cold expansion and control joints, 24 ga.						
1940	4" slab	LF	0.050	1.09	3.19		4.28
1960	5" slab	"	0.050	1.45	3.19		4.64
1980	6" slab	"	0.053	1.68	3.40		5.08
1990	8" slab	"	0.057	2.04	3.64		5.68
2000	10" slab	"	0.062	2.24	3.92		6.16
2100	Waterstops						
2120	Polyvinyl chloride						
2125	Ribbed						
2130	3/16" thick x						
2140	4" wide	LF	0.040	1.59	2.55		4.14
2160	6" wide	"	0.044	2.42	2.83		5.25
2165	1/2" thick x						
2170	9" wide	LF	0.050	6.43	3.19		9.62
2178	Ribbed with center bulb						
2180	3/16" thick x 9" wide	LF	0.050	5.40	3.19		8.59
2200	3/8" thick x 9" wide	"	0.050	6.35	3.19		9.54
2240	Dumbbell type, 3/8" thick x 6" wide	"	0.044	6.43	2.83		9.26

© 2020 BNi Publications Inc. ALL RIGHTS RESERVED

ACCESSORIES

	Description	Output		Unit Costs			
ID Code	Component Descriptions	Unit of Meas.	Manhr / Unit	Material Cost	Labor Cost	Equipment Cost	Total Cost
03 - 15001	**CONCRETE ACCESSORIES, Cont'd...**						**03 - 15001**
2260	Plain, 3/8" thick x 9" wide	LF	0.050	8.54	3.19		11.75
2280	Center bulb, 3/8" thick x 9" wide	"	0.050	10.25	3.19		13.50
2300	Rubber						
2310	Flat dumbbell						
2315	3/8" thick x						
2320	6" wide	LF	0.044	7.91	2.83		10.75
2340	9" wide	"	0.050	8.54	3.19		11.75
2350	Center bulb						
2355	3/8" thick x						
2360	6" wide	LF	0.044	8.88	2.83		11.75
2380	9" wide	"	0.050	11.00	3.19		14.25
5060	Vapor barrier						
5090	4 mil polyethylene	SF	0.003	0.05	0.17		0.22
5094	6 mil polyethylene	"	0.003	0.08	0.17		0.25
5200	Gravel porous fill, under floor slabs, 3/4" stone	CY	1.333	22.25	85.00		110
6000	Reinforcing accessories						
6010	Beam bolsters						
6020	1-1/2" high, plain	LF	0.008	0.59	0.66		1.25
6030	Galvanized	"	0.008	1.30	0.66		1.96
6035	3" high						
6040	Plain	LF	0.010	0.84	0.83		1.67
6050	Galvanized	"	0.010	2.08	0.83		2.91
6080	Slab bolsters						
6090	1" high						
6100	Plain	LF	0.004	0.63	0.33		0.96
6110	Galvanized	"	0.004	1.27	0.33		1.60
6115	2" high						
6120	Plain	LF	0.004	0.70	0.37		1.07
6130	Galvanized	"	0.004	1.49	0.37		1.86
6210	Chairs, high chairs						
6215	3" high						
6220	Plain	EA	0.020	1.70	1.66		3.36
6230	Galvanized	"	0.020	1.87	1.66		3.53
6235	5" high						
6240	Plain	EA	0.021	1.75	1.75		3.50
6250	Galvanized	"	0.021	3.52	1.75		5.27
6255	8" high						
6260	Plain	EA	0.023	2.87	1.90		4.77

© 2020 BNi Publications Inc. ALL RIGHTS RESERVED

ACCESSORIES

	Description	Output		Unit Costs			
ID Code	Component Descriptions	Unit of Meas.	Manhr / Unit	Material Cost	Labor Cost	Equipment Cost	Total Cost
03 - 15001	**CONCRETE ACCESSORIES, Cont'd...**						**03 - 15001**
6270	Galvanized	EA	0.023	4.88	1.90		6.78
6275	12" high						
6280	Plain	EA	0.027	5.20	2.22		7.42
6290	Galvanized	"	0.027	9.93	2.22		12.25
6295	Continuous, high chair						
6299	3" high						
6300	Plain	LF	0.005	2.36	0.44		2.80
6310	Galvanized	"	0.005	2.92	0.44		3.36
6315	5" high						
6320	Plain	LF	0.006	2.54	0.47		3.01
6330	Galvanized	"	0.006	3.58	0.47		4.05
6335	8" high						
6340	Plain	LF	0.006	2.92	0.51		3.43
6350	Galvanized	"	0.006	3.79	0.51		4.30
6355	12" high						
6360	Plain	LF	0.007	3.76	0.55		4.31
6370	Galvanized	"	0.007	4.88	0.55		5.43
7000	Spirals or Stirrups						
7010	8" to 24"	TON	20.000	3,050	1,660		4,710
7050	24" to 48"	"	17.778	3,050	1,480		4,530
7090	48" to 84"	"	16.000	3,050	1,330		4,380

REINFORCEMENT

03 - 21001	**BEAM REINFORCING**						**03 - 21001**
0980	Beam-girders						
1000	#3 - #4	TON	20.000	1,880	1,660		3,540
1010	#5 - #6	"	16.000	1,650	1,330		2,980
1011	#7 - #8	"	13.333	1,570	1,110		2,680
1012	#9 - #10	"	11.429	1,570	950		2,520
1013	#11	"	10.667	1,570	890		2,460
1014	#14	"	10.000	1,570	830		2,400
1018	Galvanized						
1020	#3 - #4	TON	20.000	3,190	1,660		4,850
1030	#5 - #6	"	16.000	3,020	1,330		4,350
1031	#7 - #8	"	13.333	2,900	1,110		4,010
1032	#9 - #10	"	11.429	2,900	950		3,850
1033	#11	"	10.667	2,900	890		3,790

© 2020 BNi Publications Inc. ALL RIGHTS RESERVED

REINFORCEMENT

	Description	Output		Unit Costs			
ID Code	Component Descriptions	Unit of Meas.	Manhr / Unit	Material Cost	Labor Cost	Equipment Cost	Total Cost
03 - 21001	**BEAM REINFORCING, Cont'd...**						**03 - 21001**
1034	#14	TON	10.000	2,900	830		3,730
1100	Epoxy coated						
1200	#3 - #4	TON	22.857	2,760	1,900		4,660
1210	#5 - #6	"	17.778	2,590	1,480		4,070
1220	#7 - #8	"	14.545	2,510	1,210		3,720
1230	#9 - #10	"	12.308	2,510	1,020		3,530
1240	#11	"	11.429	2,510	950		3,460
1250	#14	"	10.667	2,510	890		3,400
2000	Bond Beams						
2100	#3 - #4	TON	26.667	1,880	2,220		4,100
2110	#5 - #6	"	20.000	1,650	1,660		3,310
2120	#7 - #8	"	17.778	1,570	1,480		3,050
2200	Galvanized						
2210	#3 - #4	TON	26.667	3,060	2,220		5,280
2220	#5 - #6	"	20.000	3,020	1,660		4,680
2230	#7 - #8	"	17.778	2,900	1,480		4,380
2400	Epoxy coated						
2410	#3 - #4	TON	32.000	2,760	2,660		5,420
2420	#5 - #6	"	22.857	2,590	1,900		4,490
2430	#7 - #8	"	20.000	2,510	1,660		4,170
03 - 21002	**BOX CULVERT REINFORCING**						**03 - 21002**
0980	Box culverts						
1000	#3 - #4	TON	10.000	1,880	830		2,710
1020	#5 - #6	"	8.889	1,650	740		2,390
1040	#7 - #8	"	8.000	1,570	670		2,240
1060	#9 - #10	"	7.273	1,570	610		2,180
1080	#11	"	6.667	1,570	550		2,120
1180	Galvanized						
1200	#3 - #4	TON	10.000	3,060	830		3,890
1220	#5 - #6	"	8.889	3,020	740		3,760
1240	#7 - #8	"	8.000	2,900	670		3,570
1260	#9 - #10	"	7.273	2,900	610		3,510
1280	#11	"	6.667	2,900	550		3,450
2000	Epoxy coated						
2100	#3 - #4	TON	10.667	2,760	890		3,650
2110	#5 - #6	"	9.412	2,590	780		3,370
2120	#7 - #8	"	8.421	2,510	700		3,210

© 2020 BNi Publications Inc. ALL RIGHTS RESERVED

REINFORCEMENT

ID Code	Description: Component Descriptions	Output: Unit of Meas.	Output: Manhr / Unit	Unit Costs: Material Cost	Unit Costs: Labor Cost	Unit Costs: Equipment Cost	Unit Costs: Total Cost
03 - 21002	**BOX CULVERT REINFORCING, Cont'd...**						**03 - 21002**
2130	#9 - #10	TON	7.619	2,510	630		3,140
2140	#11	"	6.957	2,510	580		3,090
03 - 21003	**COLUMN REINFORCING**						**03 - 21003**
0980	Columns						
1000	#3 - #4	TON	22.857	1,880	1,900		3,780
1010	#5 - #6	"	17.778	1,650	1,480		3,130
1015	#7 - #8	"	16.000	1,570	1,330		2,900
1020	#9 - #10	"	14.545	1,570	1,210		2,780
1025	#11	"	13.333	1,570	1,110		2,680
1030	#14	"	12.308	1,570	1,020		2,590
1040	#18	"	11.429	1,570	950		2,520
1100	Galvanized						
1200	#3 - #4	TON	22.857	3,190	1,900		5,090
1300	#5 - #6	"	17.778	3,020	1,480		4,500
1320	#7 - #8	"	16.000	2,900	1,330		4,230
1340	#9 - #10	"	14.545	2,900	1,210		4,110
1360	#11	"	13.333	2,900	1,110		4,010
1380	#14	"	12.308	2,900	1,020		3,920
1400	#18	"	11.429	2,900	950		3,850
1500	Epoxy coated						
1510	#3 - #4	TON	26.667	2,760	2,220		4,980
1520	#5 - #6	"	20.000	2,590	1,660		4,250
1530	#7 - #8	"	17.778	2,510	1,480		3,990
1540	#9 - #10	"	16.000	2,510	1,330		3,840
1550	#11	"	14.545	2,510	1,210		3,720
1560	#14	"	13.333	2,510	1,110		3,620
1570	#18	"	12.308	2,510	1,020		3,530
03 - 21004	**ELEVATED SLAB REINFORCING**						**03 - 21004**
0980	Elevated slab						
1000	#3 - #4	TON	10.000	1,880	830		2,710
1020	#5 - #6	"	8.889	1,650	740		2,390
1040	#7 - #8	"	8.000	1,570	670		2,240
1060	#9 - #10	"	7.273	1,570	610		2,180
1080	#11	"	6.667	1,570	550		2,120
1980	Galvanized						
2000	#3 - #4	TON	10.000	3,060	830		3,890
2020	#5 - #6	"	8.889	3,020	740		3,760

© 2020 BNi Publications Inc. ALL RIGHTS RESERVED

REINFORCEMENT

ID Code	Description: Component Descriptions	Output: Unit of Meas.	Output: Manhr / Unit	Unit Costs: Material Cost	Unit Costs: Labor Cost	Unit Costs: Equipment Cost	Unit Costs: Total Cost
03 - 21004	**ELEVATED SLAB REINFORCING, Cont'd...**						**03 - 21004**
2040	#7 - #8	TON	8.000	2,900	670		3,570
2060	#9 - #10	"	7.273	2,900	610		3,510
2100	#11	"	6.667	2,900	550		3,450
3000	Epoxy coated						
3100	#3 - #4	TON	10.667	2,760	890		3,650
3110	#5 - #6	"	9.412	2,590	780		3,370
3120	#7 - #8	"	8.421	2,510	700		3,210
3130	#9 - #10	"	7.619	2,510	630		3,140
3140	#11	"	6.957	2,510	580		3,090
03 - 21005	**EQUIP. PAD REINFORCING**						**03 - 21005**
0980	Equipment pad						
1000	#3 - #4	TON	16.000	1,880	1,330		3,210
1020	#5 - #6	"	14.545	1,650	1,210		2,860
1040	#7 - #8	"	13.333	1,570	1,110		2,680
1060	#9 - #10	"	12.308	1,570	1,020		2,590
1080	#11	"	11.429	1,570	950		2,520
03 - 21006	**FOOTING REINFORCING**						**03 - 21006**
1000	Footings						
1010	Grade 50						
1020	#3 - #4	TON	13.333	1,880	1,110		2,990
1030	#5 - #6	"	11.429	1,650	950		2,600
1040	#7 - #8	"	10.000	1,570	830		2,400
1050	#9 - #10	"	8.889	1,570	740		2,310
1055	Grade 60						
1060	#3 - #4	TON	13.333	1,880	1,110		2,990
1072	#5 - #6	"	11.429	1,650	950		2,600
1074	#7 - #8	"	10.000	1,570	830		2,400
1080	#9 - #10	"	8.889	1,570	740		2,310
1090	Grade 70						
1100	#3 - #4	TON	13.333	1,880	1,110		2,990
1110	#5 - #6	"	11.429	1,650	950		2,600
1120	#7 - #8	"	10.000	1,570	830		2,400
1140	#9 - #10	"	8.889	1,570	740		2,310
1160	#11	"	8.000	1,570	670		2,240
4980	Straight dowels, 24" long						
5000	1" dia. (#8)	EA	0.020	5.82	1.66		7.48
5040	3/4" dia. (#6)	"	0.019	5.24	1.54		6.78

© 2020 BNi Publications Inc. ALL RIGHTS RESERVED

REINFORCEMENT

ID Code	Description: Component Descriptions	Output: Unit of Meas.	Output: Manhr / Unit	Unit Costs: Material Cost	Unit Costs: Labor Cost	Unit Costs: Equipment Cost	Unit Costs: Total Cost
03 - 21006	**FOOTING REINFORCING, Cont'd...**						**03 - 21006**
5050	5/8" dia. (#5)	EA	0.017	4.53	1.44		5.97
5060	1/2" dia. (#4)	"	0.016	3.41	1.33		4.74
03 - 21007	**FOUNDATION REINFORCING**						**03 - 21007**
0980	Foundations						
1000	#3 - #4	TON	13.333	1,880	1,110		2,990
1020	#5 - #6	"	11.429	1,650	950		2,600
1040	#7 - #8	"	10.000	1,570	830		2,400
1060	#9 - #10	"	8.889	1,570	740		2,310
1080	#11	"	8.000	1,570	670		2,240
1380	Galvanized						
1400	#3 - #4	TON	13.333	3,210	1,110		4,320
1410	#5 - #6	"	11.429	3,030	950		3,980
1420	#7 - #8	"	10.000	2,910	830		3,740
1430	#9 - #10	"	8.889	2,910	740		3,650
1440	#11	"	8.000	2,910	670		3,580
2000	Epoxy Coated						
2100	#3 - #4	TON	14.545	2,770	1,210		3,980
2110	#5 - #6	"	12.308	2,600	1,020		3,620
2120	#7 - #8	"	10.667	2,520	890		3,410
2130	#9 - #10	"	9.412	2,520	780		3,300
2140	#11	"	8.421	2,520	700		3,220
03 - 21008	**GRADE BEAM REINFORCING**						**03 - 21008**
0980	Grade beams						
1000	#3 - #4	TON	12.308	1,880	1,020		2,900
1020	#5 - #6	"	10.667	1,660	890		2,550
1040	#7 - #8	"	9.412	1,570	780		2,350
1060	#9 - #10	"	8.421	1,570	700		2,270
1080	#11	"	7.619	1,570	630		2,200
1090	Galvanized						
1100	#3 - #4	TON	12.308	3,210	1,020		4,230
1120	#5 - #6	"	10.667	3,030	890		3,920
1140	#7 - #8	"	9.412	2,910	780		3,690
1160	#9 - #10	"	8.421	2,910	700		3,610
1180	#11	"	7.619	2,910	630		3,540
2000	Epoxy coated						
2100	#3 - #4	TON	13.333	2,770	1,110		3,880
2110	#5 - #6	"	11.429	2,600	950		3,550

© 2020 BNi Publications Inc. ALL RIGHTS RESERVED

REINFORCEMENT

	Description	Output		Unit Costs			
ID Code	Component Descriptions	Unit of Meas.	Manhr / Unit	Material Cost	Labor Cost	Equipment Cost	Total Cost
03 - 21008	**GRADE BEAM REINFORCING, Cont'd...**						**03 - 21008**
2120	#7 - #8	TON	10.000	2,520	830		3,350
2130	#9 - #10	"	8.889	2,520	740		3,260
2140	#11	"	8.000	2,520	670		3,190
03 - 21009	**SLAB / MAT REINFORCING**						**03 - 21009**
0900	Bars, slabs						
1000	#3 - #4	TON	13.333	1,880	1,110		2,990
1020	#5 - #6	"	11.429	1,660	950		2,610
1040	#7 - #8	"	10.000	1,570	830		2,400
1060	#9 - #10	"	8.889	1,570	740		2,310
1080	#11	"	8.000	1,570	670		2,240
1090	Galvanized						
2000	#3 - #4	TON	13.333	3,210	1,110		4,320
2020	#5 - #6	"	11.429	3,030	950		3,980
2040	#7 - #8	"	10.000	2,910	830		3,740
2060	#9 - #10	"	8.889	2,910	740		3,650
2080	#11	"	8.000	2,910	670		3,580
2090	Epoxy coated						
3100	#3 - #4	TON	14.545	2,770	1,210		3,980
3110	#5 - #6	"	12.308	2,600	1,020		3,620
3120	#7 - #8	"	10.667	2,520	890		3,410
3130	#9 - #10	"	9.412	2,520	780		3,300
3140	#11	"	8.421	2,520	700		3,220
4990	Wire mesh, slabs						
5000	Galvanized						
5010	4x4						
5020	W1.4xW1.4	SF	0.005	0.46	0.44		0.90
5040	W2.0xW2.0	"	0.006	0.60	0.47		1.07
5060	W2.9xW2.9	"	0.006	0.85	0.51		1.36
5080	W4.0xW4.0	"	0.007	1.25	0.55		1.80
5090	6x6						
5100	W1.4xW1.4	SF	0.004	0.43	0.33		0.76
5120	W2.0xW2.0	"	0.004	0.60	0.37		0.97
5140	W2.9xW2.9	"	0.005	0.82	0.39		1.21
5150	W4.0xW4.0	"	0.005	0.88	0.44		1.32
5160	Standard						
5170	2x2						
5180	W.9xW.9	SF	0.005	0.46	0.44		0.90

© 2020 BNi Publications Inc. ALL RIGHTS RESERVED

REINFORCEMENT

ID Code	Description: Component Descriptions	Output: Unit of Meas.	Output: Manhr / Unit	Unit Costs: Material Cost	Unit Costs: Labor Cost	Unit Costs: Equipment Cost	Unit Costs: Total Cost
03 - 21009	**SLAB / MAT REINFORCING, Cont'd...**						**03 - 21009**
5190	4x4						
5200	W1.4xW1.4	SF	0.005	0.31	0.44		0.75
5300	W2.0xW2.0	"	0.006	0.40	0.47		0.87
5400	W2.9xW2.9	"	0.006	0.55	0.51		1.06
5500	W4.0xW4.0	"	0.007	0.85	0.55		1.40
5600	6x6						
5700	W1.4xW1.4	SF	0.004	0.20	0.33		0.53
5800	W2.0xW2.0	"	0.004	0.27	0.37		0.64
6000	W2.9xW2.9	"	0.005	0.40	0.39		0.79
6020	W4.0xW4.0	"	0.005	0.57	0.44		1.01
03 - 21010	**STAIR REINFORCING**						**03 - 21010**
0980	Stairs						
1000	#3 - #4	TON	16.000	1,880	1,330		3,210
1020	#5 - #6	"	13.333	1,660	1,110		2,770
1040	#7 - #8	"	11.429	1,570	950		2,520
1060	#9 - #10	"	10.000	1,570	830		2,400
1980	Galvanized						
2000	#3 - #4	TON	16.000	3,210	1,330		4,540
2020	#5 - #6	"	13.333	3,030	1,110		4,140
2040	#7 - #8	"	11.429	2,910	950		3,860
2060	#9 - #10	"	10.000	2,910	830		3,740
3000	Epoxy coated						
3100	#3 - #4	TON	17.778	2,770	1,480		4,250
3110	#5 - #6	"	14.545	2,600	1,210		3,810
3120	#7 - #8	"	12.308	2,520	1,020		3,540
3130	#9 - #10	"	10.667	2,520	890		3,410
4100	Stair Nosing, Steel, Galvanized						
4200	' long	EA	0.044	18.25	3.70		22.00
4210	' long	"	0.053	24.75	4.44		29.25
4230	' long	"	0.067	36.75	5.55		42.25
03 - 21011	**WALL REINFORCING**						**03 - 21011**
0980	Walls						
1000	#3 - #4	TON	11.429	1,880	950		2,830
1020	#5 - #6	"	10.000	1,660	830		2,490
1040	#7 - #8	"	8.889	1,570	740		2,310
1060	#9 - #10	"	8.000	1,570	670		2,240
1980	Galvanized						

© 2020 BNi Publications Inc. ALL RIGHTS RESERVED

REINFORCEMENT

	Description	Output		Unit Costs			
ID Code	Component Descriptions	Unit of Meas.	Manhr / Unit	Material Cost	Labor Cost	Equipment Cost	Total Cost
03 - 21011	**WALL REINFORCING, Cont'd...**						**03 - 21011**
2000	#3 - #4	TON	11.429	3,210	950		4,160
2020	#5 - #6	"	10.000	3,030	830		3,860
2040	#7 - #8	"	8.889	2,910	740		3,650
2060	#9 - #10	"	8.000	2,910	670		3,580
3000	Epoxy coated						
3100	#3 - #4	TON	12.308	2,770	1,020		3,790
3110	#5 - #6	"	10.667	2,600	890		3,490
3120	#7 - #8	"	9.412	2,520	780		3,300
3130	#9 - #10	"	8.421	2,520	700		3,220
8980	Masonry wall (horizontal)						
9000	#3 - #4	TON	32.000	1,880	2,660		4,540
9020	#5 - #6	"	26.667	1,660	2,220		3,880
9030	Galvanized						
9040	#3 - #4	TON	32.000	3,210	2,660		5,870
9060	#5 - #6	"	26.667	3,030	2,220		5,250
9180	Masonry wall (vertical)						
9200	#3 - #4	TON	40.000	1,880	3,330		5,210
9220	#5 - #6	"	32.000	1,660	2,660		4,320
9230	Galvanized						
9240	#3 - #4	TON	40.000	3,210	3,330		6,540
9260	#5 - #6	"	32.000	3,030	2,660		5,690
03 - 21016	**PILE CAP REINFORCING**						**03 - 21016**
0980	Pile caps						
1000	#3 - #4	TON	20.000	1,880	1,660		3,540
1020	#5 - #6	"	17.778	1,660	1,480		3,140
1040	#7 - #8	"	16.000	1,570	1,330		2,900
1060	#9 - #10	"	14.545	1,570	1,210		2,780
1080	#11	"	13.333	1,570	1,110		2,680
1090	Galvanized						
1100	#3 - #4	TON	20.000	3,210	1,660		4,870
1120	#5 - #6	"	17.778	3,030	1,480		4,510
1140	#7 - #8	"	16.000	2,910	1,330		4,240
1160	#9 - #10	"	14.545	2,910	1,210		4,120
1180	#11	"	13.333	2,910	1,110		4,020
2000	Epoxy coated						
2100	#3 - #4	TON	22.857	2,770	1,900		4,670
2110	#5 - #6	"	20.000	2,600	1,660		4,260

© 2020 BNi Publications Inc. ALL RIGHTS RESERVED

REINFORCEMENT

ID Code	Description / Component Descriptions	Output: Unit of Meas.	Output: Manhr / Unit	Unit Costs: Material Cost	Unit Costs: Labor Cost	Unit Costs: Equipment Cost	Unit Costs: Total Cost
03 - 21016	**PILE CAP REINFORCING, Cont'd...**						**03 - 21016**
2120	#7 - #8	TON	17.778	2,520	1,480		4,000
2130	#9 - #10	"	16.000	2,520	1,330		3,850
2140	#11	"	14.545	2,520	1,210		3,730

CAST-IN-PLACE CONCRETE

ID Code	Component Descriptions	Unit of Meas.	Manhr / Unit	Material Cost	Labor Cost	Equipment Cost	Total Cost
03 - 30531	**CONCRETE ADMIXTURES**						**03 - 30531**
1000	Concrete admixtures						
1020	Water reducing admixture	GAL					12.25
1040	Set retarder	"					26.75
1060	Air entraining agent	"					11.75

PLACING CONCRETE

ID Code	Component Descriptions	Unit of Meas.	Manhr / Unit	Material Cost	Labor Cost	Equipment Cost	Total Cost
03 - 31001	**BEAM CONCRETE**						**03 - 31001**
0960	Beams and girders						
0980	2500# or 3000# concrete						
1000	By crane	CY	0.960	140	61.00	42.00	240
1010	By pump	"	0.873	140	56.00	38.25	230
1020	By hand buggy	"	0.800	140	51.00		190
2480	3500# or 4000# concrete						
2500	By crane	CY	0.960	150	61.00	42.00	250
2520	By pump	"	0.873	150	56.00	38.25	240
2540	By hand buggy	"	0.800	150	51.00		200
4000	5000# concrete						
4010	By crane	CY	0.960	160	61.00	42.00	260
4020	By pump	"	0.873	160	56.00	38.25	250
4040	By hand buggy	"	0.800	160	51.00		210
9460	Bond beam, 3000# concrete						
9470	By pump						
9480	8" high						
9500	4" wide	LF	0.019	0.38	1.22	0.84	2.44
9520	6" wide	"	0.022	0.92	1.39	0.95	3.26
9530	8" wide	"	0.024	1.17	1.53	1.05	3.75
9540	10" wide	"	0.027	1.56	1.70	1.16	4.42
9550	12" wide	"	0.030	2.09	1.91	1.31	5.31
9555	16" high						
9560	8" wide	LF	0.030	2.88	1.91	1.31	6.10

© 2020 BNi Publications Inc. ALL RIGHTS RESERVED

PLACING CONCRETE

ID Code	Description: Component Descriptions	Output: Unit of Meas.	Output: Manhr / Unit	Unit Costs: Material Cost	Unit Costs: Labor Cost	Unit Costs: Equipment Cost	Unit Costs: Total Cost
03 - 31001	**BEAM CONCRETE, Cont'd...**						**03 - 31001**
9570	10" wide	LF	0.034	3.87	2.18	1.50	7.55
9580	12" wide	"	0.040	5.13	2.55	1.75	9.43
9585	By crane						
9590	8" high						
9600	4" wide	LF	0.021	0.47	1.33	0.91	2.71
9620	6" wide	"	0.023	0.87	1.45	1.00	3.32
9640	8" wide	"	0.024	1.11	1.53	1.05	3.69
9650	10" wide	"	0.027	1.45	1.70	1.16	4.31
9660	12" wide	"	0.030	1.97	1.91	1.31	5.19
9665	16" high						
9670	8" wide	LF	0.030	2.73	1.91	1.31	5.95
9680	10" wide	"	0.032	3.65	2.04	1.40	7.09
9690	12" wide	"	0.037	4.85	2.35	1.61	8.82
03 - 31002	**COLUMN CONCRETE**						**03 - 31002**
0980	Columns						
0990	2500# or 3000# concrete						
1000	By crane	CY	0.873	140	56.00	38.25	230
1010	By pump	"	0.800	140	51.00	35.00	230
1980	3500# or 4000# concrete						
2000	By crane	CY	0.873	150	56.00	38.25	240
2020	By pump	"	0.800	150	51.00	35.00	240
3980	5000# concrete						
4010	By crane	CY	0.873	160	56.00	38.25	250
4020	By pump	"	0.800	160	51.00	35.00	250
03 - 31003	**ELEVATED SLAB CONCRETE**						**03 - 31003**
0980	Elevated slab						
0990	2500# or 3000# concrete						
1000	By crane	CY	0.480	140	30.75	21.00	190
1010	By pump	"	0.369	140	23.50	16.25	180
1020	By hand buggy	"	0.800	140	51.00		190
1980	3500# or 4000# concrete						
2000	By crane	CY	0.480	150	30.75	21.00	200
2020	By pump	"	0.369	150	23.50	16.25	190
2040	By hand buggy	"	0.800	150	51.00		200
4000	5000# concrete						
4010	By crane	CY	0.480	160	30.75	21.00	210
4020	By pump	"	0.369	160	23.50	16.25	200

© 2020 BNi Publications Inc. ALL RIGHTS RESERVED

PLACING CONCRETE

ID Code	Description: Component Descriptions	Output: Unit of Meas.	Output: Manhr / Unit	Unit Costs: Material Cost	Unit Costs: Labor Cost	Unit Costs: Equipment Cost	Unit Costs: Total Cost
03 - 31003	**ELEVATED SLAB CONCRETE, Cont'd...**						**03 - 31003**
4040	By hand buggy	CY	0.800	160	51.00		210
8980	Topping						
8990	2500# or 3000# concrete						
9010	By crane	CY	0.480	140	30.75	21.00	190
9020	By pump	"	0.369	140	23.50	16.25	180
9040	By hand buggy	"	0.800	140	51.00		190
9080	3500# or 4000# concrete						
9100	By crane	CY	0.480	150	30.75	21.00	200
9120	By pump	"	0.369	150	23.50	16.25	190
9140	By hand buggy	"	0.800	150	51.00		200
9180	5000# concrete						
9200	By crane	CY	0.480	160	30.75	21.00	210
9210	By pump	"	0.369	160	23.50	16.25	200
9220	By hand buggy	"	0.800	160	51.00		210
03 - 31004	**EQUIPMENT PAD CONCRETE**						**03 - 31004**
0960	Equipment pad						
0980	2500# or 3000# concrete						
1000	By chute	CY	0.267	140	17.00		160
1020	By pump	"	0.686	140	43.75	30.00	210
1040	By crane	"	0.800	140	51.00	35.00	230
1050	3500# or 4000# concrete						
1060	By chute	CY	0.267	140	17.00		160
1080	By pump	"	0.686	140	43.75	30.00	210
1100	By crane	"	0.800	140	51.00	35.00	230
1110	5000# concrete						
1120	By chute	CY	0.267	160	17.00		180
1140	By pump	"	0.686	160	43.75	30.00	230
1160	By crane	"	0.800	160	51.00	35.00	250
03 - 31005	**FOOTING CONCRETE**						**03 - 31005**
0980	Continuous footing						
0990	2500# or 3000# concrete						
1000	By chute	CY	0.267	140	17.00		160
1010	By pump	"	0.600	140	38.25	26.25	200
1020	By crane	"	0.686	140	43.75	30.00	210
1980	3500# or 4000# concrete						
2000	By chute	CY	0.267	140	17.00		160
2020	By pump	"	0.600	140	38.25	26.25	200

© 2020 BNi Publications Inc. ALL RIGHTS RESERVED

PLACING CONCRETE

ID Code	Description Component Descriptions	Output Unit of Meas.	Manhr / Unit	Unit Costs Material Cost	Labor Cost	Equipment Cost	Total Cost
03 - 31005	**FOOTING CONCRETE, Cont'd...**						**03 - 31005**
2040	By crane	CY	0.686	140	43.75	30.00	210
4000	5000# concrete						
4010	By chute	CY	0.267	160	17.00		180
4020	By pump	"	0.600	160	38.25	26.25	230
4030	By crane	"	0.686	160	43.75	30.00	230
4980	Spread footing						
5000	2500# or 3000# concrete						
5010	Under 5 c.y.						
5020	By chute	CY	0.267	140	17.00		160
5040	By pump	"	0.640	140	40.75	28.00	210
5060	By crane	"	0.738	140	47.25	32.25	220
6980	Over 5 c.y.						
7000	By chute	CY	0.200	140	12.75		150
7020	By pump	"	0.565	140	36.00	24.75	200
7040	By crane	"	0.640	140	40.75	28.00	210
7060	3500# or 4000# concrete						
7070	Under 5 c.y.						
7080	By chute	CY	0.267	150	17.00		170
7100	By pump	"	0.640	140	40.75	28.00	210
7120	By crane	"	0.738	140	47.25	32.25	220
7130	Over 5 c.y.						
7140	By chute	CY	0.200	140	12.75		150
7160	By pump	"	0.565	140	36.00	24.75	200
7180	By crane	"	0.640	140	40.75	28.00	210
7200	5000# concrete						
7205	Under 5 c.y.						
7210	By chute	CY	0.267	150	17.00		170
7220	By pump	"	0.640	150	40.75	28.00	220
7230	By crane	"	0.738	150	47.25	32.25	230
7235	Over 5 c.y.						
7240	By chute	CY	0.200	150	12.75		160
7250	By pump	"	0.565	150	36.00	24.75	210
7260	By crane	"	0.640	150	40.75	28.00	220

© 2020 BNi Publications Inc. ALL RIGHTS RESERVED

PLACING CONCRETE

ID Code	Description: Component Descriptions	Output: Unit of Meas.	Output: Manhr / Unit	Unit Costs: Material Cost	Unit Costs: Labor Cost	Unit Costs: Equipment Cost	Unit Costs: Total Cost
03 - 31006	**GRADE BEAM CONCRETE**						**03 - 31006**
0960	Grade beam						
0980	2500# or 3000# concrete						
1000	By chute	CY	0.267	140	17.00		160
1020	By crane	"	0.686	140	43.75	30.00	210
1040	By pump	"	0.600	140	38.25	26.25	200
1060	By hand buggy	"	0.800	140	51.00		190
1070	3500# or 4000# concrete						
1080	By chute	CY	0.267	150	17.00		170
1100	By crane	"	0.686	150	43.75	30.00	220
1120	By pump	"	0.600	150	38.25	26.25	210
1140	By hand buggy	"	0.800	150	51.00		200
1150	5000# concrete						
1160	By chute	CY	0.267	160	17.00		180
1180	By crane	"	0.686	160	43.75	30.00	230
1190	By pump	"	0.600	160	38.25	26.25	230
1200	By hand buggy	"	0.800	160	51.00		210
03 - 31007	**PILE CAP CONCRETE**						**03 - 31007**
0970	Pile cap						
0980	2500# or 3000 concrete						
1000	By chute	CY	0.267	140	17.00		160
1005	By crane	"	0.800	140	51.00	35.00	230
1010	By pump	"	0.686	140	43.75	30.00	210
1020	By hand buggy	"	0.800	140	51.00		190
1980	3500# or 4000# concrete						
2000	By chute	CY	0.267	150	17.00		170
2010	By crane	"	0.800	150	51.00	35.00	240
2020	By pump	"	0.686	150	43.75	30.00	220
2040	By hand buggy	"	0.800	150	51.00		200
3980	5000# concrete						
4010	By chute	CY	0.267	160	17.00		180
4015	By crane	"	0.800	160	51.00	35.00	250
4020	By pump	"	0.686	160	43.75	30.00	230
4030	By hand buggy	"	0.800	160	51.00		210

© 2020 BNi Publications Inc. ALL RIGHTS RESERVED

PLACING CONCRETE

	Description	Output		Unit Costs			
ID Code	Component Descriptions	Unit of Meas.	Manhr / Unit	Material Cost	Labor Cost	Equipment Cost	Total Cost
03 - 31008	**SLAB / MAT CONCRETE**						**03 - 31008**
0960	Slab on grade						
0980	2500# or 3000# concrete						
1000	By chute	CY	0.200	140	12.75		150
1010	By crane	"	0.400	140	25.50	17.50	180
1020	By pump	"	0.343	140	22.00	15.00	180
1030	By hand buggy	"	0.533	140	34.00		170
1980	3500# or 4000# concrete						
2000	By chute	CY	0.200	150	12.75		160
2020	By crane	"	0.400	150	25.50	17.50	190
2040	By pump	"	0.343	150	22.00	15.00	190
2060	By hand buggy	"	0.533	150	34.00		180
3980	5000# concrete						
4010	By chute	CY	0.200	160	12.75		170
4020	By crane	"	0.400	160	25.50	17.50	200
4030	By pump	"	0.343	160	22.00	15.00	200
4040	By hand buggy	"	0.533	160	34.00		190
5000	Foundation mat						
5010	2500# or 3000# concrete, over 20 c.y.						
5020	By chute	CY	0.160	140	10.25		150
5040	By crane	"	0.343	140	22.00	15.00	180
5060	By pump	"	0.300	140	19.25	13.00	170
5080	By hand buggy	"	0.400	140	25.50		170
03 - 31009	**WALL CONCRETE**						**03 - 31009**
0940	Walls						
0960	2500# or 3000# concrete						
0980	To 4'						
1000	By chute	CY	0.229	140	14.50		150
1005	By crane	"	0.800	140	51.00	35.00	230
1010	By pump	"	0.738	140	47.25	32.25	220
1020	To 8'						
1030	By crane	CY	0.873	140	56.00	38.25	230
1040	By pump	"	0.800	140	51.00	35.00	230
1045	To 16'						
1050	By crane	CY	0.960	140	61.00	42.00	240
1060	By pump	"	0.873	140	56.00	38.25	230
1065	Over 16'						
1070	By crane	CY	1.067	140	68.00	46.75	250

© 2020 BNi Publications Inc. ALL RIGHTS RESERVED

PLACING CONCRETE

ID Code	Component Descriptions	Unit of Meas.	Manhr / Unit	Material Cost	Labor Cost	Equipment Cost	Total Cost
03 - 31009	**WALL CONCRETE, Cont'd...**						**03 - 31009**
1080	By pump	CY	0.960	140	61.00	42.00	240
2960	3500# or 4000# concrete						
2980	To 4'						
3000	By chute	CY	0.229	150	14.50		160
3020	By crane	"	0.800	150	51.00	35.00	240
3030	By pump	"	0.738	150	47.25	32.25	230
3060	To 8'						
3080	By crane	CY	0.873	150	56.00	38.25	240
3100	By pump	"	0.800	150	51.00	35.00	240
3105	To 16'						
3110	By crane	CY	0.960	150	61.00	42.00	250
3130	By pump	"	0.873	150	56.00	38.25	240
3135	Over 16'						
3140	By crane	CY	1.067	150	68.00	46.75	260
3150	By pump	"	0.960	150	61.00	42.00	250
3960	5000# concrete						
3980	To 4'						
4010	By chute	CY	0.229	160	14.50		170
4015	By crane	"	0.800	160	51.00	35.00	250
4020	By pump	"	0.738	160	47.25	32.25	240
4030	To 8'						
4050	By crane	CY	0.873	160	56.00	38.25	250
4070	By pump	"	0.800	160	51.00	35.00	250
4100	To 16'						
4110	By crane	CY	0.960	160	61.00	42.00	260
4150	By pump	"	0.873	160	56.00	38.25	250
8480	Filled block (CMU)						
8490	3000# concrete, by pump						
8500	4" wide	SF	0.034	0.53	2.18	1.50	4.21
8510	6" wide	"	0.040	1.19	2.55	1.75	5.49
8520	8" wide	"	0.048	1.86	3.06	2.10	7.02
8530	10" wide	"	0.056	2.49	3.60	2.47	8.56
8540	12" wide	"	0.069	3.21	4.37	3.00	10.50
8560	Pilasters, 3000# concrete	CF	0.960	7.28	61.00	42.00	110
8700	Wall cavity, 2" thick, 3000# concrete	SF	0.032	1.35	2.04	1.40	4.79

© 2020 BNi Publications Inc. ALL RIGHTS RESERVED

PLACING CONCRETE

ID Code	Description: Component Descriptions	Output: Unit of Meas.	Output: Manhr / Unit	Unit Costs: Material Cost	Unit Costs: Labor Cost	Unit Costs: Equipment Cost	Unit Costs: Total Cost
03 - 31011	**STAIR CONCRETE**						**03 - 31011**
0960	Stairs						
0980	2500# or 3000# concrete						
1000	By chute	CY	0.267	140	17.00		160
1020	By crane	"	0.800	140	51.00	35.00	230
1030	By pump	"	0.686	140	43.75	30.00	210
1040	By hand buggy	"	0.800	140	51.00		190
2100	3500# or 4000# concrete						
2120	By chute	CY	0.267	150	17.00		170
2140	By crane	"	0.800	150	51.00	35.00	240
2160	By pump	"	0.686	150	43.75	30.00	220
2180	By hand buggy	"	0.800	150	51.00		200
4000	5000# concrete						
4010	By chute	CY	0.267	160	17.00		180
4020	By crane	"	0.800	160	51.00	35.00	250
4030	By pump	"	0.686	160	43.75	30.00	230
4040	By hand buggy	"	0.800	160	51.00		210

CONCRETE FINISHING

ID Code	Component Descriptions	Unit of Meas.	Manhr / Unit	Material Cost	Labor Cost	Equipment Cost	Total Cost
03 - 35001	**CONCRETE FINISHES**						**03 - 35001**
0980	Floor finishes						
1000	Broom	SF	0.011		0.72		0.72
1020	Screed	"	0.010		0.63		0.63
1040	Darby	"	0.010		0.63		0.63
1060	Steel float	"	0.013		0.85		0.85
2000	Granolithic topping						
2020	1/2" thick	SF	0.036	0.48	2.32		2.80
2040	1" thick	"	0.040	0.88	2.55		3.43
2060	2" thick	"	0.044	1.06	2.83		3.89
4000	Wall finishes						
4020	Burlap rub, with cement paste	SF	0.013	0.13	0.85		0.98
4040	Float finish	"	0.020	0.13	1.27		1.40
4060	Etch with acid	"	0.013	0.42	0.85		1.27
4070	Sandblast						
4080	Minimum	SF	0.016	0.14	1.02	0.34	1.50
4100	Maximum	"	0.016	0.52	1.02	0.34	1.88
4110	Bush hammer						
4120	Green concrete	SF	0.040		2.55		2.55

© 2020 BNi Publications Inc. ALL RIGHTS RESERVED

CONCRETE FINISHING

ID Code	Description: Component Descriptions	Output: Unit of Meas.	Output: Manhr / Unit	Unit Costs: Material Cost	Unit Costs: Labor Cost	Unit Costs: Equipment Cost	Unit Costs: Total Cost
03 - 35001	**CONCRETE FINISHES, Cont'd...**						**03 - 35001**
4140	Cured concrete	SF	0.062		3.92		3.92
4160	Break ties and patch holes	"	0.016		1.02		1.02
4170	Carborundum						
4180	Dry rub	SF	0.027		1.70		1.70
4200	Wet rub	"	0.040		2.55		2.55
5000	Floor hardeners						
5010	Metallic						
5020	Light service	SF	0.010	0.44	0.63		1.07
5040	Heavy service	"	0.013	1.34	0.85		2.19
5050	Non-metallic						
5060	Light service	SF	0.010	0.22	0.63		0.85
5080	Heavy service	"	0.013	0.93	0.85		1.78
5200	Rusticated concrete finish						
5220	Beveled edge	LF	0.044	0.43	2.83		3.26
5240	Square edge	"	0.057	0.61	3.64		4.25
5400	Solid board concrete finish						
5420	Standard	SF	0.067	1.10	4.25		5.35
5440	Rustic	"	0.080	1.02	5.10		6.12

SPECIALTY PLACED CONCRETE

ID Code	Component Descriptions	Unit of Meas.	Manhr / Unit	Material Cost	Labor Cost	Equipment Cost	Total Cost
03 - 37190	**PNEUMATIC CONCRETE**						**03 - 37190**
0100	Pneumatic applied concrete (gunite)						
1035	2" thick	SF	0.030	6.55	1.90	1.66	10.00
1040	3" thick	"	0.040	8.04	2.53	2.21	12.75
1060	4" thick	"	0.048	9.81	3.04	2.66	15.50
1980	Finish surface						
2000	Minimum	SF	0.040		3.26		3.26
2020	Maximum	"	0.080		6.52		6.52

CONCRETE CURING

ID Code	Component Descriptions	Unit of Meas.	Manhr / Unit	Material Cost	Labor Cost	Equipment Cost	Total Cost
03 - 39001	**CURING CONCRETE**						**03 - 39001**
1000	Sprayed membrane						
1010	Slabs	SF	0.002	0.06	0.10		0.16
1020	Walls	"	0.002	0.09	0.12		0.21
1025	Curing paper						
1030	Slabs	SF	0.002	0.09	0.13		0.22

© 2020 BNi Publications Inc. ALL RIGHTS RESERVED

CONCRETE CURING

	Description	Output		Unit Costs			
ID Code	Component Descriptions	Unit of Meas.	Manhr / Unit	Material Cost	Labor Cost	Equipment Cost	Total Cost
03 - 39001	**CURING CONCRETE, Cont'd...**						**03 - 39001**
2000	Walls	SF	0.002	0.09	0.15		0.24
2010	Burlap						
2020	7.5 oz.	SF	0.003	0.07	0.17		0.24
2500	12 oz.	"	0.003	0.10	0.18		0.28

PRECAST CONCRETE

ID Code	Component Descriptions	Unit of Meas.	Manhr / Unit	Material Cost	Labor Cost	Equipment Cost	Total Cost
03 - 41001	**PRECAST BEAMS**						**03 - 41001**
0060	Prestressed, double tee, 24" deep, 8' wide						
0080	35' span						
0100	115 psf	SF	0.008	14.00	0.65	0.76	15.50
0120	140 psf	"	0.008	14.75	0.65	0.76	16.25
0130	40' span						
0140	80 psf	SF	0.009	13.25	0.69	0.82	14.75
0180	143 psf	"	0.009	14.00	0.69	0.82	15.50
0185	45' span						
0190	50 psf	SF	0.007	12.50	0.60	0.70	13.75
0200	70 psf	"	0.007	13.25	0.60	0.70	14.50
0220	100 psf	"	0.007	13.50	0.60	0.70	14.75
0230	130 psf	"	0.007	15.25	0.60	0.70	16.50
0235	50' span						
0240	75 psf	SF	0.007	12.50	0.54	0.63	13.75
0250	100 psf	"	0.007	13.50	0.54	0.63	14.75
3000	Precast beams, girders and joists						
3020	1000 lb/lf live load						
3040	10' span	LF	0.160	110	13.00	15.25	140
3060	20' span	"	0.096	120	7.82	9.20	140
3080	30' span	"	0.080	150	6.52	7.66	160
3200	3000 lb/lf live load						
3220	10' span	LF	0.160	120	13.00	15.25	150
3240	20' span	"	0.096	140	7.82	9.20	160
3260	30' span	"	0.080	180	6.52	7.66	190
3300	5000 lb/lf live load						
3320	10' span	LF	0.160	120	13.00	15.25	150
3340	20' span	"	0.096	160	7.82	9.20	180
3360	30' span	"	0.080	200	6.52	7.66	210

© 2020 BNi Publications Inc. ALL RIGHTS RESERVED

PRECAST CONCRETE

ID Code	Description: Component Descriptions	Output: Unit of Meas.	Output: Manhr / Unit	Unit Costs: Material Cost	Unit Costs: Labor Cost	Unit Costs: Equipment Cost	Unit Costs: Total Cost
03 - 41002	**PRECAST COLUMNS**						**03 - 41002**
0060	Prestressed concrete columns						
0080	10" x 10"						
0100	10' long	EA	0.960	330	78.00	92.00	500
0120	15' long	"	1.000	500	82.00	96.00	680
0140	20' long	"	1.067	660	87.00	100	850
0160	25' long	"	1.143	860	93.00	110	1,060
0180	30' long	"	1.200	1,030	98.00	110	1,240
0220	12" x 12"						
0240	20' long	EA	1.200	920	98.00	110	1,130
0300	25' long	"	1.297	1,110	110	120	1,340
0320	30' long	"	1.371	1,420	110	130	1,660
0980	16" x 16"						
1000	20' long	EA	1.200	1,420	98.00	110	1,630
1002	25' long	"	1.297	2,090	110	120	2,320
1003	30' long	"	1.371	2,470	110	130	2,710
1010	20" x 20"						
1020	20' long	EA	1.263	2,550	100	120	2,770
1022	25' long	"	1.333	3,480	110	130	3,720
1023	30' long	"	1.412	3,990	120	140	4,240
1030	24" x 24"						
1040	20' long	EA	1.333	3,890	110	130	4,130
1042	25' long	"	1.412	4,730	120	140	4,980
1043	30' long	"	1.500	5,840	120	140	6,110
1060	28" x 28"						
1080	20' long	EA	1.500	5,290	120	140	5,560
1082	25' long	"	1.600	6,400	130	150	6,680
1083	30' long	"	1.714	7,930	140	160	8,230
1089	32" x 32"						
1100	20' long	EA	1.600	6,680	130	150	6,960
1102	25' long	"	1.714	8,630	140	160	8,930
1103	30' long	"	1.846	9,880	150	180	10,210
1110	36" x 36"						
1120	20' long	EA	1.714	8,350	140	160	8,650
1122	25' long	"	1.846	10,430	150	180	10,760
1123	30' long	"	2.000	12,520	160	190	12,870

© 2020 BNi Publications Inc. ALL RIGHTS RESERVED

PRECAST CONCRETE

ID Code	Description: Component Descriptions	Output: Unit of Meas.	Output: Manhr / Unit	Unit Costs: Material Cost	Unit Costs: Labor Cost	Unit Costs: Equipment Cost	Unit Costs: Total Cost
03 - 41003	**PRECAST SLABS**						**03 - 41003**
0040	Prestressed flat slab						
0060	6" thick, 4' wide						
0080	20' span						
0100	80 psf	SF	0.020	20.25	1.63	1.91	23.75
0110	110 psf	"	0.020	20.25	1.63	1.91	23.75
0120	25' span						
0130	80 psf	SF	0.019	20.75	1.56	1.84	24.25
0940	Cored slab						
0960	6" thick, 4' wide						
0980	20' span						
2000	80 psf	SF	0.020	10.75	1.63	1.91	14.25
2030	100 psf	"	0.020	11.00	1.63	1.91	14.50
2050	130 psf	"	0.020	11.25	1.63	1.91	14.75
2060	8" thick, 4' wide						
2070	25' span						
2090	70 psf	SF	0.019	10.50	1.56	1.84	14.00
2100	125 psf	"	0.019	11.25	1.56	1.84	14.75
2110	170 psf	"	0.019	11.25	1.56	1.84	14.75
2115	30' span						
2120	70 psf	SF	0.016	10.50	1.30	1.53	13.25
2140	90 psf	"	0.016	11.00	1.30	1.53	13.75
2170	35' span						
2180	70 psf	SF	0.015	11.00	1.22	1.43	13.75
2190	10" thick, 4' wide						
2195	30' span						
2200	75 psf	SF	0.016	11.00	1.30	1.53	13.75
2220	100 psf	"	0.016	11.25	1.30	1.53	14.00
2240	130 psf	"	0.016	11.75	1.30	1.53	14.50
2260	35' span						
2280	60 psf	SF	0.015	11.25	1.22	1.43	14.00
2290	80 psf	"	0.015	11.50	1.22	1.43	14.25
2300	120 psf	"	0.015	12.00	1.22	1.43	14.75
2310	40' span						
2320	65 psf	SF	0.012	12.25	0.97	1.15	14.25
7000	Slabs, roof and floor members, 4' wide						
7020	6" thick, 25' span	SF	0.019	9.73	1.56	1.84	13.25
7040	8" thick, 30' span	"	0.015	11.25	1.18	1.39	13.75
7060	10" thick, 40' span	"	0.013	13.75	1.05	1.24	16.00

© 2020 BNi Publications Inc. ALL RIGHTS RESERVED

PRECAST CONCRETE

ID Code	Description: Component Descriptions	Output: Unit of Meas.	Output: Manhr / Unit	Unit Costs: Material Cost	Unit Costs: Labor Cost	Unit Costs: Equipment Cost	Unit Costs: Total Cost
03 - 41003	PRECAST SLABS, Cont'd...						03 - 41003
7100	Tee members						
7120	Multiple tee, roof and floor						
7140	Minimum	SF	0.012	12.50	0.97	1.15	14.50
7160	Maximum	"	0.024	15.75	1.95	2.30	20.00
7200	Double tee wall member						
7220	Minimum	SF	0.014	11.75	1.11	1.31	14.25
7240	Maximum	"	0.027	14.75	2.17	2.55	19.50
7280	Single tee						
7290	Short span, roof members						
7300	Minimum	SF	0.015	13.00	1.18	1.39	15.50
7320	Maximum	"	0.030	16.00	2.44	2.87	21.25
7400	Long span, roof members						
7420	Minimum	SF	0.012	16.50	0.97	1.15	18.50
7440	Maximum	"	0.024	19.75	1.95	2.30	24.00

PRECAST ARCHITECTURAL CONCRETE

ID Code	Component Descriptions	Unit of Meas.	Manhr / Unit	Material Cost	Labor Cost	Equipment Cost	Total Cost
03 - 45001	PRECAST WALLS						03 - 45001
0060	Wall panel, 8' x 20'						
0070	Gray cement						
0080	Liner finish						
0100	4" wall	SF	0.014	15.50	1.11	1.31	18.00
0120	5" wall	"	0.014	16.75	1.15	1.35	19.25
0140	6" wall	"	0.015	19.25	1.18	1.39	21.75
0160	8" wall	"	0.015	19.25	1.22	1.43	22.00
0180	Sandblast finish						
0200	4" wall	SF	0.014	18.00	1.11	1.31	20.50
0210	5" wall	"	0.014	19.75	1.15	1.35	22.25
0220	6" wall	"	0.015	21.50	1.18	1.39	24.00
0230	8" wall	"	0.015	22.50	1.22	1.43	25.25
0280	White cement						
0290	Liner finish						
0300	4" wall	SF	0.014	19.00	1.11	1.31	21.50
0310	5" wall	"	0.014	20.00	1.15	1.35	22.50
0320	6" wall	"	0.015	22.00	1.18	1.39	24.50
0330	8" wall	"	0.015	23.25	1.22	1.43	26.00
2000	Sandblast finish						
2010	4" wall	SF	0.014	20.25	1.11	1.31	22.75

© 2020 BNi Publications Inc. ALL RIGHTS RESERVED

PRECAST ARCHITECTURAL CONCRETE

ID Code	Description: Component Descriptions	Output: Unit of Meas.	Output: Manhr / Unit	Unit Costs: Material Cost	Unit Costs: Labor Cost	Unit Costs: Equipment Cost	Unit Costs: Total Cost
03 - 45001	PRECAST WALLS, Cont'd... 03 - 45001						
2011	5" wall	SF	0.014	21.50	1.15	1.35	24.00
2012	6" wall	"	0.015	22.00	1.18	1.39	24.50
2013	8" wall	"	0.015	24.50	1.22	1.43	27.25
2015	Double tee wall panel, 24" deep						
2018	Gray cement						
2020	Liner finish	SF	0.016	11.00	1.30	1.53	13.75
2030	Sandblast finish	"	0.016	13.50	1.30	1.53	16.25
2035	White cement						
2040	Form liner finish	SF	0.016	15.25	1.30	1.53	18.00
2050	Sandblast finish	"	0.016	19.75	1.30	1.53	22.50
3000	Partition panels						
3020	4" wall	SF	0.016	16.50	1.30	1.53	19.25
3040	5" wall	"	0.016	17.75	1.30	1.53	20.50
3060	6" wall	"	0.016	19.50	1.30	1.53	22.25
3080	8" wall	"	0.016	21.25	1.30	1.53	24.00
3200	Cladding panels						
3220	4" wall	SF	0.017	16.25	1.39	1.64	19.25
3240	5" wall	"	0.017	17.75	1.39	1.64	20.75
3260	6" wall	"	0.017	19.75	1.39	1.64	22.75
3280	8" wall	"	0.017	21.25	1.39	1.64	24.25
3400	Sandwich panel, 2.5" cladding panel, 2" insulation						
3440	5" wall	SF	0.017	24.00	1.39	1.64	27.00
3460	6" wall	"	0.017	25.25	1.39	1.64	28.25
3480	8" wall	"	0.017	26.50	1.39	1.64	29.50

PRECAST CONCRETE SPECIALTIES

ID Code	Component Descriptions	Unit of Meas.	Manhr / Unit	Material Cost	Labor Cost	Equipment Cost	Total Cost
03 - 48001	PRECAST SPECIALTIES 03 - 48001						
0980	Precast concrete, coping, 4' to 8' long						
1000	12" wide	LF	0.060	11.00	3.80	3.32	18.00
1010	10" wide	"	0.069	9.76	4.34	3.80	18.00
1520	Splash block, 30"x12"x4"	EA	0.400	16.75	25.25	22.25	64.00
2000	Stair unit, per riser	"	0.400	110	25.25	22.25	160
4000	Sun screen and trellis, 8' long, 12" high						
4020	4" thick blades	EA	0.300	120	19.00	16.50	160
4040	5" thick blades	"	0.300	140	19.00	16.50	180
4060	6" thick blades	"	0.320	180	20.25	17.75	220
4080	8" thick blades	"	0.320	230	20.25	17.75	270

© 2020 BNi Publications Inc. ALL RIGHTS RESERVED

PRECAST CONCRETE SPECIALTIES

	Description	Output		Unit Costs			
ID Code	Component Descriptions	Unit of Meas.	Manhr / Unit	Material Cost	Labor Cost	Equipment Cost	Total Cost
03 - 48001	**PRECAST SPECIALTIES, Cont'd...**						**03 - 48001**
8000	Bearing pads for precast members, 2" wide strips						
8040	1/8" thick	LF	0.003	0.39	0.20		0.59
8060	1/4" thick	"	0.003	0.51	0.20		0.71
8080	1/2" thick	"	0.003	0.58	0.20		0.78
8100	3/4" thick	"	0.004	1.14	0.23		1.37
8120	1" thick	"	0.004	1.18	0.25		1.43
8140	1-1/2" thick	"	0.004	1.51	0.25		1.76

CEMENTITIOUS TOPPINGS

03 - 53001	**CONCRETE TOPPINGS**						**03 - 53001**
1000	Gypsum fill						
1020	2" thick	SF	0.005	2.16	0.31	0.21	2.69
1040	2-1/2" thick	"	0.005	2.46	0.32	0.22	3.00
1060	3" thick	"	0.005	3.03	0.33	0.22	3.59
1080	3-1/2" thick	"	0.005	3.46	0.34	0.23	4.03
1100	4" thick	"	0.006	4.05	0.38	0.26	4.69
2000	Formboard						
2020	Mineral fiber board						
2040	1" thick	SF	0.020	1.94	1.27		3.21
2060	1-1/2" thick	"	0.023	5.09	1.45		6.54
2070	Cement fiber board						
2080	1" thick	SF	0.027	1.51	1.70		3.21
2100	1-1/2" thick	"	0.031	1.94	1.96		3.90
2110	Glass fiber board						
2120	1" thick	SF	0.020	2.38	1.27		3.65
2140	1-1/2" thick	"	0.023	3.23	1.45		4.68
4000	Poured deck						
4010	Vermiculite or perlite						
4020	1 to 4 mix	CY	0.800	210	51.00	35.00	300
4040	1 to 6 mix	"	0.738	180	47.25	32.25	260
4050	Vermiculite or perlite						
4060	2" thick						
4080	1 to 4 mix	SF	0.005	1.94	0.32	0.22	2.48
4100	1 to 6 mix	"	0.005	1.41	0.29	0.20	1.90
4200	3" thick						
4220	1 to 4 mix	SF	0.007	2.64	0.47	0.32	3.43
4240	1 to 6 mix	"	0.007	2.09	0.43	0.30	2.82

© 2020 BNi Publications Inc. ALL RIGHTS RESERVED

CEMENTITIOUS TOPPINGS

	Description	Output		Unit Costs			
ID Code	Component Descriptions	Unit of Meas.	Manhr / Unit	Material Cost	Labor Cost	Equipment Cost	Total Cost
03 - 53001	**CONCRETE TOPPINGS, Cont'd...**						**03 - 53001**
6000	Concrete plank, lightweight						
6020	2" thick	SF	0.024	10.50	1.95	2.30	14.75
6040	2-1/2" thick	"	0.024	10.75	1.95	2.30	15.00
6080	3-1/2" thick	"	0.027	11.25	2.17	2.55	16.00
6100	4" thick	"	0.027	11.75	2.17	2.55	16.50
6500	Channel slab, lightweight, straight						
6520	2-3/4" thick	SF	0.024	8.49	1.95	2.30	12.75
6540	3-1/2" thick	"	0.024	8.74	1.95	2.30	13.00
6560	3-3/4" thick	"	0.024	9.43	1.95	2.30	13.75
6580	4-3/4" thick	"	0.027	12.00	2.17	2.55	16.75
7000	Gypsum plank						
7020	2" thick	SF	0.024	3.92	1.95	2.30	8.17
7040	3" thick	"	0.024	4.12	1.95	2.30	8.37
8000	Cement fiber, T & G planks						
8020	1" thick	SF	0.022	2.16	1.77	2.09	6.03
8040	1-1/2" thick	"	0.022	2.30	1.77	2.09	6.17
8060	2" thick	"	0.024	2.74	1.95	2.30	6.99
8080	2-1/2" thick	"	0.024	2.91	1.95	2.30	7.16
8100	3" thick	"	0.024	3.78	1.95	2.30	8.03
8120	3-1/2" thick	"	0.027	4.37	2.17	2.55	9.10
8140	4" thick	"	0.027	4.80	2.17	2.55	9.53

GROUT

03 - 61001	**GROUTING**						**03 - 61001**
1000	Grouting for bases						
1010	Non-shrink						
1020	Metallic grout						
1040	1" deep	SF	0.160	8.87	10.25	3.40	22.25
1060	2" deep	"	0.178	16.75	11.25	3.77	31.75
2000	Non-metallic grout						
2020	1" deep	SF	0.160	6.68	10.25	3.40	20.25
2040	2" deep	"	0.178	12.75	11.25	3.77	27.75
2480	Fluid type						
2500	Non-metallic						
2520	1" deep	SF	0.160	6.83	10.25	3.40	20.25
2540	2" deep	"	0.178	12.00	11.25	3.77	27.00
3000	Grouting for joints						

© 2020 BNi Publications Inc. ALL RIGHTS RESERVED

GROUT

ID Code	Description: Component Descriptions	Output: Unit of Meas.	Output: Manhr / Unit	Unit Costs: Material Cost	Unit Costs: Labor Cost	Unit Costs: Equipment Cost	Unit Costs: Total Cost
03 - 61001	**GROUTING, Cont'd...**						**03 - 61001**
3020	Portland cement grout (1 cement to 3 sand)						
3040	1/2" joint thickness						
3080	6" wide joints	LF	0.027	0.22	1.70	0.56	2.48
3100	8" wide joints	"	0.032	0.25	2.04	0.68	2.97
3200	1" joint thickness						
3220	4" wide joints	LF	0.025	0.25	1.59	0.53	2.37
3240	6" wide joints	"	0.028	0.41	1.76	0.58	2.75
3260	8" wide joints	"	0.033	0.51	2.12	0.70	3.34
3400	Non-shrink, non-metallic grout						
3420	1/2" joint thickness						
3440	4" wide joint	LF	0.023	1.20	1.45	0.48	3.14
3460	6" wide joint	"	0.027	1.58	1.70	0.56	3.84
3480	8" wide joint	"	0.032	2.06	2.04	0.68	4.78
3600	1" joint thickness						
3620	4" wide joint	LF	0.025	2.06	1.59	0.53	4.18
3640	6" wide joint	"	0.028	3.12	1.76	0.58	5.46
3660	8" wide joint	"	0.033	4.23	2.12	0.70	7.06

CONCRETE BORING

ID Code	Component Descriptions	Unit of Meas.	Manhr / Unit	Material Cost	Labor Cost	Equipment Cost	Total Cost
03 - 82131	**CORE DRILLING**						**03 - 82131**
0100	Concrete						
0110	6" thick						
0120	3" dia.	EA	0.571		36.50	12.25	48.50
0140	4" dia.	"	0.667		42.50	14.25	57.00
0160	6" dia.	"	0.800		51.00	17.00	68.00
0180	8" dia.	"	1.333		85.00	28.25	110
0300	8" thick						
0320	3" dia.	EA	0.800		51.00	17.00	68.00
0360	4" dia.	"	1.000		64.00	21.25	85.00
0380	6" dia.	"	1.143		73.00	24.25	97.00
0400	8" dia.	"	1.600		100	34.00	140
0420	10" thick						
0440	3" dia.	EA	1.000		64.00	21.25	85.00
0460	4" dia.	"	1.143		73.00	24.25	97.00
0480	6" dia.	"	1.333		85.00	28.25	110
0490	8" dia.	"	2.000		130	42.50	170
0520	12" thick						

© 2020 BNi Publications Inc. ALL RIGHTS RESERVED

CONCRETE BORING

	Description	Output		Unit Costs			
ID Code	Component Descriptions	Unit of Meas.	Manhr / Unit	Material Cost	Labor Cost	Equipment Cost	Total Cost
03 - 82131	**CORE DRILLING, Cont'd...**						**03 - 82131**
0540	3" dia.	EA	1.333		85.00	28.25	110
0560	4" dia.	"	1.600		100	34.00	140
0580	6" dia.	"	2.000		130	42.50	170
0600	8" dia.	"	2.667		170	57.00	230

© 2020 BNi Publications Inc. ALL RIGHTS RESERVED

DIVISION 04
MASONRY

MASONRY RESTORATION

	Description	Output		Unit Costs			
ID Code	Component Descriptions	Unit of Meas.	Manhr / Unit	Material Cost	Labor Cost	Equipment Cost	Total Cost
04 - 01201	**RESTORATION AND CLEANING**						**04 - 01201**
1080	Masonry cleaning						
1090	Washing brick						
1120	Smooth surface	SF	0.013	0.26	1.03		1.29
1130	Rough surface	"	0.018	0.36	1.38		1.74
1140	Steam clean masonry						
1150	Smooth face						
1220	Minimum	SF	0.010	0.56	0.63	0.21	1.41
1240	Maximum	"	0.015	0.90	0.92	0.30	2.13
1250	Rough face						
1260	Minimum	SF	0.013	0.82	0.85	0.28	1.95
1270	Maximum	"	0.020	1.21	1.27	0.42	2.91
1300	Sandblast masonry						
1320	Minimum	SF	0.016	0.50	1.02	0.34	1.86
1340	Maximum	"	0.027	0.76	1.70	0.56	3.02
1360	Pointing masonry						
1420	Brick	SF	0.032	1.37	2.48		3.85
1430	Concrete block	"	0.023	0.60	1.77		2.37
1450	Cut and repoint						
1470	Brick						
2020	Minimum	SF	0.040	0.47	3.11		3.58
2030	Maximum	"	0.080	0.81	6.22		7.03
2040	Stone work	LF	0.062	1.24	4.78		6.02
2060	Cut and recaulk						
3020	Oil base caulks	LF	0.053	1.59	4.14		5.73
3030	Butyl caulks	"	0.053	1.40	4.14		5.54
3040	Polysulfides and acrylics	"	0.053	2.71	4.14		6.85
3050	Silicones	"	0.053	3.17	4.14		7.31
4000	Cement and sand grout on walls, to 1/8" thick						
4020	Minimum	SF	0.032	0.78	2.48		3.26
4040	Maximum	"	0.040	1.98	3.11		5.09
8010	Brick removal and replacement						
8020	Minimum	EA	0.100	0.82	7.77		8.59
8040	Average	"	0.133	1.07	10.25		11.25
8060	Maximum	"	0.400	2.17	31.00		33.25

© 2020 BNi Publications Inc. ALL RIGHTS RESERVED

MORTAR, GROUT AND ACCESSORIES

ID Code	Description Component Descriptions	Output Unit of Meas.	Output Manhr / Unit	Unit Costs Material Cost	Unit Costs Labor Cost	Unit Costs Equipment Cost	Unit Costs Total Cost
04 - 05161	**MASONRY GROUT**						**04 - 05161**
0100	Grout, non-shrink, non-metallic, trowelable	CF	0.016	5.94	1.01	0.88	7.84
2110	Grout door frame, hollow metal						
2120	Single	EA	0.600	14.50	38.00	33.25	86.00
2140	Double	"	0.632	20.50	40.00	35.00	96.00
2980	Grout-filled concrete block (CMU)						
3000	4" wide	SF	0.020	0.43	1.26	1.10	2.80
3020	6" wide	"	0.022	1.12	1.38	1.20	3.71
3040	8" wide	"	0.024	1.65	1.52	1.33	4.50
3060	12" wide	"	0.025	2.71	1.60	1.40	5.71
3070	Grout-filled individual CMU cells						
3090	4" wide	LF	0.012	0.36	0.76	0.66	1.78
3100	6" wide	"	0.012	0.48	0.76	0.66	1.90
3120	8" wide	"	0.012	0.64	0.76	0.66	2.06
3140	10" wide	"	0.014	0.80	0.86	0.76	2.42
3160	12" wide	"	0.014	0.97	0.86	0.76	2.59
4000	Bond beams or lintels, 8" deep						
4010	6" thick	LF	0.022	0.97	1.39	0.95	3.31
4020	8" thick	"	0.024	1.28	1.53	1.05	3.86
4040	10" thick	"	0.027	1.61	1.70	1.16	4.47
4060	12" thick	"	0.030	1.93	1.91	1.31	5.15
5000	Cavity walls						
5020	2" thick	SF	0.032	1.07	2.04	1.40	4.51
5040	3" thick	"	0.032	1.61	2.04	1.40	5.05
5060	4" thick	"	0.034	2.14	2.18	1.50	5.82
5080	6" thick	"	0.040	3.21	2.55	1.75	7.51
04 - 05231	**MASONRY ACCESSORIES**						**04 - 05231**
0200	Foundation vents	EA	0.320	26.25	24.75		51.00
1010	Bar reinforcing						
1015	Horizontal						
1020	#3 - #4	LB	0.032	0.63	2.48		3.11
1030	#5 - #6	"	0.027	0.63	2.07		2.70
1035	Vertical						
1040	#3 - #4	LB	0.040	0.63	3.11		3.74
1050	#5 - #6	"	0.032	0.63	2.48		3.11
1100	Horizontal joint reinforcing						
1105	Truss type						
1110	4" wide, 6" wall	LF	0.003	0.20	0.24		0.44

© 2020 BNi Publications Inc. ALL RIGHTS RESERVED

MORTAR, GROUT AND ACCESSORIES

	Description	Output		Unit Costs			
ID Code	Component Descriptions	Unit of Meas.	Manhr / Unit	Material Cost	Labor Cost	Equipment Cost	Total Cost
04 - 05231	**MASONRY ACCESSORIES, Cont'd...**						**04 - 05231**
1120	6" wide, 8" wall	LF	0.003	0.20	0.25		0.45
1130	8" wide, 10" wall	"	0.003	0.25	0.27		0.52
1140	10" wide, 12" wall	"	0.004	0.25	0.28		0.53
1150	12" wide, 14" wall	"	0.004	0.30	0.29		0.59
1155	Ladder type						
1160	4" wide, 6" wall	LF	0.003	0.15	0.24		0.39
1170	6" wide, 8" wall	"	0.003	0.17	0.25		0.42
1180	8" wide, 10" wall	"	0.003	0.18	0.27		0.45
1190	10" wide, 12" wall	"	0.003	0.22	0.27		0.49
2000	Rectangular wall ties						
2005	3/16" dia., galvanized						
2010	2" x 6"	EA	0.013	0.38	1.03		1.41
2020	2" x 8"	"	0.013	0.40	1.03		1.43
2040	2" x 10"	"	0.013	0.47	1.03		1.50
2050	2" x 12"	"	0.013	0.53	1.03		1.56
2060	4" x 6"	"	0.016	0.44	1.24		1.68
2070	4" x 8"	"	0.016	0.49	1.24		1.73
2080	4" x 10"	"	0.016	0.63	1.24		1.87
2090	4" x 12"	"	0.016	0.73	1.24		1.97
2095	1/4" dia., galvanized						
2100	2" x 6"	EA	0.013	0.71	1.03		1.74
2110	2" x 8"	"	0.013	0.80	1.03		1.83
2120	2" x 10"	"	0.013	0.91	1.03		1.94
2130	2" x 12"	"	0.013	1.04	1.03		2.07
2140	4" x 6"	"	0.016	0.82	1.24		2.06
2150	4" x 8"	"	0.016	0.91	1.24		2.15
2160	4" x 10"	"	0.016	1.04	1.24		2.28
2170	4" x 12"	"	0.016	1.08	1.24		2.32
2200	Z-type wall ties, galvanized						
2215	6" long						
2220	1/8" dia.	EA	0.013	0.34	1.03		1.37
2230	3/16" dia.	"	0.013	0.36	1.03		1.39
2240	1/4" dia.	"	0.013	0.38	1.03		1.41
2245	8" long						
2250	1/8" dia.	EA	0.013	0.36	1.03		1.39
2260	3/16" dia.	"	0.013	0.38	1.03		1.41
2270	1/4" dia.	"	0.013	0.40	1.03		1.43
2275	10" long						

© 2020 BNi Publications Inc. ALL RIGHTS RESERVED

MORTAR, GROUT AND ACCESSORIES

	Description	Output		Unit Costs			
ID Code	Component Descriptions	Unit of Meas.	Manhr / Unit	Material Cost	Labor Cost	Equipment Cost	Total Cost
04 - 05231	**MASONRY ACCESSORIES, Cont'd...**						**04 - 05231**
2280	1/8" dia.	EA	0.013	0.38	1.03		1.41
2290	3/16" dia.	"	0.013	0.44	1.03		1.47
2300	1/4" dia.	"	0.013	0.49	1.03		1.52
3000	Dovetail anchor slots						
3015	Galvanized steel, filled						
3020	24 ga.	LF	0.020	1.17	1.55		2.72
3040	20 ga.	"	0.020	2.46	1.55		4.01
3060	16 oz. copper, foam filled	"	0.020	3.53	1.55		5.08
3100	Dovetail anchors						
3115	16 ga.						
3120	3-1/2" long	EA	0.013	0.39	1.03		1.42
3140	5-1/2" long	"	0.013	0.48	1.03		1.51
3150	12 ga.						
3160	3-1/2" long	EA	0.013	0.52	1.03		1.55
3180	5-1/2" long	"	0.013	0.86	1.03		1.89
3200	Dovetail, triangular galvanized ties, 12 ga.						
3220	3" x 3"	EA	0.013	0.88	1.03		1.91
3240	5" x 5"	"	0.013	0.95	1.03		1.98
3260	7" x 7"	"	0.013	1.07	1.03		2.10
3280	7" x 9"	"	0.013	1.14	1.03		2.17
3400	Brick anchors						
3420	Corrugated, 3-1/2" long						
3440	16 ga.	EA	0.013	0.57	1.03		1.60
3460	12 ga.	"	0.013	0.66	1.03		1.69
3500	Non-corrugated, 3-1/2" long						
3520	16 ga.	EA	0.013	0.47	1.03		1.50
3540	12 ga.	"	0.013	0.85	1.03		1.88
3580	Cavity wall anchors, corrugated, galvanized						
3600	5" long						
3620	16 ga.	EA	0.013	0.95	1.03		1.98
3640	12 ga.	"	0.013	1.43	1.03		2.46
3660	7" long						
3680	28 ga.	EA	0.013	1.05	1.03		2.08
3700	24 ga.	"	0.013	1.33	1.03		2.36
3720	22 ga.	"	0.013	1.36	1.03		2.39
3740	16 ga.	"	0.013	1.55	1.03		2.58
3800	Mesh ties, 16 ga., 3" wide						
3820	8" long	EA	0.013	1.28	1.03		2.31

© 2020 BNi Publications Inc. ALL RIGHTS RESERVED

MORTAR, GROUT AND ACCESSORIES

ID Code	Description: Component Descriptions	Output: Unit of Meas.	Output: Manhr / Unit	Unit Costs: Material Cost	Unit Costs: Labor Cost	Unit Costs: Equipment Cost	Unit Costs: Total Cost
04 - 05231	**MASONRY ACCESSORIES, Cont'd...**						**04 - 05231**
3840	12" long	EA	0.013	1.43	1.03		2.46
3860	20" long	"	0.013	1.96	1.03		2.99
3900	24" long	"	0.013	2.16	1.03		3.19
04 - 05232	**MASONRY CONTROL JOINTS**						**04 - 05232**
1000	Control joint, cross shaped PVC	LF	0.020	2.38	1.55		3.93
1010	Closed cell joint filler						
1020	1/2"	LF	0.020	0.41	1.55		1.96
1040	3/4"	"	0.020	0.85	1.55		2.40
1070	Rubber, for						
1080	4" wall	LF	0.020	2.75	1.55		4.30
1090	6" wall	"	0.021	3.40	1.63		5.03
1100	8" wall	"	0.022	4.10	1.72		5.82
1110	PVC, for						
1120	4" wall	LF	0.020	1.43	1.55		2.98
1140	6" wall	"	0.021	2.41	1.63		4.04
1160	8" wall	"	0.022	3.65	1.72		5.37
04 - 05235	**MASONRY FLASHING**						**04 - 05235**
0080	Through-wall flashing						
1000	5 oz. coated copper	SF	0.067	4.18	5.18		9.36
1020	0.030" elastomeric	"	0.053	1.32	4.14		5.46

UNIT MASONRY

ID Code	Description	Unit of Meas.	Manhr / Unit	Material Cost	Labor Cost	Equipment Cost	Total Cost
04 - 21131	**BRICK MASONRY**						**04 - 21131**
0100	Standard size brick, running bond						
1000	Face brick, red (6.4/sf)						
1020	Veneer	SF	0.133	5.66	10.25		16.00
1030	Cavity wall	"	0.114	5.66	8.88		14.50
1040	9" solid wall	"	0.229	11.25	17.75		29.00
1200	Common brick (6.4/sf)						
1210	Select common for veneers	SF	0.133	3.69	10.25		14.00
1215	Back-up						
1220	4" thick	SF	0.100	3.32	7.77		11.00
1230	8" thick	"	0.160	6.65	12.50		19.25
1235	Firewall						
1240	12" thick	SF	0.267	10.75	20.75		31.50
1250	16" thick	"	0.364	14.25	28.25		42.50

© 2020 BNi Publications Inc. ALL RIGHTS RESERVED

UNIT MASONRY

ID Code	Description: Component Descriptions	Output: Unit of Meas.	Output: Manhr / Unit	Unit Costs: Material Cost	Unit Costs: Labor Cost	Unit Costs: Equipment Cost	Unit Costs: Total Cost
04 - 21131	**BRICK MASONRY, Cont'd...**						**04 - 21131**
1300	Glazed brick (7.4/sf)						
1310	Veneer	SF	0.145	15.25	11.25		26.50
1400	Buff or gray face brick (6.4/sf)						
1410	Veneer	SF	0.133	6.58	10.25		16.75
1420	Cavity wall	"	0.114	6.58	8.88		15.50
1500	Jumbo or oversize brick (3/sf)						
1510	4" veneer	SF	0.080	4.76	6.22		11.00
1530	4" back-up	"	0.067	4.76	5.18		9.94
1540	8" back-up	"	0.114	5.52	8.88		14.50
1550	12" firewall	"	0.200	7.42	15.50		23.00
1560	16" firewall	"	0.267	10.50	20.75		31.25
1600	Norman brick, red face (4.5/sf)						
1620	4" veneer	SF	0.100	7.88	7.77		15.75
1640	Cavity wall	"	0.089	7.88	6.91		14.75
3000	Chimney, standard brick, including flue						
3020	16" x 16"	LF	0.800	32.00	62.00		94.00
3040	16" x 20"	"	0.800	54.00	62.00		120
3060	16" x 24"	"	0.800	58.00	62.00		120
3080	20" x 20"	"	1.000	45.00	78.00		120
3100	20" x 24"	"	1.000	61.00	78.00		140
3120	20" x 32"	"	1.143	68.00	89.00		160
4000	Window sill, face brick on edge	"	0.200	3.59	15.50		19.00
04 - 21231	**STRUCTURAL TILE**						**04 - 21231**
5000	Structural glazed tile						
5010	6T series, 5-1/2" x 12"						
5020	Glazed on one side						
5040	2" thick	SF	0.080	11.00	6.22		17.25
5060	4" thick	"	0.080	13.25	6.22		19.50
5080	6" thick	"	0.089	20.75	6.91		27.75
5100	8" thick	"	0.100	25.50	7.77		33.25
5200	Glazed on two sides						
5220	4" thick	SF	0.100	19.25	7.77		27.00
5240	6" thick	"	0.114	26.50	8.88		35.50
5500	Special shapes						
5510	Group 1	SF	0.160	11.25	12.50		23.75
5520	Group 2	"	0.160	14.25	12.50		26.75
5530	Group 3	"	0.160	18.75	12.50		31.25

© 2020 BNi Publications Inc. ALL RIGHTS RESERVED

UNIT MASONRY

ID Code	Description: Component Descriptions	Output: Unit of Meas.	Output: Manhr / Unit	Unit Costs: Material Cost	Unit Costs: Labor Cost	Unit Costs: Equipment Cost	Unit Costs: Total Cost
04 - 21231	**STRUCTURAL TILE, Cont'd...**						**04 - 21231**
5540	Group 4	SF	0.160	37.75	12.50		50.00
5550	Group 5	"	0.160	46.00	12.50		59.00
5600	Fire rated						
5620	4" thick, 1 hr rating	SF	0.080	18.00	6.22		24.25
5640	6" thick, 2 hr rating	"	0.089	25.25	6.91		32.25
6000	8W series, 8" x 16"						
6010	Glazed on one side						
6020	2" thick	SF	0.053	12.50	4.14		16.75
6040	4" thick	"	0.053	13.25	4.14		17.50
6060	6" thick	"	0.062	22.00	4.78		26.75
6080	8" thick	"	0.062	24.00	4.78		28.75
6100	Glazed on two sides						
6120	4" thick	SF	0.067	21.00	5.18		26.25
6140	6" thick	"	0.080	29.25	6.22		35.50
6160	8" thick	"	0.080	35.50	6.22		41.75
6200	Special shapes						
6220	Group 1	SF	0.114	18.75	8.88		27.75
6230	Group 2	"	0.114	23.00	8.88		32.00
6240	Group 3	"	0.114	25.25	8.88		34.25
6250	Group 4	"	0.114	42.00	8.88		51.00
6260	Group 5	"	0.114	53.00	8.88		62.00
6270	Fire rated						
6290	4" thick, 1 hr rating	SF	0.114	31.50	8.88		40.50
6300	6" thick, 2 hr rating	"	0.114	43.25	8.88		52.00

CONCRETE UNIT MASONRY

ID Code	Component Descriptions	Unit of Meas.	Manhr / Unit	Material Cost	Labor Cost	Equipment Cost	Total Cost
04 - 22001	**CONCRETE MASONRY UNITS**						**04 - 22001**
0110	Hollow, load bearing						
0120	4"	SF	0.059	1.62	4.60		6.22
0140	6"	"	0.062	2.38	4.78		7.16
0160	8"	"	0.067	2.73	5.18		7.91
0180	10"	"	0.073	3.77	5.65		9.42
0190	12"	"	0.080	4.34	6.22		10.50
0280	Solid, load bearing						
0300	4"	SF	0.059	2.55	4.60		7.15
0320	6"	"	0.062	2.86	4.78		7.64
0340	8"	"	0.067	3.91	5.18		9.09

© 2020 BNi Publications Inc. ALL RIGHTS RESERVED

CONCRETE UNIT MASONRY

	Description	Output		Unit Costs			
ID Code	Component Descriptions	Unit of Meas.	Manhr / Unit	Material Cost	Labor Cost	Equipment Cost	Total Cost
04 - 22001	**CONCRETE MASONRY UNITS, Cont'd...**						**04 - 22001**
0360	10"	SF	0.073	4.16	5.65		9.81
0380	12"	"	0.080	6.19	6.22		12.50
0480	Back-up block, 8" x 16"						
0490	2"	SF	0.046	1.70	3.55		5.25
0540	4"	"	0.047	1.78	3.65		5.43
0560	6"	"	0.050	2.60	3.88		6.48
0580	8"	"	0.053	2.99	4.14		7.13
0600	10"	"	0.057	4.13	4.44		8.57
0620	12"	"	0.062	4.75	4.78		9.53
0980	Foundation wall, 8" x 16"						
1000	6"	SF	0.057	2.60	4.44		7.04
1030	8"	"	0.062	2.99	4.78		7.77
1040	10"	"	0.067	4.13	5.18		9.31
1050	12"	"	0.073	4.76	5.65		10.50
1055	Solid						
1060	6"	SF	0.062	3.15	4.78		7.93
1070	8"	"	0.067	4.30	5.18		9.48
1080	10"	"	0.073	4.57	5.65		10.25
1100	12"	"	0.080	6.79	6.22		13.00
1480	Exterior, styrofoam inserts, std weight, 8" x 16"						
1500	6"	SF	0.062	4.57	4.78		9.35
1530	8"	"	0.067	4.93	5.18		10.00
1540	10"	"	0.073	6.40	5.65		12.00
1550	12"	"	0.080	8.77	6.22		15.00
1580	Lightweight						
1600	6"	SF	0.062	5.09	4.78		9.87
1660	8"	"	0.067	5.73	5.18		11.00
1680	10"	"	0.073	6.08	5.65		11.75
1700	12"	"	0.080	8.04	6.22		14.25
1980	Acoustical slotted block						
2000	4"	SF	0.073	5.31	5.65		11.00
2020	6"	"	0.073	5.56	5.65		11.25
2040	8"	"	0.080	6.94	6.22		13.25
2050	Filled cavities						
2060	4"	SF	0.089	5.69	6.91		12.50
2070	6"	"	0.094	6.55	7.31		13.75
2080	8"	"	0.100	8.40	7.77		16.25
4000	Hollow, split face						

© 2020 BNi Publications Inc. ALL RIGHTS RESERVED

CONCRETE UNIT MASONRY

ID Code	Description: Component Descriptions	Output: Unit of Meas.	Output: Manhr / Unit	Unit Costs: Material Cost	Unit Costs: Labor Cost	Unit Costs: Equipment Cost	Unit Costs: Total Cost
04 - 22001	**CONCRETE MASONRY UNITS, Cont'd...**						**04 - 22001**
4020	4"	SF	0.059	3.64	4.60		8.24
4030	6"	"	0.062	4.21	4.78		8.99
4040	8"	"	0.067	4.42	5.18		9.60
4080	10"	"	0.073	4.95	5.65		10.50
4100	12"	"	0.080	5.28	6.22		11.50
4480	Split rib profile						
4500	4"	SF	0.073	4.42	5.65		10.00
4520	6"	"	0.073	5.13	5.65		10.75
4540	8"	"	0.080	5.58	6.22		11.75
4560	10"	"	0.080	6.12	6.22		12.25
4580	12"	"	0.080	6.63	6.22		12.75
4980	High strength block, 3500 psi						
5000	2"	SF	0.059	1.71	4.60		6.31
5020	4"	"	0.062	2.14	4.78		6.92
5030	6"	"	0.062	2.56	4.78		7.34
5040	8"	"	0.067	2.90	5.18		8.08
5050	10"	"	0.073	3.38	5.65		9.03
5060	12"	"	0.080	4.00	6.22		10.25
5500	Solar screen concrete block						
5505	4" thick						
5510	6" x 6"	SF	0.178	4.29	13.75		18.00
5520	8" x 8"	"	0.160	5.12	12.50		17.50
5530	12" x 12"	"	0.123	5.24	9.56		14.75
5540	8" thick						
5550	8" x 16"	SF	0.114	5.24	8.88		14.00
7000	Glazed block						
7020	Cove base, glazed 1 side, 2"	LF	0.089	11.00	6.91		18.00
7030	4"	"	0.089	11.25	6.91		18.25
7040	6"	"	0.100	11.50	7.77		19.25
7050	8"	"	0.100	12.50	7.77		20.25
7055	Single face						
7060	2"	SF	0.067	11.50	5.18		16.75
7080	4"	"	0.067	14.25	5.18		19.50
7090	6"	"	0.073	15.25	5.65		21.00
7100	8"	"	0.080	16.00	6.22		22.25
7105	10"	"	0.089	18.00	6.91		25.00
7110	12"	"	0.094	19.25	7.31		26.50
7115	Double face						

© 2020 BNi Publications Inc. ALL RIGHTS RESERVED

CONCRETE UNIT MASONRY

ID Code	Description: Component Descriptions	Output: Unit of Meas.	Output: Manhr / Unit	Unit Costs: Material Cost	Unit Costs: Labor Cost	Unit Costs: Equipment Cost	Unit Costs: Total Cost
04 - 22001	**CONCRETE MASONRY UNITS, Cont'd...**						**04 - 22001**
7120	4"	SF	0.084	17.25	6.54		23.75
7140	6"	"	0.089	20.25	6.91		27.25
7160	8"	"	0.100	21.25	7.77		29.00
7180	Corner or bullnose						
7200	2"	EA	0.100	18.25	7.77		26.00
7240	4"	"	0.114	23.50	8.88		32.50
7260	6"	"	0.114	28.75	8.88		37.75
7280	8"	"	0.133	31.25	10.25		41.50
7290	10"	"	0.145	34.00	11.25		45.25
7300	12"	"	0.160	36.50	12.50		49.00
9500	Gypsum unit masonry						
9510	Partition blocks (12"x30")						
9515	Solid						
9520	2"	SF	0.032	1.41	2.48		3.89
9525	Hollow						
9530	3"	SF	0.032	1.42	2.48		3.90
9540	4"	"	0.033	1.63	2.59		4.22
9550	6"	"	0.036	1.74	2.82		4.56
9900	Vertical reinforcing						
9920	4' o.c., add 5% to labor						
9940	2'8" o.c., add 15% to labor						
9960	Interior partitions, add 10% to labor						
04 - 22009	**BOND BEAMS & LINTELS**						**04 - 22009**
0980	Bond beam, no grout or reinforcement						
0990	8" x 16" x						
1000	4" thick	LF	0.062	1.85	4.78		6.63
1040	6" thick	"	0.064	2.83	4.97		7.80
1060	8" thick	"	0.067	3.24	5.18		8.42
1080	10" thick	"	0.070	4.01	5.40		9.41
1100	12" thick	"	0.073	4.56	5.65		10.25
6000	Beam lintel, no grout or reinforcement						
6010	8" x 16" x						
6020	10" thick	LF	0.080	8.85	6.22		15.00
6040	12" thick	"	0.089	9.42	6.91		16.25
6080	Precast masonry lintel						
7000	6 lf, 8" high x						
7020	4" thick	LF	0.133	7.80	10.25		18.00

© 2020 BNi Publications Inc. ALL RIGHTS RESERVED

CONCRETE UNIT MASONRY

ID Code	Description: Component Descriptions	Output: Unit of Meas.	Output: Manhr / Unit	Unit Costs: Material Cost	Unit Costs: Labor Cost	Unit Costs: Equipment Cost	Unit Costs: Total Cost
04 - 22009	**BOND BEAMS & LINTELS, Cont'd...**						**04 - 22009**
7040	6" thick	LF	0.133	9.96	10.25		20.25
7060	8" thick	"	0.145	11.25	11.25		22.50
7080	10" thick	"	0.145	13.50	11.25		24.75
7090	10 lf, 8" high x						
7100	4" thick	LF	0.080	9.80	6.22		16.00
7120	6" thick	"	0.080	12.00	6.22		18.25
7140	8" thick	"	0.089	13.50	6.91		20.50
7160	10" thick	"	0.089	18.25	6.91		25.25
8000	Steel angles and plates						
8010	Minimum	LB	0.011	1.46	0.88		2.34
8020	Maximum	"	0.020	2.15	1.55		3.70
8200	Various size angle lintels						
8205	1/4" stock						
8210	3" x 3"	LF	0.050	7.54	3.88		11.50
8220	3" x 3-1/2"	"	0.050	8.30	3.88		12.25
8225	3/8" stock						
8230	3" x 4"	LF	0.050	13.00	3.88		17.00
8240	3-1/2" x 4"	"	0.050	13.75	3.88		17.75
8250	4" x 4"	"	0.050	15.00	3.88		19.00
8260	5" x 3-1/2"	"	0.050	16.00	3.88		20.00
8262	6" x 3-1/2"	"	0.050	18.00	3.88		22.00
8265	1/2" stock						
8280	6" x 4"	LF	0.050	19.75	3.88		23.75

GLASS UNIT MASONRY

ID Code	Component Descriptions	Unit of Meas.	Manhr / Unit	Material Cost	Labor Cost	Equipment Cost	Total Cost
04 - 23001	**GLASS BLOCK**						**04 - 23001**
1000	Glass block, 4" thick						
1040	6" x 6"	SF	0.267	38.25	20.75		59.00
1060	8" x 8"	"	0.200	24.25	15.50		39.75
1080	12" x 12"	"	0.160	30.75	12.50		43.25

© 2020 BNi Publications Inc. ALL RIGHTS RESERVED

ADOBE UNIT MASONRY

	Description	Output		Unit Costs			
ID Code	Component Descriptions	Unit of Meas.	Manhr / Unit	Material Cost	Labor Cost	Equipment Cost	Total Cost
04 - 24001	**CLAY TILE**						**04 - 24001**
0100	Hollow clay tile, for back-up, 12" x 12"						
1000	Scored face						
1010	Load bearing						
1020	4" thick	SF	0.057	7.22	4.44		11.75
1040	6" thick	"	0.059	8.42	4.60		13.00
1060	8" thick	"	0.062	10.50	4.78		15.25
1080	10" thick	"	0.064	13.00	4.97		18.00
1100	12" thick	"	0.067	22.00	5.18		27.25
2000	Non-load bearing						
2020	3" thick	SF	0.055	5.85	4.28		10.25
2040	4" thick	"	0.057	6.77	4.44		11.25
2060	6" thick	"	0.059	7.84	4.60		12.50
2080	8" thick	"	0.062	9.99	4.78		14.75
2100	12" thick	"	0.067	17.75	5.18		23.00
4100	Partition, 12" x 12"						
4150	In walls						
4201	3" thick	SF	0.067	5.85	5.18		11.00
4210	4" thick	"	0.067	6.78	5.18		12.00
4220	6" thick	"	0.070	7.48	5.40		13.00
4230	8" thick	"	0.073	9.81	5.65		15.50
4240	10" thick	"	0.076	11.75	5.92		17.75
4250	12" thick	"	0.080	17.00	6.22		23.25
4300	Clay tile floors						
4320	4" thick	SF	0.044	6.77	3.45		10.25
4330	6" thick	"	0.047	8.42	3.65		12.00
4340	8" thick	"	0.050	10.50	3.88		14.50
4350	10" thick	"	0.053	13.00	4.14		17.25
4360	12" thick	"	0.057	19.25	4.44		23.75
6000	Terra cotta						
6020	Coping, 10" or 12" wide, 3" thick	LF	0.160	16.00	12.50		28.50

© 2020 BNi Publications Inc. ALL RIGHTS RESERVED

STONE

	Description	Output		Unit Costs			
ID Code	Component Descriptions	Unit of Meas.	Manhr / Unit	Material Cost	Labor Cost	Equipment Cost	Total Cost
04 - 43001	**STONE**						**04 - 43001**
0160	Rubble stone						
0180	Walls set in mortar						
0200	8" thick	SF	0.200	17.50	15.50		33.00
0220	12" thick	"	0.320	21.00	24.75		45.75
0420	18" thick	"	0.400	28.00	31.00		59.00
0440	24" thick	"	0.533	35.00	41.50		77.00
0445	Dry set wall						
0450	8" thick	SF	0.133	19.50	10.25		29.75
0455	12" thick	"	0.200	22.25	15.50		37.75
0460	18" thick	"	0.267	30.50	20.75		51.00
0465	24" thick	"	0.320	37.25	24.75		62.00
0480	Cut stone						
0490	Imported marble						
0510	Facing panels						
0520	3/4" thick	SF	0.320	45.25	24.75		70.00
0530	1-1/2" thick	"	0.364	64.00	28.25		92.00
0540	2-1/4" thick	"	0.444	77.00	34.50		110
0600	Base						
0610	1" thick						
0620	4" high	LF	0.400	20.50	31.00		52.00
0640	6" high	"	0.400	24.75	31.00		56.00
0700	Columns, solid						
0720	Plain faced	CF	5.333	150	410		560
0740	Fluted	"	5.333	420	410		830
0780	Flooring, travertine, minimum	SF	0.123	20.75	9.56		30.25
0800	Average	"	0.160	27.75	12.50		40.25
0820	Maximum	"	0.178	51.00	13.75		65.00
1000	Domestic marble						
1020	Facing panels						
1040	7/8" thick	SF	0.320	42.50	24.75		67.00
1060	1-1/2" thick	"	0.364	64.00	28.25		92.00
1080	2-1/4" thick	"	0.444	77.00	34.50		110
1500	Stairs						
1510	12" treads	LF	0.400	38.00	31.00		69.00
1520	6" risers	"	0.267	28.25	20.75		49.00
1525	Thresholds, 7/8" thick, 3' long, 4" to 6" wide						
1530	Plain	EA	0.667	34.50	52.00		87.00
1540	Beveled	"	0.667	38.25	52.00		90.00

© 2020 BNi Publications Inc. ALL RIGHTS RESERVED

STONE

ID Code	Description Component Descriptions	Output Unit of Meas.	Manhr / Unit	Unit Costs Material Cost	Labor Cost	Equipment Cost	Total Cost
04 - 43001	**STONE, Cont'd...**						**04 - 43001**
1545	Window sill						
1550	6" wide, 2" thick	LF	0.320	19.25	24.75		44.00
1555	Stools						
1560	5" wide, 7/8" thick	LF	0.320	25.75	24.75		51.00
1620	Limestone panels up to 12' x 5', smooth finish						
1630	2" thick	SF	0.096	29.50	6.08	5.32	41.00
1650	3" thick	"	0.096	34.50	6.08	5.32	46.00
1660	4" thick	"	0.096	49.25	6.08	5.32	61.00
1760	Miscellaneous limestone items						
1770	Steps, 14" wide, 6" deep	LF	0.533	63.00	41.50		100
1780	Coping, smooth finish	CF	0.267	88.00	20.75		110
1790	Sills, lintels, jambs, smooth finish	"	0.320	88.00	24.75		110
1800	Granite veneer facing panels, polished						
1810	7/8" thick						
1820	Black	SF	0.320	48.25	24.75		73.00
1840	Gray	"	0.320	38.00	24.75		63.00
1850	Base						
1860	4" high	LF	0.160	20.25	12.50		32.75
1870	6" high	"	0.178	24.50	13.75		38.25
1880	Curbing, straight, 6" x 16"	"	0.400	22.50	25.25	22.25	70.00
1890	Radius curbs, radius over 5'	"	0.533	27.50	33.75	29.50	91.00
1900	Ashlar veneer						
1905	4" thick, random	SF	0.320	34.00	24.75		59.00
1910	Pavers, 4" x 4" split						
1915	Gray	SF	0.160	33.25	12.50		45.75
1920	Pink	"	0.160	32.75	12.50		45.25
1930	Black	"	0.160	32.25	12.50		44.75
2000	Slate, panels						
2010	1" thick	SF	0.320	27.50	24.75		52.00
2020	2" thick	"	0.364	37.25	28.25		66.00
2030	Sills or stools						
2040	1" thick						
2060	6" wide	LF	0.320	12.75	24.75		37.50
2080	10" wide	"	0.348	20.75	27.00		47.75
2100	2" thick						
2120	6" wide	LF	0.364	21.00	28.25		49.25
2140	10" wide	"	0.400	34.75	31.00		66.00

© 2020 BNi Publications Inc. ALL RIGHTS RESERVED

REFRACTORIES

ID Code	Description: Component Descriptions	Output: Unit of Meas.	Output: Manhr / Unit	Unit Costs: Material Cost	Unit Costs: Labor Cost	Unit Costs: Equipment Cost	Unit Costs: Total Cost
04 - 51001	**FLUE LINERS**						**04 - 51001**
1000	Flue liners						
1020	Rectangular						
1040	8" x 12"	LF	0.133	10.50	10.25		20.75
1060	12" x 12"	"	0.145	13.25	11.25		24.50
1080	12" x 18"	"	0.160	23.25	12.50		35.75
1100	16" x 16"	"	0.178	25.00	13.75		38.75
1120	18" x 18"	"	0.190	31.00	14.75		45.75
1140	20" x 20"	"	0.200	52.00	15.50		68.00
1170	24" x 24"	"	0.229	62.00	17.75		80.00
1200	Round						
1220	18" dia.	LF	0.190	47.75	14.75		63.00
1240	24" dia.	"	0.229	94.00	17.75		110

MANUFACTURED MASONRY

ID Code	Component Descriptions	Unit of Meas.	Manhr / Unit	Material Cost	Labor Cost	Equipment Cost	Total Cost
04 - 71001	**SIMULATED BRICK AND STONE**						**04 - 71001**
1010	Brick Veneer Panel 48" x 33"						
1020	Antique	EA	0.400	140	31.00		170
1030	Baked Clay	"	0.400	120	31.00		150
1040	Brick Cream Caramel	"	0.400	140	31.00		170
1050	Dusky Evening	"	0.400	140	31.00		170
1060	Glacier	"	0.400	120	31.00		150
1070	Merlot	"	0.400	140	31.00		170
1080	Mixed Twilight	"	0.400	120	31.00		150
1090	Mocha	"	0.400	140	31.00		170
1100	Spiced	"	0.400	120	31.00		150
1200	Cast stone						
1220	6" Window/Door Molding	LF	0.080	20.75	6.22		27.00
1240	12" Dia. Column	"	0.200	88.00	15.50		100
1260	12" x 12" Quoins	EA	0.100	61.00	7.77		69.00
1280	6" Base Molding	LF	0.080	16.50	6.22		22.75
1300	10" Wallcap	"	0.080	18.75	6.22		25.00
1320	3 Piece baluster system	"	0.100	100	7.77		110
1340	24" x 24" x 1-1/2" Paver	SF	0.020	11.00	1.55		12.50
1360	Simulated Stone						
1380	Slate, 43-1/4" Wide × 8-1/2" High × approx. 1-3/4"						
1400	Arizona Red	EA	0.027	29.50	2.07		31.50
1420	Brunswick Brown	"	0.027	29.50	2.07		31.50

© 2020 BNi Publications Inc. ALL RIGHTS RESERVED

MANUFACTURED MASONRY

	Description	Output		Unit Costs			
ID Code	Component Descriptions	Unit of Meas.	Manhr / Unit	Material Cost	Labor Cost	Equipment Cost	Total Cost
04 - 71001	**SIMULATED BRICK AND STONE, Cont'd...**						**04 - 71001**
1440	Midnight Ash	EA	0.027	29.50	2.07		31.50
1460	Onyx	"	0.027	29.50	2.07		31.50
1480	Pewter	"	0.027	29.50	2.07		31.50
1500	Rocky Mountain Graphite	"	0.027	29.50	2.07		31.50
1520	Sahara	"	0.027	29.50	2.07		31.50
1540	Stacked Stone 49-1/4" Wide x 25" High 2" Thick						
1560	Birchwood	EA	0.400	110	31.00		140
1580	Earth	"	0.400	110	31.00		140
1600	Honey	"	0.400	110	31.00		140
1620	Espresso	"	0.400	110	31.00		140
1640	Spice	"	0.400	110	31.00		140
1660	Potomac	"	0.400	110	31.00		140
1680	Ponderosa	"	0.400	110	31.00		140
1700	Tudor	"	0.400	110	31.00		140
1720	Riviera	"	0.400	110	31.00		140
1740	White	"	0.400	110	31.00		140
1760	Corners 15-1/2 x 14-1/2 x 25" High x 2-1/4" Thick						
1780	Birchwood	EA	0.100	65.00	7.77		73.00
1800	Earth	"	0.100	65.00	7.77		73.00
1820	Honey	"	0.100	65.00	7.77		73.00
1840	Espresso	"	0.100	65.00	7.77		73.00
1860	Spice	"	0.100	65.00	7.77		73.00
1880	Potomac	"	0.100	65.00	7.77		73.00
1900	Ponderosa	"	0.100	65.00	7.77		73.00
1920	Tudor	"	0.100	65.00	7.77		73.00
1940	Riviera	"	0.100	65.00	7.77		73.00
1960	White	"	0.100	65.00	7.77		73.00
1980	Ledgers 48" Wide x 3-1/2" Deep x 3-5/8" High						
2000	Birchwood	EA	0.267	33.00	20.75		54.00
2020	Earth	"	0.267	33.00	20.75		54.00
2040	Honey	"	0.267	33.00	20.75		54.00
2060	Espresso	"	0.267	33.00	20.75		54.00
2080	Spice	"	0.267	33.00	20.75		54.00
2100	Potomac	"	0.267	33.00	20.75		54.00
2120	Ponderosa	"	0.267	33.00	20.75		54.00
2140	Tudor	"	0.267	33.00	20.75		54.00
2160	Riviera	"	0.267	33.00	20.75		54.00
2180	White	"	0.267	33.00	20.75		54.00

© 2020 BNi Publications Inc. ALL RIGHTS RESERVED

DIVISION 05
METALS

METAL FASTENINGS

ID Code	Description: Component Descriptions	Output: Unit of Meas.	Output: Manhr / Unit	Unit Costs: Material Cost	Unit Costs: Labor Cost	Unit Costs: Equipment Cost	Unit Costs: Total Cost
05 - 05231	**STRUCTURAL WELDING**						**05 - 05231**
0080	Welding						
0100	Single pass						
0120	1/8"	LF	0.040	0.33	3.60		3.93
0140	3/16"	"	0.053	0.55	4.80		5.35
0160	1/4"	"	0.067	0.77	6.00		6.77
0180	Miscellaneous steel shapes						
0190	Plain	LB	0.002	1.44	0.14		1.58
0200	Galvanized	"	0.003	1.80	0.24		2.04
0210	Plates						
0220	Plain	LB	0.002	1.29	0.18		1.47
0240	Galvanized	"	0.003	1.66	0.28		1.94
05 - 05239	**METAL FASTENINGS**						**05 - 05239**
0050	Powder Actuated Anchors						
0100	Loads						
0120	Single Shot						
0140	.22 Cal Green	EA					0.07
0160	.22 Cal Yellow	"					0.11
0180	.22 Cal Red	"					0.12
0200	Strip						
0220	.27 Cal Green	EA					0.18
0240	.27 Cal Yellow	"					0.17
0260	.27 Cal Red	"					0.17
0280	Pins						
0300	.145 Dia. x Length 300 Head						
0320	1/2"	EA					0.18
0340	5/8"	"					0.25
0360	3/4"	"					0.29
0380	1"	"					0.37
0400	1-1/4"	"					0.38
0420	1-1/2"	"					0.44
0440	2"	"					0.46
0460	2-1/2"	"					0.49
0480	3"	"					0.52
1000	Anchor bolts, material only						
1020	3/8" x						
1040	8" long	EA					1.11
1060	10" long	"					1.21

© 2020 BNi Publications Inc. ALL RIGHTS RESERVED

METAL FASTENINGS

	Description	Output		Unit Costs			
ID Code	Component Descriptions	Unit of Meas.	Manhr / Unit	Material Cost	Labor Cost	Equipment Cost	Total Cost
05 - 05239	**METAL FASTENINGS, Cont'd...**						**05 - 05239**
1080	12" long	EA					1.31
1090	1/2" x						
1100	8" long	EA					1.65
1120	10" long	"					1.76
1140	12" long	"					1.93
1160	18" long	"					2.10
1170	5/8" x						
1180	8" long	EA					1.54
1200	10" long	"					1.70
1220	12" long	"					1.81
1240	18" long	"					1.93
1260	24" long	"					2.10
1270	3/4" x						
1280	8" long	EA					2.20
1300	12" long	"					2.48
1320	18" long	"					3.41
1340	24" long	"					4.52
1350	7/8" x						
1360	8" long	EA					2.20
1380	12" long	"					2.48
1400	18" long	"					3.41
1420	24" long	"					4.52
1430	1" x						
1440	12" long	EA					4.41
1460	18" long	"					5.51
1480	24" long	"					6.61
1500	36" long	"					9.93
3980	Expansion shield						
4000	1/4"	EA					0.68
4020	3/8"	"					1.14
4040	1/2"	"					2.25
4060	5/8"	"					3.26
4080	3/4"	"					3.99
4100	1"	"					5.40
4480	Non-drilling anchor						
4500	1/4"	EA					0.71
4540	3/8"	"					0.88
4560	1/2"	"					1.35

© 2020 BNi Publications Inc. ALL RIGHTS RESERVED

METAL FASTENINGS

ID Code	Description: Component Descriptions	Output: Unit of Meas.	Output: Manhr / Unit	Unit Costs: Material Cost	Unit Costs: Labor Cost	Unit Costs: Equipment Cost	Unit Costs: Total Cost
05 - 05239	**METAL FASTENINGS, Cont'd...**						**05 - 05239**
4580	5/8"	EA					2.22
4600	3/4"	"					3.81
7000	Self-drilling anchor						
7020	1/4"	EA					1.79
7040	5/16"	"					2.23
7060	3/8"	"					2.68
7080	1/2"	"					3.58
7100	5/8"	"					6.80
7120	3/4"	"					8.92
7140	7/8"	"					12.50
8020	Add 25% for galvanized anchor bolts						
8040	Channel door frame, with anchors	LB	0.009	1.96	0.80		2.76
8060	Corner guard angle, with anchors	"	0.013	1.75	1.20		2.95
05 - 05240	**METAL LINTELS**						**05 - 05240**
0080	Lintels, steel						
0100	Plain	LB	0.020	1.33	1.80		3.13
0120	Galvanized	"	0.020	2.00	1.80		3.80

STRUCTURAL METAL FRAMING

ID Code	Component Descriptions	Unit of Meas.	Manhr / Unit	Material Cost	Labor Cost	Equipment Cost	Total Cost
05 - 12001	**BEAMS, GIRDERS, COLUMNS, TRUSSES**						**05 - 12001**
0100	Beams and girders, A-36						
0120	Welded	TON	4.800	3,160	390	460	4,010
0140	Bolted	"	4.364	3,070	360	420	3,840
0180	Columns						
0185	Pipe						
0190	6" dia.	LB	0.005	1.61	0.39	0.46	2.46
0200	12" dia.	"	0.004	1.37	0.32	0.38	2.07
0220	Purlins and girts						
0230	Welded	TON	8.000	3,010	650	770	4,430
0240	Bolted	"	6.857	2,960	560	660	4,180
1000	Column base plates						
1020	Up to 150 lb each	LB	0.005	1.73	0.48		2.21
1040	Over 150 lb each	"	0.007	1.41	0.60		2.01
1200	Structural pipe						
1220	3" to 5" o.d.	TON	9.600	3,250	780	920	4,950
1240	6" to 12" o.d.	"	6.857	3,010	560	660	4,230
1300	Structural tube						

© 2020 BNi Publications Inc. ALL RIGHTS RESERVED

STRUCTURAL METAL FRAMING

	Description	Output		Unit Costs			
ID Code	Component Descriptions	Unit of Meas.	Manhr / Unit	Material Cost	Labor Cost	Equipment Cost	Total Cost
05 - 12001	**BEAMS, GIRDERS, COLUMNS, TRUSSES, Cont'd...**						**05 - 12001**
1310	6" square						
1320	Light sections	TON	9.600	3,770	780	920	5,470
1340	Heavy sections	"	6.857	3,530	560	660	4,750
1350	6" wide rectangular						
1360	Light sections	TON	8.000	3,770	650	770	5,190
1380	Heavy sections	"	6.000	3,530	490	580	4,590
1390	Greater than 6" wide rectangular						
1400	Light sections	TON	8.000	4,010	650	770	5,430
1420	Heavy sections	"	6.000	3,780	490	580	4,840
1500	Miscellaneous structural shapes						
1520	Steel angle	TON	12.000	2,680	980	1,150	4,810
1540	Steel plate	"	8.000	3,040	650	770	4,460
5980	Trusses, field welded						
6000	60 lb/lf	TON	6.000	4,090	490	580	5,150
6020	100 lb/lf	"	4.800	3,580	390	460	4,430
6040	150 lb/lf	"	4.000	3,370	330	380	4,080
6050	Bolted						
6060	60 lb/lf	TON	5.333	4,040	430	510	4,990
6080	100 lb/lf	"	4.364	3,530	360	420	4,300
6100	150 lb/lf	"	3.692	3,350	300	350	4,000
9100	Add for galvanizing	"					830

WIRE ROPE ASSEMBLIES

ID Code	Component Descriptions	Unit of Meas.	Manhr / Unit	Material Cost	Labor Cost	Equipment Cost	Total Cost
05 - 15011	**WIRE ROPE**						**05 - 15011**
0100	Galvanized Cable						
0120	1/16"	LF					0.30
0130	1/8"	"					1.01
0140	3/16"	"					1.33
0160	1/4"	"					1.67
0180	5/16"	"					2.33
0200	3/8"	"					2.78
0220	1/2"	"					3.03
0240	5/8"	"					3.41
0260	3/4"	"					5.62
0280	1-1/4"	"					14.75
0300	Stainless						
0320	1/16"	LF					0.34

© 2020 BNi Publications Inc. ALL RIGHTS RESERVED

WIRE ROPE ASSEMBLIES

	Description	Output		Unit Costs			
ID Code	Component Descriptions	Unit of Meas.	Manhr / Unit	Material Cost	Labor Cost	Equipment Cost	Total Cost
05 - 15011	**WIRE ROPE, Cont'd...**						**05 - 15011**
0340	1/8"	LF					1.12
0360	3/16"	"					1.47
0380	1/4"	"					1.85
0400	5/16"	"					2.58
0420	3/8"	"					3.09
0440	1/2"	"					3.36
0460	5/8"	"					4.11
0480	3/4"	"					6.23
0490	1-1/4"	"					16.25
0520	Coated						
0540	1/16"	LF					0.58
0560	1/8"	"					1.25
0580	3/16"	"					2.00
1300	Clips						
1320	1/16"	EA					1.37
1340	1/8"	"					1.52
1360	3/16"	"					1.57
1380	1/4"	"					1.56
1400	5/16"	"					2.68
1420	3/8"	"					1.89
1440	1/2"	"					2.32
1460	5/8"	"					4.04
1480	3/4"	"					6.30
1500	1-1/4"	"					10.50
1520	Thimbles						
1580	3/16"	EA					0.75
1600	1/4"	"					1.00
1620	5/16"	"					1.07
1640	3/8"	"					1.47
1660	1/2"	"					1.83
1680	5/8"	"					4.95
1700	3/4"	"					4.85
1720	1-1/4"	"					13.50
1740	Sleeves (Swage type)						
1760	1/16"	EA					0.29
1780	1/8"	"					0.74
1800	3/16"	"					0.69
1820	1/4"	"					0.85

© 2020 BNi Publications Inc. ALL RIGHTS RESERVED

WIRE ROPE ASSEMBLIES

ID Code	Description: Component Descriptions	Output: Unit of Meas.	Output: Manhr / Unit	Unit Costs: Material Cost	Unit Costs: Labor Cost	Unit Costs: Equipment Cost	Unit Costs: Total Cost
05 - 15011	**WIRE ROPE, Cont'd...**						**05 - 15011**
1840	5/16"	EA					1.76
1860	3/8"	"					1.90
1880	1/2"	"					1.95
1900	5/8"	"					2.11
1920	3/4"	"					2.21
1940	1-1/4"	"					2.32

STEEL JOIST FRAMING

ID Code	Component Descriptions	Unit of Meas.	Manhr / Unit	Material Cost	Labor Cost	Equipment Cost	Total Cost
05 - 21001	**METAL JOISTS**						**05 - 21001**
0090	Joist						
0100	DLH series	TON	3.200	1,960	260	310	2,530
0120	K series	"	3.200	2,030	260	310	2,600
0140	LH series	"	3.200	1,960	260	310	2,530

STEEL DECKING

ID Code	Component Descriptions	Unit of Meas.	Manhr / Unit	Material Cost	Labor Cost	Equipment Cost	Total Cost
05 - 31001	**METAL DECKING**						**05 - 31001**
0090	Roof, 1-1/2" deep, non-composite						
0095	16 ga.						
0100	Primed	SF	0.008	4.91	0.65	0.76	6.32
0120	Galvanized	"	0.008	5.15	0.65	0.76	6.56
0130	18 ga.						
0140	Primed	SF	0.008	3.84	0.65	0.76	5.25
0200	Galvanized	"	0.008	4.21	0.65	0.76	5.62
0210	20 ga.						
0220	Primed	SF	0.008	2.82	0.65	0.76	4.23
0240	Galvanized	"	0.008	3.35	0.65	0.76	4.76
0250	22 ga.						
0260	Primed	SF	0.008	2.29	0.65	0.76	3.70
0270	Galvanized	"	0.008	2.57	0.65	0.76	3.98
1000	Open type decking, galvanized						
1010	1-1/2" deep						
1020	18 ga.	SF	0.008	3.69	0.65	0.76	5.10
1040	20 ga.	"	0.008	2.82	0.65	0.76	4.23
1060	22 ga.	"	0.008	2.43	0.65	0.76	3.84
1070	3" deep						
1080	16 ga.	SF	0.009	5.73	0.71	0.83	7.27

© 2020 BNi Publications Inc. ALL RIGHTS RESERVED

STEEL DECKING

ID Code	Description: Component Descriptions	Output: Unit of Meas.	Output: Manhr / Unit	Unit Costs: Material Cost	Unit Costs: Labor Cost	Unit Costs: Equipment Cost	Unit Costs: Total Cost
05 - 31001	**METAL DECKING, Cont'd...**						**05 - 31001**
1100	18 ga.	SF	0.009	5.39	0.71	0.83	6.93
1120	20 ga.	"	0.009	4.08	0.71	0.83	5.62
1140	22 ga.	"	0.009	3.96	0.71	0.83	5.50
1150	4-1/2" deep						
1160	16 ga.	SF	0.010	8.50	0.78	0.92	10.25
1180	18 ga.	"	0.010	7.03	0.78	0.92	8.73
1200	6" deep						
1220	16 ga.	SF	0.011	11.75	0.86	1.02	13.75
1240	18 ga.	"	0.011	10.25	0.86	1.02	12.25
1250	7-1/2" deep						
1260	16 ga.	SF	0.011	12.75	0.91	1.06	14.75
1280	18 ga.	"	0.011	12.00	0.91	1.06	14.00
2480	Cellular type						
2500	1-1/2" deep, galvanized						
2520	18-18 ga.	SF	0.010	10.25	0.78	0.92	12.00
2530	22-18 ga.	"	0.010	9.23	0.78	0.92	11.00
2535	3" deep, galvanized						
2540	16-16 ga.	SF	0.011	14.50	0.86	1.02	16.50
2550	18-16 ga.	"	0.011	13.50	0.86	1.02	15.50
2560	18-18 ga.	"	0.011	12.50	0.86	1.02	14.50
2570	20-18 ga.	"	0.011	11.00	0.86	1.02	13.00
2575	4-1/2" deep, galvanized						
2580	16-16 ga.	SF	0.011	21.75	0.93	1.09	23.75
2590	18-16 ga.	"	0.011	20.00	0.93	1.09	22.00
2600	18-18 ga.	"	0.011	19.00	0.93	1.09	21.00
2610	20-18 ga.	"	0.011	17.25	0.93	1.09	19.25
3500	Composite deck, non-cellular, galvanized						
3520	1-1/2" deep						
3530	18 ga.	SF	0.009	3.41	0.71	0.83	4.95
3540	20 ga.	"	0.009	3.11	0.71	0.83	4.65
3550	22 ga.	"	0.009	2.52	0.71	0.83	4.06
3560	3" deep						
3570	18 ga.	SF	0.009	4.80	0.75	0.88	6.43
3580	20 ga.	"	0.009	3.11	0.75	0.88	4.74
3590	22 ga.	"	0.009	2.52	0.75	0.88	4.15
3600	Slab form floor deck						
3610	9/16" deep						
3620	28 ga.	SF	0.009	1.53	0.71	0.83	3.07

© 2020 BNi Publications Inc. ALL RIGHTS RESERVED

STEEL DECKING

ID Code	Component Descriptions	Unit of Meas.	Manhr / Unit	Material Cost	Labor Cost	Equipment Cost	Total Cost
05 - 31001	**METAL DECKING, Cont'd...**						**05 - 31001**
3630	1-5/16" deep						
3640	24 ga.	SF	0.009	2.24	0.73	0.86	3.84
3650	22 ga.	"	0.009	2.69	0.73	0.86	4.29

COLD FORMED FRAMING

ID Code	Component Descriptions	Unit of Meas.	Manhr / Unit	Material Cost	Labor Cost	Equipment Cost	Total Cost
05 - 41001	**METAL FRAMING**						**05 - 41001**
0100	Furring channel, galvanized						
0110	Beams and columns, 3/4"						
0120	12" o.c.	SF	0.080	0.44	7.20		7.64
0140	16" o.c.	"	0.073	0.34	6.54		6.88
0150	Walls, 3/4"						
0160	12" o.c.	SF	0.040	0.44	3.60		4.04
0170	16" o.c.	"	0.033	0.34	3.00		3.34
0172	24" o.c.	"	0.027	0.24	2.40		2.64
0173	1-1/2"						
0174	12" o.c.	SF	0.040	0.72	3.60		4.32
0175	16" o.c.	"	0.033	0.55	3.00		3.55
0176	24" o.c.	"	0.027	0.37	2.40		2.77
0177	Stud, load bearing						
0178	16" o.c.						
0179	16 ga.						
0180	2-1/2"	SF	0.036	1.33	3.20		4.53
0190	3-5/8"	"	0.036	1.57	3.20		4.77
0200	4"	"	0.036	1.63	3.20		4.83
0220	6"	"	0.040	2.05	3.60		5.65
0280	18 ga.						
0300	2-1/2"	SF	0.036	1.08	3.20		4.28
0310	3-5/8"	"	0.036	1.33	3.20		4.53
0320	4"	"	0.036	1.39	3.20		4.59
0330	6"	"	0.040	1.76	3.60		5.36
0350	8"	"	0.040	2.12	3.60		5.72
0360	20 ga.						
0370	2-1/2"	SF	0.036	0.60	3.20		3.80
0390	3-5/8"	"	0.036	0.72	3.20		3.92
0400	4"	"	0.036	0.79	3.20		3.99
0420	6"	"	0.040	0.96	3.60		4.56
0450	8"	"	0.040	1.15	3.60		4.75

© 2020 BNi Publications Inc. ALL RIGHTS RESERVED

COLD FORMED FRAMING

	Description	Output		Unit Costs			
ID Code	Component Descriptions	Unit of Meas.	Manhr / Unit	Material Cost	Labor Cost	Equipment Cost	Total Cost
05 - 41001	**METAL FRAMING, Cont'd...**						**05 - 41001**
0460	24" o.c.						
0470	16 ga.						
0480	2-1/2"	SF	0.031	0.91	2.76		3.67
0510	3-5/8"	"	0.031	1.08	2.76		3.84
0520	4"	"	0.031	1.15	2.76		3.91
0530	6"	"	0.033	1.39	3.00		4.39
0540	8"	"	0.033	1.76	3.00		4.76
0545	18 ga.						
0550	2-1/2"	SF	0.031	0.72	2.76		3.48
0560	3-5/8"	"	0.031	0.84	2.76		3.60
0570	4"	"	0.031	0.91	2.76		3.67
0580	6"	"	0.033	1.15	3.00		4.15
0590	8"	"	0.033	1.39	3.00		4.39
0595	20 ga.						
0600	2-1/2"	SF	0.031	0.44	2.76		3.20
0610	3-5/8"	"	0.031	0.49	2.76		3.25
0620	4"	"	0.031	0.55	2.76		3.31
0630	6"	"	0.033	0.71	3.00		3.71
0640	8"	"	0.033	0.88	3.00		3.88

METAL FABRICATIONS

05 - 51001	**STAIRS**						**05 - 51001**
1000	Stock unit, steel, complete, per riser						
1010	Tread						
1020	3'-6" wide	EA	1.000	240	90.00		330
1040	4' wide	"	1.143	280	100		380
1060	5' wide	"	1.333	330	120		450
1200	Metal pan stair, cement filled, per riser						
1220	3'-6" wide	EA	0.800	260	72.00		330
1240	4' wide	"	0.889	300	80.00		380
1260	5' wide	"	1.000	340	90.00		430
1280	Landing, steel pan	SF	0.200	100	18.00		120
1300	Cast iron tread, steel stringers, stock units, per riser						
1310	Tread						
1320	3'-6" wide	EA	1.000	460	90.00		550
1340	4' wide	"	1.143	530	100		630
1360	5' wide	"	1.333	640	120		760

© 2020 BNi Publications Inc. ALL RIGHTS RESERVED

METAL FABRICATIONS

ID Code	Description: Component Descriptions	Output: Unit of Meas.	Output: Manhr / Unit	Unit Costs: Material Cost	Unit Costs: Labor Cost	Unit Costs: Equipment Cost	Unit Costs: Total Cost
05 - 51001	**STAIRS, Cont'd...**						**05 - 51001**
1400	Stair treads, abrasive, 12" x 3'-6"						
1410	Cast iron						
1420	3/8"	EA	0.400	220	36.00		260
1440	1/2"	"	0.400	280	36.00		320
1450	Cast aluminum						
1460	5/16"	EA	0.400	250	36.00		290
1480	3/8"	"	0.400	270	36.00		310
1500	1/2"	"	0.400	320	36.00		360
05 - 51331	**LADDERS**						**05 - 51331**
0100	Ladder, 18" wide						
0110	With cage	LF	0.533	110	48.00		160
0120	Without cage	"	0.400	70.00	36.00		110

METAL RAILINGS

ID Code	Description	Unit of Meas.	Manhr / Unit	Material Cost	Labor Cost	Equipment Cost	Total Cost
05 - 52131	**RAILINGS**						**05 - 52131**
0080	Railing, pipe						
0090	1-1/4" diameter, welded steel						
0095	2-rail						
0100	Primed	LF	0.160	31.75	14.50		46.25
0120	Galvanized	"	0.160	40.75	14.50		55.00
0130	3-rail						
0140	Primed	LF	0.200	40.75	18.00		59.00
0160	Galvanized	"	0.200	53.00	18.00		71.00
0170	Wall mounted, single rail, welded steel						
0180	Primed	LF	0.123	21.25	11.00		32.25
0200	Galvanized	"	0.123	27.50	11.00		38.50
0210	1-1/2" diameter, welded steel						
0215	2-rail						
0220	Primed	LF	0.160	34.50	14.50		49.00
0240	Galvanized	"	0.160	44.75	14.50		59.00
0245	3-rail						
0250	Primed	LF	0.200	43.25	18.00		61.00
0260	Galvanized	"	0.200	56.00	18.00		74.00
0270	Wall mounted, single rail, welded steel						
0280	Primed	LF	0.123	21.75	11.00		32.75
0300	Galvanized	"	0.123	28.50	11.00		39.50
0960	2" diameter, welded steel						

© 2020 BNi Publications Inc. ALL RIGHTS RESERVED

METAL RAILINGS

ID Code	Description: Component Descriptions	Output: Unit of Meas.	Output: Manhr / Unit	Unit Costs: Material Cost	Unit Costs: Labor Cost	Unit Costs: Equipment Cost	Unit Costs: Total Cost
05 - 52131	**RAILINGS, Cont'd...**						**05 - 52131**
0980	2-rail						
1000	Primed	LF	0.178	41.25	16.00		57.00
1020	Galvanized	"	0.178	54.00	16.00		70.00
1030	3-rail						
1040	Primed	LF	0.229	52.00	20.50		73.00
1070	Galvanized	"	0.229	68.00	20.50		89.00
1075	Wall mounted, single rail, welded steel						
1080	Primed	LF	0.133	23.75	12.00		35.75
1100	Galvanized	"	0.133	30.75	12.00		42.75

METAL GRATINGS

ID Code	Component Descriptions	Unit of Meas.	Manhr / Unit	Material Cost	Labor Cost	Equipment Cost	Total Cost
05 - 53001	**METAL GRATING**						**05 - 53001**
0200	Floor plate, checkered, steel						
0220	1/4"						
0240	Primed	SF	0.011	11.75	1.02		12.75
0260	Galvanized	"	0.011	18.00	1.02		19.00
0270	3/8"						
0280	Primed	SF	0.012	17.25	1.10		18.25
0300	Galvanized	"	0.012	27.25	1.10		28.25
1000	Aluminum grating, pressure-locked bearing bars						
1020	3/4" x 1/8"	SF	0.020	28.75	1.80		30.50
1030	1" x 1/8"	"	0.020	32.25	1.80		34.00
1040	1-1/4" x 1/8"	"	0.020	46.00	1.80		47.75
1050	1-1/4" x 3/16"	"	0.020	48.25	1.80		50.00
1060	1-1/2" x 1/8"	"	0.020	63.00	1.80		65.00
1070	1-3/4" x 3/16"	"	0.020	35.75	1.80		37.50
2000	Miscellaneous expenses						
2010	Cutting						
2020	Minimum	LF	0.053		4.80		4.80
2040	Maximum	"	0.080		7.20		7.20
2050	Banding						
2060	Minimum	LF	0.133		12.00		12.00
2080	Maximum	"	0.160		14.50		14.50
2090	Toe plates						
2100	Minimum	LF	0.160		14.50		14.50
2120	Maximum	"	0.200		18.00		18.00
3000	Steel grating, primed						

© 2020 BNi Publications Inc. ALL RIGHTS RESERVED

METAL GRATINGS

	Description	Output		Unit Costs			
ID Code	Component Descriptions	Unit of Meas.	Manhr / Unit	Material Cost	Labor Cost	Equipment Cost	Total Cost
05 - 53001	**METAL GRATING, Cont'd...**						**05 - 53001**
3020	3/4" x 1/8"	SF	0.027	9.98	2.40		12.50
3040	1" x 1/8"	"	0.027	10.25	2.40		12.75
3060	1-1/4" x 1/8"	"	0.027	11.50	2.40		14.00
3080	1-1/4" x 3/16"	"	0.027	15.25	2.40		17.75
3100	1-1/2" x 1/8"	"	0.027	14.00	2.40		16.50
3120	1-3/4" x 3/16"	"	0.027	20.00	2.40		22.50
3140	Galvanized						
3160	3/4" x 1/8"	SF	0.027	12.50	2.40		15.00
3180	1" x 1/8"	"	0.027	13.00	2.40		15.50
3200	1-1/4" x 1/8"	"	0.027	14.25	2.40		16.75
3220	1-1/4" x 3/16"	"	0.027	19.00	2.40		21.50
3240	1-1/2" x 1/8"	"	0.027	17.50	2.40		20.00
3320	1-3/4" x 3/16"	"	0.027	24.75	2.40		27.25
3400	Miscellaneous expenses						
3410	Cutting						
3420	Minimum	LF	0.057		5.14		5.14
3440	Maximum	"	0.089		8.00		8.00
3450	Banding						
3460	Minimum	LF	0.145		13.00		13.00
3480	Maximum	"	0.178		16.00		16.00
3490	Toe plates						
3500	Minimum	LF	0.178		16.00		16.00
3520	Maximum	"	0.229		20.50		20.50

METAL CASTINGS

05 - 56001	**CASTINGS**						**05 - 56001**
1000	Miscellaneous castings						
1020	Light sections	LB	0.016	8.84	1.44		10.25
1040	Heavy sections	"	0.011	6.47	1.02		7.49
1060	Manhole covers and frames						
1080	Regular, city type						
1090	18" dia.						
1100	100 lb	EA	1.600	410	140		550
1110	24" dia.						
1120	200 lb	EA	1.600	380	140		520
1130	300 lb	"	1.778	390	160		550
1140	400 lb	"	1.778	410	160		570

© 2020 BNi Publications Inc. ALL RIGHTS RESERVED

METAL CASTINGS

ID Code	Description: Component Descriptions	Output: Unit of Meas.	Output: Manhr / Unit	Unit Costs: Material Cost	Unit Costs: Labor Cost	Unit Costs: Equipment Cost	Unit Costs: Total Cost
05 - 56001	CASTINGS, Cont'd...						05 - 56001
1160	26" dia., 475 lb	EA	2.000	500	180		680
1180	30" dia., 600 lb	"	2.286	640	210		850
1200	8" square, 75 lb	"	0.320	180	28.75		210
1210	24" square						
1220	126 lb	EA	1.600	400	140		540
1240	500 lb	"	2.000	640	180		820
1400	Watertight type						
1420	20" dia., 200 lb	EA	2.000	330	180		510
1440	24" dia., 350 lb	"	2.667	570	240		810
1500	Steps, cast iron						
1520	7" x 9"	EA	0.160	18.75	14.50		33.25
1540	8" x 9"	"	0.178	26.50	16.00		42.50
1600	Manhole covers and frames, aluminum						
1620	12" x 12"	EA	0.320	91.00	28.75		120
1640	18" x 18"	"	0.320	94.00	28.75		120
1660	24" x 24"	"	0.400	100	36.00		140
1800	Corner protection						
1820	Steel angle guard with anchors						
1840	2" x 2" x 3/16"	LF	0.114	17.50	10.25		27.75
1860	2" x 3" x 1/4"	"	0.114	19.75	10.25		30.00
1880	3" x 3" x 5/16"	"	0.114	23.00	10.25		33.25
1900	3" x 4" x 5/16"	"	0.123	27.75	11.00		38.75
1920	4" x 4" x 5/16"	"	0.123	28.75	11.00		39.75

MISC. FABRICATIONS

ID Code	Description	Unit of Meas.	Manhr / Unit	Material Cost	Labor Cost	Equipment Cost	Total Cost
05 - 59001	METAL SPECIALTIES						05 - 59001
0060	Kick plate						
0080	4" high x 1/4" thick						
0100	Primed	LF	0.160	8.89	14.50		23.50
0120	Galvanized	"	0.160	10.00	14.50		24.50
0130	6" high x 1/4" thick						
0140	Primed	LF	0.178	10.00	16.00		26.00
0160	Galvanized	"	0.178	12.00	16.00		28.00

© 2020 BNi Publications Inc. ALL RIGHTS RESERVED

DECORATIVE METAL RAILINGS

	Description	Output		Unit Costs			
ID Code	Component Descriptions	Unit of Meas.	Manhr / Unit	Material Cost	Labor Cost	Equipment Cost	Total Cost
05 - 73001	**ORNAMENTAL METAL**						**05 - 73001**
1030	Railings, square bars, 6" o.c., shaped top rails						
1040	Steel	LF	0.400	97.00	36.00		130
1060	Aluminum	"	0.400	120	36.00		160
1080	Bronze	"	0.533	240	48.00		290
1100	Stainless steel	"	0.533	250	48.00		300
1200	Laminated metal or wood handrails						
1220	2-1/2" round or oval shape	LF	0.400	290	36.00		330

© 2020 BNi Publications Inc. ALL RIGHTS RESERVED

DIVISION 06
WOOD AND PLASTICS

FASTENERS AND ADHESIVES

	Description	Output		Unit Costs			
ID Code	Component Descriptions	Unit of Meas.	Manhr / Unit	Material Cost	Labor Cost	Equipment Cost	Total Cost
06 - 05231	**ACCESSORIES**						**06 - 05231**
0080	Column/post base, cast aluminum						
0100	4" x 4"	EA	0.200	19.75	16.25		36.00
0120	6" x 6"	"	0.200	27.50	16.25		43.75
0130	Bridging, metal, per pair						
0140	12" o.c.	EA	0.080	2.68	6.52		9.20
0160	16" o.c.	"	0.073	2.47	5.93		8.40
1000	Anchors						
1020	Bolts, threaded two ends, with nuts and washers						
1030	1/2" dia.						
1040	4" long	EA	0.050	3.13	4.07		7.20
1060	7-1/2" long	"	0.050	3.64	4.07		7.71
1070	3/4" dia.						
1080	7-1/2" long	EA	0.050	6.91	4.07		11.00
1100	15" long	"	0.050	10.50	4.07		14.50
1200	Framing anchors						
1202	10 gauge	EA	0.067	1.23	5.43		6.66
1210	Bolts, carriage						
1212	1/4 x 4	EA	0.080	0.81	6.52		7.33
1214	5/16 x 6	"	0.084	1.84	6.86		8.70
1216	3/8 x 6	"	0.084	3.72	6.86		10.50
1218	1/2 x 6	"	0.084	5.19	6.86		12.00
1240	Joist and beam hangers						
1250	18 ga.						
1260	2 x 4	EA	0.080	1.50	6.52		8.02
1280	2 x 6	"	0.080	1.80	6.52		8.32
1282	2 x 8	"	0.080	2.10	6.52		8.62
1284	2 x 10	"	0.089	2.26	7.24		9.50
1286	2 x 12	"	0.100	2.93	8.15		11.00
1288	16 ga.						
1290	3 x 6	EA	0.089	5.26	7.24		12.50
1292	3 x 8	"	0.089	6.39	7.24		13.75
1300	3 x 10	"	0.094	7.21	7.67		15.00
1302	3 x 12	"	0.107	8.12	8.69		16.75
1304	3 x 14	"	0.114	8.79	9.32		18.00
1320	4 x 6	"	0.089	9.02	7.24		16.25
1322	4 x 8	"	0.089	10.50	7.24		17.75
1324	4 x 10	"	0.094	12.00	7.67		19.75
1326	4 x 12	"	0.107	15.50	8.69		24.25

© 2020 BNi Publications Inc. ALL RIGHTS RESERVED

FASTENERS AND ADHESIVES

ID Code	Description: Component Descriptions	Output: Unit of Meas.	Output: Manhr / Unit	Unit Costs: Material Cost	Unit Costs: Labor Cost	Unit Costs: Equipment Cost	Unit Costs: Total Cost
06 - 05231	**ACCESSORIES, Cont'd...**						**06 - 05231**
1328	4 x 14	EA	0.114	16.25	9.32		25.50
1520	Rafter anchors, 18 ga., 1-1/2" wide						
1540	5-1/4" long	EA	0.067	1.13	5.43		6.56
1560	10-3/4" long	"	0.067	1.68	5.43		7.11
1600	Shear plates						
1620	2-5/8" dia.	EA	0.062	4.02	5.01		9.03
1640	4" dia.	"	0.067	8.37	5.43		13.75
1700	Sill anchors						
1720	Embedded in concrete	EA	0.080	2.96	6.52		9.48
1800	Split rings						
1820	2-1/2" dia.	EA	0.089	2.43	7.24		9.67
1840	4" dia.	"	0.100	4.48	8.15		12.75
1900	Strap ties, 14 ga., 1-3/8" wide						
1920	12" long	EA	0.067	3.04	5.43		8.47
1940	18" long	"	0.073	3.27	5.93		9.20
1960	24" long	"	0.080	4.86	6.52		11.50
1980	36" long	"	0.089	6.69	7.24		14.00
2000	Toothed rings						
2020	2-5/8" dia.	EA	0.133	2.81	10.75		13.50
2040	4" dia.	"	0.160	3.27	13.00		16.25
06 - 05731	**WOOD TREATMENT**						**06 - 05731**
1000	Creosote preservative treatment						
1020	8 lb/cf	BF					0.74
1040	10 lb/cf	"					0.88
1060	Salt preservative treatment						
1070	Oil borne						
1080	Minimum	BF					0.67
1100	Maximum	"					0.95
1120	Water borne						
1140	Minimum	BF					0.47
1150	Maximum	"					0.74
1200	Fire retardant treatment						
1220	Minimum	BF					0.95
1240	Maximum	"					1.15
1300	Kiln dried, softwood, add to framing costs						
1320	1" thick	BF					0.34
1340	2" thick	"					0.47

© 2020 BNi Publications Inc. ALL RIGHTS RESERVED

FASTENERS AND ADHESIVES

ID Code	Component Descriptions	Unit of Meas.	Manhr / Unit	Material Cost	Labor Cost	Equipment Cost	Total Cost
06 - 05731	**WOOD TREATMENT, Cont'd...**						**06 - 05731**
1360	3" thick	BF					0.60
1380	4" thick	"					0.74

ROUGH CARPENTRY

ID Code	Component Descriptions	Unit of Meas.	Manhr / Unit	Material Cost	Labor Cost	Equipment Cost	Total Cost
06 - 11001	**BLOCKING**						**06 - 11001**
1100	Steel construction						
1105	Walls						
1110	2x4	LF	0.053	0.54	4.34		4.88
1120	2x6	"	0.062	0.82	5.01		5.83
1130	2x8	"	0.067	1.08	5.43		6.51
1140	2x10	"	0.073	1.44	5.93		7.37
1150	2x12	"	0.080	1.86	6.52		8.38
1160	Ceilings						
1170	2x4	LF	0.062	0.54	5.01		5.55
1180	2x6	"	0.073	0.82	5.93		6.75
1190	2x8	"	0.080	1.08	6.52		7.60
1200	2x10	"	0.089	1.44	7.24		8.68
1210	2x12	"	0.100	1.86	8.15		10.00
1215	Wood construction						
1220	Walls						
1230	2x4	LF	0.044	0.60	3.62		4.22
1240	2x6	"	0.050	0.91	4.07		4.98
1250	2x8	"	0.053	1.20	4.34		5.54
1260	2x10	"	0.057	1.60	4.66		6.26
1270	2x12	"	0.062	2.07	5.01		7.08
1280	Ceilings						
1290	2x4	LF	0.050	0.60	4.07		4.67
1300	2x6	"	0.057	0.91	4.66		5.57
1310	2x8	"	0.062	1.20	5.01		6.21
1320	2x10	"	0.067	1.60	5.43		7.03
1330	2x12	"	0.073	2.07	5.93		8.00

© 2020 BNi Publications Inc. ALL RIGHTS RESERVED

ROUGH CARPENTRY

	Description	Output		Unit Costs			
ID Code	Component Descriptions	Unit of Meas.	Manhr / Unit	Material Cost	Labor Cost	Equipment Cost	Total Cost
06 - 11002	**CEILING FRAMING**						**06 - 11002**
1000	Ceiling joists						
1010	12" o.c.						
1020	2x4	SF	0.019	0.89	1.55		2.44
1030	2x6	"	0.020	1.28	1.63		2.91
1040	2x8	"	0.021	1.89	1.71		3.60
1050	2x10	"	0.022	2.15	1.81		3.96
1060	2x12	"	0.024	3.96	1.91		5.87
1070	16" o.c.						
1080	2x4	SF	0.015	0.72	1.25		1.97
1090	2x6	"	0.016	1.08	1.30		2.38
1100	2x8	"	0.017	1.53	1.35		2.88
1110	2x10	"	0.017	1.72	1.41		3.13
1120	2x12	"	0.018	3.23	1.48		4.71
1130	24" o.c.						
1140	2x4	SF	0.013	0.52	1.03		1.55
1150	2x6	"	0.013	0.86	1.08		1.94
1160	2x8	"	0.014	1.28	1.14		2.42
1170	2x10	"	0.015	1.53	1.20		2.73
1180	2x12	"	0.016	3.88	1.27		5.15
1200	Headers and nailers						
1210	2x4	LF	0.026	0.60	2.10		2.70
1220	2x6	"	0.027	0.91	2.17		3.08
1230	2x8	"	0.029	1.20	2.33		3.53
1240	2x10	"	0.031	1.60	2.50		4.10
1250	2x12	"	0.033	1.97	2.71		4.68
1300	Sister joists for ceilings						
1310	2x4	LF	0.057	0.60	4.66		5.26
1320	2x6	"	0.067	0.91	5.43		6.34
1330	2x8	"	0.080	1.20	6.52		7.72
1340	2x10	"	0.100	1.60	8.15		9.75
1350	2x12	"	0.133	1.97	10.75		12.75
06 - 11003	**FLOOR FRAMING**						**06 - 11003**
1000	Floor joists						
1010	12" o.c.						
1020	2x6	SF	0.016	1.09	1.30		2.39
1030	2x8	"	0.016	1.61	1.33		2.94
1040	2x10	"	0.017	2.23	1.35		3.58

© 2020 BNi Publications Inc. ALL RIGHTS RESERVED

ROUGH CARPENTRY

ID Code	Description: Component Descriptions	Output: Unit of Meas.	Output: Manhr / Unit	Unit Costs: Material Cost	Unit Costs: Labor Cost	Unit Costs: Equipment Cost	Unit Costs: Total Cost
06 - 11003	FLOOR FRAMING, Cont'd...						06 - 11003
1050	2x12	SF	0.017	3.27	1.41		4.68
1060	2x14	"	0.017	4.99	1.35		6.34
1070	3x6	"	0.017	3.70	1.38		5.08
1080	3x8	"	0.017	4.82	1.41		6.23
1090	3x10	"	0.018	6.03	1.48		7.51
1100	3x12	"	0.019	7.24	1.55		8.79
1120	3x14	"	0.020	8.26	1.63		9.89
1130	4x6	"	0.017	4.82	1.35		6.17
1140	4x8	"	0.017	6.19	1.41		7.60
1150	4x10	"	0.018	7.92	1.48		9.40
1160	4x12	"	0.019	9.64	1.55		11.25
1170	4x14	"	0.020	11.25	1.63		13.00
1180	16" o.c.						
1190	2x6	SF	0.013	0.94	1.08		2.02
1200	2x8	"	0.014	1.33	1.10		2.43
1220	2x10	"	0.014	1.61	1.12		2.73
1230	2x12	"	0.014	2.01	1.16		3.17
1240	2x14	"	0.015	4.48	1.20		5.68
1250	3x6	"	0.014	3.09	1.12		4.21
1260	3x8	"	0.014	3.96	1.16		5.12
1270	3x10	"	0.015	4.99	1.20		6.19
1280	3x12	"	0.015	6.03	1.25		7.28
1290	3x14	"	0.016	7.14	1.30		8.44
1300	4x6	"	0.014	3.96	1.12		5.08
1310	4x8	"	0.014	5.43	1.16		6.59
1320	4x10	"	0.015	6.71	1.20		7.91
1330	4x12	"	0.015	7.92	1.25		9.17
1340	4x14	"	0.016	9.47	1.30		10.75
2000	Sister joists for floors						
2010	2x4	LF	0.050	0.60	4.07		4.67
2020	2x6	"	0.057	0.91	4.66		5.57
2030	2x8	"	0.067	1.20	5.43		6.63
2040	2x10	"	0.080	1.60	6.52		8.12
2050	2x12	"	0.100	2.07	8.15		10.25
2060	3x6	"	0.080	3.01	6.52		9.53
2070	3x8	"	0.089	3.70	7.24		11.00
2080	3x10	"	0.100	4.90	8.15		13.00
2090	3x12	"	0.114	5.93	9.32		15.25

© 2020 BNi Publications Inc. ALL RIGHTS RESERVED

ROUGH CARPENTRY

	Description	Output		Unit Costs			
ID Code	Component Descriptions	Unit of Meas.	Manhr / Unit	Material Cost	Labor Cost	Equipment Cost	Total Cost
06 - 11003	**FLOOR FRAMING, Cont'd...**						**06 - 11003**
2100	4x6	LF	0.080	3.88	6.52		10.50
2110	4x8	"	0.089	5.16	7.24		12.50
2120	4x10	"	0.100	6.71	8.15		14.75
2130	4x12	"	0.114	7.48	9.32		16.75
3000	Plywood Web Joists, 16" o.c.						
3010	3/8" plywood web and 1-3/4" top & bottom chord						
3020	9-1/2" depth	SF	0.014	1.78	1.12		2.90
3030	11-7/8" depth	"	0.016	1.84	1.30		3.14
3040	14" depth	"	0.018	2.26	1.44		3.70
3050	16" depth	"	0.020	3.39	1.63		5.02
3100	Wood Trussed Floor Joists						
3110	2" x 4" chord and truss members						
3130	11-1/4" depth	SF	0.016	5.73	1.30		7.03
3140	14" depth	"	0.018	6.15	1.44		7.59
3150	16" depth	"	0.020	6.31	1.63		7.94
3160	18" depth	"	0.023	6.63	1.86		8.49
06 - 11004	**FURRING**						**06 - 11004**
1100	Furring, wood strips						
1102	Walls						
1105	On masonry or concrete walls						
1107	1x2 furring						
1110	12" o.c.	SF	0.025	0.48	2.03		2.51
1120	16" o.c.	"	0.023	0.41	1.86		2.27
1130	24" o.c.	"	0.021	0.39	1.71		2.10
1135	1x3 furring						
1140	12" o.c.	SF	0.025	0.60	2.03		2.63
1150	16" o.c.	"	0.023	0.54	1.86		2.40
1160	24" o.c.	"	0.021	0.42	1.71		2.13
1165	On wood walls						
1167	1x2 furring						
1170	12" o.c.	SF	0.018	0.48	1.44		1.92
1180	16" o.c.	"	0.016	0.41	1.30		1.71
1190	24" o.c.	"	0.015	0.38	1.18		1.56
1195	1x3 furring						
1200	12" o.c.	SF	0.018	0.61	1.44		2.05
1210	16" o.c.	"	0.016	0.52	1.30		1.82
1220	24" o.c.	"	0.015	0.42	1.18		1.60

© 2020 BNi Publications Inc. ALL RIGHTS RESERVED

ROUGH CARPENTRY

ID Code	Component Descriptions	Unit of Meas.	Manhr / Unit	Material Cost	Labor Cost	Equipment Cost	Total Cost
06 - 11004	**FURRING, Cont'd...**						**06 - 11004**
1224	Ceilings						
1226	On masonry or concrete ceilings						
1228	1x2 furring						
1230	12" o.c.	SF	0.044	0.48	3.62		4.10
1240	16" o.c.	"	0.040	0.41	3.26		3.67
1250	24" o.c.	"	0.036	0.38	2.96		3.34
1254	1x3 furring						
1260	12" o.c.	SF	0.044	0.60	3.62		4.22
1270	16" o.c.	"	0.040	0.52	3.26		3.78
1280	24" o.c.	"	0.036	0.42	2.96		3.38
1286	On wood ceilings						
1288	1x2 furring						
1290	12" o.c.	SF	0.030	0.48	2.41		2.89
1300	16" o.c.	"	0.027	0.41	2.17		2.58
1310	24" o.c.	"	0.024	0.38	1.97		2.35
1316	1x3						
1320	12" o.c.	SF	0.030	0.60	2.41		3.01
1330	16" o.c.	"	0.027	0.52	2.17		2.69
1340	24" o.c.	"	0.024	0.42	1.97		2.39
06 - 11005	**ROOF FRAMING**						**06 - 11005**
1000	Roof framing						
1005	Rafters, gable end						
1008	0-2 pitch (flat to 2-in-12)						
1010	12" o.c.						
1020	2x4	SF	0.017	0.86	1.35		2.21
1030	2x6	"	0.017	1.20	1.41		2.61
1040	2x8	"	0.018	1.72	1.48		3.20
1050	2x10	"	0.019	2.15	1.55		3.70
1060	2x12	"	0.020	3.96	1.63		5.59
1070	16" o.c.						
1080	2x6	SF	0.014	1.08	1.16		2.24
1090	2x8	"	0.015	1.52	1.20		2.72
1100	2x10	"	0.015	1.72	1.25		2.97
1110	2x12	"	0.016	3.18	1.30		4.48
1120	24" o.c.						
1130	2x6	SF	0.012	0.60	0.98		1.58
1140	2x8	"	0.013	1.26	1.01		2.27

© 2020 BNi Publications Inc. ALL RIGHTS RESERVED

ROUGH CARPENTRY

	Description	Output		Unit Costs			
ID Code	Component Descriptions	Unit of Meas.	Manhr / Unit	Material Cost	Labor Cost	Equipment Cost	Total Cost
06 - 11005	**ROOF FRAMING, Cont'd...**						**06 - 11005**
1150	2x10	SF	0.013	1.46	1.05		2.51
1160	2x12	"	0.013	2.57	1.08		3.65
1165	4-6 pitch (4-in-12 to 6-in-12)						
1170	12" o.c.						
1175	2x4	SF	0.017	0.86	1.41		2.27
1180	2x6	"	0.018	1.28	1.48		2.76
1190	2x8	"	0.019	1.97	1.55		3.52
1200	2x10	"	0.020	2.23	1.63		3.86
1210	2x12	"	0.021	3.44	1.71		5.15
1220	16" o.c.						
1230	2x6	SF	0.015	1.08	1.20		2.28
1240	2x8	"	0.015	1.72	1.25		2.97
1250	2x10	"	0.016	1.97	1.30		3.27
1260	2x12	"	0.017	2.93	1.35		4.28
1270	24" o.c.						
1280	2x6	SF	0.013	0.86	1.01		1.87
1290	2x8	"	0.013	1.46	1.05		2.51
1300	2x10	"	0.014	1.54	1.12		2.66
1310	2x12	"	0.015	2.41	1.25		3.66
1315	8-12 pitch (8-in-12 to 12-in-12)						
1320	12" o.c.						
1330	2x4	SF	0.018	0.94	1.48		2.42
1340	2x6	"	0.019	1.46	1.55		3.01
1350	2x8	"	0.020	2.07	1.63		3.70
1360	2x10	"	0.021	2.41	1.71		4.12
1370	2x12	"	0.022	3.70	1.81		5.51
1380	16" o.c.						
1390	2x6	SF	0.015	1.20	1.25		2.45
1400	2x8	"	0.016	1.93	1.30		3.23
1410	2x10	"	0.017	2.15	1.35		3.50
1420	2x12	"	0.017	3.09	1.41		4.50
1430	24" o.c.						
1440	2x6	SF	0.013	0.94	1.05		1.99
1450	2x8	"	0.013	1.53	1.08		2.61
1460	2x10	"	0.014	1.72	1.12		2.84
1470	2x12	"	0.014	2.75	1.16		3.91
2000	Ridge boards						
2010	2x6	LF	0.040	0.91	3.26		4.17

© 2020 BNi Publications Inc. ALL RIGHTS RESERVED

ROUGH CARPENTRY

	Description	Output		Unit Costs			
ID Code	Component Descriptions	Unit of Meas.	Manhr / Unit	Material Cost	Labor Cost	Equipment Cost	Total Cost
06 - 11005	**ROOF FRAMING, Cont'd...**						**06 - 11005**
2020	2x8	LF	0.044	1.20	3.62		4.82
2030	2x10	"	0.050	1.60	4.07		5.67
2040	2x12	"	0.057	2.07	4.66		6.73
3000	Hip rafters						
3010	2x6	LF	0.029	0.91	2.33		3.24
3020	2x8	"	0.030	1.20	2.41		3.61
3030	2x10	"	0.031	1.60	2.50		4.10
3040	2x12	"	0.032	2.07	2.60		4.67
3180	Jack rafters						
3190	4-6 pitch (4-in-12 to 6-in-12)						
3200	16" o.c.						
3210	2x6	SF	0.024	1.12	1.91		3.03
3220	2x8	"	0.024	1.72	1.97		3.69
3230	2x10	"	0.026	1.97	2.10		4.07
3240	2x12	"	0.027	2.93	2.17		5.10
3250	24" o.c.						
3260	2x6	SF	0.018	0.86	1.48		2.34
3270	2x8	"	0.019	1.46	1.51		2.97
3280	2x10	"	0.020	1.72	1.59		3.31
3290	2x12	"	0.020	2.49	1.63		4.12
3295	8-12 pitch (8-in-12 to 12-in-12)						
3300	16" o.c.						
3310	2x6	SF	0.025	1.72	2.03		3.75
3320	2x8	"	0.026	2.15	2.10		4.25
3330	2x10	"	0.027	3.09	2.17		5.26
3340	2x12	"	0.028	4.30	2.24		6.54
3350	24" o.c.						
3360	2x6	SF	0.019	1.37	1.55		2.92
3370	2x8	"	0.020	1.72	1.59		3.31
3380	2x10	"	0.020	2.75	1.63		4.38
3390	2x12	"	0.021	3.96	1.67		5.63
4980	Sister rafters						
5000	2x4	LF	0.057	0.60	4.66		5.26
5010	2x6	"	0.067	0.91	5.43		6.34
5020	2x8	"	0.080	1.20	6.52		7.72
5030	2x10	"	0.100	1.60	8.15		9.75
5040	2x12	"	0.133	2.07	10.75		12.75
5050	Fascia boards						

© 2020 BNi Publications Inc. ALL RIGHTS RESERVED

ROUGH CARPENTRY

	Description	Output		Unit Costs			
ID Code	Component Descriptions	Unit of Meas.	Manhr / Unit	Material Cost	Labor Cost	Equipment Cost	Total Cost
06 - 11005	**ROOF FRAMING, Cont'd...**						**06 - 11005**
5060	2x4	LF	0.040	0.60	3.26		3.86
5070	2x6	"	0.040	0.91	3.26		4.17
5080	2x8	"	0.044	1.20	3.62		4.82
5090	2x10	"	0.044	1.60	3.62		5.22
5100	2x12	"	0.050	2.07	4.07		6.14
7980	Cant strips						
7985	Fiber						
8000	3x3	LF	0.023	0.48	1.86		2.34
8020	4x4	"	0.024	0.67	1.97		2.64
8030	Wood						
8040	3x3	LF	0.024	2.49	1.97		4.46
06 - 11006	**SLEEPERS**						**06 - 11006**
0960	Sleepers, over concrete						
0980	12" o.c.						
1000	1x2	SF	0.018	0.29	1.48		1.77
1020	1x3	"	0.019	0.43	1.55		1.98
1060	2x4	"	0.022	0.94	1.81		2.75
1080	2x6	"	0.024	1.37	1.91		3.28
1090	16" o.c.						
1100	1x2	SF	0.016	0.26	1.30		1.56
1120	1x3	"	0.016	0.37	1.30		1.67
1140	2x4	"	0.019	0.78	1.55		2.33
1160	2x6	"	0.020	1.15	1.63		2.78
06 - 11007	**SOFFITS**						**06 - 11007**
0980	Soffit framing						
1000	2x3	LF	0.057	0.41	4.66		5.07
1020	2x4	"	0.062	0.50	5.01		5.51
1030	2x6	"	0.067	0.74	5.43		6.17
1040	2x8	"	0.073	1.05	5.93		6.98
06 - 11008	**WALL FRAMING**						**06 - 11008**
0960	Framing wall, studs						
0980	12" o.c.						
1000	2x3	SF	0.015	0.53	1.20		1.73
1040	2x4	"	0.015	0.74	1.20		1.94
1080	2x6	"	0.016	1.08	1.30		2.38
1100	2x8	"	0.017	1.44	1.35		2.79

© 2020 BNi Publications Inc. ALL RIGHTS RESERVED

ROUGH CARPENTRY

	Description	Output		Unit Costs			
ID Code	Component Descriptions	Unit of Meas.	Manhr / Unit	Material Cost	Labor Cost	Equipment Cost	Total Cost
06 - 11008	**WALL FRAMING, Cont'd...**						**06 - 11008**
1110	16" o.c.						
1120	2x3	SF	0.013	0.43	1.01		1.44
1140	2x4	"	0.013	0.60	1.01		1.61
1150	2x6	"	0.013	0.86	1.08		1.94
1160	2x8	"	0.014	1.36	1.12		2.48
1165	24" o.c.						
1170	2x3	SF	0.011	0.33	0.88		1.21
1180	2x4	"	0.011	0.45	0.88		1.33
1190	2x6	"	0.011	0.72	0.93		1.65
1200	2x8	"	0.012	0.94	0.95		1.89
1480	Plates, top or bottom						
1500	2x3	LF	0.024	0.41	1.91		2.32
1510	2x4	"	0.025	0.50	2.03		2.53
1520	2x6	"	0.027	0.74	2.17		2.91
1530	2x8	"	0.029	1.05	2.33		3.38
2000	Headers, door or window						
2005	2x6						
2008	Single						
2010	3' long	EA	0.400	2.44	32.50		35.00
2020	6' long	"	0.500	4.88	40.75		45.75
2025	Double						
2030	3' long	EA	0.444	4.89	36.25		41.25
2040	6' long	"	0.571	9.79	46.50		56.00
2044	2x8						
2046	Single						
2050	4' long	EA	0.500	4.46	40.75		45.25
2060	8' long	"	0.615	8.92	50.00		59.00
2065	Double						
2070	4' long	EA	0.571	8.92	46.50		55.00
2080	8' long	"	0.727	18.00	59.00		77.00
2085	2x10						
2088	Single						
2090	5' long	EA	0.615	6.75	50.00		57.00
2100	10' long	"	0.800	13.50	65.00		79.00
2110	Double						
2120	5' long	EA	0.667	13.50	54.00		68.00
2130	10' long	"	0.800	27.00	65.00		92.00
2134	2x12						

© 2020 BNi Publications Inc. ALL RIGHTS RESERVED

ROUGH CARPENTRY

ID Code	Description: Component Descriptions	Output: Unit of Meas.	Output: Manhr / Unit	Unit Costs: Material Cost	Unit Costs: Labor Cost	Unit Costs: Equipment Cost	Unit Costs: Total Cost
06 - 11008	**WALL FRAMING, Cont'd...**						**06 - 11008**
2138	Single						
2140	6' long	EA	0.615	9.79	50.00		60.00
2150	12' long	"	0.800	19.25	65.00		84.00
2155	Double						
2160	6' long	EA	0.727	19.25	59.00		78.00
2170	12' long	"	0.889	38.50	72.00		110

TIMBER

ID Code	Description	Unit of Meas.	Manhr / Unit	Material Cost	Labor Cost	Equipment Cost	Total Cost
06 - 13001	**HEAVY TIMBER**						**06 - 13001**
1000	Mill framed structures						
1010	Beams to 20' long						
1020	Douglas fir						
1040	6x8	LF	0.080	8.27	5.10	3.50	16.75
1042	6x10	"	0.083	9.77	5.28	3.62	18.75
1044	6x12	"	0.089	11.75	5.67	3.88	21.25
1046	6x14	"	0.092	14.00	5.89	4.03	24.00
1048	6x16	"	0.096	15.25	6.12	4.20	25.50
1060	8x10	"	0.083	13.00	5.28	3.62	22.00
1070	8x12	"	0.089	15.25	5.67	3.88	24.75
1080	8x14	"	0.092	17.50	5.89	4.03	27.50
1090	8x16	"	0.096	20.00	6.12	4.20	30.25
1200	Southern yellow pine						
1220	6x8	LF	0.080	6.56	5.10	3.50	15.25
1222	6x10	"	0.083	7.97	5.28	3.62	16.75
1224	6x12	"	0.089	10.25	5.67	3.88	19.75
1226	6x14	"	0.092	11.75	5.89	4.03	21.75
1228	6x16	"	0.096	13.00	6.12	4.20	23.25
1240	8x10	"	0.083	10.75	5.28	3.62	19.75
1242	8x12	"	0.089	13.25	5.67	3.88	22.75
1244	8x14	"	0.092	15.00	5.89	4.03	25.00
1246	8x16	"	0.096	17.50	6.12	4.20	27.75
1380	Columns to 12' high						
1400	Douglas fir						
1420	6x6	LF	0.120	5.95	7.66	5.25	19.00
1440	8x8	"	0.120	10.25	7.66	5.25	23.25
1460	10x10	"	0.133	17.75	8.51	5.83	32.00
1480	12x12	"	0.133	22.00	8.51	5.83	36.25

© 2020 BNi Publications Inc. ALL RIGHTS RESERVED

TIMBER

ID Code	Component Descriptions	Unit of Meas.	Manhr / Unit	Material Cost	Labor Cost	Equipment Cost	Total Cost
06 - 13001	**HEAVY TIMBER, Cont'd...**						**06 - 13001**
1500	Southern yellow pine						
1520	6x6	LF	0.120	5.12	7.66	5.25	18.00
1540	8x8	"	0.120	8.61	7.66	5.25	21.50
1560	10x10	"	0.133	13.50	8.51	5.83	27.75
1580	12x12	"	0.133	18.75	8.51	5.83	33.00
2000	Posts, treated						
2100	4x4	LF	0.032	2.05	2.60		4.65
2120	6x6	"	0.040	5.95	3.26		9.21

WOOD DECKING

ID Code	Component Descriptions	Unit of Meas.	Manhr / Unit	Material Cost	Labor Cost	Equipment Cost	Total Cost
06 - 15001	**WOOD DECKING**						**06 - 15001**
0090	Decking, T&G solid						
0095	Cedar						
0100	3" thick	SF	0.020	11.50	1.63		13.25
0120	4" thick	"	0.021	14.50	1.73		16.25
1030	Fir						
1040	3" thick	SF	0.020	5.05	1.63		6.68
1060	4" thick	"	0.021	6.13	1.73		7.86
1080	Southern yellow pine						
2000	3" thick	SF	0.023	5.05	1.86		6.91
2020	4" thick	"	0.025	5.34	2.00		7.34
3120	White pine						
3140	3" thick	SF	0.020	6.13	1.63		7.76
3160	4" thick	"	0.021	8.30	1.73		10.00

SHEATHING

ID Code	Component Descriptions	Unit of Meas.	Manhr / Unit	Material Cost	Labor Cost	Equipment Cost	Total Cost
06 - 16001	**FLOOR SHEATHING**						**06 - 16001**
1980	Sub-flooring, plywood, CDX						
2000	1/2" thick	SF	0.010	0.60	0.81		1.41
2020	5/8" thick	"	0.011	0.87	0.93		1.80
2080	3/4" thick	"	0.013	1.60	1.08		2.68
2090	Structural plywood						
2100	1/2" thick	SF	0.010	0.95	0.81		1.76
2120	5/8" thick	"	0.011	1.52	0.93		2.45
2140	3/4" thick	"	0.012	1.60	1.00		2.60
3100	Board type sub-flooring						

© 2020 BNi Publications Inc. ALL RIGHTS RESERVED

SHEATHING

ID Code	Description: Component Descriptions	Output: Unit of Meas.	Output: Manhr / Unit	Unit Costs: Material Cost	Unit Costs: Labor Cost	Unit Costs: Equipment Cost	Unit Costs: Total Cost
06 - 16001	**FLOOR SHEATHING, Cont'd...**						**06 - 16001**
3105	1x8						
3110	Minimum	SF	0.018	1.45	1.44		2.89
3115	Maximum	"	0.020	1.84	1.63		3.47
3117	1x8						
3120	Minimum	SF	0.017	1.60	1.37		2.97
3140	Maximum	"	0.019	1.88	1.53		3.41
3150	1x10						
3160	Minimum	SF	0.016	2.24	1.30		3.54
3180	Maximum	"	0.018	2.41	1.44		3.85
5990	Underlayment						
6000	Hardboard, 1/4" tempered	SF	0.010	0.89	0.81		1.70
6010	Plywood, CDX						
6020	3/8" thick	SF	0.010	0.93	0.81		1.74
6040	1/2" thick	"	0.011	1.11	0.86		1.97
6060	5/8" thick	"	0.011	1.28	0.93		2.21
6080	3/4" thick	"	0.012	1.60	1.00		2.60
06 - 16002	**ROOF SHEATHING**						**06 - 16002**
0080	Sheathing						
0090	Plywood, CDX						
1000	3/8" thick	SF	0.010	0.93	0.84		1.77
1020	1/2" thick	"	0.011	1.11	0.86		1.97
1040	5/8" thick	"	0.011	1.28	0.93		2.21
1060	3/4" thick	"	0.012	1.60	1.00		2.60
1080	Structural plywood						
2040	3/8" thick	SF	0.010	0.59	0.84		1.43
2060	1/2" thick	"	0.011	0.77	0.86		1.63
2080	5/8" thick	"	0.011	0.93	0.93		1.86
2100	3/4" thick	"	0.012	1.12	1.00		2.12
06 - 16003	**WALL SHEATHING**						**06 - 16003**
0980	Sheathing						
0990	Plywood, CDX						
1000	3/8" thick	SF	0.012	0.93	0.96		1.89
1020	1/2" thick	"	0.012	1.11	1.00		2.11
1040	5/8" thick	"	0.013	1.28	1.08		2.36
1060	3/4" thick	"	0.015	1.60	1.18		2.78
3000	Waferboard						
3020	3/8" thick	SF	0.012	0.59	0.96		1.55

© 2020 BNi Publications Inc. ALL RIGHTS RESERVED

SHEATHING

ID Code	Component Descriptions	Unit of Meas.	Manhr / Unit	Material Cost	Labor Cost	Equipment Cost	Total Cost
06 - 16003	**WALL SHEATHING, Cont'd...**						**06 - 16003**
3040	1/2" thick	SF	0.012	0.77	1.00		1.77
3060	5/8" thick	"	0.013	0.93	1.08		2.01
3080	3/4" thick	"	0.015	1.02	1.18		2.20
4100	Structural plywood						
4120	3/8" thick	SF	0.012	0.93	0.96		1.89
4140	1/2" thick	"	0.012	1.11	1.00		2.11
4160	5/8" thick	"	0.013	1.28	1.08		2.36
4180	3/4" thick	"	0.015	1.11	1.18		2.29
7000	Gypsum, 1/2" thick	"	0.012	0.59	1.00		1.59
8000	Asphalt impregnated fiberboard, 1/2" thick	"	0.012	1.02	1.00		2.02

TRUSSES

ID Code	Component Descriptions	Unit of Meas.	Manhr / Unit	Material Cost	Labor Cost	Equipment Cost	Total Cost
06 - 17531	**WOOD TRUSSES**						**06 - 17531**
0960	Truss, fink, 2x4 members						
0980	3-in-12 slope						
1000	24' span	EA	0.686	120	43.75	30.00	190
1020	26' span	"	0.686	130	43.75	30.00	200
1021	28' span	"	0.727	140	46.50	31.75	220
1022	30' span	"	0.727	140	46.50	31.75	220
1024	34' span	"	0.774	150	49.50	33.75	230
1025	38' span	"	0.774	150	49.50	33.75	230
1030	5-in-12 slope						
1040	24' span	EA	0.706	130	45.00	31.00	210
1050	28' span	"	0.727	140	46.50	31.75	220
1055	30' span	"	0.750	150	48.00	32.75	230
1060	32' span	"	0.750	160	48.00	32.75	240
1070	40' span	"	0.800	210	51.00	35.00	300
1074	Gable, 2x4 members						
1078	5-in-12 slope						
1080	24' span	EA	0.706	150	45.00	31.00	230
1090	26' span	"	0.706	160	45.00	31.00	240
1100	28' span	"	0.727	180	46.50	31.75	260
1120	30' span	"	0.750	190	48.00	32.75	270
1140	32' span	"	0.750	190	48.00	32.75	270
1160	36' span	"	0.774	210	49.50	33.75	290
1180	40' span	"	0.800	220	51.00	35.00	310
1190	King post type, 2x4 members						

© 2020 BNi Publications Inc. ALL RIGHTS RESERVED

TRUSSES

ID Code	Description: Component Descriptions	Output: Unit of Meas.	Output: Manhr / Unit	Unit Costs: Material Cost	Unit Costs: Labor Cost	Unit Costs: Equipment Cost	Unit Costs: Total Cost
06 - 17531	**WOOD TRUSSES, Cont'd...**						**06 - 17531**
2000	4-in-12 slope						
2040	16' span	EA	0.649	90.00	41.50	28.25	160
2060	18' span	"	0.667	97.00	42.50	29.25	170
2080	24' span	"	0.706	100	45.00	31.00	180
2100	26' span	"	0.706	110	45.00	31.00	190
2120	30' span	"	0.750	140	48.00	32.75	220
2140	34' span	"	0.750	150	48.00	32.75	230
2160	38' span	"	0.774	180	49.50	33.75	260
2180	42' span	"	0.828	210	53.00	36.25	300

GLUED-LAMINATED CONSTRUCTION

ID Code	Component Descriptions	Unit of Meas.	Manhr / Unit	Material Cost	Labor Cost	Equipment Cost	Total Cost
06 - 18131	**LAMINATED BEAMS**						**06 - 18131**
0010	Parallel strand beams 3-1/2" wide x						
0020	9-1/2"	LF	0.034	12.25	2.18	1.50	16.00
0030	11-1/4"	"	0.036	13.00	2.27	1.55	16.75
0040	11-7/8"	"	0.037	13.75	2.35	1.61	17.75
0050	14"	"	0.044	17.25	2.78	1.90	22.00
0060	16"	"	0.048	20.50	3.06	2.10	25.75
0070	18"	"	0.053	24.25	3.40	2.33	30.00
1000	Laminated veneer beams, 1-3/4" wide x						
1010	11-7/8"	LF	0.037	8.27	2.35	1.61	12.25
1020	14"	"	0.044	10.25	2.78	1.90	15.00
1030	16"	"	0.048	10.25	3.06	2.10	15.50
1040	18"	"	0.053	13.00	3.40	2.33	18.75
2000	Laminated strand beams, 1-3/4" wide x						
2010	9-1/2"	LF	0.034	5.31	2.18	1.50	8.99
2020	11-7/8"	"	0.037	6.07	2.35	1.61	10.00
2030	14"	"	0.044	7.14	2.78	1.90	11.75
2040	16"	"	0.048	8.19	3.06	2.10	13.25
2050	3-1/2" wide x						
2060	9-1/2"	LF	0.034	9.34	2.18	1.50	13.00
2070	11-7/8"	"	0.037	12.25	2.35	1.61	16.25
2080	14"	"	0.044	14.25	2.78	1.90	19.00
2090	16"	"	0.048	17.00	3.06	2.10	22.25
3000	Gluelam beam, 3-1/2" wide x						
3010	10"	LF	0.034	14.25	2.18	1.50	18.00
3020	12"	"	0.040	16.75	2.55	1.75	21.00

© 2020 BNi Publications Inc. ALL RIGHTS RESERVED

GLUED-LAMINATED CONSTRUCTION

ID Code	Description: Component Descriptions	Output: Unit of Meas.	Output: Manhr / Unit	Unit Costs: Material Cost	Unit Costs: Labor Cost	Unit Costs: Equipment Cost	Unit Costs: Total Cost
06 - 18131	**LAMINATED BEAMS, Cont'd...**						**06 - 18131**
3030	15"	LF	0.046	19.75	2.91	2.00	24.75
3040	5-1/2" wide x						
3050	10"	LF	0.034	22.75	2.18	1.50	26.50
3060	16"	"	0.048	35.75	3.06	2.10	41.00
3070	20"	"	0.056	42.00	3.60	2.47	48.00
3100	7-1/2" wide x						
3150	10"	LF	0.037	30.50	2.35	1.61	34.50
3200	12"	"	0.051	48.00	3.22	2.21	53.00
3250	15"	"	0.060	57.00	3.83	2.62	63.00
3280	18"	"	0.064	68.00	4.08	2.80	75.00
3300	9-1/2" wide x						
3350	10"	LF	0.037	30.50	2.35	1.61	34.50
3400	16"	"	0.051	49.00	3.22	2.21	54.00
3450	20"	"	0.060	61.00	3.83	2.62	67.00
3500	24"	"	0.064	77.00	4.08	2.80	84.00
3550	30"	"	0.074	96.00	4.71	3.23	100

FINISH CARPENTRY

ID Code	Component Descriptions	Unit of Meas.	Manhr / Unit	Material Cost	Labor Cost	Equipment Cost	Total Cost
06 - 20231	**FINISH CARPENTRY**						**06 - 20231**
0070	Mouldings and trim						
0980	Apron, flat						
1000	9/16 x 2	LF	0.040	1.97	3.26		5.23
1010	9/16 x 3-1/2	"	0.042	4.55	3.43		7.98
1015	Base						
1020	Colonial						
1022	7/16 x 2-1/4	LF	0.040	2.35	3.26		5.61
1024	7/16 x 3	"	0.040	3.04	3.26		6.30
1026	7/16 x 3-1/4	"	0.040	3.11	3.26		6.37
1028	9/16 x 3	"	0.042	3.04	3.43		6.47
1030	9/16 x 3-1/4	"	0.042	3.18	3.43		6.61
1034	11/16 x 2-1/4	"	0.044	3.34	3.62		6.96
1035	Ranch						
1036	7/16 x 2-1/4	LF	0.040	2.57	3.26		5.83
1038	7/16 x 3-1/4	"	0.040	3.04	3.26		6.30
1039	9/16 x 2-1/4	"	0.042	2.80	3.43		6.23
1041	9/16 x 3	"	0.042	3.04	3.43		6.47
1043	9/16 x 3-1/4	"	0.042	3.11	3.43		6.54

© 2020 BNi Publications Inc. ALL RIGHTS RESERVED

FINISH CARPENTRY

	Description	Output		Unit Costs			
ID Code	Component Descriptions	Unit of Meas.	Manhr / Unit	Material Cost	Labor Cost	Equipment Cost	Total Cost
06 - 20231	**FINISH CARPENTRY, Cont'd...**						**06 - 20231**
1050	Casing						
1060	11/16 x 2-1/2	LF	0.036	2.42	2.96		5.38
1070	11/16 x 3-1/2	"	0.038	2.73	3.10		5.83
1180	Chair rail						
1200	9/16 x 2-1/2	LF	0.040	2.57	3.26		5.83
1210	9/16 x 3-1/2	"	0.040	3.56	3.26		6.82
1250	Closet pole						
1300	1-1/8" dia.	LF	0.053	1.74	4.34		6.08
1310	1-5/8" dia.	"	0.053	2.57	4.34		6.91
1340	Cove						
1500	9/16 x 1-3/4	LF	0.040	1.97	3.26		5.23
1510	11/16 x 2-3/4	"	0.040	3.04	3.26		6.30
1550	Crown						
1600	9/16 x 1-5/8	LF	0.053	2.57	4.34		6.91
1610	9/16 x 2-5/8	"	0.062	2.80	5.01		7.81
1620	11/16 x 3-5/8	"	0.067	3.04	5.43		8.47
1630	11/16 x 4-1/4	"	0.073	4.55	5.93		10.50
1640	11/16 x 5-1/4	"	0.080	5.08	6.52		11.50
1680	Drip cap						
1700	1-1/16 x 1-5/8	LF	0.040	2.73	3.26		5.99
1780	Glass bead						
1800	3/8 x 3/8	LF	0.050	0.98	4.07		5.05
1820	1/2 x 9/16	"	0.050	1.21	4.07		5.28
1840	5/8 x 5/8	"	0.050	1.29	4.07		5.36
1860	3/4 x 3/4	"	0.050	1.52	4.07		5.59
1880	Half round						
1900	1/2	LF	0.032	1.14	2.60		3.74
1910	5/8	"	0.032	1.52	2.60		4.12
1920	3/4	"	0.032	2.05	2.60		4.65
1980	Lattice						
2000	1/4 x 7/8	LF	0.032	0.91	2.60		3.51
2010	1/4 x 1-1/8	"	0.032	0.98	2.60		3.58
2020	1/4 x 1-3/8	"	0.032	1.05	2.60		3.65
2030	1/4 x 1-3/4	"	0.032	1.18	2.60		3.78
2040	1/4 x 2	"	0.032	1.36	2.60		3.96
2080	Ogee molding						
2100	5/8 x 3/4	LF	0.040	1.81	3.26		5.07
2110	11/16 x 1-1/8	"	0.040	4.25	3.26		7.51

© 2020 BNi Publications Inc. ALL RIGHTS RESERVED

FINISH CARPENTRY

ID Code	Description: Component Descriptions	Output: Unit of Meas.	Output: Manhr / Unit	Unit Costs: Material Cost	Unit Costs: Labor Cost	Unit Costs: Equipment Cost	Unit Costs: Total Cost
06 - 20231	**FINISH CARPENTRY, Cont'd...**						**06 - 20231**
2120	11/16 x 1-3/8	LF	0.040	3.34	3.26		6.60
2180	Parting bead						
2200	3/8 x 7/8	LF	0.050	1.52	4.07		5.59
2300	Quarter round						
2301	1/4 x 1/4	LF	0.032	0.53	2.60		3.13
2303	3/8 x 3/8	"	0.032	0.76	2.60		3.36
2305	1/2 x 1/2	"	0.032	0.98	2.60		3.58
2307	11/16 x 11/16	"	0.035	0.98	2.83		3.81
2309	3/4 x 3/4	"	0.035	1.81	2.83		4.64
2311	1-1/16 x 1-1/16	"	0.036	1.43	2.96		4.39
2380	Railings, balusters						
2400	1-1/8 x 1-1/8	LF	0.080	4.86	6.52		11.50
2410	1-1/2 x 1-1/2	"	0.073	5.69	5.93		11.50
2480	Screen moldings						
2500	1/4 x 3/4	LF	0.067	1.21	5.43		6.64
2510	5/8 x 5/16	"	0.067	1.52	5.43		6.95
2580	Shoe						
2600	7/16 x 11/16	LF	0.032	1.52	2.60		4.12
2605	Sash beads						
2610	1/2 x 3/4	LF	0.067	1.74	5.43		7.17
2620	1/2 x 7/8	"	0.067	1.97	5.43		7.40
2630	1/2 x 1-1/8	"	0.073	2.12	5.93		8.05
2640	5/8 x 7/8	"	0.073	2.12	5.93		8.05
2760	Stop						
2780	5/8 x 1-5/8						
2800	Colonial	LF	0.050	1.05	4.07		5.12
2810	Ranch	"	0.050	1.05	4.07		5.12
2880	Stools						
2900	11/16 x 2-1/4	LF	0.089	4.63	7.24		11.75
2910	11/16 x 2-1/2	"	0.089	4.86	7.24		12.00
2920	11/16 x 5-1/4	"	0.100	5.01	8.15		13.25
4000	Exterior trim, casing, select pine, 1x3	"	0.040	3.34	3.26		6.60
4010	Douglas fir						
4020	1x3	LF	0.040	1.59	3.26		4.85
4040	1x4	"	0.040	1.97	3.26		5.23
4060	1x6	"	0.044	2.57	3.62		6.19
4100	1x8	"	0.050	3.56	4.07		7.63
5000	Cornices, white pine, #2 or better						

© 2020 BNi Publications Inc. ALL RIGHTS RESERVED

FINISH CARPENTRY

ID Code	Description: Component Descriptions	Output: Unit of Meas.	Output: Manhr / Unit	Unit Costs: Material Cost	Unit Costs: Labor Cost	Unit Costs: Equipment Cost	Unit Costs: Total Cost
06 - 20231	**FINISH CARPENTRY, Cont'd...**						**06 - 20231**
5020	1x2	LF	0.040	0.98	3.26		4.24
5040	1x4	"	0.040	1.21	3.26		4.47
5060	1x6	"	0.044	1.97	3.62		5.59
5080	1x8	"	0.047	2.42	3.83		6.25
5100	1x10	"	0.050	3.11	4.07		7.18
5120	1x12	"	0.053	3.87	4.34		8.21
8600	Shelving, pine						
8620	1x8	LF	0.062	1.74	5.01		6.75
8640	1x10	"	0.064	2.28	5.21		7.49
8660	1x12	"	0.067	2.88	5.43		8.31
8800	Plywood shelf, 3/4", with edge band, 12" wide	"	0.080	3.11	6.52		9.63
8840	Adjustable shelf, and rod, 12" wide						
8860	3' to 4' long	EA	0.200	24.75	16.25		41.00
8880	5' to 8' long	"	0.267	46.50	21.75		68.00
8900	Prefinished wood shelves with brackets and supports						
8905	8" wide						
8910	3' long	EA	0.200	73.00	16.25		89.00
8922	4' long	"	0.200	84.00	16.25		100
8924	6' long	"	0.200	120	16.25		140
8930	10" wide						
8940	3' long	EA	0.200	81.00	16.25		97.00
8942	4' long	"	0.200	120	16.25		140
8946	6' long	"	0.200	130	16.25		150

MILLWORK

ID Code	Component Descriptions	Unit of Meas.	Manhr / Unit	Material Cost	Labor Cost	Equipment Cost	Total Cost
06 - 22001	**MILLWORK**						**06 - 22001**
0070	Countertop, laminated plastic						
0080	25" x 7/8" thick						
0099	Minimum	LF	0.200	18.50	16.25		34.75
0100	Average	"	0.267	34.75	21.75		57.00
0110	Maximum	"	0.320	51.00	26.00		77.00
0115	25" x 1-1/4" thick						
0120	Minimum	LF	0.267	22.25	21.75		44.00
0130	Average	"	0.320	44.50	26.00		71.00
0140	Maximum	"	0.400	67.00	32.50		100
0160	Add for cutouts	EA	0.500		40.75		40.75
0165	Backsplash, 4" high, 7/8" thick	LF	0.160	24.50	13.00		37.50

© 2020 BNi Publications Inc. ALL RIGHTS RESERVED

MILLWORK

ID Code	Description: Component Descriptions	Output: Unit of Meas.	Output: Manhr / Unit	Unit Costs: Material Cost	Unit Costs: Labor Cost	Unit Costs: Equipment Cost	Unit Costs: Total Cost
06 - 22001	**MILLWORK, Cont'd...**						**06 - 22001**
2000	Plywood, sanded, A-C						
2020	1/4" thick	SF	0.027	1.58	2.17		3.75
2040	3/8" thick	"	0.029	1.71	2.33		4.04
2060	1/2" thick	"	0.031	1.94	2.50		4.44
2070	A-D						
2080	1/4" thick	SF	0.027	1.50	2.17		3.67
2090	3/8" thick	"	0.029	1.71	2.33		4.04
2100	1/2" thick	"	0.031	1.86	2.50		4.36
2500	Base cabinet, 34-1/2" high, 24" deep, hardwood						
2540	Minimum	LF	0.320	240	26.00		270
2560	Average	"	0.400	270	32.50		300
2580	Maximum	"	0.533	310	43.50		350
2600	Wall cabinets						
2640	Minimum	LF	0.267	73.00	21.75		95.00
2660	Average	"	0.320	99.00	26.00		130
2680	Maximum	"	0.400	120	32.50		150

ARCHITECTURAL WOODWORK

ID Code	Component Descriptions	Unit of Meas.	Manhr / Unit	Material Cost	Labor Cost	Equipment Cost	Total Cost
06 - 26001	**PANEL WORK**						**06 - 26001**
1020	Hardboard, tempered, 1/4" thick						
1040	Natural faced	SF	0.020	1.14	1.63		2.77
1060	Plastic faced	"	0.023	1.71	1.86		3.57
1080	Pegboard, natural	"	0.020	1.43	1.63		3.06
1100	Plastic faced	"	0.023	1.71	1.86		3.57
1200	Untempered, 1/4" thick						
1220	Natural faced	SF	0.020	1.07	1.63		2.70
1240	Plastic faced	"	0.023	1.86	1.86		3.72
1260	Pegboard, natural	"	0.020	1.14	1.63		2.77
1280	Plastic faced	"	0.023	1.65	1.86		3.51
1300	Plywood, unfinished, 1/4" thick						
1320	Birch						
1330	Natural	SF	0.027	1.22	2.17		3.39
1340	Select	"	0.027	1.79	2.17		3.96
1400	Knotty pine	"	0.027	2.37	2.17		4.54
1500	Cedar (closet lining)						
1520	Standard boards T&G	SF	0.027	2.94	2.17		5.11
1540	Particle board	"	0.027	1.79	2.17		3.96

© 2020 BNi Publications Inc. ALL RIGHTS RESERVED

ARCHITECTURAL WOODWORK

ID Code	Component Descriptions	Unit of Meas.	Manhr / Unit	Material Cost	Labor Cost	Equipment Cost	Total Cost
06 - 26001	**PANEL WORK, Cont'd...**						**06 - 26001**
2000	Plywood, prefinished, 1/4" thick, premium grade						
2020	Birch veneer	SF	0.032	4.30	2.60		6.90
2040	Cherry veneer	"	0.032	5.02	2.60		7.62
2060	Chestnut veneer	"	0.032	9.69	2.60		12.25
2080	Lauan veneer	"	0.032	1.86	2.60		4.46
2100	Mahogany veneer	"	0.032	4.95	2.60		7.55
2120	Oak veneer (red)	"	0.032	4.95	2.60		7.55
2140	Pecan veneer	"	0.032	6.24	2.60		8.84
2160	Rosewood veneer	"	0.032	9.69	2.60		12.25
2180	Teak veneer	"	0.032	6.39	2.60		8.99
2200	Walnut veneer	"	0.032	5.52	2.60		8.12

WOOD STAIRS AND RAILINGS

ID Code	Component Descriptions	Unit of Meas.	Manhr / Unit	Material Cost	Labor Cost	Equipment Cost	Total Cost
06 - 43131	**STAIRWORK**						**06 - 43131**
0080	Risers, 1x8, 42" wide						
0100	White oak	EA	0.400	52.00	32.50		85.00
0120	Pine	"	0.400	46.00	32.50		79.00
0130	Treads, 1-1/16" x 9-1/2" x 42"						
0140	White oak	EA	0.500	62.00	40.75		100

ORNAMENTAL WOODWORK

ID Code	Component Descriptions	Unit of Meas.	Manhr / Unit	Material Cost	Labor Cost	Equipment Cost	Total Cost
06 - 44001	**COLUMNS**						**06 - 44001**
0980	Column, hollow, round wood						
0990	12" diameter						
1000	10' high	EA	0.800	920	51.00	44.25	1,020
1040	12' high	"	0.857	1,130	54.00	47.50	1,230
1060	14' high	"	0.960	1,350	61.00	53.00	1,460
1080	16' high	"	1.200	1,680	76.00	67.00	1,820
2000	24" diameter						
2020	16' high	EA	1.200	3,830	76.00	67.00	3,970
2040	18' high	"	1.263	4,350	80.00	70.00	4,500
2060	20' high	"	1.263	5,350	80.00	70.00	5,500
2080	22' high	"	1.333	5,630	84.00	74.00	5,790
2100	24' high	"	1.333	6,150	84.00	74.00	6,310

© 2020 BNi Publications Inc. ALL RIGHTS RESERVED

DIVISION 07 THERMAL AND MOISTURE

MOISTURE PROTECTION

ID Code	Description: Component Descriptions	Output: Unit of Meas.	Output: Manhr / Unit	Unit Costs: Material Cost	Unit Costs: Labor Cost	Unit Costs: Equipment Cost	Unit Costs: Total Cost
07 - 11001	**DAMPPROOFING**						**07 - 11001**
1000	Silicone dampproofing, sprayed on						
1020	Concrete surface						
1040	1 coat	SF	0.004	0.64	0.28		0.92
1060	2 coats	"	0.006	1.06	0.39		1.45
1070	Concrete block						
1080	1 coat	SF	0.005	0.64	0.34		0.98
1100	2 coats	"	0.007	1.06	0.46		1.52
1110	Brick						
1120	1 coat	SF	0.006	0.64	0.39		1.03
1140	2 coats	"	0.008	1.06	0.51		1.57
07 - 11131	**BITUMINOUS DAMPPROOFING**						**07 - 11131**
0100	Building paper, asphalt felt						
0120	15 lb	SF	0.032	0.19	2.04		2.23
0140	30 lb	"	0.033	0.37	2.12		2.49
1000	Asphalt, troweled, cold, primer plus						
1020	1 coat	SF	0.027	0.68	1.70		2.38
1040	2 coats	"	0.040	1.43	2.55		3.98
1060	3 coats	"	0.050	2.04	3.19		5.23
1200	Fibrous asphalt, hot troweled, primer plus						
1220	1 coat	SF	0.032	0.68	2.04		2.72
1240	2 coats	"	0.044	1.43	2.83		4.26
1260	3 coats	"	0.057	2.04	3.64		5.68
1400	Asphaltic paint dampproofing, per coat						
1420	Brush on	SF	0.011	0.35	0.72		1.07
1440	Spray on	"	0.009	0.49	0.56		1.05
07 - 11161	**PARGING / MASONRY PLASTER**						**07 - 11161**
0080	Parging						
0100	1/2" thick	SF	0.053	0.44	4.14		4.58
0200	3/4" thick	"	0.067	0.48	5.18		5.66
0300	1" thick	"	0.080	0.63	6.22		6.85

© 2020 BNi Publications Inc. ALL RIGHTS RESERVED

SHEET WATERPROOFING

	Description	Output		Unit Costs			
ID Code	Component Descriptions	Unit of Meas.	Manhr / Unit	Material Cost	Labor Cost	Equipment Cost	Total Cost
07 - 13001	**WATERPROOFING**						**07 - 13001**
1000	Membrane waterproofing, elastomeric						
1020	Butyl						
1040	1/32" thick	SF	0.032	1.54	2.04		3.58
1060	1/16" thick	"	0.033	2.01	2.12		4.13
1080	Butyl with nylon						
1100	1/32" thick	SF	0.032	1.80	2.04		3.84
1120	1/16" thick	"	0.033	2.17	2.12		4.29
1140	Neoprene						
1160	1/32" thick	SF	0.032	2.64	2.04		4.68
1180	1/16" thick	"	0.033	3.78	2.12		5.90
1190	Neoprene with nylon						
1220	1/32" thick	SF	0.032	2.74	2.04		4.78
1240	1/16" thick	"	0.033	4.42	2.12		6.54
1420	Bituminous membrane, asphalt felt, 15 lb.						
1440	One ply	SF	0.020	0.91	1.27		2.18
1460	Two ply	"	0.024	1.08	1.54		2.62
1480	Three ply	"	0.029	1.33	1.82		3.15
1500	Four ply	"	0.033	1.54	2.12		3.66
1520	Five ply	"	0.042	1.64	2.68		4.32
1620	Modified asphalt membrane, fibrous asphalt						
1630	One ply	SF	0.033	0.57	2.12		2.69
1640	Two ply	"	0.040	1.17	2.55		3.72
1650	Three ply	"	0.044	1.82	2.83		4.65
1669	Four ply	"	0.053	2.42	3.40		5.82
1670	Five ply	"	0.064	2.91	4.08		6.99
1700	Asphalt coated protective board						
1710	1/8" thick	SF	0.020	0.54	1.27		1.81
1720	1/4" thick	"	0.020	0.73	1.27		2.00
1730	3/8" thick	"	0.020	0.82	1.27		2.09
1740	1/2" thick	"	0.021	0.99	1.34		2.33
1800	Cement protective board						
1820	3/8" thick	SF	0.027	1.52	1.70		3.22
1840	1/2" thick	"	0.027	2.13	1.70		3.83
2000	Fluid applied, neoprene						
2040	50 mil	SF	0.027	2.13	1.70		3.83
2060	90 mil	"	0.027	3.53	1.70		5.23
2100	Tab extended polyurethane						
2120	.050" thick	SF	0.020	1.97	1.27		3.24

© 2020 BNi Publications Inc. ALL RIGHTS RESERVED

SHEET WATERPROOFING

	Description	Output		Unit Costs			
ID Code	Component Descriptions	Unit of Meas.	Manhr / Unit	Material Cost	Labor Cost	Equipment Cost	Total Cost
07 - 13001	**WATERPROOFING, Cont'd...**						**07 - 13001**
2140	Fluid applied rubber based polyurethane						
2160	6 mil	SF	0.025	1.26	1.59		2.85
2200	15 mil	"	0.020	2.39	1.27		3.66
2300	Bentonite waterproofing, panels						
2320	3/16" thick	SF	0.020	2.14	1.27		3.41
2330	1/4" thick	"	0.020	2.43	1.27		3.70
2340	5/8" thick	"	0.021	3.62	1.34		4.96
2350	Granular admixtures, trowel on, 3/8" thick	"	0.020	2.00	1.27		3.27
2410	Metallic oxide, iron compound, troweled						
2420	5/8" thick	SF	0.020	1.82	1.27		3.09
2440	3/4" thick	"	0.023	2.20	1.45		3.65

INSULATION

07 - 21131	**BOARD INSULATION**						**07 - 21131**
1000	Insulation, rigid						
1010	Fiberglass, roof						
1020	0.75" thick, R2.78	SF	0.007	0.55	0.46		1.01
1040	1.06" thick, R4.17	"	0.008	0.84	0.48		1.32
1060	1.31" thick, R5.26	"	0.008	1.13	0.51		1.64
1080	1.63" thick, R6.67	"	0.008	1.39	0.53		1.92
1100	2.25" thick, R8.33	"	0.009	1.53	0.56		2.09
2000	Composite board, roof						
2020	1-1/2" thick, R6.67	SF	0.008	1.20	0.51		1.71
2040	1-5/8" thick, R7.69	"	0.008	1.27	0.53		1.80
2060	2" thick, R10.0	"	0.009	2.29	0.56		2.85
2080	2-1/4" thick, R12.50	"	0.009	2.54	0.60		3.14
2100	2-1/2" thick, R14.29	"	0.010	2.76	0.63		3.39
2120	2-3/4" thick, R16.67	"	0.011	3.03	0.68		3.71
2140	3-1/4" thick, R20.00	"	0.011	3.90	0.72		4.62
2200	Perlite board, roof						
2220	1.00" thick, R2.78	SF	0.007	0.57	0.42		0.99
2240	1.50" thick, R4.17	"	0.007	0.90	0.44		1.34
2260	2.00" thick, R5.92	"	0.007	1.10	0.46		1.56
2280	2.50" thick, R6.67	"	0.008	1.35	0.48		1.83
2290	3.00" thick, R8.33	"	0.008	1.69	0.51		2.20
2300	4.00" thick, R10.00	"	0.008	1.88	0.53		2.41
2320	5.25" thick, R14.29	"	0.009	2.07	0.56		2.63

© 2020 BNi Publications Inc. ALL RIGHTS RESERVED

INSULATION

	Description	Output		Unit Costs			
ID Code	Component Descriptions	Unit of Meas.	Manhr / Unit	Material Cost	Labor Cost	Equipment Cost	Total Cost
07 - 21131	BOARD INSULATION, Cont'd...						07 - 21131
2580	Rigid urethane						
2590	Roof						
2600	1" thick, R6.67	SF	0.007	1.09	0.42		1.51
2620	1.20" thick, R8.33	"	0.007	1.25	0.43		1.68
2640	1.50" thick, R11.11	"	0.007	1.49	0.44		1.93
2660	2" thick, R14.29	"	0.007	1.92	0.46		2.38
2680	2.25" thick, R16.67	"	0.008	2.52	0.48		3.00
2685	Wall						
2690	1" thick, R6.67	SF	0.008	1.09	0.53		1.62
2700	1.5" thick, R11.11	"	0.009	1.49	0.56		2.05
2720	2" thick, R14.29	"	0.009	1.97	0.60		2.57
2780	Polystyrene						
2790	Roof						
2800	1.0" thick, R4.17	SF	0.007	0.41	0.42		0.83
2820	1.5" thick, R6.26	"	0.007	0.63	0.44		1.07
2840	2.0" thick, R8.33	"	0.007	0.78	0.46		1.24
2880	Wall						
2900	1.0" thick, R4.17	SF	0.008	0.41	0.53		0.94
2920	1.5" thick, R6.26	"	0.009	0.63	0.56		1.19
2940	2.0" thick, R8.33	"	0.009	0.78	0.60		1.38
4020	Rigid board insulation, deck						
4025	Mineral fiberboard						
4030	1" thick, R3.0	SF	0.007	0.60	0.42		1.02
4040	2" thick, R5.26	"	0.007	1.33	0.46		1.79
4045	Fiberglass						
4050	1" thick, R4.3	SF	0.007	1.03	0.42		1.45
4060	2" thick, R8.5	"	0.007	1.52	0.46		1.98
4065	Polystyrene						
4070	1" thick, R5.4	SF	0.007	0.41	0.42		0.83
4080	2" thick, R10.8	"	0.007	1.05	0.46		1.51
4090	Urethane						
4100	.75" thick, R5.4	SF	0.007	0.95	0.42		1.37
4120	1" thick, R6.4	"	0.007	1.13	0.42		1.55
4140	1.5" thick, R10.7	"	0.007	1.36	0.44		1.80
4160	2" thick, R14.3	"	0.007	1.55	0.46		2.01
4170	Foamglass						
4180	1" thick, R1.8	SF	0.007	1.54	0.42		1.96
4220	2" thick, R5.26	"	0.007	1.96	0.46		2.42

© 2020 BNi Publications Inc. ALL RIGHTS RESERVED

INSULATION

	Description	Output		Unit Costs			
ID Code	Component Descriptions	Unit of Meas.	Manhr / Unit	Material Cost	Labor Cost	Equipment Cost	Total Cost
07 - 21131	**BOARD INSULATION, Cont'd...**						**07 - 21131**
4230	Wood fiber						
4240	1" thick, R3.85	SF	0.007	1.74	0.42		2.16
4260	2" thick, R7.7	"	0.007	2.10	0.46		2.56
4270	Particle board						
4280	3/4" thick, R2.08	SF	0.007	0.92	0.42		1.34
4300	1" thick, R2.77	"	0.007	0.96	0.42		1.38
4320	2" thick, R5.50	"	0.007	1.27	0.46		1.73
07 - 21161	**BATT INSULATION**						**07 - 21161**
0980	Ceiling, fiberglass, unfaced						
1000	3-1/2" thick, R11	SF	0.009	0.45	0.60		1.05
1020	6" thick, R19	"	0.011	0.58	0.68		1.26
1030	9" thick, R30	"	0.012	1.15	0.78		1.93
1035	Suspended ceiling, unfaced						
1040	3-1/2" thick, R11	SF	0.009	0.45	0.56		1.01
1060	6" thick, R19	"	0.010	0.58	0.63		1.21
1070	9" thick, R30	"	0.011	1.15	0.72		1.87
1075	Crawl space, unfaced						
1080	3-1/2" thick, R11	SF	0.012	0.45	0.78		1.23
1100	6" thick, R19	"	0.013	0.58	0.85		1.43
1120	9" thick, R30	"	0.015	1.15	0.92		2.07
2000	Wall, fiberglass						
2010	Paper backed						
2020	2" thick, R7	SF	0.008	0.34	0.53		0.87
2040	3" thick, R8	"	0.009	0.38	0.56		0.94
2060	4" thick, R11	"	0.009	0.62	0.60		1.22
2080	6" thick, R19	"	0.010	0.92	0.63		1.55
2090	Foil backed, 1 side						
2100	2" thick, R7	SF	0.008	0.66	0.53		1.19
2120	3" thick, R11	"	0.009	0.71	0.56		1.27
2140	4" thick, R14	"	0.009	0.75	0.60		1.35
2160	6" thick, R21	"	0.010	0.98	0.63		1.61
2170	Foil backed, 2 sides						
2180	2" thick, R7	SF	0.009	0.76	0.60		1.36
2200	3" thick, R11	"	0.010	0.97	0.63		1.60
2220	4" thick, R14	"	0.011	1.14	0.68		1.82
2240	6" thick, R21	"	0.011	1.23	0.72		1.95
2250	Unfaced						

© 2020 BNi Publications Inc. ALL RIGHTS RESERVED

INSULATION

	Description	Output		Unit Costs			
ID Code	Component Descriptions	Unit of Meas.	Manhr / Unit	Material Cost	Labor Cost	Equipment Cost	Total Cost
07 - 21161	**BATT INSULATION, Cont'd...**						**07 - 21161**
2260	2" thick, R7	SF	0.008	0.42	0.53		0.95
2280	3" thick, R9	"	0.009	0.48	0.56		1.04
2300	4" thick, R11	"	0.009	0.51	0.60		1.11
2320	6" thick, R19	"	0.010	0.66	0.63		1.29
2400	Mineral wool batts						
2410	Paper backed						
2420	2" thick, R6	SF	0.008	0.36	0.53		0.89
2440	4" thick, R12	"	0.009	0.83	0.56		1.39
2460	6" thick, R19	"	0.010	1.03	0.63		1.66
8980	Fasteners, self adhering, attached to ceiling deck						
9000	2-1/2" long	EA	0.013	0.26	0.85		1.11
9020	4-1/2" long	"	0.015	0.29	0.92		1.21
9060	Capped, self-locking washers	"	0.008	0.26	0.51		0.77
07 - 21231	**LOOSE FILL INSULATION**						**07 - 21231**
1000	Blown-in type						
1010	Fiberglass						
1020	5" thick, R11	SF	0.007	0.41	0.42		0.83
1040	6" thick, R13	"	0.008	0.48	0.51		0.99
1060	9" thick, R19	"	0.011	0.58	0.72		1.30
2000	Rockwool, attic application						
2040	6" thick, R13	SF	0.008	0.38	0.51		0.89
2060	8" thick, R19	"	0.010	0.45	0.63		1.08
2080	10" thick, R22	"	0.012	0.53	0.78		1.31
2100	12" thick, R26	"	0.013	0.68	0.85		1.53
2120	15" thick, R30	"	0.016	0.82	1.02		1.84
6200	Poured type						
6210	Fiberglass						
6220	1" thick, R4	SF	0.005	0.45	0.31		0.76
6222	2" thick, R8	"	0.006	0.84	0.36		1.20
6224	3" thick, R12	"	0.007	1.24	0.42		1.66
6226	4" thick, R16	"	0.008	1.63	0.51		2.14
6230	Mineral wool						
6240	1" thick, R3	SF	0.005	0.50	0.31		0.81
6242	2" thick, R6	"	0.006	0.92	0.36		1.28
6244	3" thick, R9	"	0.007	1.40	0.42		1.82
6246	4" thick, R12	"	0.008	1.63	0.51		2.14
6300	Vermiculite or perlite						

© 2020 BNi Publications Inc. ALL RIGHTS RESERVED

INSULATION

	Description	Output		Unit Costs			
ID Code	Component Descriptions	Unit of Meas.	Manhr / Unit	Material Cost	Labor Cost	Equipment Cost	Total Cost
07 - 21231	**LOOSE FILL INSULATION, Cont'd...**						**07 - 21231**
6310	2" thick, R4.8	SF	0.006	0.98	0.36		1.34
6320	3" thick, R7.2	"	0.007	1.39	0.42		1.81
6330	4" thick, R9.6	"	0.008	1.81	0.51		2.32
8000	Masonry, poured vermiculite or perlite						
8020	4" block	SF	0.004	0.57	0.25		0.82
8040	6" block	"	0.005	0.86	0.31		1.17
8060	8" block	"	0.006	1.25	0.36		1.61
8100	10" block	"	0.006	1.65	0.39		2.04
8120	12" block	"	0.007	2.05	0.42		2.47
07 - 21291	**SPRAYED INSULATION**						**07 - 21291**
1000	Foam, sprayed on						
1010	Polystyrene						
1020	1" thick, R4	SF	0.008	0.69	0.51		1.20
1040	2" thick, R8	"	0.011	1.34	0.68		2.02
1050	Urethane						
1060	1" thick, R4	SF	0.008	0.65	0.51		1.16
1080	2" thick, R8	"	0.011	1.24	0.68		1.92

EXTERIOR INSULATION AND FINISH SYSTEMS

07 - 24001	**AGGREGATE COATED PANELS**						**07 - 24001**
0980	Dryvit type system						
1000	1" thick	SF	0.027	3.31	2.40		5.71
1020	1-1/2" thick	"	0.029	3.45	2.57		6.02
1040	2" thick	"	0.033	3.91	3.00		6.91
07 - 24011	**EXTERIOR INSULATION FINISH SYSTEM, EIFS**						**07 - 24011**
1000	3" Thick	SF	0.010	5.08	0.63		5.71
2000	For Base & Finish Only	"					1.83

© 2020 BNi Publications Inc. ALL RIGHTS RESERVED

VAPOR RETARDERS

ID Code	Description: Component Descriptions	Output: Unit of Meas.	Output: Manhr / Unit	Unit Costs: Material Cost	Unit Costs: Labor Cost	Unit Costs: Equipment Cost	Unit Costs: Total Cost
07 - 26001	**VAPOR BARRIERS**						**07 - 26001**
0980	Vapor barrier, polyethylene						
1000	2 mil	SF	0.004	0.02	0.25		0.27
1010	6 mil	"	0.004	0.07	0.25		0.32
1020	8 mil	"	0.004	0.08	0.28		0.36
1040	10 mil	"	0.004	0.09	0.28		0.37

SHINGLES AND TILES

ID Code	Component Descriptions	Unit of Meas.	Manhr / Unit	Material Cost	Labor Cost	Equipment Cost	Total Cost
07 - 31131	**ASPHALT SHINGLES**						**07 - 31131**
1000	Standard asphalt shingles, strip shingles						
1020	210 lb/square	SQ	0.800	95.00	62.00		160
1040	235 lb/square	"	0.889	99.00	69.00		170
1060	240 lb/square	"	1.000	100	78.00		180
1080	260 lb/square	"	1.143	150	89.00		240
1100	300 lb/square	"	1.333	160	100		260
1120	385 lb/square	"	1.600	220	120		340
5980	Roll roofing, mineral surface						
6000	90 lb	SQ	0.571	55.00	44.50		100
6020	110 lb	"	0.667	92.00	52.00		140
6040	140 lb	"	0.800	95.00	62.00		160
07 - 31161	**METAL SHINGLES**						**07 - 31161**
0980	Aluminum, .020" thick						
1000	Plain	SQ	1.600	280	120		400
1020	Colors	"	1.600	310	120		430
1960	Steel, galvanized						
1980	26 ga.						
2000	Plain	SQ	1.600	350	120		470
2020	Colors	"	1.600	440	120		560
2030	24 ga.						
2040	Plain	SQ	1.600	400	120		520
2060	Colors	"	1.600	510	120		630
2960	Porcelain enamel, 22 ga.						
3000	Minimum	SQ	2.000	870	160		1,030
3020	Average	"	2.000	1,000	160		1,160
3040	Maximum	"	2.000	1,120	160		1,280

© 2020 BNi Publications Inc. ALL RIGHTS RESERVED

SHINGLES AND TILES

Description		Output		Unit Costs			
ID Code	Component Descriptions	Unit of Meas.	Manhr / Unit	Material Cost	Labor Cost	Equipment Cost	Total Cost
07 - 31261	**SLATE SHINGLES**						**07 - 31261**
0960	Slate shingles						
0980	Pennsylvania						
1000	Ribbon	SQ	4.000	600	310		910
1020	Clear	"	4.000	770	310		1,080
1030	Vermont						
1040	Black	SQ	4.000	710	310		1,020
1060	Gray	"	4.000	780	310		1,090
1070	Green	"	4.000	800	310		1,110
1080	Red	"	4.000	1,440	310		1,750
1980	Replacement shingles						
2000	Small jobs	EA	0.267	12.75	20.75		33.50
2020	Large jobs	SF	0.133	9.94	10.25		20.25
07 - 31291	**WOOD SHINGLES**						**07 - 31291**
1000	Wood shingles, on roofs						
1010	White cedar, #1 shingles						
1020	4" exposure	SQ	2.667	270	210		480
1040	5" exposure	"	2.000	240	160		400
1050	#2 shingles						
1060	4" exposure	SQ	2.667	190	210		400
1080	5" exposure	"	2.000	160	160		320
1090	Resquared and rebutted						
1100	4" exposure	SQ	2.667	240	210		450
1120	5" exposure	"	2.000	200	160		360
1140	On walls						
1150	White cedar, #1 shingles						
1160	4" exposure	SQ	4.000	270	310		580
1180	5" exposure	"	3.200	240	250		490
1200	6" exposure	"	2.667	200	210		410
1210	#2 shingles						
1220	4" exposure	SQ	4.000	190	310		500
1240	5" exposure	"	3.200	160	250		410
1260	6" exposure	"	2.667	130	210		340
1300	Add for fire retarding	"					120

© 2020 BNi Publications Inc. ALL RIGHTS RESERVED

SHINGLES AND TILES

ID Code	Description: Component Descriptions	Output: Unit of Meas.	Output: Manhr / Unit	Unit Costs: Material Cost	Unit Costs: Labor Cost	Unit Costs: Equipment Cost	Unit Costs: Total Cost
07 - 31292	**WOOD SHAKES**						**07 - 31292**
2010	Shakes, hand split, 24" red cedar, on roofs						
2020	5" exposure	SQ	4.000	300	310		610
2040	7" exposure	"	3.200	280	250		530
2060	9" exposure	"	2.667	260	210		470
2080	On walls						
2100	6" exposure	SQ	4.000	280	310		590
2120	8" exposure	"	3.200	270	250		520
2140	10" exposure	"	2.667	260	210		470
3000	Add for fire retarding	"					78.00

ROOF TILES

ID Code	Component Descriptions	Unit of Meas.	Manhr / Unit	Material Cost	Labor Cost	Equipment Cost	Total Cost
07 - 32161	**CONCRETE ROOF TILE**						**07 - 32161**
0100	Concrete Roof Tile, Corrugated						
0120	13"x16-1/2", 90/SQ, 950 LB/SQ Earthtone Colors	SQ	4.000	180	310		490
0140	Custom Blues	"	4.000	390	310		700
0160	Custom Greens	"	4.000	230	310		540
0180	Concrete Roof Tile, Shakes						
0200	13"x16-1/2", 90/SQ,950 LB/SQ,Colors	SQ	4.000	240	310		550

ROOFING AND SIDING

ID Code	Component Descriptions	Unit of Meas.	Manhr / Unit	Material Cost	Labor Cost	Equipment Cost	Total Cost
07 - 41001	**MANUFACTURED ROOFS**						**07 - 41001**
1020	Aluminum roof panels, for steel framing						
1040	Corrugated						
1045	Unpainted finish						
1050	.024"	SF	0.020	2.06	1.55		3.61
1060	.030"	"	0.020	3.35	1.55		4.90
1065	Painted finish						
1070	.024"	SF	0.020	2.61	1.55		4.16
1080	.030"	"	0.020	3.52	1.55		5.07
1100	V-beam						
1110	Unpainted finish						
1120	.032"	SF	0.020	2.96	1.55		4.51
1140	.040"	"	0.020	3.56	1.55		5.11
1160	.050"	"	0.020	4.48	1.55		6.03
1170	Painted finish						
1180	.032"	SF	0.020	3.85	1.55		5.40

© 2020 BNi Publications Inc. ALL RIGHTS RESERVED

ROOFING AND SIDING

	Description	Output		Unit Costs			
ID Code	Component Descriptions	Unit of Meas.	Manhr / Unit	Material Cost	Labor Cost	Equipment Cost	Total Cost
07 - 41001	**MANUFACTURED ROOFS, Cont'd...**						**07 - 41001**
1200	.040"	SF	0.020	4.61	1.55		6.16
1240	.050"	"	0.020	5.51	1.55		7.06
2020	Steel roof panels, for structural steel framing						
2040	Corrugated, painted						
2100	18 ga.	SF	0.020	4.26	1.55		5.81
2120	20 ga.	"	0.020	3.97	1.55		5.52
2140	22 ga.	"	0.020	3.50	1.55		5.05
2160	Box rib, painted						
2180	18 ga.	SF	0.021	4.97	1.63		6.60
2200	20 ga.	"	0.021	4.09	1.63		5.72
2220	22 ga.	"	0.021	3.64	1.63		5.27
2240	4" rib, painted						
2250	18 ga.	SF	0.022	5.73	1.73		7.46
2260	20 ga.	"	0.022	4.85	1.73		6.58
2280	22 ga.	"	0.022	4.34	1.73		6.07
2300	Standing seam roof						
2320	2" high seam, painted						
2360	22 ga.	SF	0.032	5.55	2.49		8.04
2380	24 ga.	"	0.032	5.43	2.49		7.92
2400	26 ga.	"	0.032	5.26	2.49		7.75

WALL PANELS

07 - 42003	**MANUFACTURED WALLS**						**07 - 42003**
2090	Sandwich panels with 1-1/2" fiberglass insulation						
2091	Galvanized 18 ga. steel interior panels						
2095	Exterior panels						
2100	16 ga. aluminum	SF	0.107	7.26	9.60		16.75
2120	18 ga. galvanized steel	"	0.107	10.25	9.60		19.75
2140	20 ga. painted steel	"	0.107	10.50	9.60		20.00
2160	20 ga. stainless steel	"	0.107	10.75	9.60		20.25
3000	Metal liner panels, 1-3/8" thick, 24" wide						
3020	Galvanized						
3040	22 ga.	SF	0.027	3.64	2.40		6.04
3060	20 ga.	"	0.027	4.01	2.40		6.41
3080	18 ga.	"	0.027	4.93	2.40		7.33
3100	Primed						
3120	22 ga.	SF	0.027	2.83	2.40		5.23

© 2020 BNi Publications Inc. ALL RIGHTS RESERVED

WALL PANELS

	Description	Output		Unit Costs			
ID Code	Component Descriptions	Unit of Meas.	Manhr / Unit	Material Cost	Labor Cost	Equipment Cost	Total Cost
07 - 42003	**MANUFACTURED WALLS, Cont'd...**						**07 - 42003**
3140	20 ga.	SF	0.027	3.26	2.40		5.66
3160	18 ga.	"	0.027	4.01	2.40		6.41
07 - 42005	**METAL SIDING PANELS**						**07 - 42005**
1000	Aluminum siding panels						
1020	Corrugated						
1030	Plain finish						
1040	.024"	SF	0.032	2.23	2.88		5.11
1060	.032"	"	0.032	2.62	2.88		5.50
1070	Painted finish						
1080	.024"	SF	0.032	2.78	2.88		5.66
1100	.032"	"	0.032	3.19	2.88		6.07
1120	V-beam						
1130	Plain finish						
1140	.032"	SF	0.032	3.37	2.88		6.25
1160	.040"	"	0.032	3.89	2.88		6.77
1180	.050"	"	0.032	4.93	2.88		7.81
1190	Painted finish						
1200	.032"	SF	0.032	4.15	2.88		7.03
1220	.040"	"	0.032	4.90	2.88		7.78
1240	.050"	"	0.032	6.97	2.88		9.85
1260	4" rib						
1270	Plain finish						
1280	.032"	SF	0.036	3.09	3.27		6.36
1300	.040"	"	0.036	3.44	3.27		6.71
1320	.050"	"	0.036	4.19	3.27		7.46
1330	Painted finish						
1340	.032"	SF	0.036	4.01	3.27		7.28
1360	.040"	"	0.036	4.18	3.27		7.45
1380	.050"	"	0.036	4.78	3.27		8.05
2000	Steel siding panels						
2040	Corrugated						
2080	22 ga.	SF	0.053	2.73	4.80		7.53
2100	24 ga.	"	0.053	2.49	4.80		7.29
2120	26 ga.	"	0.053	2.26	4.80		7.06
2140	Box rib						
2160	20 ga.	SF	0.053	4.08	4.80		8.88
2180	22 ga.	"	0.053	3.42	4.80		8.22

© 2020 BNi Publications Inc. ALL RIGHTS RESERVED

WALL PANELS

ID Code	Component Descriptions	Unit of Meas.	Manhr / Unit	Material Cost	Labor Cost	Equipment Cost	Total Cost
07 - 42005	**METAL SIDING PANELS, Cont'd...**						**07 - 42005**
2200	24 ga.	SF	0.053	3.01	4.80		7.81
2220	26 ga.	"	0.053	2.46	4.80		7.26
3000	Ribbed, sheets, galvanized						
3020	22 ga.	SF	0.032	3.42	2.88		6.30
3040	24 ga.	"	0.032	3.01	2.88		5.89
3060	26 ga.	"	0.032	2.46	2.88		5.34
3080	28 ga.	"	0.032	2.25	2.88		5.13
3200	Primed						
3220	24 ga.	SF	0.032	3.01	2.88		5.89
3240	26 ga.	"	0.032	2.46	2.88		5.34
3260	28 ga.	"	0.032	2.25	2.88		5.13

SIDING

ID Code	Component Descriptions	Unit of Meas.	Manhr / Unit	Material Cost	Labor Cost	Equipment Cost	Total Cost
07 - 46231	**WOOD SIDING**						**07 - 46231**
1000	Beveled siding, cedar						
1010	A grade						
1020	1/2 x 6	SF	0.040	4.89	3.26		8.15
1040	1/2 x 8	"	0.032	5.00	2.60		7.60
1060	3/4 x 10	"	0.027	6.43	2.17		8.60
1070	Clear						
1080	1/2 x 6	SF	0.040	5.44	3.26		8.70
1100	1/2 x 8	"	0.032	5.56	2.60		8.16
1120	3/4 x 10	"	0.027	7.45	2.17		9.62
1130	B grade						
1140	1/2 x 6	SF	0.040	5.26	3.26		8.52
1160	1/2 x 8	"	0.032	5.94	2.60		8.54
1180	3/4 x 10	"	0.027	5.60	2.17		7.77
2000	Board and batten						
2010	Cedar						
2020	1x6	SF	0.040	6.84	3.26		10.00
2040	1x8	"	0.032	6.22	2.60		8.82
2060	1x10	"	0.029	5.62	2.33		7.95
2080	1x12	"	0.026	5.04	2.10		7.14
2090	Pine						
2100	1x6	SF	0.040	1.73	3.26		4.99
2120	1x8	"	0.032	1.69	2.60		4.29
2140	1x10	"	0.029	1.62	2.33		3.95

© 2020 BNi Publications Inc. ALL RIGHTS RESERVED

SIDING

	Description	Output		Unit Costs			
ID Code	Component Descriptions	Unit of Meas.	Manhr / Unit	Material Cost	Labor Cost	Equipment Cost	Total Cost
07 - 46231	**WOOD SIDING, Cont'd...**						**07 - 46231**
2100	1x12	SF	0.020	1.49	2.10		3.59
2170	Redwood						
2180	1x6	SF	0.040	7.44	3.26		10.75
2200	1x8	"	0.032	6.93	2.60		9.53
2220	1x10	"	0.029	6.43	2.33		8.76
2240	1x12	"	0.026	5.93	2.10		8.03
3000	Tongue and groove						
3010	Cedar						
3020	1x4	SF	0.044	6.43	3.62		10.00
3040	1x6	"	0.042	6.18	3.43		9.61
3060	1x8	"	0.040	5.80	3.26		9.06
3080	1x10	"	0.038	5.69	3.10		8.79
3090	Pine						
3100	1x4	SF	0.044	1.93	3.62		5.55
3120	1x6	"	0.042	1.82	3.43		5.25
3140	1x8	"	0.040	1.70	3.26		4.96
3160	1x10	"	0.038	1.62	3.10		4.72
3170	Redwood						
3180	1x4	SF	0.044	6.80	3.62		10.50
3200	1x6	"	0.042	6.55	3.43		9.98
3220	1x8	"	0.040	6.33	3.26		9.59
3240	1x10	"	0.038	6.04	3.10		9.14
07 - 46291	**PLYWOOD SIDING**						**07 - 46291**
1000	Rough sawn cedar, 3/8" thick	SF	0.027	2.16	2.17		4.33
1020	Fir, 3/8" thick	"	0.027	1.19	2.17		3.36
1980	Texture 1-11, 5/8" thick						
2000	Cedar	SF	0.029	2.92	2.33		5.25
2020	Fir	"	0.029	2.04	2.33		4.37
2040	Redwood	"	0.029	3.14	2.22		5.36
2060	Southern Yellow Pine	"	0.029	1.66	2.33		3.99
07 - 46331	**PLASTIC SIDING**						**07 - 46331**
1000	Horizontal vinyl siding, solid						
1010	8" wide						
1020	Standard	SF	0.031	1.35	2.50		3.85
1040	Insulated	"	0.031	1.64	2.50		4.14
1050	10" wide						
1060	Standard	SF	0.029	1.40	2.33		3.73

© 2020 BNi Publications Inc. ALL RIGHTS RESERVED

SIDING

ID Code	Component Descriptions	Unit of Meas.	Manhr / Unit	Material Cost	Labor Cost	Equipment Cost	Total Cost
07 - 46331	**PLASTIC SIDING, Cont'd...**						**07 - 46331**
1080	Insulated	SF	0.029	1.68	2.33		4.01
8500	Vinyl moldings for doors and windows	LF	0.032	0.87	2.60		3.47
07 - 46461	**WOOD FIBER CEMENT SIDING**						**07 - 46461**
1000	Lap siding 5/16" x 8.25 x 12', 7" exposure	SF	0.031	1.32	2.50		3.82
2000	Panel siding 4' x 8'	"	0.029	0.99	2.33		3.32
3000	Shingle siding 14-5/8" x 25-5/32" (per bundle)	EA					130

MEMBRANE ROOFING

ID Code	Component Descriptions	Unit of Meas.	Manhr / Unit	Material Cost	Labor Cost	Equipment Cost	Total Cost
07 - 51131	**BUILT-UP ASPHALT ROOFING**						**07 - 51131**
0980	Built-up roofing, asphalt felt, including gravel						
1000	2 ply	SQ	2.000	88.00	160		250
1500	3 ply	"	2.667	120	210		330
2000	4 ply	"	3.200	170	250		420
2090	Walkway, for built-up roofs						
2095	3' x 3' x						
2100	1/2" thick	SF	0.027	2.46	2.07		4.53
2110	3/4" thick	"	0.027	3.81	2.07		5.88
2120	1" thick	"	0.027	4.12	2.07		6.19
2150	Roof bonds						
2155	10 yrs	SQ					37.25
2160	20 yrs	"					42.50
2195	Cant strip, 4" x 4"						
2200	Treated wood	LF	0.023	2.56	1.77		4.33
2260	Foamglass	"	0.020	2.20	1.55		3.75
2280	Mineral fiber	"	0.020	0.43	1.55		1.98
8000	New gravel for built-up roofing, 400 lb/sq	SQ	1.600	44.50	120		160
8220	Roof gravel (ballast)	CY	4.000	32.75	310		340
9000	Aluminum coating, top surfacing, for built-up roofing	SQ	1.333	49.00	100		150
9500	Remove 4-ply built-up roof (includes gravel)	"	4.000		310		310
9600	Remove & replace gravel, includes flood coat	"	2.667	59.00	210		270

© 2020 BNi Publications Inc. ALL RIGHTS RESERVED

ELASTOMERIC MEMBRANE ROOFING

	Description	Output		Unit Costs			
ID Code	Component Descriptions	Unit of Meas.	Manhr / Unit	Material Cost	Labor Cost	Equipment Cost	Total Cost
07 - 53001	**SINGLE-PLY ROOFING**						**07 - 53001**
2000	Elastic sheet roofing						
2060	Neoprene, 1/16" thick	SF	0.010	2.83	0.77		3.60
2080	EPDM rubber						
2100	45 mil	SF	0.010	1.78	0.77		2.55
2110	60 mil	"	0.010	2.44	0.77		3.21
2115	PVC						
2120	45 mil	SF	0.010	2.34	0.77		3.11
2140	60 mil	"	0.010	2.78	0.77		3.55
2200	Flashing						
2220	Pipe flashing, 90 mil thick						
2260	1" pipe	EA	0.200	34.00	15.50		49.50
2280	2" pipe	"	0.200	36.50	15.50		52.00
2290	3" pipe	"	0.211	36.75	16.50		53.00
2300	4" pipe	"	0.211	40.00	16.50		57.00
2310	5" pipe	"	0.222	42.75	17.25		60.00
2320	6" pipe	"	0.222	46.50	17.25		64.00
2330	8" pipe	"	0.235	53.00	18.25		71.00
2340	10" pipe	"	0.267	61.00	20.75		82.00
2350	12" pipe	"	0.267	74.00	20.75		95.00
2360	Neoprene flashing, 60 mil thick strip						
2380	6" wide	LF	0.067	1.72	5.19		6.91
2390	12" wide	"	0.100	3.38	7.78		11.25
2400	18" wide	"	0.133	4.98	10.25		15.25
2420	24" wide	"	0.200	6.55	15.50		22.00
2500	Adhesives						
2520	Mastic sealer, applied at joints only						
2540	1/4" bead	LF	0.004	0.19	0.31		0.50
3000	Fluid applied roofing						
3100	Urethane, 2 part, elastomeric membrane						
3101	1" thick	SF	0.013	4.97	1.03		6.00
3200	Vinyl liquid roofing, 2 coats, 2 mils per coat	"	0.011	7.07	0.88		7.95
3300	Silicone roofing, 2 coats sprayed, 16 mil per coat	"	0.013	5.28	1.03		6.31
3400	Inverted roof system						
3420	Insulated membrane with coarse gravel ballast						
3421	3 ply with 2" polystyrene	SF	0.013	8.47	1.03		9.50
8000	Ballast, 3/4" through 1-1/2" gravel, 100lb/sf	"	0.008	0.52	0.62		1.14
8100	Walkway for membrane roofs, 1/2" thick	"	0.027	2.69	2.07		4.76

© 2020 BNi Publications Inc. ALL RIGHTS RESERVED

FLASHING AND SHEET METAL

ID Code	Description: Component Descriptions	Output: Unit of Meas.	Output: Manhr / Unit	Unit Costs: Material Cost	Unit Costs: Labor Cost	Unit Costs: Equipment Cost	Unit Costs: Total Cost
07 - 61001	**METAL ROOFING**						**07 - 61001**
1000	Sheet metal roofing, copper, 16 oz, batten seam	SQ	5.333	1,800	420		2,220
1020	Standing seam	"	5.000	1,760	390		2,150
2000	Aluminum roofing, natural finish						
2005	Corrugated, on steel frame						
2010	.0175" thick	SQ	2.286	140	180		320
2040	.0215" thick	"	2.286	180	180		360
2060	.024" thick	"	2.286	210	180		390
2080	.032" thick	"	2.286	260	180		440
2100	V-beam, on steel frame						
2120	.032" thick	SQ	2.286	270	180		450
2130	.040" thick	"	2.286	290	180		470
2140	.050" thick	"	2.286	370	180		550
2200	Ridge cap						
2220	.019" thick	LF	0.027	4.25	2.07		6.32
2500	Corrugated galvanized steel roofing, on steel frame						
2520	28 ga.	SQ	2.286	230	180		410
2540	26 ga.	"	2.286	270	180		450
2550	24 ga.	"	2.286	300	180		480
2560	22 ga.	"	2.286	330	180		510
2580	26 ga., factory insulated with 1" polystyrene	"	3.200	510	250		760
2600	Ridge roll						
2620	10" wide	LF	0.027	2.31	2.07		4.38
2640	20" wide	"	0.032	4.70	2.49		7.19

SHEET METAL FLASHING AND TRIM

ID Code	Component Descriptions	Unit of Meas.	Manhr / Unit	Material Cost	Labor Cost	Equipment Cost	Total Cost
07 - 62001	**FLASHING AND TRIM**						**07 - 62001**
0050	Counter flashing						
0060	Aluminum, .032"	SF	0.080	2.09	6.22		8.31
0100	Stainless steel, .015"	"	0.080	6.69	6.22		13.00
0105	Copper						
0110	16 oz.	SF	0.080	9.36	6.22		15.50
0112	20 oz.	"	0.080	11.00	6.22		17.25
0114	24 oz.	"	0.080	13.50	6.22		19.75
0116	32 oz.	"	0.080	16.50	6.22		22.75
0118	Valley flashing						
0120	Aluminum, .032"	SF	0.050	1.74	3.89		5.63
0130	Stainless steel, .015	"	0.050	5.56	3.89		9.45

© 2020 BNi Publications Inc. ALL RIGHTS RESERVED

SHEET METAL FLASHING AND TRIM

ID Code	Description: Component Descriptions	Output: Unit of Meas.	Output: Manhr / Unit	Unit Costs: Material Cost	Unit Costs: Labor Cost	Unit Costs: Equipment Cost	Unit Costs: Total Cost
07 - 62001	**FLASHING AND TRIM, Cont'd...**						**07 - 62001**
0135	Copper						
0140	16 oz.	SF	0.050	9.36	3.89		13.25
0160	20 oz.	"	0.067	11.00	5.19		16.25
0180	24 oz.	"	0.050	13.50	3.89		17.50
0200	32 oz.	"	0.050	16.50	3.89		20.50
0380	Base flashing						
0400	Aluminum, .040"	SF	0.067	2.60	5.19		7.79
0410	Stainless steel, .018"	"	0.067	6.65	5.19		11.75
0415	Copper						
0420	16 oz.	SF	0.067	9.36	5.19		14.50
0422	20 oz.	"	0.050	11.00	3.89		15.00
0424	24 oz.	"	0.067	13.50	5.19		18.75
0426	32 oz.	"	0.067	16.50	5.19		21.75
2040	Waterstop, T section, 22 ga.						
2050	1-1/2" x 3"	LF	0.040	3.25	3.11		6.36
2060	2" x 2"	"	0.040	3.60	3.11		6.71
2070	4" x 3"	"	0.040	4.41	3.11		7.52
2080	6" x 4"	"	0.040	4.67	3.11		7.78
2090	8" x 4"	"	0.040	5.79	3.11		8.90
2500	Scupper outlets						
2520	10" x 10" x 4"	EA	0.200	58.00	15.50		74.00
2530	22" x 4" x 4"	"	0.200	62.00	15.50		78.00
2540	8" x 8" x 5"	"	0.200	57.00	15.50		73.00
3201	Flashing and trim, aluminum						
3221	.019" thick	SF	0.057	1.28	4.44		5.72
3231	.032" thick	"	0.057	1.57	4.44		6.01
3241	.040" thick	"	0.062	2.69	4.79		7.48
3310	Neoprene sheet flashing, .060" thick	"	0.050	2.14	3.89		6.03
3320	Copper, paper backed						
3330	2 oz.	SF	0.080	2.75	6.22		8.97
3340	5 oz.	"	0.080	3.55	6.22		9.77
3400	Drainage boots, roof, cast iron						
3420	2 x 3	LF	0.100	120	7.78		130
3430	3 x 4	"	0.100	140	7.78		150
3440	4 x 5	"	0.107	160	8.30		170
3450	4 x 6	"	0.107	170	8.30		180
3460	5 x 7	"	0.114	190	8.89		200
7980	Pitch pocket, copper, 16 oz.						

© 2020 BNi Publications Inc. ALL RIGHTS RESERVED

SHEET METAL FLASHING AND TRIM

ID Code	Component Descriptions	Unit of Meas.	Manhr / Unit	Material Cost	Labor Cost	Equipment Cost	Total Cost
07 - 62001	**FLASHING AND TRIM, Cont'd...**						**07 - 62001**
8120	4 x 4	EA	0.200	140	15.50		160
8140	6 x 6	"	0.200	150	15.50		170
8160	8 x 8	"	0.200	200	15.50		220
8180	8 x 10	"	0.200	230	15.50		250
8200	8 x 12	"	0.200	270	15.50		290
8400	Reglets, copper 10 oz.	LF	0.053	7.76	4.15		12.00
8420	Stainless steel, .020"	"	0.053	3.51	4.15		7.66
8480	Gravel stop						
8500	Aluminum, .032"						
8600	4"	LF	0.027	2.04	2.07		4.11
8620	6"	"	0.027	3.01	2.07		5.08
8640	8"	"	0.031	4.04	2.39		6.43
8660	10"	"	0.031	5.05	2.39		7.44
8670	Copper, 16 oz.						
8680	4"	LF	0.027	6.52	2.07		8.59
8700	6"	"	0.027	9.72	2.07		11.75
8720	8"	"	0.031	13.00	2.39		15.50
8740	10"	"	0.031	16.25	2.39		18.75

ROOFING SPECIALTIES

ID Code	Component Descriptions	Unit of Meas.	Manhr / Unit	Material Cost	Labor Cost	Equipment Cost	Total Cost
07 - 71001	**MANUFACTURED SPECIALTIES**						**07 - 71001**
0080	Moisture relief vent						
0100	Aluminum	EA	0.114	23.25	8.89		32.25
0120	Copper	"	0.114	63.00	8.89		72.00
0130	Expansion joint						
0135	Aluminum						
0140	Opening to 2.5"	LF	0.057	17.75	4.44		22.25
0160	Opening to 3.5"	"	0.062	17.50	4.79		22.25
0170	Copper, 16 oz.						
0180	Opening to 2.5"	LF	0.057	33.25	4.44		37.75
0200	Opening to 3.5"	"	0.062	41.25	4.79		46.00
0210	Butyl or neoprene						
0215	4" wide						
0220	16 oz. copper bellows	LF	0.067	32.00	5.19		37.25
0240	28 ga. stainless steel bellows	"	0.067	20.00	5.19		25.25
0260	6" wide						
0280	Copper bellows	LF	0.073	35.25	5.66		41.00

© 2020 BNi Publications Inc. ALL RIGHTS RESERVED

ROOFING SPECIALTIES

	Description	Output		Unit Costs			
ID Code	Component Descriptions	Unit of Meas.	Manhr / Unit	Material Cost	Labor Cost	Equipment Cost	Total Cost
07 - 71001	**MANUFACTURED SPECIALTIES, Cont'd...**						**07 - 71001**
0290	Stainless steel						
0300	Opening to 2.5"	LF	0.057	20.50	4.44		25.00
0320	Opening to 3.5"	"	0.062	26.00	4.79		30.75
0800	Smoke vent, 48" x 48"						
0820	Aluminum	EA	2.000	2,310	160		2,470
0860	Galvanized steel	"	2.000	2,030	160		2,190
0900	Heat/smoke vent, 48" x 96"						
1000	Aluminum	EA	2.667	3,240	210		3,450
1020	Galvanized steel	"	2.667	2,760	210		2,970
2020	Ridge vent strips						
2040	Mill finish	LF	0.053	3.90	4.15		8.05
2050	Connectors	EA	0.200	3.85	15.50		19.25
2060	End cap	"	0.229	1.92	17.75		19.75
2080	Soffit vents						
2085	Mill finish						
2090	2-1/2" wide	LF	0.032	0.49	2.49		2.98
2100	3" wide	"	0.032	0.60	2.49		3.09
2125	6" wide	"	0.032	1.04	2.49		3.53
3000	Roof hatches						
3020	Steel, plain, primed						
3040	2'6" x 3'0"	EA	2.000	800	160		960
3060	2'6" x 4'6"	"	2.667	1,180	210		1,390
3080	2'6" x 8'0"	"	4.000	1,820	310		2,130
3100	Galvanized steel						
3120	2'6" x 3'0"	EA	2.000	820	160		980
3140	2'6" x 4'6"	"	2.667	1,240	210		1,450
3160	2'6" x 8'0"	"	4.000	1,940	310		2,250
3180	Aluminum						
3200	2'6" x 3'0"	EA	2.000	950	160		1,110
3220	2'6" x 4'6"	"	2.667	1,430	210		1,640
3240	2'6" x 8'0"	"	4.000	2,660	310		2,970
3520	Ceiling access doors						
3540	Swing up model, metal frame						
3550	Steel door						
3560	2'6" x 2'6"	EA	0.800	690	62.00		750
3580	2'6" x 3'0"	"	0.800	740	62.00		800
3590	Aluminum door						
3600	2'6" x 2'6"	EA	0.800	850	62.00		910

© 2020 BNi Publications Inc. ALL RIGHTS RESERVED

ROOFING SPECIALTIES

ID Code	Description Component Descriptions	Output Unit of Meas.	Manhr / Unit	Unit Costs Material Cost	Labor Cost	Equipment Cost	Total Cost
07 - 71001	**MANUFACTURED SPECIALTIES, Cont'd...**						**07 - 71001**
3620	2'6" x 3'0"	EA	0.800	920	62.00		980
3640	Swing down model, metal frame						
3650	Steel door						
3660	2'6" x 2'6"	EA	0.800	1,040	62.00		1,100
3670	2'6" x 3'0"	"	0.800	1,120	62.00		1,180
3680	Aluminum door						
3690	2'6" x 2'6"	EA	0.800	1,240	62.00		1,300
3700	2'6" x 3'0"	"	0.800	1,330	62.00		1,390
3800	Gravity ventilators, with curb, base, damper and screen						
3820	Stationary siphon						
3830	6" dia.	EA	0.533	54.00	41.50		96.00
3840	12" dia.	"	0.533	93.00	41.50		130
3850	24" dia.	"	0.800	340	62.00		400
3860	36" dia.	"	0.800	720	62.00		780
3900	Wind driven spinner						
3920	6" dia.	EA	0.533	82.00	41.50		120
3940	12" dia.	"	0.533	110	41.50		150
3960	24" dia.	"	0.800	410	62.00		470
3980	36" dia.	"	0.800	840	62.00		900
4000	Stationary mushroom						
4020	16" dia.	EA	0.800	670	62.00		730
4060	30" dia.	"	1.000	1,510	78.00		1,590
4080	36" dia.	"	1.333	1,940	100		2,040
4100	42" dia.	"	1.600	2,900	120		3,020
07 - 71231	**GUTTERS AND DOWNSPOUTS**						**07 - 71231**
1500	Copper gutter and downspout						
1520	Downspouts, 16 oz. copper						
1530	Round						
1540	3" dia.	LF	0.053	12.50	4.15		16.75
1550	4" dia.	"	0.053	15.50	4.15		19.75
1560	Rectangular, corrugated						
1570	2" x 3"	LF	0.050	12.00	3.89		16.00
1580	3" x 4"	"	0.050	14.75	3.89		18.75
1585	Rectangular, flat surface						
1590	2" x 3"	LF	0.053	13.75	4.15		18.00
1600	3" x 4"	"	0.053	19.50	4.15		23.75
1620	Lead-coated copper downspouts						

© 2020 BNi Publications Inc. ALL RIGHTS RESERVED

ROOFING SPECIALTIES

ID Code	Description: Component Descriptions	Output: Unit of Meas.	Output: Manhr / Unit	Unit Costs: Material Cost	Unit Costs: Labor Cost	Unit Costs: Equipment Cost	Unit Costs: Total Cost
07 - 71231	GUTTERS AND DOWNSPOUTS, Cont'd...						07 - 71231
1625	Round						
1630	3" dia.	LF	0.050	16.25	3.89		20.25
1650	4" dia.	"	0.057	19.75	4.44		24.25
1670	Rectangular, corrugated						
1680	2" x 3"	LF	0.053	16.25	4.15		20.50
1690	3" x 4"	"	0.053	19.50	4.15		23.75
1695	Rectangular, plain						
1700	2" x 3"	LF	0.053	11.25	4.15		15.50
1750	3" x 4"	"	0.053	13.25	4.15		17.50
1800	Gutters, 16 oz. copper						
1810	Half round						
1820	4" wide	LF	0.080	11.25	6.22		17.50
1840	5" wide	"	0.089	13.75	6.92		20.75
1860	Type K						
1880	4" wide	LF	0.080	12.50	6.22		18.75
1890	5" wide	"	0.089	13.00	6.92		20.00
1900	Lead-coated copper gutters						
1905	Half round						
1910	4" wide	LF	0.080	13.50	6.22		19.75
1920	6" wide	"	0.089	18.50	6.92		25.50
1925	Type K						
1930	4" wide	LF	0.080	14.75	6.22		21.00
1940	5" wide	"	0.089	19.25	6.92		26.25
3000	Aluminum gutter and downspout						
3005	Downspouts						
3010	2" x 3"	LF	0.053	1.45	4.15		5.60
3030	3" x 4"	"	0.057	2.00	4.44		6.44
3035	4" x 5"	"	0.062	2.31	4.79		7.10
3038	Round						
3040	3" dia.	LF	0.053	2.44	4.15		6.59
3050	4" dia.	"	0.057	3.13	4.44		7.57
3240	Gutters, stock units						
3260	4" wide	LF	0.084	2.26	6.55		8.81
3270	5" wide	"	0.089	2.69	6.92		9.61
4101	Galvanized steel gutter and downspout						
4111	Downspouts, round corrugated						
4121	3" dia.	LF	0.053	2.09	4.15		6.24
4131	4" dia.	"	0.053	2.80	4.15		6.95

© 2020 BNi Publications Inc. ALL RIGHTS RESERVED

ROOFING SPECIALTIES

ID Code	Description: Component Descriptions	Output: Unit of Meas.	Output: Manhr / Unit	Unit Costs: Material Cost	Unit Costs: Labor Cost	Unit Costs: Equipment Cost	Unit Costs: Total Cost
07 - 71231	**GUTTERS AND DOWNSPOUTS, Cont'd...**						**07 - 71231**
4141	5" dia.	LF	0.057	4.18	4.44		8.62
4151	6" dia.	"	0.057	5.54	4.44		9.98
4161	Rectangular						
4171	2" x 3"	LF	0.053	1.89	4.15		6.04
4191	3" x 4"	"	0.050	2.70	3.89		6.59
4201	4" x 4"	"	0.050	3.38	3.89		7.27
4300	Gutters, stock units						
4310	5" wide						
4320	Plain	LF	0.089	1.82	6.92		8.74
4330	Painted	"	0.089	1.98	6.92		8.90
4335	6" wide						
4340	Plain	LF	0.094	2.55	7.32		9.87
4360	Painted	"	0.094	2.86	7.32		10.25

FIREPROOFING

ID Code	Component Descriptions	Unit of Meas.	Manhr / Unit	Material Cost	Labor Cost	Equipment Cost	Total Cost
07 - 81001	**FIREPROOFING**						**07 - 81001**
0980	Sprayed on						
1000	1" thick						
1020	On beams	SF	0.018	0.89	1.13		2.02
1040	On columns	"	0.016	0.91	1.02		1.93
1050	On decks						
1060	Flat surface	SF	0.008	0.91	0.51		1.42
1080	Fluted surface	"	0.010	1.15	0.63		1.78
1100	1-1/2" thick						
1120	On beams	SF	0.023	1.60	1.45		3.05
1140	On columns	"	0.020	1.81	1.27		3.08
1150	On decks						
1160	Flat surface	SF	0.010	1.36	0.63		1.99
1170	Fluted surface	"	0.013	1.60	0.85		2.45

© 2020 BNi Publications Inc. ALL RIGHTS RESERVED

JOINT SEALANTS

	Description	Output		Unit Costs			
ID Code	Component Descriptions	Unit of Meas.	Manhr / Unit	Material Cost	Labor Cost	Equipment Cost	Total Cost
07 - 92001	**CAULKING**						**07 - 92001**
0100	Caulk exterior, two component						
0120	1/4 x 1/2	LF	0.040	0.43	3.26		3.69
0140	3/8 x 1/2	"	0.044	0.66	3.62		4.28
0160	1/2 x 1/2	"	0.050	0.90	4.07		4.97
0220	Caulk interior, single component						
0240	1/4 x 1/2	LF	0.038	0.29	3.10		3.39
0260	3/8 x 1/2	"	0.042	0.41	3.43		3.84
0280	1/2 x 1/2	"	0.047	0.54	3.83		4.37
1000	Butyl rubber fillers						
1010	1/4" x 1/4"	LF	0.016	0.83	1.30		2.13
1020	1/2" x 1/2"	"	0.027	1.21	2.17		3.38
1030	1/2" x 3/4"	"	0.032	1.85	2.60		4.45
1040	3/4" x 3/4"	"	0.032	2.43	2.60		5.03
1060	1" x 1"	"	0.036	2.85	2.89		5.74
1400	Seals, O-ring type cord						
1410	1/4" dia.	LF	0.020	0.91	1.63		2.54
1420	1/2" dia.	"	0.021	2.72	1.71		4.43
1440	1" dia.	"	0.022	9.64	1.81		11.50
1450	1-1/4" dia.	"	0.024	12.25	1.91		14.25
1460	1-1/2" dia.	"	0.025	16.25	2.03		18.25
1470	1-3/4" dia.	"	0.026	24.00	2.10		26.00
1480	2" dia.	"	0.027	31.00	2.17		33.25
1500	Polyvinyl chloride, closed cell						
1520	1/4" x 2"	LF	0.029	0.62	2.33		2.95
1540	1/4" x 6"	"	0.036	1.96	2.96		4.92
1600	Silicone foam penetration seal						
1620	1/4" x 1/2"	LF	0.010	0.24	0.81		1.05
1640	1/2" x 1/2"	"	0.013	0.41	1.08		1.49
1660	1/2" x 3/4"	"	0.016	0.64	1.30		1.94
1680	3/4" x 3/4"	"	0.020	0.96	1.63		2.59
1700	1/8" x 1"	"	0.010	0.24	0.81		1.05
1720	1/8" x 3"	"	0.016	0.64	1.30		1.94
1740	1/4" x 3"	"	0.020	1.27	1.63		2.90
1760	1/4" x 6"	"	0.027	2.61	2.17		4.78
1780	1/2" x 6"	"	0.062	5.14	5.01		10.25
1800	1/2" x 9"	"	0.100	7.72	8.15		15.75
1820	1/2" x 12"	"	0.145	10.25	11.75		22.00
2020	Oil base sealants and caulking						

© 2020 BNi Publications Inc. ALL RIGHTS RESERVED

JOINT SEALANTS

ID Code	Description Component Descriptions	Output Unit of Meas.	Manhr / Unit	Unit Costs Material Cost	Labor Cost	Equipment Cost	Total Cost
07 - 92001	**CAULKING, Cont'd...**						**07 - 92001**
2040	1/4" x 1/4"	LF	0.020	0.05	1.63		1.68
2060	1/4" x 3/8"	"	0.021	0.11	1.69		1.80
2080	1/4" x 1/2"	"	0.022	0.12	1.76		1.88
2100	3/8" x 3/8"	"	0.023	0.15	1.86		2.01
2120	3/8" x 1/2"	"	0.024	0.17	1.97		2.14
2140	3/8" x 5/8"	"	0.026	0.31	2.10		2.41
2160	3/8" x 3/4"	"	0.028	0.35	2.24		2.59
2180	1/2" x 1/2"	"	0.031	0.31	2.50		2.81
2200	1/2" x 5/8"	"	0.035	0.39	2.83		3.22
2220	1/2" x 3/4"	"	0.040	0.45	3.26		3.71
2240	1/2" x 7/8"	"	0.041	0.53	3.34		3.87
2260	1/2" x 1"	"	0.042	0.58	3.43		4.01
2280	3/4" x 3/4"	"	0.043	0.68	3.52		4.20
2300	1" x 1"	"	0.044	1.19	3.62		4.81
2400	Polyurethane compounds						
2420	1/4" x 1/4"	LF	0.020	0.24	1.63		1.87
2440	1/4" x 3/8"	"	0.021	0.43	1.69		2.12
2460	1/4" x 1/2"	"	0.022	0.54	1.76		2.30
2480	3/8" x 3/8"	"	0.023	0.60	1.86		2.46
2500	3/8" x 1/2"	"	0.024	0.77	1.97		2.74
2520	3/8" x 5/8"	"	0.026	0.96	2.10		3.06
2540	3/8" x 3/4"	"	0.028	1.16	2.24		3.40
2560	1/2" x 1/2"	"	0.031	1.11	2.50		3.61
2580	1/2" x 5/8"	"	0.035	1.21	2.83		4.04
2590	1/2" x 3/4"	"	0.040	1.39	3.26		4.65
2600	1/2" x 7/8"	"	0.041	1.77	3.34		5.11
2620	1/2" x 1"	"	0.044	2.24	3.62		5.86
2640	3/4" x 3/4"	"	0.043	2.28	3.52		5.80
2660	3/4" x 1"	"	0.044	2.49	3.62		6.11
4000	Backer rod, polyethylene						
4100	1/4"	LF	0.020	0.06	1.63		1.69
4120	1/2"	"	0.021	0.10	1.71		1.81
4140	3/4"	"	0.022	0.13	1.81		1.94
4160	1"	"	0.024	0.19	1.91		2.10

© 2020 BNi Publications Inc. ALL RIGHTS RESERVED

EXPANSION CONTROL

ID Code	Description: Component Descriptions	Output: Unit of Meas.	Output: Manhr / Unit	Unit Costs: Material Cost	Unit Costs: Labor Cost	Unit Costs: Equipment Cost	Unit Costs: Total Cost
07 - 95001	**EXPANSION JOINTS**						**07 - 95001**
1000	Expansion joints with covers, floor assembly type						
1040	With 1" space						
1060	Aluminum	LF	0.133	31.25	12.00		43.25
1080	Bronze	"	0.133	65.00	12.00		77.00
1100	Stainless steel	"	0.133	51.00	12.00		63.00
1200	With 2" space						
1210	Aluminum	LF	0.133	33.75	12.00		45.75
1220	Bronze	"	0.133	60.00	12.00		72.00
1240	Stainless steel	"	0.133	53.00	12.00		65.00
1400	Ceiling and wall assembly type						
1420	With 1" space						
1430	Aluminum	LF	0.160	17.50	14.50		32.00
1440	Bronze	"	0.160	60.00	14.50		75.00
1450	Stainless steel	"	0.160	54.00	14.50		69.00
1500	With 2" space						
1520	Aluminum	LF	0.160	19.25	14.50		33.75
1540	Bronze	"	0.160	65.00	14.50		80.00
1560	Stainless steel	"	0.160	57.00	14.50		72.00
1600	Exterior roof and wall, aluminum						
1640	Roof to roof						
1660	With 1" space	LF	0.133	51.00	12.00		63.00
1680	With 2" space	"	0.133	55.00	12.00		67.00
1700	Roof to wall						
1720	With 1" space	LF	0.145	39.75	13.00		53.00
1740	With 2" space	"	0.145	47.25	13.00		60.00
1760	Flat wall to wall						
1780	With 1" space	LF	0.133	19.25	12.00		31.25
1790	With 2" space	"	0.133	21.00	12.00		33.00
1800	Corner to flat wall						
1820	With 1" space	LF	0.160	23.25	14.50		37.75
1840	With 2" in space	"	0.160	23.50	14.50		38.00

© 2020 BNi Publications Inc. ALL RIGHTS RESERVED

DIVISION 08
DOORS AND WINDOWS

METAL DOORS & TRANSOMS

	Description	Output		Unit Costs			
ID Code	Component Descriptions	Unit of Meas.	Manhr / Unit	Material Cost	Labor Cost	Equipment Cost	Total Cost
08 - 11131	**METAL DOORS**						**08 - 11131**
1000	Flush hollow metal, std. duty, 20 ga., 1-3/8" thick						
1020	2-6 x 6-8	EA	0.889	370	72.00		440
1040	2-8 x 6-8	"	0.889	390	72.00		460
1080	3-0 x 6-8	"	0.889	420	72.00		490
1090	1-3/4" thick						
1100	2-6 x 6-8	EA	0.889	410	72.00		480
1120	2-8 x 6-8	"	0.889	510	72.00		580
1150	3-0 x 6-8	"	0.889	470	72.00		540
1200	2-6 x 7-0	"	0.889	450	72.00		520
1210	2-8 x 7-0	"	0.889	470	72.00		540
1240	3-0 x 7-0	"	0.889	500	72.00		570
2110	Heavy duty, 20 ga., unrated, 1-3/4"						
2130	2-8 x 6-8	EA	0.889	450	72.00		520
2135	3-0 x 6-8	"	0.889	490	72.00		560
2140	2-8 x 7-0	"	0.889	520	72.00		590
2150	3-0 x 7-0	"	0.889	500	72.00		570
2170	3-4 x 7-0	"	0.889	520	72.00		590
2200	18 ga., 1-3/4", unrated door						
2210	2-0 x 7-0	EA	0.889	480	72.00		550
2230	2-4 x 7-0	"	0.889	480	72.00		550
2235	2-6 x 7-0	"	0.889	480	72.00		550
2240	2-8 x 7-0	"	0.889	530	72.00		600
2260	3-0 x 7-0	"	0.889	490	72.00		560
2270	3-4 x 7-0	"	0.889	600	72.00		670
2310	2", unrated door						
2320	2-0 x 7-0	EA	1.000	530	82.00		610
2330	2-4 x 7-0	"	1.000	530	82.00		610
2340	2-6 x 7-0	"	1.000	530	82.00		610
2350	2-8 x 7-0	"	1.000	580	82.00		660
2360	3-0 x 7-0	"	1.000	600	82.00		680
2370	3-4 x 7-0	"	1.000	610	82.00		690
2400	Galvanized metal door						
2410	3-0 x 7-0	EA	1.000	620	82.00		700
2450	For lead lining in doors	"					1,120
2460	For sound attenuation	"					100
4280	Vision glass						
4300	8" x 8"	EA	1.000	130	82.00		210
4320	8" x 48"	"	1.000	200	82.00		280

© 2020 BNi Publications Inc. ALL RIGHTS RESERVED

METAL DOORS & TRANSOMS

ID Code	Description: Component Descriptions	Output: Unit of Meas.	Output: Manhr / Unit	Unit Costs: Material Cost	Unit Costs: Labor Cost	Unit Costs: Equipment Cost	Unit Costs: Total Cost
08 - 11131	**METAL DOORS, Cont'd...**						**08 - 11131**
4340	Fixed metal louver	EA	0.800	290	65.00		350
4350	For fire rating, add						
4370	3 hr door	EA					490
4380	1-1/2 hr door	"					220
4400	3/4 hr door	"					110
4430	1' extra height, add to material, 20%						
4440	1'6" extra height, add to material, 60%						
4470	For dutch doors with shelf, add to material, 100%						
08 - 11134	**METAL DOOR FRAMES**						**08 - 11134**
1000	Hollow metal, stock, 18 ga., 4-3/4" x 1-3/4"						
1020	2-0 x 7-0	EA	1.000	160	82.00		240
1040	2-4 x 7-0	"	1.000	190	82.00		270
1060	2-6 x 7-0	"	1.000	190	82.00		270
1080	2-8 x 7-0	"	1.000	190	82.00		270
1100	3-0 x 7-0	"	1.000	190	82.00		270
1120	4-0 x 7-0	"	1.333	210	110		320
1140	5-0 x 7-0	"	1.333	220	110		330
1160	6-0 x 7-0	"	1.333	260	110		370
1500	16 ga., 6-3/4" x 1-3/4"						
1520	2-0 x 7-0	EA	1.000	190	90.00		280
1530	2-4 x 7-0	"	1.000	180	90.00		270
1535	2-6 x 7-0	"	1.000	180	90.00		270
1540	2-8 x 7-0	"	1.000	180	90.00		270
1550	3-0 x 7-0	"	1.000	200	90.00		290
1560	4-0 x 7-0	"	1.333	230	120		350
1580	6-0 x 7-0	"	1.333	260	120		380
1600	Transom frame						
1620	3-4 x 1 6	EA	1.000	130	90.00		220
1640	3-8 x 1-6	"	1.000	130	90.00		220
1660	6-4 x 1-6	"	1.000	190	90.00		280
1680	Transom sash						
1690	3-0 x 1-4	EA	1.000	120	90.00		210
1700	3-4 x 1-4	"	1.000	130	90.00		220
1720	6-0 x 1-4	"	1.000	190	90.00		280
1760	1' extension of frame, add	"					24.75
1770	14 ga. frame, add	"					24.75
1775	For fire rating, add						

© 2020 BNi Publications Inc. ALL RIGHTS RESERVED

METAL DOORS & TRANSOMS

	Description	Output		Unit Costs			
ID Code	Component Descriptions	Unit of Meas.	Manhr / Unit	Material Cost	Labor Cost	Equipment Cost	Total Cost
08 - 11134	**METAL DOOR FRAMES, Cont'd...**						**08 - 11134**
1780	3 hour	EA					67.00
1790	1-1/2 hour	"					54.00
1800	3/4 hour	"					47.00
1810	Lead lining in frame, add	"					140
1900	Sidelights, complete						
1920	1-0 x 7-2	EA	1.000	460	90.00		550
1940	1-4 x 7-2	"	1.000	510	90.00		600
1960	1-0 x 8-8	"	1.000	540	90.00		630
1970	1-6 x 8-8	"	1.000	560	90.00		650
2000	16 ga., 4-3/4" x 1-3/4"						
2020	2-0 x 7-0	EA	1.000	170	90.00		260
2030	2-4 x 7-0	"	1.000	170	90.00		260
2035	2-6 x 7-0	"	1.000	170	90.00		260
2040	2-8 x 7-0	"	1.000	170	90.00		260
2050	3-0 x 7-0	"	1.000	180	90.00		270
2060	4-0 x 7-0	"	1.333	180	120		300
2070	6-0 x 7-0	"	1.333	190	120		310
2100	Transom frame						
2120	3-4 x 1-6	EA	1.000	120	90.00		210
2140	3-8 x 1-6	"	1.000	120	90.00		210
2160	6-4 x 1-6	"	1.000	160	90.00		250
2180	Transom sash						
2200	3-0 x 1-4	EA	1.000	100	90.00		190
2220	3-4 x 1-4	"	1.000	110	90.00		200
2240	6-0 x 1-4	"	1.000	160	90.00		250
2320	1' extension of door frame, add	"					24.75
2330	14 ga., metal frame, add	"					24.75
2335	For fire rating, add						
2340	3 hour	EA					64.00
2360	1-1/2 hour	"					51.00
2380	3/4 hour	"					44.75
2390	Lead lining in frame, add	"					140
2400	Sidelights, complete						
2420	1-0 x 7-2	EA	1.000	440	90.00		530
2440	1-4 x 7-2	"	1.000	460	90.00		550
2460	1-0 x 8-8	"	1.000	500	90.00		590
2480	1-4 x 8-8	"	1.000	510	90.00		600
2500	16 ga., 5-3/4" x 1-3/4"						

© 2020 BNi Publications Inc. ALL RIGHTS RESERVED

METAL DOORS & TRANSOMS

ID Code	Description: Component Descriptions	Output: Unit of Meas.	Output: Manhr / Unit	Unit Costs: Material Cost	Unit Costs: Labor Cost	Unit Costs: Equipment Cost	Unit Costs: Total Cost
08 - 11134	**METAL DOOR FRAMES, Cont'd...**						**08 - 11134**
2520	2-0 x 7-0	EA	1.000	170	82.00		250
2530	2-4 x 7-0	"	1.000	180	82.00		260
2540	2-6 x 7-0	"	1.000	180	82.00		260
2550	2-8 x 7-0	"	1.000	190	82.00		270
2560	3-0 x 7-0	"	1.000	190	82.00		270
2580	4-0 x 7-0	"	1.333	200	110		310
2590	5-0 x 7-0	"	1.333	210	110		320
2600	6-0 x 7-0	"	1.333	220	110		330
2610	Mullions, vertical						
2640	5-1/4" x 1-3/4"	LF	0.100	19.25	8.15		27.50
2650	5-1/4" x 2"	"	0.100	24.00	8.15		32.25
2700	Horizontal						
2730	5-1/4" x 1-3/4"	LF	0.100	19.25	8.15		27.50
2740	5-1/4" x 2"	"	0.100	24.00	8.15		32.25
08 - 11161	**ALUMINUM DOORS**						**08 - 11161**
1490	Aluminum doors, commercial						
1500	Narrow stile						
1520	2-6 x 7-0	EA	4.000	940	360		1,300
1540	3-0 x 7-0	"	4.000	990	360		1,350
1560	3-6 x 7-0	"	4.000	1,010	360		1,370
1570	Pair						
1580	5-0 x 7-0	EA	8.000	1,570	720		2,290
1600	6-0 x 7-0	"	8.000	1,590	720		2,310
1620	7-0 x 7-0	"	8.000	1,660	720		2,380
1700	Wide stile						
1720	2-6 x 7-0	EA	4.000	1,320	360		1,680
1740	3-0 x 7-0	"	4.000	1,360	360		1,720
1760	3-6 x 7-0	"	4.000	1,400	360		1,760
1770	Pair						
1780	5-0 x 7-0	EA	8.000	2,410	720		3,130
1800	6-0 x 7-0	"	8.000	2,520	720		3,240
1820	7-0 x 7-0	"	8.000	2,570	720		3,290

© 2020 BNi Publications Inc. ALL RIGHTS RESERVED

WOOD AND PLASTIC

	Description	Output		Unit Costs			
ID Code	Component Descriptions	Unit of Meas.	Manhr / Unit	Material Cost	Labor Cost	Equipment Cost	Total Cost
08 - 14001	**WOOD DOORS**						**08 - 14001**
0980	Solid core, 1-3/8" thick						
1000	Birch faced						
1020	2-4 x 7-0	EA	1.000	180	82.00		260
1040	2-8 x 7-0	"	1.000	180	82.00		260
1060	3-0 x 7-0	"	1.000	180	82.00		260
1070	3-4 x 7-0	"	1.000	370	82.00		450
1080	2-4 x 6-8	"	1.000	180	82.00		260
1090	2-6 x 6-8	"	1.000	180	82.00		260
1095	2-8 x 6-8	"	1.000	180	82.00		260
1100	3-0 x 6-8	"	1.000	180	82.00		260
1120	Lauan faced						
1140	2-4 x 6-8	EA	1.000	160	82.00		240
1160	2-8 x 6-8	"	1.000	170	82.00		250
1180	3-0 x 6-8	"	1.000	180	82.00		260
1200	3-4 x 6-8	"	1.000	190	82.00		270
1300	Tempered hardboard faced						
1320	2-4 x 7-0	EA	1.000	200	82.00		280
1340	2-8 x 7-0	"	1.000	210	82.00		290
1360	3-0 x 7-0	"	1.000	240	82.00		320
1380	3-4 x 7-0	"	1.000	250	82.00		330
1420	Hollow core, 1-3/8" thick						
1440	Birch faced						
1460	2-4 x 7-0	EA	1.000	160	82.00		240
1480	2-8 x 7-0	"	1.000	160	82.00		240
1500	3-0 x 7-0	"	1.000	170	82.00		250
1520	3-4 x 7-0	"	1.000	180	82.00		260
1600	Lauan faced						
1620	2-4 x 6-8	EA	1.000	69.00	82.00		150
1630	2-6 x 6-8	"	1.000	74.00	82.00		160
1640	2-8 x 6-8	"	1.000	93.00	82.00		180
1660	3-0 x 6-8	"	1.000	97.00	82.00		180
1680	3-4 x 6-8	"	1.000	110	82.00		190
1740	Tempered hardboard faced						
1760	2-4 x 7-0	EA	1.000	84.00	82.00		170
1770	2-6 x 7-0	"	1.000	90.00	82.00		170
1780	2-8 x 7-0	"	1.000	100	82.00		180
1800	3-0 x 7-0	"	1.000	110	82.00		190
1820	3-4 x 7-0	"	1.000	120	82.00		200

© 2020 BNi Publications Inc. ALL RIGHTS RESERVED

WOOD AND PLASTIC

	Description	Output		Unit Costs			
ID Code	Component Descriptions	Unit of Meas.	Manhr / Unit	Material Cost	Labor Cost	Equipment Cost	Total Cost
08 - 14001	**WOOD DOORS, Cont'd...**						**08 - 14001**
1000	Solid core, 1 3/4" thick						
1920	Birch faced						
1940	2-4 x 7-0	EA	1.000	280	82.00		360
1950	2-6 x 7-0	"	1.000	280	82.00		360
1960	2-8 x 7-0	"	1.000	290	82.00		370
1970	3-0 x 7-0	"	1.000	270	82.00		350
1980	3-4 x 7-0	"	1.000	280	82.00		360
2000	Lauan faced						
2020	2-4 x 7-0	EA	1.000	190	82.00		270
2030	2-6 x 7-0	"	1.000	220	82.00		300
2040	2-8 x 7-0	"	1.000	230	82.00		310
2060	3-4 x 7-0	"	1.000	240	82.00		320
2080	3-0 x 7-0	"	1.000	260	82.00		340
2140	Tempered hardboard faced						
2160	2-4 x 7-0	EA	1.000	250	82.00		330
2170	2-6 x 7-0	"	1.000	280	82.00		360
2180	2-8 x 7-0	"	1.000	300	82.00		380
2190	3-0 x 7-0	"	1.000	320	82.00		400
2200	3-4 x 7-0	"	1.000	340	82.00		420
2250	Hollow core, 1-3/4" thick						
2270	Birch faced						
2290	2-4 x 7-0	EA	1.000	190	82.00		270
2295	2-6 x 7-0	"	1.000	190	82.00		270
2300	2-8 x 7-0	"	1.000	200	82.00		280
2320	3-0 x 7-0	"	1.000	200	82.00		280
2340	3-4 x 7-0	"	1.000	220	82.00		300
2400	Lauan faced						
2420	2-4 x 6-8	EA	1.000	110	82.00		190
2430	2-6 x 6-8	"	1.000	130	82.00		210
2440	2-8 x 6-8	"	1.000	110	82.00		190
2460	3-0 x 6-8	"	1.000	120	82.00		200
2480	3-4 x 6-8	"	1.000	120	82.00		200
2520	Tempered hardboard						
2540	2-4 x 7-0	EA	1.000	100	82.00		180
2550	2-6 x 7-0	"	1.000	110	82.00		190
2560	2-8 x 7-0	"	1.000	110	82.00		190
2580	3-0 x 7-0	"	1.000	120	82.00		200
2600	3-4 x 7-0	"	1.000	130	82.00		210

© 2020 BNi Publications Inc. ALL RIGHTS RESERVED

WOOD AND PLASTIC

ID Code	Description: Component Descriptions	Output: Unit of Meas.	Output: Manhr / Unit	Unit Costs: Material Cost	Unit Costs: Labor Cost	Unit Costs: Equipment Cost	Unit Costs: Total Cost
08 - 14001	**WOOD DOORS, Cont'd...**						**08 - 14001**
2620	Add-on, louver	EA	0.800	35.00	65.00		100
2640	Glass	"	0.800	110	65.00		180
2700	Exterior doors, 3-0 x 7-0 x 2-1/2", solid core						
2710	Carved						
2720	One face	EA	2.000	1,460	160		1,620
2740	Two faces	"	2.000	2,020	160		2,180
3000	Closet doors, 1-3/4" thick						
3001	Bi-fold or bi-passing, includes frame and trim						
3020	Paneled						
3040	4-0 x 6-8	EA	1.333	510	110		620
3060	6-0 x 6-8	"	1.333	580	110		690
3070	Louvered						
3080	4-0 x 6-8	EA	1.333	350	110		460
3100	6-0 x 6-8	"	1.333	420	110		530
3130	Flush						
3140	4-0 x 6-8	EA	1.333	260	110		370
3160	6-0 x 6-8	"	1.333	330	110		440
3170	Primed						
3180	4-0 x 6-8	EA	1.333	280	110		390
3200	6-0 x 6-8	"	1.333	310	110		420
08 - 14009	**WOOD FRAMES**						**08 - 14009**
0080	Frame, interior, pine						
0100	2-6 x 6-8	EA	1.143	100	93.00		190
0140	2-8 x 6-8	"	1.143	110	93.00		200
0160	3-0 x 6-8	"	1.143	120	93.00		210
0180	5-0 x 6-8	"	1.143	120	93.00		210
0200	6-0 x 6-8	"	1.143	130	93.00		220
0220	2-6 x 7-0	"	1.143	120	93.00		210
0240	2-8 x 7-0	"	1.143	140	93.00		230
0260	3-0 x 7-0	"	1.143	140	93.00		230
0280	5-0 x 7-0	"	1.600	150	130		280
0300	6-0 x 7-0	"	1.600	160	130		290
1000	Exterior, custom, with threshold, including trim						
1040	Walnut						
1060	3-0 x 7-0	EA	2.000	420	160		580
1080	6-0 x 7-0	"	2.000	480	160		640
1090	Oak						

© 2020 BNi Publications Inc. ALL RIGHTS RESERVED

WOOD AND PLASTIC

ID Code	Description: Component Descriptions	Output: Unit of Meas.	Output: Manhr / Unit	Unit Costs: Material Cost	Unit Costs: Labor Cost	Unit Costs: Equipment Cost	Unit Costs: Total Cost
08 - 14009	**WOOD FRAMES, Cont'd...**						**08 - 14009**
1100	3-0 x 7-0	EA	2.000	380	160		540
1120	6-0 x 7-0	"	2.000	430	160		590
1200	Pine						
1220	2-4 x 7-0	EA	1.600	160	130		290
1240	2-6 x 7-0	"	1.600	160	130		290
1280	2-8 x 7-0	"	1.600	170	130		300
1300	3-0 x 7-0	"	1.600	180	130		310
1320	3-4 x 7-0	"	1.600	200	130		330
1340	6-0 x 7-0	"	2.667	210	220		430

SPECIALTY DOORS AND FRAMES

ID Code	Component Descriptions	Unit of Meas.	Manhr / Unit	Material Cost	Labor Cost	Equipment Cost	Total Cost
08 - 31131	**CONTROL**						**08 - 31131**
1020	Access control, 7' high, indoor or outdoor						
1040	Remote or card control, type B	EA	10.667	1,880	890		2,770
1060	Free passage, type B	"	10.667	1,530	890		2,420
1080	Remote or card control, type AA	"	10.667	3,000	890		3,890
1100	Free passage, type AA	"	10.667	2,720	890		3,610

SPECIAL FUNCTION DOORS

ID Code	Component Descriptions	Unit of Meas.	Manhr / Unit	Material Cost	Labor Cost	Equipment Cost	Total Cost
08 - 34001	**SPECIAL DOORS**						**08 - 34001**
1000	Vault door and frame, class 5, steel	EA	8.000	7,940	650		8,590
1400	Overhead door, coiling insulated						
1500	Chain gear, no frame, 12' x 12'	EA	10.000	3,410	820		4,230
2000	Aluminum, bronze glass panels, 12-9 x 13-0	"	8.000	3,850	650		4,500
2200	Garage, flush, ins. metal, primed, 9-0 x 7-0	"	2.667	1,100	220		1,320
3020	Sliding fire doors, motorized, fusible link, 3 hr.						
3040	3-0 x 6-8	EA	16.000	5,620	1,300		6,920
3060	3-8 x 6-8	"	16.000	5,690	1,300		6,990
3080	4-0 x 8-0	"	16.000	5,790	1,300		7,090
3100	5-0 x 8-0	"	16.000	5,900	1,300		7,200
3180	Metal clad doors, including electric motor						
3200	Light duty						
3220	Minimum	SF	0.133	45.00	10.75		56.00
3240	Maximum	"	0.320	73.00	26.00		99.00
3250	Heavy duty						
3260	Minimum	SF	0.400	70.00	32.50		100

© 2020 BNi Publications Inc. ALL RIGHTS RESERVED

SPECIAL FUNCTION DOORS

	Description	Output		Unit Costs			
ID Code	Component Descriptions	Unit of Meas.	Manhr / Unit	Material Cost	Labor Cost	Equipment Cost	Total Cost
08 - 34001	**SPECIAL DOORS, Cont'd...**						**08 - 34001**
3280	Maximum	SF	0.500	110	40.75		150
3300	Hangar doors, based on 150' openings						
3320	To 20' high	SF	0.096	61.00	6.12	4.20	71.00
3340	20' to 40' high	"	0.060	67.00	3.83	2.62	73.00
3360	40' to 60' high	"	0.040	70.00	2.55	1.75	74.00
3380	60' to 80' high	"	0.024	73.00	1.53	1.05	76.00
3400	Over 80' high	"	0.019	100	1.22	0.84	100
3480	Counter doors (roll-up shutters), std, manual						
3500	Opening, 4' high						
3520	4' wide	EA	6.667	1,300	540		1,840
3540	6' wide	"	6.667	1,760	540		2,300
3560	8' wide	"	7.273	1,980	590		2,570
3580	10' wide	"	10.000	2,200	820		3,020
3590	14' wide	"	10.000	2,750	820		3,570
3600	6' high						
3610	4' wide	EA	6.667	1,540	540		2,080
3620	6' wide	"	7.273	2,010	590		2,600
3630	8' wide	"	8.000	2,200	650		2,850
3640	10' wide	"	10.000	2,480	820		3,300
3650	14' wide	"	11.429	2,810	930		3,740
3660	For stainless steel, add to material, 40%						
3670	For motor operator, add	EA					1,520
3700	Service doors (roll-up shutters), std, manual						
3800	Opening						
3820	8' high x 8' wide	EA	4.444	1,650	360		2,010
3830	10' high x 10' wide	"	6.667	2,060	540		2,600
3840	12' high x 12' wide	"	10.000	2,310	820		3,130
3850	14' high x 14' wide	"	13.333	3,030	1,090		4,120
3860	16' high x 14' wide	"	13.333	4,240	1,090		5,330
3870	20' high x 14' wide	"	20.000	4,790	1,630		6,420
3880	24' high x 16' wide	"	17.778	7,810	1,450		9,260
3890	For motor operator						
3900	Up to 12-0 x 12-0, add	EA					1,550
3920	Over 12-0 x 12-0, add	"					1,980
4000	Roll-up doors						
4050	13-0 high x 14-0 wide	EA	11.429	1,560	930		2,490
4060	12-0 high x 14-0 wide	"	11.429	1,980	930		2,910
4100	Top coiling grilles, manual, steel or aluminum						

© 2020 BNi Publications Inc. ALL RIGHTS RESERVED

SPECIAL FUNCTION DOORS

ID Code	Description: Component Descriptions	Output: Unit of Meas.	Output: Manhr / Unit	Unit Costs: Material Cost	Unit Costs: Labor Cost	Unit Costs: Equipment Cost	Unit Costs: Total Cost
08 - 34001	**SPECIAL DOORS, Cont'd...**						**08 - 34001**
4120	Opening, 4' high x						
4140	4' wide	EA	3.200	1,830	260		2,090
4160	6' wide	"	3.200	1,890	260		2,150
4180	8' wide	"	4.444	2,190	360		2,550
4200	12' wide	"	4.444	2,530	360		2,890
4220	16' wide	"	6.667	2,900	540		3,440
4230	6' high x						
4240	4' wide	EA	6.667	1,920	540		2,460
4260	6' wide	"	7.273	2,140	590		2,730
4280	8' wide	"	8.000	2,220	650		2,870
4300	12' wide	"	8.889	2,690	720		3,410
4320	16' wide	"	11.429	3,420	930		4,350
4400	Side coiling grilles, manual, aluminum						
4430	Opening, 8' high x						
4440	18' wide	EA	60.000	4,370	3,830	2,630	10,830
4460	24' wide	"	68.571	5,640	4,380	3,000	13,020
4470	12' high x						
4480	12' wide	EA	60.000	4,430	3,830	2,630	10,890
4490	18' wide	"	68.571	5,640	4,380	3,000	13,020
4500	24' wide	"	80.000	8,060	5,110	3,500	16,670
5000	Accordion folding, tracks and fittings included						
5020	Vinyl covered, 2 layers	SF	0.320	14.50	26.00		40.50
5040	Woven mahogany and vinyl	"	0.320	18.25	26.00		44.25
5060	Economy vinyl	"	0.320	12.25	26.00		38.25
5080	Rigid polyvinyl chloride	"	0.320	20.00	26.00		46.00
5200	Sectional wood overhead, frames not incl.						
5220	Commercial grade, HD, 1-3/4" thick, manual						
5240	8' x 8'	EA	6.667	1,080	540		1,620
5260	10' x 10'	"	7.273	1,560	590		2,150
5280	12' x 12'	"	8.000	2,030	650		2,680
5290	Chain hoist						
5300	12' x 16' high	EA	13.333	3,000	1,090		4,090
5320	14' x 14' high	"	10.000	3,290	820		4,110
5340	20' x 8' high	"	16.000	2,830	1,300		4,130
5360	16' high	"	20.000	6,350	1,630		7,980
5800	Sectional metal overhead doors, complete						
5900	Residential grade, manual						
6020	9' x 7'	EA	3.200	740	260		1,000

© 2020 BNi Publications Inc. ALL RIGHTS RESERVED

SPECIAL FUNCTION DOORS

	Description	Output		Unit Costs			
ID Code	Component Descriptions	Unit of Meas.	Manhr / Unit	Material Cost	Labor Cost	Equipment Cost	Total Cost
08 - 34001	**SPECIAL DOORS, Cont'd...**						**08 - 34001**
6040	16' x 7'	EA	4.000	1,330	330		1,660
6100	Commercial grade						
6120	8' x 8'	EA	6.667	860	540		1,400
6140	10' x 10'	"	7.273	1,150	590		1,740
6160	12' x 12'	"	8.000	1,910	650		2,560
6180	20' x 14', with chain hoist	"	16.000	4,560	1,300		5,860
6400	Sliding glass doors						
6410	Tempered plate glass, 1/4" thick						
6420	6' wide						
6440	Economy grade	EA	2.667	1,180	220		1,400
6450	Premium grade	"	2.667	1,350	220		1,570
6455	12' wide						
6460	Economy grade	EA	4.000	1,650	330		1,980
6465	Premium grade	"	4.000	2,480	330		2,810
6470	Insulating glass, 5/8" thick						
6475	6' wide						
6480	Economy grade	EA	2.667	1,450	220		1,670
6490	Premium grade	"	2.667	1,860	220		2,080
6500	12' wide						
6510	Economy grade	EA	4.000	1,800	330		2,130
6515	Premium grade	"	4.000	2,890	330		3,220
6520	1" thick						
6525	6' wide						
6530	Economy grade	EA	2.667	1,820	220		2,040
6535	Premium grade	"	2.667	2,100	220		2,320
6540	12' wide						
6550	Economy grade	EA	4.000	2,830	330		3,160
6560	Premium grade	"	4.000	4,140	330		4,470
6600	Added costs						
6610	Custom quality, add to material, 30%						
6630	Tempered glass, 6' wide, add	SF					5.08
6650	Vertical lift doors, channel frame construction						
6670	20' high x						
6720	10' wide	EA	9.600	41,820	610	530	42,960
6730	15' wide	"	9.600	52,280	610	530	53,420
6740	20' wide	"	17.143	57,500	1,090	950	59,540
6750	25' wide	"	17.143	62,730	1,090	950	64,770
6755	25' high x						

© 2020 BNi Publications Inc. ALL RIGHTS RESERVED

SPECIAL FUNCTION DOORS

Description		Output		Unit Costs			
ID Code	Component Descriptions	Unit of Meas.	Manhr / Unit	Material Cost	Labor Cost	Equipment Cost	Total Cost
08 - 34001	**SPECIAL DOORS, Cont'd...**						**08 - 34001**
6760	20' wide	EA	17.143	65,340	1,090	950	67,380
6770	25' wide	"	20.000	73,190	1,270	1,110	75,570
6775	30' high x						
6780	25' wide	EA	20.000	78,420	1,270	1,110	80,800
6790	30' wide	"	20.000	86,260	1,270	1,110	88,640
6800	35' wide	"	20.000	101,940	1,270	1,110	104,320
6850	Residential storm door						
6900	Minimum	EA	1.333	180	110		290
6920	Average	"	1.333	240	110		350
6940	Maximum	"	2.000	530	160		690

ENTRANCES AND STOREFRONTS

ID Code	Component Descriptions	Unit of Meas.	Manhr / Unit	Material Cost	Labor Cost	Equipment Cost	Total Cost
08 - 41001	**STOREFRONTS**						**08 - 41001**
0135	Storefront, aluminum and glass						
0140	Minimum	SF	0.100	29.25	9.00		38.25
0150	Average	"	0.114	43.75	10.25		54.00
0160	Maximum	"	0.133	87.00	12.00		99.00
1020	Entrance doors, premium, closers, panic dev.,etc.						
1030	1/2" thick glass						
1040	3' x 7'	EA	6.667	3,890	600		4,490
1060	6' x 7'	"	10.000	6,650	900		7,550
1065	3/4" thick glass						
1070	3' x 7'	EA	6.667	4,030	600		4,630
1080	6' x 7'	"	10.000	6,720	900		7,620
1085	1" thick glass						
1090	3' x 7'	EA	6.667	4,370	600		4,970
1100	6' x 7'	"	10.000	7,720	900		8,620
1150	Revolving doors						
1151	7' diameter, 7' high						
1160	Minimum	EA	60.000	27,660	3,800	3,330	34,790
1170	Average	"	96.000	34,800	6,080	5,320	46,200
1180	Maximum	"	120.000	44,910	7,600	6,650	59,160

© 2020 BNi Publications Inc. ALL RIGHTS RESERVED

GLAZED CURTAIN WALLS

ID Code	Description: Component Descriptions	Output: Unit of Meas.	Output: Manhr / Unit	Unit Costs: Material Cost	Unit Costs: Labor Cost	Unit Costs: Equipment Cost	Unit Costs: Total Cost
08 - 44001	GLAZED CURTAIN WALLS						08 - 44001
1000	Curtain wall, aluminum system, framing sections						
1005	2" x 3"						
1010	Jamb	LF	0.067	19.50	6.00		25.50
1020	Horizontal	"	0.067	19.75	6.00		25.75
1030	Mullion	"	0.067	26.50	6.00		32.50
1035	2" x 4"						
1040	Jamb	LF	0.100	26.50	9.00		35.50
1060	Horizontal	"	0.100	27.25	9.00		36.25
1070	Mullion	"	0.100	26.50	9.00		35.50
1080	3" x 5-1/2"						
1090	Jamb	LF	0.100	35.00	9.00		44.00
1100	Horizontal	"	0.100	38.75	9.00		47.75
1110	Mullion	"	0.100	35.25	9.00		44.25
1115	4" corner mullion	"	0.133	46.75	12.00		59.00
1120	Coping sections						
1130	1/8" x 8"	LF	0.133	44.25	12.00		56.00
1140	1/8" x 9"	"	0.133	44.50	12.00		57.00
1150	1/8" x 12-1/2"	"	0.160	45.75	14.50		60.00
1160	Sill section						
1170	1/8" x 6"	LF	0.080	43.75	7.20		51.00
1180	1/8" x 7"	"	0.080	44.25	7.20		51.00
1190	1/8" x 8-1/2"	"	0.080	45.00	7.20		52.00
1200	Column covers, aluminum						
1210	1/8" x 26"	LF	0.200	43.75	18.00		62.00
1220	1/8" x 34"	"	0.211	48.50	19.00		68.00
1230	1/8" x 38"	"	0.211	49.00	19.00		68.00
1500	Doors						
1600	Aluminum framed, standard hardware						
1620	Narrow stile						
1630	2-6 x 7-0	EA	4.000	810	360		1,170
1640	3-0 x 7-0	"	4.000	810	360		1,170
1660	3-6 x 7-0	"	4.000	840	360		1,200
1700	Wide stile						
1720	2-6 x 7-0	EA	4.000	1,380	360		1,740
1730	3-0 x 7-0	"	4.000	1,490	360		1,850
1750	3-6 x 7-0	"	4.000	1,600	360		1,960
1800	Flush panel doors, to match adjacent wall panels						
1810	2-6 x 7-0	EA	5.000	1,170	450		1,620

© 2020 BNi Publications Inc. ALL RIGHTS RESERVED

GLAZED CURTAIN WALLS

	Description	Output		Unit Costs			
ID Code	Component Descriptions	Unit of Meas.	Manhr / Unit	Material Cost	Labor Cost	Equipment Cost	Total Cost
08 - 44001	**GLAZED CURTAIN WALLS, Cont'd...**						**08 - 44001**
1820	3-0 x 7-0	EA	5.000	1,230	450		1,680
1840	3-6 x 7-0	"	5.000	1,270	450		1,720
2100	Wall panel, insulated						
2120	U=.08	SF	0.067	14.75	6.00		20.75
2140	U=.10	"	0.067	14.00	6.00		20.00
2160	U=.15	"	0.067	12.50	6.00		18.50
3000	Window wall system, complete						
3010	Minimum	SF	0.080	42.75	7.20		50.00
3030	Average	"	0.089	68.00	8.00		76.00
3050	Maximum	"	0.114	160	10.25		170
4860	Added costs						
4870	For bronze, add 20% to material						
4880	For stainless steel, add 50% to material						

METAL WINDOWS

ID Code	Component Descriptions	Unit of Meas.	Manhr / Unit	Material Cost	Labor Cost	Equipment Cost	Total Cost
08 - 51131	**ALUMINUM WINDOWS**						**08 - 51131**
0110	Jalousie						
0120	3-0 x 4-0	EA	1.000	390	90.00		480
0140	3-0 x 5-0	"	1.000	450	90.00		540
0220	Fixed window						
0240	6 sf to 8 sf	SF	0.114	19.25	10.25		29.50
0250	12 sf to 16 sf	"	0.089	17.00	8.00		25.00
0255	Projecting window						
0260	6 sf to 8 sf	SF	0.200	42.50	18.00		61.00
0270	12 sf to 16 sf	"	0.133	38.25	12.00		50.00
0275	Horizontal sliding						
0280	6 sf to 8 sf	SF	0.100	27.75	9.00		36.75
0290	12 sf to 16 sf	"	0.080	25.50	7.20		32.75
1140	Double hung						
1160	6 sf to 8 sf	SF	0.160	38.25	14.50		53.00
1180	10 sf to 12 sf	"	0.133	34.00	12.00		46.00
3010	Storm window, 0.5 cfm, up to						
3020	60 u.i. (united inches)	EA	0.400	89.00	36.00		130
3040	70 u.i.	"	0.400	92.00	36.00		130
3060	80 u.i.	"	0.400	100	36.00		140
3080	90 u.i.	"	0.444	100	40.00		140
3100	100 u.i.	"	0.444	110	40.00		150

© 2020 BNi Publications Inc. ALL RIGHTS RESERVED

METAL WINDOWS

ID Code	Description Component Descriptions	Output Unit of Meas.	Manhr / Unit	Unit Costs Material Cost	Labor Cost	Equipment Cost	Total Cost
08 - 51131	**ALUMINUM WINDOWS, Cont'd...**						**08 - 51131**
3110	2.0 cfm, up to						
3120	60 u.i.	EA	0.400	110	36.00		150
3140	70 u.i.	"	0.400	120	36.00		160
3160	80 u.i.	"	0.400	120	36.00		160
3180	90 u.i.	"	0.444	130	40.00		170
3200	100 u.i.	"	0.444	130	40.00		170
08 - 51231	**STEEL WINDOWS**						**08 - 51231**
0100	Steel windows, primed						
1000	Casements						
1010	Operable						
1020	Minimum	SF	0.047	54.00	4.23		58.00
1040	Maximum	"	0.053	81.00	4.80		86.00
1060	Fixed sash	"	0.040	43.25	3.60		46.75
1080	Double hung	"	0.044	81.00	4.00		85.00
1100	Industrial windows						
1120	Horizontally pivoted sash	SF	0.053	69.00	4.80		74.00
1130	Fixed sash	"	0.044	54.00	4.00		58.00
1135	Security sash						
1140	Operable	SF	0.053	86.00	4.80		91.00
1150	Fixed	"	0.044	76.00	4.00		80.00
1155	Picture window	"	0.044	37.00	4.00		41.00
1160	Projecting sash						
1170	Minimum	SF	0.050	64.00	4.50		69.00
1180	Maximum	"	0.050	79.00	4.50		84.00
1930	Mullions	LF	0.040	16.75	3.60		20.25

WOOD & PLASTIC

ID Code	Component Descriptions	Unit of Meas.	Manhr / Unit	Material Cost	Labor Cost	Equipment Cost	Total Cost
08 - 52001	**WOOD WINDOWS**						**08 - 52001**
0980	Double hung						
0990	24" x 36"						
1000	Minimum	EA	0.800	240	65.00		300
1002	Average	"	1.000	350	82.00		430
1004	Maximum	"	1.333	470	110		580
1010	24" x 48"						
1020	Minimum	EA	0.800	280	65.00		350
1022	Average	"	1.000	410	82.00		490
1024	Maximum	"	1.333	570	110		680

© 2020 BNi Publications Inc. ALL RIGHTS RESERVED

WOOD & PLASTIC

	Description	Output		Unit Costs			
ID Code	Component Descriptions	Unit of Meas.	Manhr / Unit	Material Cost	Labor Cost	Equipment Cost	Total Cost
08 - 52001	**WOOD WINDOWS, Cont'd...**						**08 - 52001**
1030	30" x 48"						
1040	Minimum	EA	0.889	290	72.00		360
1042	Average	"	1.143	410	93.00		500
1044	Maximum	"	1.600	590	130		720
1050	30" x 60"						
1060	Minimum	EA	0.889	320	72.00		390
1062	Average	"	1.143	510	93.00		600
1064	Maximum	"	1.600	630	130		760
1160	Casement						
1180	1 leaf, 22" x 38" high						
1220	Minimum	EA	0.800	350	65.00		420
1222	Average	"	1.000	430	82.00		510
1224	Maximum	"	1.333	500	110		610
1230	2 leaf, 50" x 50" high						
1240	Minimum	EA	1.000	940	82.00		1,020
1242	Average	"	1.333	1,230	110		1,340
1244	Maximum	"	2.000	1,410	160		1,570
1250	3 leaf, 71" x 62" high						
1260	Minimum	EA	1.000	1,550	82.00		1,630
1262	Average	"	1.333	1,580	110		1,690
1264	Maximum	"	2.000	1,890	160		2,050
1270	4 leaf, 95" x 75" high						
1280	Minimum	EA	1.143	2,060	93.00		2,150
1282	Average	"	1.600	2,350	130		2,480
1284	Maximum	"	2.667	3,000	220		3,220
1290	5 leaf, 119" x 75" high						
1300	Minimum	EA	1.143	2,670	93.00		2,760
1302	Average	"	1.600	2,880	130		3,010
1304	Maximum	"	2.667	3,680	220		3,900
1360	Picture window, fixed glass, 54" x 54" high						
1400	Minimum	EA	1.000	550	82.00		630
1422	Average	"	1.143	620	93.00		710
1424	Maximum	"	1.333	1,100	110		1,210
1430	68" x 55" high						
1440	Minimum	EA	1.000	990	82.00		1,070
1442	Average	"	1.143	1,140	93.00		1,230
1444	Maximum	"	1.333	1,490	110		1,600
1480	Sliding, 40" x 31" high						

© 2020 BNi Publications Inc. ALL RIGHTS RESERVED

WOOD & PLASTIC

ID Code	Description: Component Descriptions	Output: Unit of Meas.	Output: Manhr / Unit	Unit Costs: Material Cost	Unit Costs: Labor Cost	Unit Costs: Equipment Cost	Unit Costs: Total Cost
08 - 52001	**WOOD WINDOWS, Cont'd...**						**08 - 52001**
1520	Minimum	EA	0.800	330	65.00		400
1522	Average	"	1.000	500	82.00		580
1524	Maximum	"	1.333	600	110		710
1530	52" x 39" high						
1540	Minimum	EA	1.000	410	82.00		490
1542	Average	"	1.143	610	93.00		700
1544	Maximum	"	1.333	650	110		760
1550	64" x 72" high						
1560	Minimum	EA	1.000	630	82.00		710
1562	Average	"	1.333	1,010	110		1,120
1564	Maximum	"	1.600	1,110	130		1,240
1760	Awning windows						
1780	34" x 21" high						
1800	Minimum	EA	0.800	330	65.00		400
1822	Average	"	1.000	380	82.00		460
1824	Maximum	"	1.333	440	110		550
1840	40" x 21" high						
1860	Minimum	EA	0.889	390	72.00		460
1862	Average	"	1.143	430	93.00		520
1864	Maximum	"	1.600	480	130		610
1880	48" x 27" high						
1900	Minimum	EA	0.889	410	72.00		480
1902	Average	"	1.143	490	93.00		580
1904	Maximum	"	1.600	570	130		700
1920	60" x 36" high						
1940	Minimum	EA	1.000	430	82.00		510
1942	Average	"	1.333	760	110		870
1944	Maximum	"	1.600	860	130		990
8000	Window frame, milled						
8010	Minimum	LF	0.160	6.09	13.00		19.00
8020	Average	"	0.200	6.80	16.25		23.00
8030	Maximum	"	0.267	10.25	21.75		32.00

© 2020 BNi Publications Inc. ALL RIGHTS RESERVED

SKYLIGHTS

	Description	Output		Unit Costs			
ID Code	Component Descriptions	Unit of Meas.	Manhr / Unit	Material Cost	Labor Cost	Equipment Cost	Total Cost
08 - 62001	**PLASTIC SKYLIGHTS**						**08 - 62001**
1020	Single thickness, not including mounting curb						
1040	2' x 4'	EA	1.000	430	78.00		510
1050	4' x 4'	"	1.333	580	100		680
1060	5' x 5'	"	2.000	770	160		930
1070	6' x 8'	"	2.667	1,630	210		1,840
1200	Double thickness, not including mounting curb						
1220	2' x 4'	EA	1.000	560	78.00		640
1240	4' x 4'	"	1.333	710	100		810
1260	5' x 5'	"	2.000	1,040	160		1,200
1270	6' x 8'	"	2.667	1,820	210		2,030
1290	Metal framed skylights						
1420	Translucent panels, 2-1/2" thick	SF	0.080	46.00	6.22		52.00
1490	Continuous vaults, 8' wide						
1500	Single glazed	SF	0.100	62.00	7.78		70.00
1560	Double glazed	"	0.114	100	8.89		110
08 - 62101	**SOLAR SKYLIGHTS**						**08 - 62101**
0050	Tubular solar skylight, basic kit						
0100	Minimum	EA	2.667	380	220		600
0200	Average	"	4.000	430	330		760
0300	Maximum	"	8.000	480	650		1,130
0400	Tubular solar skylight dome, 10" Diameter						
0450	Minimum	EA	0.800	79.00	65.00		140
0600	Average	"	1.000	85.00	82.00		170
0700	Maximum	"	1.333	91.00	110		200
0800	14" Diameter						
0900	Minimum	EA	0.800	91.00	65.00		160
1000	Average	"	1.000	97.00	82.00		180
1100	Maximum	"	1.333	100	110		210
1200	Straight extension tube, 10" Diameter x 12" long						
1300	Minimum	EA	0.667	52.00	54.00		110
1400	Average	"	0.800	61.00	65.00		130
1500	Maximum	"	1.000	67.00	82.00		150
1700	24" long						
1800	Minimum	EA	0.667	61.00	54.00		110
1900	Average	"	0.800	67.00	65.00		130
2000	Maximum	"	1.000	73.00	82.00		160
2200	36" long						

© 2020 BNi Publications Inc. ALL RIGHTS RESERVED

SKYLIGHTS

ID Code	Description Component Descriptions	Output Unit of Meas.	Manhr / Unit	Unit Costs Material Cost	Labor Cost	Equipment Cost	Total Cost
08 - 62101	**SOLAR SKYLIGHTS, Cont'd...**						**08 - 62101**
2300	Minimum	EA	0.667	90.00	54.00		140
2400	Average	"	0.800	97.00	65.00		160
2500	Maximum	"	1.000	100	82.00		180
2700	48" long						
2800	Minimum	EA	0.800	110	65.00		180
2900	Average	"	1.000	110	82.00		190
3000	Maximum	"	1.333	120	110		230
3200	14" Diameter x 12" long						
3300	Minimum	EA	0.667	67.00	54.00		120
3400	Average	"	0.800	73.00	65.00		140
3500	Maximum	"	1.000	79.00	82.00		160
3700	24" long						
3800	Minimum	EA	0.667	85.00	54.00		140
3900	Average	"	0.800	97.00	65.00		160
4000	Maximum	"	1.000	110	82.00		190
4200	36" long						
4300	Minimum	EA	0.667	110	54.00		160
4400	Average	"	0.800	120	65.00		190
4500	Maximum	"	1.000	130	82.00		210
4700	90 Degree extension tubes, 10" Diameter						
4800	Minimum	EA	0.444	61.00	36.25		97.00
4900	Average	"	0.500	70.00	40.75		110
5000	Maximum	"	0.571	79.00	46.50		130
5100	14" Diameter						
5200	Minimum	EA	0.444	75.00	36.25		110
5300	Average	"	0.500	83.00	40.75		120
5400	Maximum	"	0.571	91.00	46.50		140
5500	Bottom tube adaptor, 10" Diameter						
5600	Minimum	EA	0.444	64.00	36.25		100
5700	Average	"	0.500	70.00	40.75		110
5800	Maximum	"	0.571	75.00	46.50		120
5900	14" Diameter						
6000	Minimum	EA	0.444	79.00	36.25		120
6100	Average	"	0.500	85.00	40.75		130
6200	Maximum	"	0.571	91.00	46.50		140
6300	Top tube adaptor, 10" Diameter						
6400	Minimum	EA	0.444	64.00	36.25		100
6500	Average	"	0.500	70.00	40.75		110

© 2020 BNi Publications Inc. ALL RIGHTS RESERVED

SKYLIGHTS

ID Code	Component Descriptions	Unit of Meas.	Manhr / Unit	Material Cost	Labor Cost	Equipment Cost	Total Cost
08 - 62101	**SOLAR SKYLIGHTS, Cont'd...**						**08 - 62101**
6600	Maximum	EA	0.571	75.00	46.50		120
6700	14" Diameter						
6800	Minimum	EA	0.444	79.00	36.25		120
6900	Average	"	0.500	85.00	40.75		130
7000	Maximum	"	0.571	91.00	46.50		140
7100	Tube flashing						
7200	Minimum	EA	0.667	85.00	54.00		140
7300	Average	"	0.800	97.00	65.00		160
7400	Maximum	"	1.000	110	82.00		190
7500	Daylight dimmer switch						
7600	Minimum	EA	0.444	63.00	36.25		99.00
7700	Average	"	0.500	68.00	40.75		110
7800	Maximum	"	0.571	73.00	46.50		120
7900	Dimmer						
8000	Minimum	EA	0.444	250	36.25		290
8100	Average	"	0.500	280	40.75		320
8300	Maximum	"	0.571	300	46.50		350

METAL-FRAMED SKYLIGHTS

ID Code	Component Descriptions	Unit of Meas.	Manhr / Unit	Material Cost	Labor Cost	Equipment Cost	Total Cost
08 - 63011	**METAL-FRAMED SKYLIGHTS**						**08 - 63011**
1600	Metal Framed, 2-1/4" Thick, Translucent, <5,000 SF	SF	0.400	30.75	31.25		62.00
1700	>5,000 SF	"	0.444	27.75	34.50		62.00
2000	Continous Vaulted - Semi-Circular						
2020	Skylight, 2-1/4" Thick, to 8', Single Glazed	SF	0.400	36.25	31.25		68.00
2040	Double Glazed	"	0.400	44.25	31.25		76.00
2060	Skylight, 2-1/4" Thick, to 9', Single Glazed	"	0.444	62.00	34.50		97.00
2080	Double Glazed	"	0.444	70.00	34.50		100
2500	Pyramid Type, Self Supporting, Clear Opening						
2520	Minimum	SF	0.348	49.50	27.00		77.00
2540	Average	"	0.400	56.00	31.25		87.00
2560	Maximum	"	0.533	72.00	41.50		110
3000	Grid Type, 4' x 10' Modlule						
3020	Minimum	SF	0.320	30.25	25.00		55.00
3040	Maximum	"	0.533	56.00	41.50		98.00
3060	Preformed Acrylic Skylight						
3080	Minimum	SF	0.320	24.25	25.00		49.25
4000	Maximum	"	0.533	40.25	41.50		82.00

© 2020 BNi Publications Inc. ALL RIGHTS RESERVED

HARDWARE

	Description	Output		Unit Costs			
ID Code	Component Descriptions	Unit of Meas.	Manhr / Unit	Material Cost	Labor Cost	Equipment Cost	Total Cost
08 - 71001	**HINGES**						**08 - 71001**
1200	Hinges, material only						
1250	3 x 3 butts, steel, interior, plain bearing	PAIR					20.75
1260	4 x 4 butts, steel, standard	"					30.50
1270	5 x 4-1/2 butts, bronze/s. steel, heavy duty	"					79.00
1290	Pivot hinges						
1300	Top pivot	EA					88.00
1310	Intermediate pivot	"					94.00
1320	Bottom pivot	"					180
1500	BHMA specifications						
1520	3-1/2 x 3-1/2, full mortise butts						
1540	Plain bearing	PAIR					24.75
1550	Ball bearing	"					29.75
1560	Half surface butts	"					42.25
1580	4 x 4						
1600	Full mortise butts, plain bearing, standard duty	PAIR					27.00
1640	Full mortise butts, ball bearing	"					32.50
1645	Half surface butts						
1650	Standard duty	PAIR					42.50
1660	Ball bearing	"					42.50
1670	4-1/2 x 4-1/2						
1680	Full mortise butts, plain bearing	PAIR					35.75
1690	Ball bearing, heavy duty	"					73.00
1695	Half mortise and half surface butts						
1700	Plain bearing	PAIR					42.25
1720	Full surface and half surface butts						
1740	Standard duty	PAIR					72.00
1780	Heavy duty	"					130
1785	Full mortise and full slide-in butts, ball bearing	"					33.00
1786	Half mortise butts, ball bearing						
1788	Standard duty	PAIR					150
1790	Heavy duty	"					170
1795	5 x 5, ball bearing						
1800	Full mortise butts	PAIR					69.00
1820	Half mortise, full & half surface butts	"					150
1830	Full mortise, full surface and half surface butts	"					200
1910	4 x 4						
1930	Full mortise butts, plain bearing, standard duty	PAIR					16.50
2020	5 x 4-1/2						

© 2020 BNi Publications Inc. ALL RIGHTS RESERVED

HARDWARE

	Description	Output		Unit Costs			
ID Code	Component Descriptions	Unit of Meas.	Manhr / Unit	Material Cost	Labor Cost	Equipment Cost	Total Cost
08 - 71001	**HINGES, Cont'd...**						**08 - 71001**
2040	Full mortise butts, ball bearing, heavy duty	PAIR					77.00
08 - 71002	**LOCKSETS**						**08 - 71002**
1280	Latchset, heavy duty						
1300	Cylindrical	EA	0.500	190	40.75		230
1320	Mortise	"	0.800	200	65.00		260
1325	Lockset, heavy duty						
1330	Cylindrical	EA	0.500	310	40.75		350
1350	Mortise	"	0.800	350	65.00		420
2000	Mortise locks and latchsets, chrome						
2020	Latchset passage or closet latch	EA	0.667	240	54.00		290
2030	Privacy (bath or bedroom)	"	0.667	250	54.00		300
2040	Entry lockset	"	0.667	310	54.00		360
2050	Classroom lockset (outside key operated)	"	0.667	310	54.00		360
2060	Storeroom lock	"	0.667	310	54.00		360
2070	Front door lock	"	0.667	310	54.00		360
2080	Dormitory or exit lock	"	0.667	310	54.00		360
2200	Preassembled locks and latches, brass						
2220	Latchset, passage or closet latch	EA	0.667	280	54.00		330
2225	Lockset						
2230	Privacy (bath or bathroom)	EA	0.667	340	54.00		390
2240	Entry lock	"	0.667	490	54.00		540
2250	Classroom lock (outside key operated)	"	0.667	490	54.00		540
2260	Storeroom lock	"	0.667	540	54.00		590
2270	Bored locks and latches, satin chrome plated						
2280	Latchset passage or closet latch	EA	0.667	160	54.00		210
2285	Lockset						
2290	Privacy (bath or bedroom)	EA	0.667	220	54.00		270
2300	Entry lock	"	0.667	240	54.00		290
2320	Classroom lock	"	0.667	240	54.00		290
2330	Corridor lock	"	0.667	240	54.00		290
4000	Miscellaneous locks						
4020	Exit lock with alarm, single door	EA	3.200	780	260		1,040
4040	Electric strike						
4050	Rim mounted wrought steel	EA	2.000	580	160		740
4060	Mortised, wrought steel with bronze plating	"	3.200	230	260		490
4065	Dead bolt						
4070	Bored, wrought brass, keyed both sides	EA	1.333	120	110		230

© 2020 BNi Publications Inc. ALL RIGHTS RESERVED

HARDWARE

ID Code	Component Descriptions	Unit of Meas.	Manhr / Unit	Material Cost	Labor Cost	Equipment Cost	Total Cost
08 - 71002	**LOCKSETS, Cont'd...**						**08 - 71002**
4080	Mortised, cast brass	EA	1.333	330	110		440
4090	Lockset, cipher, mechanical	"	0.800	2,020	65.00		2,090
08 - 71003	**CLOSERS**						**08 - 71003**
2600	Door closers						
2605	Surface mounted, traditional type, parallel arm						
2610	Standard	EA	1.000	240	82.00		320
2620	Heavy duty	"	1.000	280	82.00		360
2630	Modern type, parallel arm, standard duty	"	1.000	290	82.00		370
2640	Overhead, concealed, pivot hung, single acting						
2650	Interior	EA	1.000	450	82.00		530
2660	Exterior	"	1.000	670	82.00		750
2665	Floor concealed, single acting, offset, pivoted						
2670	Interior	EA	2.667	730	220		950
2680	Exterior	"	2.667	930	220		1,150
08 - 71004	**DOOR TRIM**						**08 - 71004**
1100	Door bumper, bronze, wall type	EA	0.160	6.35	13.00		19.25
1105	Wall type, 4" dia. with convex rubber pad, aluminum	"	0.160	11.75	13.00		24.75
1108	Floor type						
1110	Aluminum	EA	0.160	5.37	13.00		18.25
1120	Brass	"	0.160	6.46	13.00		19.50
1130	Door holders						
1140	Wall type, bronze	EA	0.160	32.00	13.00		45.00
1160	Overhead	"	0.400	26.75	32.50		59.00
1180	Floor type	"	0.400	26.75	32.50		59.00
1200	Plunger type	"	0.400	26.00	32.50		59.00
1240	Wall type, aluminum	"	0.400	25.25	32.50		58.00
1520	Surface bolt	"	0.160	22.50	13.00		35.50
1600	Panic device						
1601	Rim type with thumb piece	EA	2.000	650	160		810
1610	Mortise	"	2.000	810	160		970
1620	Vertical rod	"	2.000	1,230	160		1,390
1630	Labeled, rim type	"	2.000	850	160		1,010
1640	Mortise	"	2.000	1,110	160		1,270
1650	Vertical rod	"	2.000	1,180	160		1,340
2070	Silencers, rubber type	"	0.016	3.06	1.30		4.36
2080	Dust proof strike with plate, brass	"	0.267	18.75	21.75		40.50
2090	Flush bolt, lever extension, brass, rated	"	0.160	34.25	13.00		47.25

© 2020 BNi Publications Inc. ALL RIGHTS RESERVED

HARDWARE

ID Code	Description Component Descriptions	Output Unit of Meas.	Manhr / Unit	Unit Costs Material Cost	Labor Cost	Equipment Cost	Total Cost
08 - 71004	**DOOR TRIM, Cont'd...**						**08 - 71004**
2100	Surface bolt with strike, brass, 6" long	EA	0.160	26.00	13.00		39.00
2250	Door coordinator, labeled, brass, satin chrome	"	0.571	120	46.50		170
2300	Door plates						
2305	Kick plate, aluminum, 3 beveled edges						
2310	10" x 28"	EA	0.400	30.50	32.50		63.00
2320	10" x 30"	"	0.400	33.50	32.50		66.00
2330	10" x 34"	"	0.400	36.50	32.50		69.00
2340	10" x 38"	"	0.400	39.75	32.50		72.00
2350	Push plate, 4" x 16"						
2360	Aluminum	EA	0.160	27.75	13.00		40.75
2371	Bronze	"	0.160	88.00	13.00		100
2380	Stainless steel	"	0.160	70.00	13.00		83.00
2385	Armor plate, 40" x 34"	"	0.320	81.00	26.00		110
2388	Pull handle, 4" x 16"						
2390	Aluminum	EA	0.160	98.00	13.00		110
2400	Bronze	"	0.160	190	13.00		200
2420	Stainless steel	"	0.160	140	13.00		150
2425	Hasp assembly						
2430	3"	EA	0.133	4.62	10.75		15.25
2440	4-1/2"	"	0.178	5.77	14.50		20.25
2450	6"	"	0.229	9.18	18.75		28.00
2720	Electro-magnetic door holder						
2730	Wall mounted	EA	2.667	200	220		420
2740	Floor mounted	"	2.667	350	220		570
2780	Smoke detector door holder						
2800	Photoelectric type	EA	2.667	300	220		520
2810	Ionization type	"	2.667	300	220		520
5100	Pneumatic operators, activated by rubber mats						
5105	Swing						
5110	Single	EA	6.667	4,330	540		4,870
5120	Double	"	10.000	7,160	820		7,980
5125	Sliding						
5130	Single	EA	6.667	4,790	540		5,330
5140	Double	"	10.000	8,320	820		9,140

© 2020 BNi Publications Inc. ALL RIGHTS RESERVED

HARDWARE

	Description	Output		Unit Costs			
ID Code	Component Descriptions	Unit of Meas.	Manhr / Unit	Material Cost	Labor Cost	Equipment Cost	Total Cost
08 - 71006	**WEATHERSTRIPPING**						**08 - 71006**
0100	Weatherstrip, head and jamb, metal strip, neoprene						
0140	Standard duty	LF	0.044	5.19	3.62		8.81
0160	Heavy duty	"	0.050	5.77	4.07		9.84
3980	Spring type						
4000	Metal doors	EA	2.000	57.00	160		220
4010	Wood doors	"	2.667	57.00	220		280
4020	Sponge type with adhesive backing	"	0.800	54.00	65.00		120
4025	Astragal						
4030	1-3/4" x 13 ga., aluminum	LF	0.067	7.10	5.43		12.50
4040	1-3/8" x 5/8", oak	"	0.053	5.77	4.34		10.00
4500	Thresholds						
4510	Bronze	LF	0.200	56.00	16.25		72.00
4515	Aluminum						
4520	Plain	LF	0.200	40.25	16.25		57.00
4525	Vinyl insert	"	0.200	41.25	16.25		58.00
4530	Aluminum with grit	"	0.200	39.25	16.25		56.00
4533	Steel						
4535	Plain	LF	0.200	31.00	16.25		47.25
4540	Interlocking	"	0.667	41.50	54.00		96.00

GLAZING

08 - 81001	**GLASS GLAZING**						**08 - 81001**
0800	Sheet glass, 1/8" thick	SF	0.044	8.91	4.00		13.00
1020	Plate glass, bronze or grey, 1/4" thick	"	0.073	13.00	6.54		19.50
1040	Clear	"	0.073	10.25	6.54		16.75
1060	Polished	"	0.073	12.00	6.54		18.50
1800	Plexiglass						
2000	1/8" thick	SF	0.073	5.73	6.54		12.25
2020	1/4" thick	"	0.044	10.25	4.00		14.25
3000	Float glass, clear						
3010	3/16" thick	SF	0.067	6.93	6.00		13.00
3020	1/4" thick	"	0.073	7.07	6.54		13.50
3030	5/16" thick	"	0.080	13.25	7.20		20.50
3040	3/8" thick	"	0.100	14.25	9.00		23.25
3050	1/2" thick	"	0.133	24.00	12.00		36.00
3060	5/8" thick	"	0.160	31.75	14.50		46.25
3070	3/4" thick	"	0.200	34.50	18.00		53.00

© 2020 BNi Publications Inc. ALL RIGHTS RESERVED

GLAZING

	Description	Output		Unit Costs			
ID Code	Component Descriptions	Unit of Meas.	Manhr / Unit	Material Cost	Labor Cost	Equipment Cost	Total Cost
08 - 81001	**GLASS GLAZING, Cont'd...**						**08 - 81001**
3080	1" thick	SF	0.267	61.00	24.00		85.00
3100	Tinted glass, polished plate, twin ground						
3120	3/16" thick	SF	0.067	9.55	6.00		15.50
3130	1/4" thick	"	0.073	9.55	6.54		16.00
3140	3/8" thick	"	0.100	15.25	9.00		24.25
3150	1/2" thick	"	0.133	24.75	12.00		36.75
3190	Total, full vision, all glass window system						
3200	To 10' high						
3220	Minimum	SF	0.200	64.00	18.00		82.00
3222	Average	"	0.200	84.00	18.00		100
3224	Maximum	"	0.200	100	18.00		120
3225	10' to 20' high						
3230	Minimum	SF	0.200	77.00	18.00		95.00
3240	Average	"	0.200	94.00	18.00		110
3250	Maximum	"	0.200	120	18.00		140
5000	Insulated glass, bronze or gray						
5020	1/2" thick	SF	0.133	19.50	12.00		31.50
5040	1" thick	"	0.200	23.25	18.00		41.25
5100	Spandrel, polished, 1 side, 1/4" thick	"	0.073	15.50	6.54		22.00
5900	Tempered glass (safety)						
6000	Clear sheet glass						
6020	1/8" thick	SF	0.044	10.75	4.00		14.75
6030	3/16" thick	"	0.062	13.00	5.53		18.50
6040	Clear float glass						
6050	1/4" thick	SF	0.067	11.25	6.00		17.25
6060	5/16" thick	"	0.080	20.00	7.20		27.25
6070	3/8" thick	"	0.100	24.50	9.00		33.50
6080	1/2" thick	"	0.133	33.50	12.00		45.50
6090	5/8" thick	"	0.160	38.00	14.50		53.00
6100	3/4" thick	"	0.267	47.00	24.00		71.00
6160	Tinted float glass						
6180	3/16" thick	SF	0.062	13.50	5.53		19.00
6200	1/4" thick	"	0.067	14.75	6.00		20.75
6210	3/8" thick	"	0.100	26.75	9.00		35.75
6220	1/2" thick	"	0.133	35.75	12.00		47.75
6490	Laminated glass						
6500	Float safety glass with polyvinyl plastic layer						
6510	1/4", sheet or float						

© 2020 BNi Publications Inc. ALL RIGHTS RESERVED

GLAZING

	Description	Output		Unit Costs			
ID Code	Component Descriptions	Unit of Meas.	Manhr / Unit	Material Cost	Labor Cost	Equipment Cost	Total Cost
08 - 81001	**GLASS GLAZING, Cont'd...**						**08 - 81001**
6530	Two lites, 1/8" thick, clear glass	SF	0.067	14.25	6.00		20.25
6540	1/2" thick, float glass						
6550	Two lites, 1/4" thick, clear glass	SF	0.133	21.75	12.00		33.75
6570	Tinted glass	"	0.133	25.50	12.00		37.50
6800	Insulating glass, two lites, clear float glass						
6840	1/2" thick	SF	0.133	13.75	12.00		25.75
6850	5/8" thick	"	0.160	16.00	14.50		30.50
6860	3/4" thick	"	0.200	17.50	18.00		35.50
6870	7/8" thick	"	0.229	18.50	20.50		39.00
6880	1" thick	"	0.267	24.75	24.00		48.75
6885	Glass seal edge						
6890	3/8" thick	SF	0.133	11.75	12.00		23.75
6895	Tinted glass						
6900	1/2" thick	SF	0.133	23.75	12.00		35.75
6910	1" thick	"	0.267	25.50	24.00		49.50
6920	Tempered, clear						
6930	1" thick	SF	0.267	46.50	24.00		71.00
6950	Wire reinforced	"	0.267	59.00	24.00		83.00
7100	Plate mirror glass						
7200	1/4" thick						
7210	15 sf	SF	0.080	11.75	7.20		19.00
7220	Over 15 sf	"	0.073	10.75	6.54		17.25
7230	Door type, 1/4" thick	"	0.080	12.00	7.20		19.25
7240	Transparent, one-way vision, 1/4" thick	"	0.080	25.75	7.20		33.00
7250	Sheet mirror glass						
7260	3/16" thick	SF	0.080	10.75	7.20		18.00
7270	1/4" thick	"	0.067	11.25	6.00		17.25
7300	Wall tiles, 12" x 12"						
7310	Clear glass	SF	0.044	3.64	4.00		7.64
7320	Veined glass	"	0.044	4.62	4.00		8.62
8800	Wire glass, 1/4" thick						
8810	Clear	SF	0.267	21.75	24.00		45.75
8820	Hammered	"	0.267	21.75	24.00		45.75
8840	Obscure	"	0.267	25.25	24.00		49.25
8900	Bullet resistant, plate, with inter-leaved vinyl						
8910	1-3/16" thick						
8930	To 15 sf	SF	0.400	110	36.00		150
8940	Over 15 sf	"	0.400	120	36.00		160

© 2020 BNi Publications Inc. ALL RIGHTS RESERVED

GLAZING

ID Code	Description: Component Descriptions	Output: Unit of Meas.	Output: Manhr / Unit	Unit Costs: Material Cost	Unit Costs: Labor Cost	Unit Costs: Equipment Cost	Unit Costs: Total Cost
08 - 81001	GLASS GLAZING, Cont'd...						08 - 81001
8945	2" thick						
8950	To 15 sf	SF	0.667	150	60.00		210
8960	Over 15 sf	"	0.667	160	60.00		220
9500	Glazing accessories						
9510	Neoprene glazing gaskets						
9530	1/4" glass	LF	0.032	2.28	2.88		5.16
9540	3/8" glass	"	0.033	2.54	3.00		5.54
9550	1/2" glass	"	0.035	2.67	3.13		5.80
9560	3/4" glass	"	0.036	3.81	3.27		7.08
9570	1" glass	"	0.040	4.44	3.60		8.04
9580	Mullion section						
9590	1/4" glass	LF	0.016	0.70	1.44		2.14
9600	3/8" glass	"	0.020	0.89	1.80		2.69
9610	1/2" glass	"	0.023	1.27	2.05		3.32
9620	3/4" glass	"	0.027	1.90	2.40		4.30
9630	1" glass	"	0.032	2.54	2.88		5.42
9640	Molded corners	EA	0.533	2.72	48.00		51.00

LOUVERS AND VENTS

ID Code	Component Descriptions	Unit of Meas.	Manhr / Unit	Material Cost	Labor Cost	Equipment Cost	Total Cost
08 - 91001	VENTS AND WALL LOUVERS						08 - 91001
0100	Block vent, 8"x16"x4" alum., w/screen, mill finish	EA	0.267	180	24.00		200
1200	Standard	"	0.250	98.00	22.50		120
1210	Vents w/screen, 4" deep, 8" wide, 5" high						
1220	Modular	EA	0.250	110	22.50		130
1230	Grilles and louvers						
2000	Aluminum gable louvers	SF	0.133	20.75	12.00		32.75
2020	Vent screen aluminum, 4" wide, continuous	LF	0.027	6.02	2.40		8.42
2040	Fixed type louvers						
2060	4 through 10 sf	SF	0.133	35.00	12.00		47.00
2080	Over 10 sf	"	0.100	41.50	9.00		51.00
2090	Movable type louvers						
2220	4 through 10 sf	SF	0.133	41.50	12.00		54.00
2240	Over 10 sf	"	0.100	45.75	9.00		55.00
2260	Aluminum louvers						
2980	Louvers, aluminum, anodized, fixed blade						
3000	Horizontal line	SF	0.200	61.00	18.00		79.00
3020	Vertical line	"	0.200	61.00	18.00		79.00

© 2020 BNi Publications Inc. ALL RIGHTS RESERVED

LOUVERS AND VENTS

ID Code	Component Descriptions	Unit of Meas.	Manhr / Unit	Material Cost	Labor Cost	Equipment Cost	Total Cost
	Description	**Output**		**Unit Costs**			
08 - 91001	**VENTS AND WALL LOUVERS, Cont'd...**						**08 - 91001**
3040	Wall louver, aluminum mill finish						
3060	Under, 2 sf	SF	0.100	46.25	9.00		55.00
3080	2 to 4 sf	"	0.089	40.50	8.00		48.50
3090	5 to 10 sf	"	0.089	38.00	8.00		46.00
3110	Galvanized steel						
3120	Under 2 sf	SF	0.100	41.75	9.00		51.00
3140	2 to 4 sf	"	0.089	28.75	8.00		36.75
3160	5 to 10 sf	"	0.089	27.00	8.00		35.00
4000	Residential use, fixed type, with screen						
4050	8" x 8"	EA	0.400	21.25	36.00		57.00
4060	12" x 12"	"	0.400	23.25	36.00		59.00
4080	12" x 18"	"	0.400	28.00	36.00		64.00
4100	14" x 24"	"	0.400	40.25	36.00		76.00
4120	18" x 24"	"	0.400	45.25	36.00		81.00
4140	30" x 24"	"	0.444	62.00	40.00		100
08 - 91261	**DOOR LOUVERS**						**08 - 91261**
0110	Fixed, 1" thick, enameled steel						
0120	8"x8"	EA	0.100	64.00	8.15		72.00
0140	12"x8"	"	0.100	73.00	8.15		81.00
0160	12"x12"	"	0.114	82.00	9.32		91.00
0180	16"x12"	"	0.123	120	10.00		130
0200	18"x12"	"	0.200	120	16.25		140
0220	20"x8"	"	0.114	130	9.32		140
0240	20"x12"	"	0.229	150	18.75		170
0260	20"x16"	"	0.267	150	21.75		170
0270	20"x20"	"	0.320	160	26.00		190
0280	24"x12"	"	0.267	130	21.75		150
0290	24"x16"	"	0.286	140	23.25		160
0300	24"x18"	"	0.308	160	25.00		190
0320	24"x20"	"	0.333	170	27.25		200
0340	24"x24"	"	0.364	170	29.75		200
0390	26"x26"	"	0.500	200	40.75		240

© 2020 BNi Publications Inc. ALL RIGHTS RESERVED

DIVISION 09
FINISHES

SUPPORT SYSTEMS

	Description	Output		Unit Costs			
ID Code	Component Descriptions	Unit of Meas.	Manhr / Unit	Material Cost	Labor Cost	Equipment Cost	Total Cost
09 - 21161	**METAL STUDS**						**09 - 21161**
0060	Studs, non load bearing, galvanized						
0080	2-1/2", 20 ga.						
0100	12" o.c.	SF	0.017	0.68	1.35		2.03
0102	16" o.c.	"	0.013	0.52	1.08		1.60
0110	25 ga.						
0120	12" o.c.	SF	0.017	0.46	1.35		1.81
0122	16" o.c.	"	0.013	0.36	1.08		1.44
0124	24" o.c.	"	0.011	0.28	0.90		1.18
0130	3-5/8", 20 ga.						
0140	12" o.c.	SF	0.020	0.81	1.63		2.44
0142	16" o.c.	"	0.016	0.62	1.30		1.92
0144	24" o.c.	"	0.013	0.47	1.08		1.55
0170	25 ga.						
0180	12" o.c.	SF	0.020	0.53	1.63		2.16
0182	16" o.c.	"	0.016	0.44	1.30		1.74
0184	24" o.c.	"	0.013	0.33	1.08		1.41
0188	4", 20 ga.						
0190	12" o.c.	SF	0.020	0.89	1.63		2.52
0192	16" o.c.	"	0.016	0.68	1.30		1.98
0194	24" o.c.	"	0.013	0.52	1.08		1.60
0198	25 ga.						
0200	12" o.c.	SF	0.020	0.60	1.63		2.23
0202	16" o.c.	"	0.016	0.47	1.30		1.77
0204	24" o.c.	"	0.013	0.35	1.08		1.43
0210	6", 20 ga.						
0220	12" o.c.	SF	0.025	1.14	2.03		3.17
0222	16" o.c.	"	0.020	0.83	1.63		2.46
0224	24" o.c.	"	0.017	0.68	1.35		2.03
0230	25 ga.						
0240	12" o.c.	SF	0.025	0.73	2.03		2.76
0242	16" o.c.	"	0.020	0.58	1.63		2.21
0244	24" o.c.	"	0.017	0.44	1.35		1.79
0980	Load bearing studs, galvanized						
0990	3-5/8", 16 ga.						
1000	12" o.c.	SF	0.020	1.47	1.63		3.10
1020	16" o.c.	"	0.016	1.36	1.30		2.66
1110	18 ga.						
1130	12" o.c.	SF	0.013	1.15	1.08		2.23

© 2020 BNi Publications Inc. ALL RIGHTS RESERVED

SUPPORT SYSTEMS

ID Code	Description: Component Descriptions	Output: Unit of Meas.	Output: Manhr / Unit	Unit Costs: Material Cost	Unit Costs: Labor Cost	Unit Costs: Equipment Cost	Unit Costs: Total Cost
09 - 21161	**METAL STUDS, Cont'd...**						**09 - 21161**
1140	16" o.c.	SF	0.016	1.05	1.30		2.35
1145	4", 16 ga.						
1150	12" o.c.	SF	0.020	1.55	1.63		3.18
1160	16" o.c.	"	0.016	1.40	1.30		2.70
1980	6", 16 ga.						
2000	12" o.c.	SF	0.025	1.98	2.03		4.01
2001	16" o.c.	"	0.020	1.78	1.63		3.41
3000	Furring						
3160	On beams and columns						
3170	7/8" channel	LF	0.053	0.52	4.34		4.86
3180	1-1/2" channel	"	0.062	0.62	5.01		5.63
4460	On ceilings						
4470	3/4" furring channels						
4480	12" o.c.	SF	0.033	0.37	2.71		3.08
4490	16" o.c.	"	0.032	0.29	2.60		2.89
4495	24" o.c.	"	0.029	0.20	2.33		2.53
4500	1-1/2" furring channels						
4520	12" o.c.	SF	0.036	0.62	2.96		3.58
4540	16" o.c.	"	0.033	0.47	2.71		3.18
4560	24" o.c.	"	0.031	0.31	2.50		2.81
5000	On walls						
5020	3/4" furring channels						
5050	12" o.c.	SF	0.027	0.37	2.17		2.54
5100	16" o.c.	"	0.025	0.29	2.03		2.32
5150	24" o.c.	"	0.024	0.20	1.91		2.11
5200	1-1/2" furring channels						
5210	12" o.c.	SF	0.029	0.62	2.33		2.95
5220	16" o.c.	"	0.027	0.47	2.17		2.64
5230	24" o.c.	"	0.025	0.31	2.03		2.34

© 2020 BNi Publications Inc. ALL RIGHTS RESERVED

LATH AND PLASTER

ID Code	Description: Component Descriptions	Output: Unit of Meas.	Output: Manhr / Unit	Unit Costs: Material Cost	Unit Costs: Labor Cost	Unit Costs: Equipment Cost	Unit Costs: Total Cost
09 - 22361	**GYPSUM LATH**						**09 - 22361**
1070	Gypsum lath, 1/2" thick						
1090	Clipped	SY	0.044	4.92	3.62		8.54
1110	Nailed	"	0.050	4.92	4.07		8.99
09 - 22362	**METAL LATH**						**09 - 22362**
0960	Diamond expanded, galvanized						
0980	2.5 lb., on walls						
1010	Nailed	SY	0.100	4.22	8.15		12.25
1030	Wired	"	0.114	4.22	9.32		13.50
1040	On ceilings						
1050	Nailed	SY	0.114	4.22	9.32		13.50
1070	Wired	"	0.133	4.22	10.75		15.00
1980	3.4 lb., on walls						
2000	Nailed	SY	0.100	5.73	8.15		14.00
2020	Wired	"	0.114	5.73	9.32		15.00
2030	On ceilings						
2040	Nailed	SY	0.114	5.73	9.32		15.00
2060	Wired	"	0.133	5.73	10.75		16.50
2064	Flat rib						
2068	2.75 lb., on walls						
2070	Nailed	SY	0.100	3.99	8.15		12.25
2100	Wired	"	0.114	3.99	9.32		13.25
2110	On ceilings						
2120	Nailed	SY	0.114	3.99	9.32		13.25
2140	Wired	"	0.133	3.99	10.75		14.75
2150	3.4 lb., on walls						
2160	Nailed	SY	0.100	4.80	8.15		13.00
2180	Wired	"	0.114	4.80	9.32		14.00
2190	On ceilings						
2200	Nailed	SY	0.114	4.80	9.32		14.00
2220	Wired	"	0.133	4.80	10.75		15.50
2230	Stucco lath						
2240	1.8 lb.	SY	0.100	4.96	8.15		13.00
2300	3.6 lb.	"	0.100	5.56	8.15		13.75
2310	Paper backed						
2320	Minimum	SY	0.080	3.85	6.52		10.25
2400	Maximum	"	0.114	6.21	9.32		15.50

© 2020 BNi Publications Inc. ALL RIGHTS RESERVED

LATH AND PLASTER

ID Code	Description Component Descriptions	Output Unit of Meas.	Manhr / Unit	Unit Costs Material Cost	Labor Cost	Equipment Cost	Total Cost
09 - 22366	**PLASTER ACCESSORIES**						**09 - 22366**
0120	Expansion joint, 3/4", 26 ga., galv.	LF	0.020	1.63	1.63		3.26
2000	Plaster corner beads, 3/4", galvanized	"	0.023	0.45	1.86		2.31
2020	Casing bead, expanded flange, galvanized	"	0.020	0.56	1.63		2.19
2100	Expanded wing, 1-1/4" wide, galvanized	"	0.020	0.72	1.63		2.35
2500	Joint clips for lath	EA	0.004	0.19	0.32		0.51
2580	Metal base, galvanized, 2-1/2" high	LF	0.027	0.83	2.17		3.00
2600	Stud clips for gypsum lath	EA	0.004	0.19	0.32		0.51
2700	Tie wire galvanized, 18 ga., 25 lb. hank	"					52.00
8000	Sound deadening board, 1/4"	SF	0.013	0.35	1.08		1.43

SUPPORTS FOR PLASTER GYPSUM BOARD

ID Code	Component Descriptions	Unit of Meas.	Manhr / Unit	Material Cost	Labor Cost	Equipment Cost	Total Cost
09 - 23001	**PLASTER**						**09 - 23001**
0980	Gypsum plaster, trowel finish, 2 coats						
1000	Ceilings	SY	0.250	4.30	19.00		23.25
1020	Walls	"	0.235	4.30	17.75		22.00
1030	3 coats						
1040	Ceilings	SY	0.348	5.96	26.50		32.50
1060	Walls	"	0.308	5.96	23.25		29.25
1960	Vermiculite plaster						
1980	2 coats						
2000	Ceilings	SY	0.381	4.89	29.00		34.00
2020	Walls	"	0.348	4.89	26.50		31.50
2030	3 coats						
2040	Ceilings	SY	0.471	7.68	35.75		43.50
2060	Walls	"	0.421	7.68	32.00		39.75
5960	Keenes cement plaster						
5980	2 coats						
6000	Ceilings	SY	0.308	2.09	23.25		25.25
6020	Walls	"	0.267	2.09	20.25		22.25
6030	3 coats						
6040	Ceilings	SY	0.348	2.14	26.50		28.75
6060	Walls	"	0.308	2.14	23.25		25.50
7000	On columns, add to installation, 50%	"					
7020	Chases, fascia, and soffits, add to installation, 50%	"					
7040	Beams, add to installation, 50%	"					

© 2020 BNi Publications Inc. ALL RIGHTS RESERVED

CEMENT PLASTERING

ID Code	Component Descriptions	Unit of Meas.	Manhr / Unit	Material Cost	Labor Cost	Equipment Cost	Total Cost
09 - 24001	**PORTLAND CEMENT PLASTER**						**09 - 24001**
2980	Stucco, portland, gray, 3 coat, 1" thick						
3000	Sand finish	SY	0.348	8.43	26.50		35.00
3020	Trowel finish	"	0.364	8.43	27.50		36.00
3030	White cement						
3040	Sand finish	SY	0.364	9.63	27.50		37.25
3060	Trowel finish	"	0.400	9.63	30.50		40.25
3980	Scratch coat						
4000	For ceramic tile	SY	0.080	3.05	6.07		9.12
4020	For quarry tile	"	0.080	3.05	6.07		9.12
5000	Portland cement plaster						
5020	2 coats, 1/2"	SY	0.160	6.07	12.25		18.25
5040	3 coats, 7/8"	"	0.200	7.25	15.25		22.50

GYPSUM BOARD

ID Code	Component Descriptions	Unit of Meas.	Manhr / Unit	Material Cost	Labor Cost	Equipment Cost	Total Cost
09 - 29001	**GYPSUM BOARD**						**09 - 29001**
0080	Drywall, plasterboard, 3/8" clipped to						
0100	Metal furred ceiling	SF	0.009	0.41	0.72		1.13
0120	Columns and beams	"	0.020	0.41	1.63		2.04
0140	Walls	"	0.008	0.41	0.65		1.06
0150	Nailed or screwed to						
0160	Wood or metal framed ceiling	SF	0.008	0.41	0.65		1.06
0180	Columns and beams	"	0.018	0.41	1.44		1.85
0190	Walls	"	0.007	0.41	0.59		1.00
0220	1/2", clipped to						
0240	Metal furred ceiling	SF	0.009	0.42	0.72		1.14
0260	Columns and beams	"	0.020	0.38	1.63		2.01
0270	Walls	"	0.008	0.38	0.65		1.03
0280	Nailed or screwed to						
0290	Wood or metal framed ceiling	SF	0.008	0.38	0.65		1.03
0300	Columns and beams	"	0.018	0.38	1.44		1.82
0400	Walls	"	0.007	0.38	0.59		0.97
1000	5/8", clipped to						
1020	Metal furred ceiling	SF	0.010	0.42	0.81		1.23
1040	Columns and beams	"	0.022	0.42	1.81		2.23
1060	Walls	"	0.009	0.42	0.72		1.14
1070	Nailed or screwed to						
1080	Wood or metal framed ceiling	SF	0.010	0.42	0.81		1.23

© 2020 BNi Publications Inc. ALL RIGHTS RESERVED

GYPSUM BOARD

	Description	Output		Unit Costs			
ID Code	Component Descriptions	Unit of Meas.	Manhr / Unit	Material Cost	Labor Cost	Equipment Cost	Total Cost
09 - 29001	**GYPSUM BOARD, Cont'd...**						**09 - 29001**
1100	Columns and beams	SF	0.022	0.42	1.81		2.23
1120	Walls	"	0.009	0.42	0.72		1.14
1122	Vinyl faced, clipped to metal studs						
1124	1/2"	SF	0.010	1.19	0.81		2.00
1126	5/8"	"	0.010	1.13	0.81		1.94
1130	Add for						
1140	Fire resistant	SF					0.12
1180	Water resistant	"					0.19
1200	Water and fire resistant	"					0.24
1220	Taping and finishing joints						
1222	Minimum	SF	0.005	0.04	0.43		0.47
1224	Average	"	0.007	0.07	0.54		0.61
1226	Maximum	"	0.008	0.10	0.65		0.75
5020	Casing bead						
5022	Minimum	LF	0.023	0.16	1.86		2.02
5024	Average	"	0.027	0.18	2.17		2.35
5026	Maximum	"	0.040	0.22	3.26		3.48
5040	Corner bead						
5042	Minimum	LF	0.023	0.18	1.86		2.04
5044	Average	"	0.027	0.22	2.17		2.39
5046	Maximum	"	0.040	0.27	3.26		3.53

TILE

09 - 30131	**CERAMIC TILE**						**09 - 30131**
0980	Glazed wall tile, 4-1/4" x 4-1/4"						
1000	Minimum	SF	0.057	2.32	4.44		6.76
1020	Average	"	0.067	3.68	5.18		8.86
1040	Maxlmum	"	0.080	13.25	6.22		19.50
1042	6" x 6"						
1044	Minimum	SF	0.050	1.65	3.88		5.53
1046	Average	"	0.057	2.22	4.44		6.66
1048	Maximum	"	0.067	2.77	5.18		7.95
2960	Base, 4-1/4" high						
2980	Minimum	LF	0.100	4.47	7.77		12.25
3000	Average	"	0.100	5.20	7.77		13.00
3040	Maximum	"	0.100	6.87	7.77		14.75
3042	Glazed moldings and trim, 12" x 12"						

© 2020 BNi Publications Inc. ALL RIGHTS RESERVED

TILE

ID Code	Component Descriptions	Unit of Meas.	Manhr / Unit	Material Cost	Labor Cost	Equipment Cost	Total Cost
09 - 30131	**CERAMIC TILE, Cont'd...**						**09 - 30131**
3044	Minimum	LF	0.080	2.46	6.22		8.68
3046	Average	"	0.080	3.75	6.22		9.97
3048	Maximum	"	0.080	5.04	6.22		11.25
6100	Unglazed floor tile						
6120	Portland cem., cushion edge, face mtd						
6140	1" x 1"	SF	0.073	8.87	5.65		14.50
6150	2" x 2"	"	0.067	9.38	5.18		14.50
6162	4" x 4"	"	0.067	8.73	5.18		14.00
6164	6" x 6"	"	0.057	3.12	4.44		7.56
6166	12" x 12"	"	0.050	2.75	3.88		6.63
6168	16" x 16"	"	0.044	2.38	3.45		5.83
6170	18" x 18"	"	0.040	2.31	3.11		5.42
6200	Adhesive bed, with white grout						
6220	1" x 1"	SF	0.073	7.38	5.65		13.00
6230	2" x 2"	"	0.067	7.81	5.18		13.00
6260	4" x 4"	"	0.067	7.81	5.18		13.00
6262	6" x 6"	"	0.057	2.60	4.44		7.04
6264	12" x 12"	"	0.050	2.28	3.88		6.16
6266	16" x 16"	"	0.044	1.98	3.45		5.43
6268	18" x 18"	"	0.040	1.92	3.11		5.03
6300	Organic adhesive bed, thin set, back mounted						
6320	1" x 1"	SF	0.073	7.38	5.65		13.00
6350	2" x 2"	"	0.067	8.59	5.18		13.75
6360	For group 2 colors, add to material, 10%						
6370	For group 3 colors, add to material, 20%						
6380	For abrasive surface, add to material, 25%						
6382	Porcelain floor tile						
6384	1" x 1"	SF	0.073	9.90	5.65		15.50
6386	2" x 2"	"	0.070	9.05	5.40		14.50
6388	4" x 4"	"	0.067	8.41	5.18		13.50
6390	6" x 6"	"	0.057	3.02	4.44		7.46
6392	12" x 12"	"	0.050	2.72	3.88		6.60
6394	16" x 16"	"	0.044	2.16	3.45		5.61
6396	18" x 18"	"	0.040	2.04	3.11		5.15
6400	Unglazed wall tile						
6420	Organic adhesive, face mounted cushion edge						
6425	1" x 1"						
6430	Minimum	SF	0.067	4.65	5.18		9.83

© 2020 BNi Publications Inc. ALL RIGHTS RESERVED

TILE

	Description	Output		Unit Costs			
ID Code	Component Descriptions	Unit of Meas.	Manhr / Unit	Material Cost	Labor Cost	Equipment Cost	Total Cost
09 - 30131	**CERAMIC TILE, Cont'd...**						**09 - 30131**
6432	Average	SF	0.073	6.09	5.65		11.75
6434	Maximum	"	0.080	9.09	6.22		15.25
6448	2" x 2"						
6450	Minimum	SF	0.062	5.37	4.78		10.25
6452	Average	"	0.067	6.09	5.18		11.25
6454	Maximum	"	0.073	9.96	5.65		15.50
6500	Back mounted						
6510	1" x 1"						
6520	Minimum	SF	0.067	4.65	5.18		9.83
6522	Average	"	0.073	6.09	5.65		11.75
6524	Maximum	"	0.080	9.09	6.22		15.25
6538	2" x 2"						
6540	Minimum	SF	0.062	5.37	4.78		10.25
6542	Average	"	0.067	6.09	5.18		11.25
6544	Maximum	"	0.073	9.96	5.65		15.50
6600	For glazed finish, add to material, 25%						
6620	For glazed mosaic, add to material, 100%						
6630	For metallic colors, add to material, 125%						
6640	For exterior wall use, add to total, 25%						
6650	For exterior soffit, add to total, 25%						
6660	For portland cement bed, add to total, 25%						
6670	For dry set portland cement bed, add to total, 10%						
8020	Conductive floor tile, unglazed square edged						
8040	Portland cement bed						
8060	1 x 1	SF	0.100	6.93	7.77		14.75
8080	1-9/16 x 1-9/16	"	0.100	6.38	7.77		14.25
8100	Dry set						
8120	1 x 1	SF	0.100	6.93	7.77		14.75
8140	1-9/16 x 1-9/16	"	0.100	6.38	7.77		14.25
8160	Epoxy bed with epoxy joints						
8180	1 x 1	SF	0.100	6.93	7.77		14.75
8200	1-9/16 x 1-9/16	"	0.100	6.38	7.77		14.25
8400	For WWF in bed add to total, 15%						
8420	For abrasive surface, add to material, 40%						
8990	Ceramic accessories						
9000	Towel bar, 24" long						
9002	Minimum	EA	0.320	18.00	24.75		42.75
9004	Average	"	0.400	22.25	31.00		53.00

© 2020 BNi Publications Inc. ALL RIGHTS RESERVED

TILE

ID Code	Description: Component Descriptions	Output: Unit of Meas.	Output: Manhr / Unit	Unit Costs: Material Cost	Unit Costs: Labor Cost	Unit Costs: Equipment Cost	Unit Costs: Total Cost
09 - 30131	**CERAMIC TILE, Cont'd...**						**09 - 30131**
9006	Maximum	EA	0.533	59.00	41.50		100
9020	Soap dish						
9022	Minimum	EA	0.533	8.47	41.50		50.00
9024	Average	"	0.667	11.50	52.00		64.00
9026	Maximum	"	0.800	30.25	62.00		92.00
09 - 30161	**QUARRY TILE**						**09 - 30161**
1060	Floor						
1080	4 x 4 x 1/2"	SF	0.107	6.57	8.29		14.75
1100	6 x 6 x 1/2"	"	0.100	6.44	7.77		14.25
1120	6 x 6 x 3/4"	"	0.100	7.99	7.77		15.75
1122	12 x 12 x 3/4"	"	0.089	11.25	6.91		18.25
1124	16 x 1 6 x 3/4"	"	0.080	7.83	6.22		14.00
1126	18 x 18 x 3/4"	"	0.067	5.52	5.18		10.75
1150	Medallion						
1160	36" dia.	EA	2.000	350	160		510
1162	48" dia.	"	2.000	410	160		570
1200	Wall, applied to 3/4" portland cement bed						
1220	4 x 4 x 1/2"	SF	0.160	5.84	12.50		18.25
1240	6 x 6 x 3/4"	"	0.133	6.53	10.25		16.75
1320	Cove base						
1330	5 x 6 x 1/2" straight top	LF	0.133	6.66	10.25		17.00
1340	6 x 6 x 3/4" round top	"	0.133	6.18	10.25		16.50
1345	Moldings						
1350	2 x 12	LF	0.080	10.50	6.22		16.75
1352	4 x 12	"	0.080	16.50	6.22		22.75
1360	Stair treads 6 x 6 x 3/4"	"	0.200	9.13	15.50		24.75
1380	Window sill 6 x 8 x 3/4"	"	0.160	8.33	12.50		20.75
1400	For abrasive surface, add to material, 25%						

ACOUSTICAL TREATMENT

ID Code	Component Descriptions	Unit of Meas.	Manhr / Unit	Material Cost	Labor Cost	Equipment Cost	Total Cost
09 - 51001	**CEILINGS AND WALLS**						**09 - 51001**
1400	Acoustical panels, suspension system not included						
1420	Fiberglass panels						
1500	5/8" thick						
1560	2' x 2'	SF	0.011	1.61	0.93		2.54
1580	2' x 4'	"	0.009	1.34	0.72		2.06
1590	3/4" thick						

© 2020 BNi Publications Inc. ALL RIGHTS RESERVED

ACOUSTICAL TREATMENT

ID Code	Description	Output		Unit Costs			
	Component Descriptions	Unit of Meas.	Manhr / Unit	Material Cost	Labor Cost	Equipment Cost	Total Cost
09 - 51001	**CEILINGS AND WALLS, Cont'd...**						**09 - 51001**
1600	2' x 2'	SF	0.011	2.14	0.93		3.07
1620	2' x 4'	"	0.009	2.07	0.72		2.79
1630	Glass cloth faced fiberglass panels						
1660	3/4" thick	SF	0.013	3.05	1.08		4.13
1680	1" thick	"	0.013	3.41	1.08		4.49
1690	Mineral fiber panels						
1700	5/8" thick						
1720	2' x 2'	SF	0.011	1.37	0.93		2.30
1740	2' x 4'	"	0.009	1.37	0.72		2.09
1750	3/4" thick						
1760	2' x 2'	SF	0.011	2.14	0.93		3.07
1780	2' x 4'	"	0.009	2.07	0.72		2.79
1790	For aluminum faced panels, add to material, 80%						
1800	For vinyl faced panels, add to total, 125%						
1810	For fire rated panels, add to material, 75%						
1820	Wood fiber panels						
1840	1/2" thick						
1850	2' x 2'	SF	0.011	1.77	0.93		2.70
1860	2' x 4'	"	0.009	1.77	0.72		2.49
1870	5/8" thick						
1880	2' x 2'	SF	0.011	2.03	0.93		2.96
1890	2' x 4'	"	0.009	2.03	0.72		2.75
1900	3/4" thick						
1910	2' x 2'	SF	0.011	2.49	0.93		3.42
1920	2' x 4'	"	0.009	2.49	0.72		3.21
1930	2" thick						
1940	2' x 2'	SF	0.013	2.91	1.08		3.99
1950	2' x 4'	"	0.010	2.91	0.81		3.72
2000	For flameproofing, add to material, 10%						
2010	For sculptured finish, add to material, 15%						
2020	Air distributing panels						
2060	3/4" thick	SF	0.020	2.64	1.63		4.27
2080	5/8" thick	"	0.016	2.26	1.30		3.56
2090	Acoustical tiles, suspension system not included						
2100	Fiberglass tile, 12" x 12"						
3040	5/8" thick	SF	0.015	2.00	1.18		3.18
3060	3/4" thick	"	0.018	2.32	1.44		3.76
3080	Glass cloth faced fiberglass tile						

© 2020 BNi Publications Inc. ALL RIGHTS RESERVED

ACOUSTICAL TREATMENT

	Description	Output		Unit Costs			
ID Code	Component Descriptions	Unit of Meas.	Manhr / Unit	Material Cost	Labor Cost	Equipment Cost	Total Cost
09 - 51001	**CEILINGS AND WALLS, Cont'd...**						**09 - 51001**
3100	3/4" thick	SF	0.018	3.73	1.44		5.17
3120	3" thick	"	0.020	4.17	1.63		5.80
3130	Mineral fiber tile, 12" x 12"						
3140	5/8" thick						
3160	Standard	SF	0.016	1.05	1.30		2.35
3170	Vinyl faced	"	0.016	2.08	1.30		3.38
3180	3/4" thick						
3190	Standard	SF	0.016	1.53	1.30		2.83
3200	Vinyl faced	"	0.016	2.66	1.30		3.96
3240	Fire rated	"	0.016	3.39	1.30		4.69
3260	Aluminum or mylar faced	"	0.016	6.47	1.30		7.77
3280	Wood fiber tile, 12" x 12"						
3300	1/2" thick	SF	0.016	1.68	1.30		2.98
3320	3/4" thick	"	0.016	2.44	1.30		3.74
3340	For flameproofing, add to material, 10%						
3360	For sculptured 3 dimensional, add to material, 50%						
3380	Metal pan units, 24 ga. steel						
3700	12" x 12"	SF	0.032	6.00	2.60		8.60
3710	12" x 24"	"	0.027	6.83	2.17		9.00
3720	Aluminum, .025" thick						
3740	12" x 12"	SF	0.032	7.03	2.60		9.63
3750	12" x 24"	"	0.027	7.24	2.17		9.41
3760	Anodized aluminum, 0.25" thick						
3770	12" x 12"	SF	0.032	7.72	2.60		10.25
3775	12" x 24"	"	0.027	9.17	2.17		11.25
3780	Stainless steel, 24 ga.						
3790	12" x 12"	SF	0.032	17.75	2.60		20.25
3800	12" x 24"	"	0.027	15.25	2.17		17.50
3840	For flameproof sound absorbing pads, add to material	"					2.28
3860	Metal ceiling systems						
3880	.020" thick panels						
4030	10', 12', and 16' lengths	SF	0.023	5.49	1.86		7.35
4040	Custom lengths, 3' to 20'	"	0.023	5.55	1.86		7.41
4050	.025" thick panels						
4080	32 sf, 38 sf, and 52 sf pieces	SF	0.027	5.51	2.17		7.68
4100	Custom lengths, 10 sf to 65 sf	"	0.027	6.43	2.17		8.60
4140	Carriers, black, add	"					3.31
4160	Recess filler strip, add	"					1.11

© 2020 BNi Publications Inc. ALL RIGHTS RESERVED

ACOUSTICAL TREATMENT

	Description	Output		Unit Costs			
ID Code	Component Descriptions	Unit of Meas.	Manhr / Unit	Material Cost	Labor Cost	Equipment Cost	Total Cost
09 - 51001	**CEILINGS AND WALLS, Cont'd...**						**09 - 51001**
4180	Custom lengths, add	SF					1.65
4190	Sound absorption walls, with fabric cover						
5000	2-6" x 9' x 3/4"	SF	0.027	10.25	2.17		12.50
5020	2' x 9' x 1"	"	0.027	11.25	2.17		13.50
5040	Starter spline	LF	0.020	1.65	1.63		3.28
5060	Internal spline	"	0.020	1.44	1.63		3.07
5080	Acoustical treatment						
5100	Barriers for plenums						
5120	Leaded vinyl						
5140	0.48 lb per sf	SF	0.038	4.23	3.10		7.33
5160	0.87 lb per sf	"	0.040	5.11	3.26		8.37
5170	Aluminum foil, fiberglass reinforcement						
5180	Minimum	SF	0.027	1.16	2.17		3.33
5200	Maximum	"	0.040	1.32	3.26		4.58
5220	Aluminum mesh, paper backed	"	0.027	1.09	2.17		3.26
5240	Fibered cement sheet, 3/16" thick	"	0.029	2.25	2.33		4.58
5260	Sheet lead, 1/64" thick	"	0.020	3.83	1.63		5.46
5300	Sound attenuation blanket						
5360	1" thick	SF	0.080	0.43	6.52		6.95
5380	1-1/2" thick	"	0.080	0.60	6.52		7.12
5390	2" thick	"	0.080	0.75	6.52		7.27
5400	3" thick	"	0.089	0.90	7.24		8.14
5420	Ceiling suspension systems						
5440	T-bar system						
5510	2' x 4'	SF	0.008	1.27	0.65		1.92
5520	2' x 2'	"	0.009	1.37	0.72		2.09
5530	Concealed Z-bar suspension system, 12" module	"	0.013	1.30	1.08		2.38
5550	For 1-1/2" carrier channels, 4' o.c., add	"					0.41
5560	Carrier channel for recessed light fixtures	"					0.75

© 2020 BNi Publications Inc. ALL RIGHTS RESERVED

FLOORING

	Description	Output		Unit Costs			
ID Code	Component Descriptions	Unit of Meas.	Manhr / Unit	Material Cost	Labor Cost	Equipment Cost	Total Cost
09 - 63000	**FLOOR LEVELING**						**09 - 63000**
0980	Repair and level floors to receive new flooring						
1000	Minimum	SY	0.027	1.65	2.17		3.82
1020	Average	"	0.067	3.92	5.43		9.35
1030	Maximum	"	0.080	5.81	6.52		12.25
09 - 63161	**UNIT MASONRY FLOORING**						**09 - 63161**
1000	Clay brick						
1020	9 x 4-1/2 x 3" thick						
1040	Glazed	SF	0.067	8.43	5.43		13.75
1060	Unglazed	"	0.067	8.08	5.43		13.50
1070	8 x 4 x 3/4" thick						
1080	Glazed	SF	0.070	7.62	5.67		13.25
1100	Unglazed	"	0.070	7.27	5.67		13.00
1140	For herringbone pattern, add to labor, 15%						

WOOD FLOORING

ID Code	Component Descriptions	Unit of Meas.	Manhr / Unit	Material Cost	Labor Cost	Equipment Cost	Total Cost
09 - 64001	**WOOD FLOORING**						**09 - 64001**
0100	Wood strip flooring, unfinished						
1000	Fir floor						
1010	C and better						
1020	Vertical grain	SF	0.027	3.52	2.17		5.69
1040	Flat grain	"	0.027	4.40	2.17		6.57
1060	Oak floor						
1080	Minimum	SF	0.038	3.72	3.10		6.82
1100	Average	"	0.038	5.13	3.10		8.23
1120	Maximum	"	0.038	7.42	3.10		10.50
1200	Maple floor						
1220	25/32" x 2-1/4"						
1240	Minimum	SF	0.038	5.32	3.10		8.42
1260	Maximum	"	0.038	7.54	3.10		10.75
1280	33/32" x 3-1/4"						
1300	Minimum	SF	0.038	7.42	3.10		10.50
1320	Maximum	"	0.038	8.38	3.10		11.50
1340	Added costs						
1350	For factory finish, add to material, 10%						
1355	For random width floor, add to total, 20%						
1360	For simulated pegs, add to total, 10%						
1500	Wood block industrial flooring						

© 2020 BNi Publications Inc. ALL RIGHTS RESERVED

WOOD FLOORING

ID Code	Description: Component Descriptions	Output: Unit of Meas.	Output: Manhr / Unit	Unit Costs: Material Cost	Unit Costs: Labor Cost	Unit Costs: Equipment Cost	Unit Costs: Total Cost
09 - 64001	**WOOD FLOORING, Cont'd...**						**09 - 64001**
1510	Creosoted						
1520	2" thick	SF	0.021	4.18	1.71		5.89
1540	2-1/2" thick	"	0.025	4.34	2.03		6.37
1560	3" thick	"	0.027	4.51	2.17		6.68
2500	Parquet, 5/16", white oak						
2520	Finished	SF	0.040	10.00	3.26		13.25
2540	Unfinished	"	0.040	4.84	3.26		8.10
3000	Gym floor, 2 ply felt, 25/32" maple, finished, in mastic	"	0.044	8.54	3.62		12.25
3020	Over wood sleepers	"	0.050	8.69	4.07		12.75
9020	Finishing, sand, fill, finish, and wax	"	0.020	0.66	1.63		2.29
9100	Refinish sand, seal, and 2 coats of polyurethane	"	0.027	1.16	2.17		3.33
9540	Clean and wax floors	"	0.004	0.24	0.32		0.56

RESILIENT FLOORING

ID Code	Component Descriptions	Unit of Meas.	Manhr / Unit	Material Cost	Labor Cost	Equipment Cost	Total Cost
09 - 65131	**RESILIENT BASE AND ACCESSORIES**						**09 - 65131**
1000	Wall base, vinyl						
1120	Group 1						
1130	4" high	LF	0.027	1.41	2.17		3.58
1140	6" high	"	0.027	1.92	2.17		4.09
1160	Group 2						
1180	4" high	LF	0.027	1.24	2.17		3.41
1200	6" high	"	0.027	1.97	2.17		4.14
1220	Group 3						
1230	4" high	LF	0.027	2.80	2.17		4.97
1240	6" high	"	0.027	3.15	2.17		5.32
6000	Stair accessories						
6010	Treads, 1/4" x 12", rubber diamond surface						
6020	Marbled	LF	0.067	17.00	5.43		22.50
6040	Plain	"	0.067	17.50	5.43		23.00
6080	Grit strip safety tread, 12" wide, colors						
6100	3/16" thick	LF	0.067	17.50	5.43		23.00
6120	5/16" thick	"	0.067	23.50	5.43		29.00
6140	Risers, 7" high, 1/8" thick, colors						
6160	Flat	LF	0.040	6.57	3.26		9.83
6180	Coved	"	0.040	4.53	3.26		7.79
6300	Nosing, rubber						
6310	3/16" thick, 3" wide						

© 2020 BNi Publications Inc. ALL RIGHTS RESERVED

RESILIENT FLOORING

ID Code	Description: Component Descriptions	Output: Unit of Meas.	Output: Manhr / Unit	Unit Costs: Material Cost	Unit Costs: Labor Cost	Unit Costs: Equipment Cost	Unit Costs: Total Cost
09 - 65131	**RESILIENT BASE AND ACCESSORIES, Cont'd...**						**09 - 65131**
6320	Black	LF	0.040	5.69	3.26		8.95
6340	Colors	"	0.040	6.41	3.26		9.67
6350	6" wide						
6360	Black	LF	0.067	6.99	5.43		12.50
6380	Colors	"	0.067	7.29	5.43		12.75
09 - 65161	**RESILIENT SHEET FLOORING**						**09 - 65161**
0980	Vinyl sheet flooring						
1000	Minimum	SF	0.008	4.22	0.65		4.87
1002	Average	"	0.010	6.81	0.77		7.58
1004	Maximum	"	0.013	11.50	1.08		12.50
1020	Cove, to 6"	LF	0.016	2.51	1.30		3.81
2000	Fluid applied resilient flooring						
2020	Polyurethane, poured in place, 3/8" thick	SF	0.067	10.50	5.43		16.00
6200	Vinyl sheet goods, backed						
6220	0.070" thick	SF	0.010	4.10	0.81		4.91
6240	0.093" thick	"	0.010	6.35	0.81		7.16
6260	0.125" thick	"	0.010	7.33	0.81		8.14
6280	0.250" thick	"	0.010	8.43	0.81		9.24
09 - 65191	**RESILIENT TILE FLOORING**						**09 - 65191**
1020	Solid vinyl tile, 1/8" thick, 12" x 12"						
1040	Marble patterns	SF	0.020	5.13	1.63		6.76
1060	Solid colors	"	0.020	6.65	1.63		8.28
1080	Travertine patterns	"	0.020	7.47	1.63		9.10
2000	Conductive resilient flooring, vinyl tile						
2040	1/8" thick, 12" x 12"	SF	0.023	8.11	1.86		9.97

TERRAZZO FLOORING

ID Code	Component Descriptions	Unit of Meas.	Manhr / Unit	Material Cost	Labor Cost	Equipment Cost	Total Cost
09 - 66131	**TERRAZZO**						**09 - 66131**
1100	Floors on concrete, 1-3/4" thick, 5/8" topping						
1120	Gray cement	SF	0.114	5.67	8.68		14.25
1140	White cement	"	0.114	5.92	8.68		14.50
1200	Sand cushion, 3" thick, 5/8" top, 1/4"						
1220	Gray cement	SF	0.133	6.69	10.00		16.75
1240	White cement	"	0.133	6.96	10.00		17.00
1260	Monolithic terrazzo, 3-1/2" base slab, 5/8" topping	"	0.100	4.75	7.59		12.25
1280	Terrazzo wainscot, cast-in-place, 1/2" thick	"	0.200	5.77	15.25		21.00

© 2020 BNi Publications Inc. ALL RIGHTS RESERVED

TERRAZZO FLOORING

ID Code	Description: Component Descriptions	Output: Unit of Meas.	Output: Manhr / Unit	Unit Costs: Material Cost	Unit Costs: Labor Cost	Unit Costs: Equipment Cost	Unit Costs: Total Cost
09 - 66131	**TERRAZZO, Cont'd...**						**09 - 66131**
1300	Base, cast-in-place, terrazzo cove type, 6" high	LF	0.114	7.26	8.68		16.00
1320	Curb, cast-in-place, 6" wide x 6" high, polished top	"	0.400	6.60	30.50		37.00
1340	For venetian type terrazzo, add to material, 10%						
1360	For abrasive heavy duty terrazzo, add to material,						
1400	Divider strips						
1500	Zinc	LF					1.43
1510	Brass	"					2.66
1560	Stairs, cast-in-place, topping on concrete or metal						
1620	1-1/2" thick treads, 12" wide	LF	0.400	5.61	30.50		36.00
1640	Combined tread and riser	"	1.000	8.42	76.00		84.00
1680	Precast terrazzo, thin set						
1690	Terrazzo tiles, non-slip surface						
2120	9" x 9" x 1" thick	SF	0.114	16.00	8.68		24.75
2130	12" x 12"						
2140	1" thick	SF	0.107	17.25	8.10		25.25
2160	1-1/2" thick	"	0.114	18.00	8.68		26.75
2180	18" x 18" x 1-1/2" thick	"	0.114	23.50	8.68		32.25
2200	24" x 24" x 1-1/2" thick	"	0.094	30.25	7.14		37.50
2400	For white cement, add to material, 10%						
2800	For venetian type terrazzo, add to material, 25%						
3000	Terrazzo wainscot						
3020	12" x 12" x 1" thick	SF	0.200	8.83	15.25		24.00
3040	18" x 18" x 1-1/2" thick	"	0.229	14.50	17.25		31.75
3060	Base						
3080	6" high						
3220	Straight	LF	0.062	12.75	4.67		17.50
3240	Coved	"	0.062	15.00	4.67		19.75
3260	8" high						
3280	Straight	LF	0.067	14.25	5.06		19.25
3300	Coved	"	0.067	16.75	5.06		21.75
3310	Terrazzo curbs						
3320	8" wide x 8" high	LF	0.320	33.00	24.25		57.00
3340	6" wide x 6" high	"	0.267	29.75	20.25		50.00
3400	Precast terrazzo stair treads, 12" wide						
3410	1-1/2" thick						
3420	Diamond pattern	LF	0.145	39.75	11.00		51.00
3430	Non-slip surface	"	0.145	41.75	11.00		53.00
3440	2" thick						

© 2020 BNi Publications Inc. ALL RIGHTS RESERVED

TERRAZZO FLOORING

ID Code	Description: Component Descriptions	Output: Unit of Meas.	Output: Manhr / Unit	Unit Costs: Material Cost	Unit Costs: Labor Cost	Unit Costs: Equipment Cost	Unit Costs: Total Cost
09 - 66131	**TERRAZZO, Cont'd...**						**09 - 66131**
3450	Diamond pattern	LF	0.145	41.75	11.00		53.00
3460	Non-slip surface	"	0.160	44.00	12.25		56.00
3480	Stair risers, 1" thick to 6" high						
3520	Straight sections	LF	0.080	13.25	6.07		19.25
3530	Cove sections	"	0.080	15.75	6.07		21.75
3600	Combined tread and riser						
3620	Straight sections						
3640	1-1/2" tread, 3/4" riser	LF	0.229	57.00	17.25		74.00
3660	3" tread, 1" riser	"	0.229	69.00	17.25		86.00
3680	Curved sections						
3700	2" tread, 1" riser	LF	0.267	73.00	20.25		93.00
3720	3" tread, 1" riser	"	0.267	76.00	20.25		96.00
3800	Stair stringers, notched for treads and risers						
3820	1" thick	LF	0.200	34.50	15.25		49.75
3840	2" thick	"	0.267	36.00	20.25		56.00
3860	Landings, structural, nonslip						
3870	1-1/2" thick	SF	0.133	32.75	10.00		42.75
3880	3" thick	"	0.160	45.75	12.25		58.00
4000	Conductive terrazzo, spark proof industrial floor						
4020	Epoxy terrazzo						
4040	Floor	SF	0.050	6.89	3.79		10.75
4060	Base	"	0.067	7.70	5.06		12.75
4070	Polyacrylate						
4080	Floor	SF	0.050	8.85	3.79		12.75
4100	Base	"	0.067	10.00	5.06		15.00
4110	Polyester						
4120	Floor	SF	0.032	4.12	2.43		6.55
4140	Base	"	0.040	4.40	3.03		7.43
4150	Synthetic latex mastic						
4170	Floor	SF	0.050	7.15	3.79		11.00
4180	Base	"	0.067	7.15	5.06		12.25

© 2020 BNi Publications Inc. ALL RIGHTS RESERVED

FLUID APPLIED FLOORING

	Description	Output		Unit Costs			
ID Code	Component Descriptions	Unit of Meas.	Manhr / Unit	Material Cost	Labor Cost	Equipment Cost	Total Cost
09 - 67001	**SPECIAL FLOORING**						**09 - 67001**
1020	Epoxy flooring, marble chips						
1040	Epoxy with colored quartz chips in 1/4" base	SF	0.044	4.48	3.62		8.10
1060	Heavy duty epoxy topping, 3/16" thick	"	0.044	3.62	3.62		7.24
1080	Epoxy terrazzo						
1090	1/4" thick chemical resistant	SF	0.050	6.58	4.07		10.75

CARPET

ID Code	Component Descriptions	Unit of Meas.	Manhr / Unit	Material Cost	Labor Cost	Equipment Cost	Total Cost
09 - 68001	**CARPET PADDING**						**09 - 68001**
1000	Carpet padding						
1005	Foam rubber, waffle type, 0.3" thick	SY	0.040	6.74	3.26		10.00
1010	Jute padding						
1020	Minimum	SY	0.036	4.57	2.96		7.53
1022	Average	"	0.040	5.95	3.26		9.21
1024	Maximum	"	0.044	8.98	3.62		12.50
1030	Sponge rubber cushion						
1040	Minimum	SY	0.036	5.42	2.96		8.38
1042	Average	"	0.040	7.22	3.26		10.50
1044	Maximum	"	0.044	10.25	3.62		13.75
1050	Urethane cushion, 3/8" thick						
1060	Minimum	SY	0.036	5.42	2.96		8.38
1062	Average	"	0.040	6.32	3.26		9.58
1064	Maximum	"	0.044	8.24	3.62		11.75
09 - 68002	**CARPET**						**09 - 68002**
0990	Carpet, acrylic						
1000	24 oz., light traffic	SY	0.089	17.75	7.24		25.00
1020	28 oz., medium traffic	"	0.089	21.25	7.24		28.50
2000	Residential						
2010	Nylon						
2020	15 oz., light traffic	SY	0.089	24.50	7.24		31.75
2040	28 oz., medium traffic	"	0.089	32.00	7.24		39.25
2100	Commercial						
2110	Nylon						
2120	28 oz., medium traffic	SY	0.089	30.50	7.24		37.75
2140	35 oz., heavy traffic	"	0.089	37.25	7.24		44.50
2145	Wool						
2150	30 oz., medium traffic	SY	0.089	51.00	7.24		58.00
2160	36 oz., medium traffic	"	0.089	53.00	7.24		60.00

© 2020 BNi Publications Inc. ALL RIGHTS RESERVED

CARPET

ID Code	Component Descriptions	Unit of Meas.	Manhr / Unit	Material Cost	Labor Cost	Equipment Cost	Total Cost
	09 - 68002 CARPET, Cont'd... 09 - 68002						
2180	42 oz., heavy traffic	SY	0.089	71.00	7.24		78.00
3000	Carpet tile						
3020	Foam backed						
3022	Minimum	SF	0.016	4.07	1.30		5.37
3024	Average	"	0.018	4.71	1.44		6.15
3026	Maximum	"	0.020	7.47	1.63		9.10
3040	Tufted loop or shag						
3042	Minimum	SF	0.016	4.41	1.30		5.71
3044	Average	"	0.018	5.32	1.44		6.76
3046	Maximum	"	0.020	8.56	1.63		10.25
8980	Clean and vacuum carpet						
9000	Minimum	SY	0.004	0.36	0.25		0.61
9020	Average	"	0.005	0.56	0.43		0.99
9040	Maximum	"	0.008	0.77	0.65		1.42

ACCESS FLOORING

ID Code	Component Descriptions	Unit of Meas.	Manhr / Unit	Material Cost	Labor Cost	Equipment Cost	Total Cost
	09 - 69001 ACCESS & PEDESTAL FLOOR 09 - 69001						
0980	Panels, no covering, 2'x2'						
1000	Plain	SF	0.010	11.75	0.81		12.50
1040	Perforated	"	0.400	16.25	32.50		48.75
1100	Pedestals						
1120	For 6" to 12" clearance	EA	0.080	9.40	6.52		16.00
1200	Stringers						
1220	2'	LF	0.038	2.59	3.10		5.69
1240	6'	"	0.027	2.59	2.17		4.76
1300	Accessories						
1320	Ramp assembly	SF	0.032	54.00	2.60		57.00
1330	Elevated floor assembly	"	0.030	86.00	2.41		88.00
1340	Handrail	LF	0.400	67.00	32.50		100
1360	Fascia plate	"	0.200	32.75	16.25		49.00
1400	For carpet tiles, add	SF					9.08
1420	For vinyl flooring, add	"					10.25
1500	RF shielding components, floor liner						
1520	Hot rolled steel sheet						
1540	14 ga.	SF	0.020	15.50	1.63		17.25
1560	11 ga.	"	0.062	21.50	5.01		26.50

© 2020 BNi Publications Inc. ALL RIGHTS RESERVED

WALL COVERING

	Description	Output		Unit Costs			
ID Code	Component Descriptions	Unit of Meas.	Manhr / Unit	Material Cost	Labor Cost	Equipment Cost	Total Cost
09 - 72001	**WALL COVERING**						**09 - 72001**
0900	Vinyl wall covering						
1000	Medium duty	SF	0.011	1.10	0.77		1.87
1010	Heavy duty	"	0.013	2.26	0.90		3.16
1020	Over pipes and irregular shapes						
1030	Lightweight, 13 oz.	SF	0.016	1.89	1.08		2.97
1040	Medium weight, 25 oz.	"	0.018	2.26	1.20		3.46
1060	Heavyweight, 34 oz.	"	0.020	2.77	1.36		4.13
1080	Cork wall covering						
1100	1' x 1' squares						
1140	1/4" thick	SF	0.020	5.66	1.36		7.02
1160	1/2" thick	"	0.020	7.19	1.36		8.55
1180	3/4" thick	"	0.020	8.10	1.36		9.46
1190	Wall fabrics						
1200	Natural fabrics, grass cloths						
1220	Minimum	SF	0.012	1.65	0.83		2.48
1240	Average	"	0.013	1.83	0.90		2.73
1260	Maximum	"	0.016	6.16	1.08		7.24
1280	Flexible gypsum coated wall fabric, fire resistant	"	0.008	1.85	0.54		2.39
2000	Vinyl corner guards						
2020	3/4" x 3/4" x 8'	EA	0.100	8.70	6.80		15.50
2040	2-3/4" x 2-3/4" x 4'	"	0.100	5.14	6.80		12.00

PAINT

09 - 91001	**PAINTING PREPARATION**						**09 - 91001**
1000	Dropcloths						
1050	Minimum	SF	0.001	0.16	0.03		0.19
1100	Average	"	0.001	0.18	0.04		0.22
1150	Maximum	"	0.001	0.37	0.06		0.43
1200	Masking						
1250	Paper and tape						
1300	Minimum	LF	0.008	0.04	0.54		0.58
1350	Average	"	0.010	0.06	0.68		0.74
1400	Maximum	"	0.013	0.07	0.90		0.97
1450	Doors						
1500	Minimum	EA	0.100	0.05	6.80		6.85
1550	Average	"	0.133	0.06	9.07		9.13
1600	Maximum	"	0.178	0.07	12.00		12.00

© 2020 BNi Publications Inc. ALL RIGHTS RESERVED

PAINT

ID Code	Description: Component Descriptions	Output: Unit of Meas.	Output: Manhr / Unit	Unit Costs: Material Cost	Unit Costs: Labor Cost	Unit Costs: Equipment Cost	Unit Costs: Total Cost
09 - 91001	**PAINTING PREPARATION, Cont'd...**						**09 - 91001**
1650	Windows						
1700	Minimum	EA	0.100	0.05	6.80		6.85
1750	Average	"	0.133	0.06	9.07		9.13
1800	Maximum	"	0.178	0.07	12.00		12.00
2000	Sanding						
2050	Walls and flat surfaces						
2100	Minimum	SF	0.005		0.36		0.36
2150	Average	"	0.007		0.45		0.45
2200	Maximum	"	0.008		0.54		0.54
2250	Doors and windows						
2300	Minimum	EA	0.133		9.07		9.07
2350	Average	"	0.200		13.50		13.50
2400	Maximum	"	0.267		18.25		18.25
2450	Trim						
2500	Minimum	LF	0.010		0.68		0.68
2550	Average	"	0.013		0.90		0.90
2600	Maximum	"	0.018		1.20		1.20
2650	Puttying						
2700	Minimum	SF	0.012	0.01	0.83		0.84
2750	Average	"	0.016	0.02	1.08		1.10
2800	Maximum	"	0.020	0.03	1.36		1.39
09 - 91009	**PAINT**						**09 - 91009**
0830	Paint, enamel						
0850	600 sf per gal.	GAL					54.00
0900	550 sf per gal.	"					50.00
1000	500 sf per gal.	"					36.00
1020	450 sf per gal.	"					33.75
1060	350 sf per gal.	"					32.50
1100	Filler, 60 sf per gal.	"					38.50
1160	Latex, 400 sf per gal.	"					36.00
1170	Aluminum						
1180	400 sf per gal.	GAL					48.00
1190	500 sf per gal.	"					77.00
1200	Red lead, 350 sf per gal.	"					67.00
1220	Primer						
1240	400 sf per gal.	GAL					32.50
1250	300 sf per gal.	"					32.50

© 2020 BNi Publications Inc. ALL RIGHTS RESERVED

PAINT

	Description	Output		Unit Costs			
ID Code	Component Descriptions	Unit of Meas.	Manhr / Unit	Material Cost	Labor Cost	Equipment Cost	Total Cost
09 - 91009	**PAINT, Cont'd...**						**09 - 91009**
1280	Latex base, interior, white	GAL					36.00
1480	Sealer and varnish						
1500	400 sf per gal.	GAL					33.75
1520	425 sf per gal.	"					48.00
1540	600 sf per gal.	"					62.00
09 - 91130	**EXT. PAINTING, SITEWORK**						**09 - 91130**
1020	Benches						
1040	Brush						
1060	First Coat						
1080	Minimum	SF	0.008	0.19	0.54		0.73
1100	Average	"	0.010	0.21	0.68		0.89
1120	Maximum	"	0.013	0.22	0.90		1.12
1140	Second Coat						
1160	Minimum	SF	0.005	0.18	0.34		0.52
1180	Average	"	0.006	0.19	0.38		0.57
1200	Maximum	"	0.007	0.21	0.45		0.66
1220	Roller						
1240	First Coat						
1260	Minimum	SF	0.004	0.19	0.27		0.46
1280	Average	"	0.004	0.21	0.30		0.51
1300	Maximum	"	0.005	0.22	0.34		0.56
1320	Second Coat						
1340	Minimum	SF	0.003	0.18	0.19		0.37
1360	Average	"	0.003	0.19	0.22		0.41
1380	Maximum	"	0.004	0.21	0.24		0.45
2000	Brickwork						
2020	Brush						
2040	First Coat						
2060	Minimum	SF	0.005	0.19	0.34		0.53
2080	Average	"	0.007	0.21	0.45		0.66
2100	Maximum	"	0.010	0.22	0.68		0.90
2120	Second Coat						
2140	Minimum	SF	0.004	0.19	0.30		0.49
2160	Average	"	0.005	0.21	0.36		0.57
2180	Maximum	"	0.007	0.22	0.45		0.67
2380	Spray						
2400	First Coat						

© 2020 BNi Publications Inc. ALL RIGHTS RESERVED

PAINT

ID Code	Description: Component Descriptions	Output: Unit of Meas.	Output: Manhr / Unit	Unit Costs: Material Cost	Unit Costs: Labor Cost	Unit Costs: Equipment Cost	Unit Costs: Total Cost
09 - 91130	**EXT. PAINTING, SITEWORK, Cont'd...**						**09 - 91130**
2420	Minimum	SF	0.002	0.15	0.15		0.30
2440	Average	"	0.003	0.17	0.19		0.36
2460	Maximum	"	0.004	0.18	0.24		0.42
2480	Second Coat						
2500	Minimum	SF	0.002	0.15	0.14		0.29
2520	Average	"	0.003	0.17	0.17		0.34
2540	Maximum	"	0.003	0.18	0.22		0.40
3000	Concrete Block						
3020	Roller						
3040	First Coat						
3060	Minimum	SF	0.004	0.19	0.27		0.46
3080	Average	"	0.005	0.21	0.36		0.57
3100	Maximum	"	0.008	0.22	0.54		0.76
3120	Second Coat						
3140	Minimum	SF	0.003	0.19	0.22		0.41
3160	Average	"	0.004	0.21	0.30		0.51
3180	Maximum	"	0.007	0.22	0.45		0.67
3200	Spray						
3220	First Coat						
3240	Minimum	SF	0.002	0.15	0.15		0.30
3260	Average	"	0.003	0.17	0.18		0.35
3280	Maximum	"	0.003	0.18	0.20		0.38
3300	Second Coat						
3320	Minimum	SF	0.001	0.15	0.09		0.24
3340	Average	"	0.002	0.17	0.12		0.29
3360	Maximum	"	0.003	0.18	0.17		0.35
3500	Fences, Chain Link						
3520	Brush						
3540	First Coat						
3560	Minimum	SF	0.008	0.13	0.54		0.67
3580	Average	"	0.009	0.14	0.60		0.74
3600	Maximum	"	0.010	0.15	0.68		0.83
3620	Second Coat						
3640	Minimum	SF	0.005	0.13	0.36		0.49
3660	Average	"	0.006	0.14	0.41		0.55
3680	Maximum	"	0.007	0.15	0.49		0.64
3700	Roller						
3720	First Coat						

© 2020 BNi Publications Inc. ALL RIGHTS RESERVED

PAINT

ID Code	Description: Component Descriptions	Output: Unit of Meas.	Output: Manhr / Unit	Unit Costs: Material Cost	Unit Costs: Labor Cost	Unit Costs: Equipment Cost	Unit Costs: Total Cost
09 - 91130	**EXT. PAINTING, SITEWORK, Cont'd...**						**09 - 91130**
3740	Minimum	SF	0.006	0.13	0.38		0.51
3760	Average	"	0.007	0.14	0.45		0.59
3780	Maximum	"	0.008	0.15	0.51		0.66
3800	Second Coat						
3820	Minimum	SF	0.003	0.13	0.22		0.35
3840	Average	"	0.004	0.14	0.27		0.41
3860	Maximum	"	0.005	0.15	0.34		0.49
3880	Spray						
3900	First Coat						
3920	Minimum	SF	0.003	0.10	0.17		0.27
3940	Average	"	0.003	0.11	0.19		0.30
3960	Maximum	"	0.003	0.13	0.22		0.35
3980	Second Coat						
4000	Minimum	SF	0.002	0.10	0.12		0.22
4060	Average	"	0.002	0.11	0.15		0.26
4080	Maximum	"	0.003	0.13	0.17		0.30
4200	Fences, Wood or Masonry						
4220	Brush						
4240	First Coat						
4260	Minimum	SF	0.008	0.19	0.57		0.76
4280	Average	"	0.010	0.21	0.68		0.89
4300	Maximum	"	0.013	0.22	0.90		1.12
4320	Second Coat						
4340	Minimum	SF	0.005	0.19	0.34		0.53
4360	Average	"	0.006	0.21	0.41		0.62
4380	Maximum	"	0.008	0.22	0.54		0.76
4400	Roller						
4420	First Coat						
4440	Minimum	SF	0.004	0.19	0.30		0.49
4460	Average	"	0.005	0.21	0.36		0.57
4480	Maximum	"	0.006	0.22	0.41		0.63
4500	Second Coat						
4520	Minimum	SF	0.003	0.19	0.20		0.39
4540	Average	"	0.004	0.21	0.25		0.46
4560	Maximum	"	0.005	0.22	0.34		0.56
4580	Spray						
4600	First Coat						
4620	Minimum	SF	0.003	0.15	0.19		0.34

© 2020 BNi Publications Inc. ALL RIGHTS RESERVED

PAINT

ID Code	Description: Component Descriptions	Output: Unit of Meas.	Output: Manhr / Unit	Unit Costs: Material Cost	Unit Costs: Labor Cost	Unit Costs: Equipment Cost	Unit Costs: Total Cost
09 - 91130	**EXT. PAINTING, SITEWORK, Cont'd...**						**09 - 91130**
4640	Average	SF	0.004	0.17	0.24		0.41
4660	Maximum	"	0.005	0.18	0.34		0.52
4680	Second Coat						
4700	Minimum	SF	0.002	0.15	0.13		0.28
4760	Average	"	0.003	0.17	0.17		0.34
4780	Maximum	"	0.003	0.18	0.22		0.40
4800	Storage Tanks						
4820	Roller						
4840	First Coat						
4860	Minimum	SF	0.003	0.15	0.22		0.37
4880	Average	"	0.004	0.17	0.27		0.44
4900	Maximum	"	0.005	0.18	0.34		0.52
4920	Second Coat						
4940	Minimum	SF	0.003	0.15	0.18		0.33
4960	Average	"	0.003	0.17	0.21		0.38
4980	Maximum	"	0.004	0.18	0.27		0.45
5000	Spray						
5020	First Coat						
5040	Minimum	SF	0.002	0.13	0.13		0.26
5060	Average	"	0.002	0.14	0.16		0.30
5080	Maximum	"	0.003	0.15	0.19		0.34
5100	Second Coat						
5160	Minimum	SF	0.002	0.13	0.10		0.23
5180	Average	"	0.002	0.14	0.12		0.26
5200	Maximum	"	0.002	0.15	0.13		0.28
09 - 91131	**EXT. PAINTING, BUILDINGS**						**09 - 91131**
1000	Decks, Metal						
1020	Spray						
1040	First Coat						
1060	Minimum	SF	0.004	0.13	0.24		0.37
1080	Average	"	0.004	0.14	0.27		0.41
1100	Maximum	"	0.004	0.15	0.30		0.45
1120	Second Coat						
1140	Minimum	SF	0.003	0.11	0.17		0.28
1160	Average	"	0.003	0.13	0.19		0.32
1180	Maximum	"	0.003	0.14	0.22		0.36
1200	Decks, Wood, Stained						

© 2020 BNi Publications Inc. ALL RIGHTS RESERVED

PAINT

	Description	Output		Unit Costs			
ID Code	Component Descriptions	Unit of Meas.	Manhr / Unit	Material Cost	Labor Cost	Equipment Cost	Total Cost
09 - 91131	**EXT. PAINTING, BUILDINGS, Cont'd...**						**09 - 91131**
1220	Brush						
1240	First Coat						
1260	Minimum	SF	0.004	0.15	0.27		0.42
1280	Average	"	0.004	0.17	0.30		0.47
1300	Maximum	"	0.005	0.18	0.34		0.52
1320	Second Coat						
1340	Minimum	SF	0.003	0.15	0.19		0.34
1360	Average	"	0.003	0.17	0.20		0.37
1380	Maximum	"	0.003	0.18	0.22		0.40
1400	Roller						
1420	First Coat						
1440	Minimum	SF	0.003	0.15	0.19		0.34
1460	Average	"	0.003	0.17	0.20		0.37
1480	Maximum	"	0.003	0.18	0.22		0.40
1500	Second Coat						
1520	Minimum	SF	0.003	0.15	0.17		0.32
1540	Average	"	0.003	0.17	0.18		0.35
1560	Maximum	"	0.003	0.18	0.20		0.38
1580	Spray						
1600	First Coat						
1620	Minimum	SF	0.003	0.13	0.17		0.30
1640	Average	"	0.003	0.14	0.18		0.32
1660	Maximum	"	0.003	0.15	0.20		0.35
1680	Second Coat						
1700	Minimum	SF	0.002	0.13	0.15		0.28
1720	Average	"	0.002	0.14	0.16		0.30
1740	Maximum	"	0.003	0.15	0.18		0.33
1760	Doors, Metal						
1780	Roller						
1800	First Coat						
1820	Minimum	SF	0.006	0.15	0.38		0.53
1840	Average	"	0.007	0.17	0.45		0.62
1860	Maximum	"	0.008	0.18	0.54		0.72
1880	Second Coat						
1900	Minimum	SF	0.004	0.15	0.27		0.42
1920	Average	"	0.004	0.17	0.30		0.47
1940	Maximum	"	0.005	0.18	0.34		0.52
1960	Spray						

© 2020 BNi Publications Inc. ALL RIGHTS RESERVED

PAINT

ID Code	Component Descriptions	Unit of Meas.	Manhr / Unit	Material Cost	Labor Cost	Equipment Cost	Total Cost
09 - 91131	**EXT. PAINTING, BUILDINGS, Cont'd...**						**09 - 91131**
1980	First Coat						
2000	Minimum	SF	0.005	0.13	0.34		0.47
2020	Average	"	0.006	0.14	0.38		0.52
2040	Maximum	"	0.007	0.15	0.45		0.60
2060	Second Coat						
2080	Minimum	SF	0.004	0.13	0.24		0.37
2100	Average	"	0.004	0.14	0.27		0.41
2120	Maximum	"	0.004	0.15	0.30		0.45
2140	Door Frames, Metal						
2160	Brush						
2180	First Coat						
2200	Minimum	LF	0.010	0.19	0.68		0.87
2220	Average	"	0.013	0.21	0.85		1.06
2240	Maximum	"	0.015	0.22	0.98		1.20
2260	Second Coat						
2280	Minimum	LF	0.006	0.19	0.38		0.57
2300	Average	"	0.007	0.21	0.45		0.66
2320	Maximum	"	0.008	0.22	0.54		0.76
2340	Spray						
2360	First Coat						
2380	Minimum	LF	0.004	0.13	0.30		0.43
2400	Average	"	0.006	0.14	0.38		0.52
2420	Maximum	"	0.008	0.15	0.54		0.69
2440	Second Coat						
2460	Minimum	LF	0.004	0.13	0.24		0.37
2480	Average	"	0.004	0.14	0.27		0.41
2500	Maximum	"	0.004	0.15	0.30		0.45
2520	Doors, Wood						
2540	Brush						
2560	First Coat						
2580	Minimum	SF	0.012	0.15	0.83		0.98
2600	Average	"	0.016	0.17	1.08		1.25
2620	Maximum	"	0.020	0.18	1.36		1.54
2640	Second Coat						
2660	Minimum	SF	0.010	0.15	0.68		0.83
2680	Average	"	0.011	0.17	0.77		0.94
2700	Maximum	"	0.013	0.18	0.90		1.08
2720	Roller						

© 2020 BNi Publications Inc. ALL RIGHTS RESERVED

PAINT

ID Code	Description: Component Descriptions	Output: Unit of Meas.	Output: Manhr / Unit	Unit Costs: Material Cost	Unit Costs: Labor Cost	Unit Costs: Equipment Cost	Unit Costs: Total Cost
09 - 91131	**EXT. PAINTING, BUILDINGS, Cont'd...**						**09 - 91131**
2740	First Coat						
2760	Minimum	SF	0.006	0.15	0.36		0.51
2780	Average	"	0.007	0.17	0.45		0.62
2800	Maximum	"	0.010	0.18	0.68		0.86
2820	Second Coat						
2840	Minimum	SF	0.004	0.15	0.27		0.42
2860	Average	"	0.004	0.17	0.30		0.47
2880	Maximum	"	0.007	0.18	0.45		0.63
2900	Spray						
2920	First Coat						
2940	Minimum	SF	0.003	0.13	0.17		0.30
2960	Average	"	0.003	0.14	0.20		0.34
2980	Maximum	"	0.004	0.15	0.27		0.42
3000	Second Coat						
3020	Minimum	SF	0.002	0.13	0.13		0.26
3040	Average	"	0.002	0.14	0.15		0.29
3060	Maximum	"	0.003	0.15	0.18		0.33
3080	Gutters and Downspouts						
3100	Brush						
3120	First Coat						
3140	Minimum	LF	0.010	0.19	0.68		0.87
3160	Average	"	0.011	0.21	0.77		0.98
3180	Maximum	"	0.013	0.22	0.90		1.12
3200	Second Coat						
3220	Minimum	LF	0.007	0.19	0.45		0.64
3240	Average	"	0.008	0.21	0.54		0.75
3260	Maximum	"	0.010	0.22	0.68		0.90
3300	Siding, Metal						
3320	Roller						
3340	First Coat						
3360	Minimum	SF	0.003	0.15	0.22		0.37
3380	Average	"	0.004	0.17	0.24		0.41
3400	Maximum	"	0.004	0.18	0.27		0.45
3420	Second Coat						
3440	Minimum	SF	0.003	0.15	0.20		0.35
3460	Average	"	0.003	0.17	0.22		0.39
3480	Maximum	"	0.004	0.18	0.24		0.42
3500	Spray						

© 2020 BNi Publications Inc. ALL RIGHTS RESERVED

PAINT

ID Code	Description: Component Descriptions	Output: Unit of Meas.	Output: Manhr / Unit	Unit Costs: Material Cost	Unit Costs: Labor Cost	Unit Costs: Equipment Cost	Unit Costs: Total Cost
09 - 91131	**EXT. PAINTING, BUILDINGS, Cont'd...**						**09 - 91131**
3520	First Coat						
3540	Minimum	SF	0.003	0.13	0.17		0.30
3560	Average	"	0.003	0.14	0.19		0.33
3580	Maximum	"	0.003	0.15	0.22		0.37
3600	Second Coat						
3620	Minimum	SF	0.002	0.13	0.10		0.23
3640	Average	"	0.002	0.14	0.13		0.27
3660	Maximum	"	0.003	0.15	0.18		0.33
3680	Siding, Wood						
3700	Roller						
3720	First Coat						
3740	Minimum	SF	0.003	0.13	0.19		0.32
3760	Average	"	0.003	0.14	0.22		0.36
3780	Maximum	"	0.004	0.15	0.24		0.39
3800	Second Coat						
3820	Minimum	SF	0.003	0.13	0.22		0.35
3840	Average	"	0.004	0.14	0.24		0.38
3860	Maximum	"	0.004	0.15	0.27		0.42
3880	Spray						
3900	First Coat						
3920	Minimum	SF	0.003	0.13	0.18		0.31
3940	Average	"	0.003	0.14	0.19		0.33
3960	Maximum	"	0.003	0.15	0.20		0.35
3980	Second Coat						
4000	Minimum	SF	0.002	0.13	0.13		0.26
4020	Average	"	0.003	0.14	0.18		0.32
4040	Maximum	"	0.004	0.15	0.27		0.42
4060	Stucco						
4080	Roller						
4100	First Coat						
4120	Minimum	SF	0.004	0.19	0.24		0.43
4140	Average	"	0.004	0.21	0.28		0.49
4160	Maximum	"	0.005	0.22	0.34		0.56
4180	Second Coat						
4200	Minimum	SF	0.003	0.19	0.20		0.39
4220	Average	"	0.003	0.21	0.23		0.44
4240	Maximum	"	0.004	0.22	0.27		0.49
4260	Spray						

© 2020 BNi Publications Inc. ALL RIGHTS RESERVED

PAINT

ID Code	Description: Component Descriptions	Output: Unit of Meas.	Output: Manhr / Unit	Unit Costs: Material Cost	Unit Costs: Labor Cost	Unit Costs: Equipment Cost	Unit Costs: Total Cost
09 - 91131	**EXT. PAINTING, BUILDINGS, Cont'd...**						**09 - 91131**
4280	First Coat						
4300	Minimum	SF	0.003	0.15	0.17		0.32
4320	Average	"	0.003	0.17	0.19		0.36
4340	Maximum	"	0.003	0.18	0.22		0.40
4360	Second Coat						
4380	Minimum	SF	0.002	0.15	0.13		0.28
4400	Average	"	0.002	0.17	0.15		0.32
4420	Maximum	"	0.003	0.18	0.18		0.36
4440	Trim						
4460	Brush						
4480	First Coat						
4500	Minimum	LF	0.003	0.19	0.22		0.41
4520	Average	"	0.004	0.21	0.27		0.48
4540	Maximum	"	0.005	0.22	0.34		0.56
4560	Second Coat						
4580	Minimum	LF	0.003	0.19	0.17		0.36
4600	Average	"	0.003	0.21	0.22		0.43
4620	Maximum	"	0.005	0.22	0.34		0.56
4640	Walls						
4660	Roller						
4680	First Coat						
4700	Minimum	SF	0.003	0.15	0.19		0.34
4720	Average	"	0.003	0.17	0.20		0.37
4740	Maximum	"	0.003	0.18	0.21		0.39
4760	Second Coat						
4780	Minimum	SF	0.003	0.15	0.17		0.32
4800	Average	"	0.003	0.17	0.18		0.35
4820	Maximum	"	0.003	0.18	0.20		0.38
4840	Spray						
4860	First Coat						
4880	Minimum	SF	0.001	0.11	0.08		0.19
4900	Average	"	0.002	0.13	0.10		0.23
4920	Maximum	"	0.002	0.14	0.13		0.27
4940	Second Coat						
4960	Minimum	SF	0.001	0.11	0.07		0.18
4980	Average	"	0.001	0.13	0.09		0.22
5000	Maximum	"	0.002	0.14	0.12		0.26
5020	Windows						

© 2020 BNi Publications Inc. ALL RIGHTS RESERVED

PAINT

ID Code	Description: Component Descriptions	Output: Unit of Meas.	Output: Manhr / Unit	Unit Costs: Material Cost	Unit Costs: Labor Cost	Unit Costs: Equipment Cost	Unit Costs: Total Cost
09 - 91131	**EXT. PAINTING, BUILDINGS, Cont'd...**						**09 - 91131**
5040	Brush						
5060	First Coat						
5080	Minimum	SF	0.013	0.13	0.90		1.03
5100	Average	"	0.016	0.14	1.08		1.22
5120	Maximum	"	0.020	0.15	1.36		1.51
5140	Second Coat						
5160	Minimum	SF	0.011	0.13	0.77		0.90
5180	Average	"	0.013	0.14	0.90		1.04
5200	Maximum	"	0.016	0.15	1.08		1.23
09 - 91132	**EXT. PAINTING, MISC.**						**09 - 91132**
1000	Gratings, Metal						
1020	Roller						
1040	First Coat						
1060	Minimum	SF	0.023	0.15	1.55		1.70
1080	Average	"	0.027	0.17	1.81		1.98
1100	Maximum	"	0.032	0.18	2.17		2.35
1120	Second Coat						
1140	Minimum	SF	0.016	0.15	1.08		1.23
1160	Average	"	0.020	0.17	1.36		1.53
1180	Maximum	"	0.027	0.18	1.81		1.99
1200	Spray						
1220	First Coat						
1240	Minimum	SF	0.011	0.13	0.77		0.90
1260	Average	"	0.013	0.14	0.90		1.04
1280	Maximum	"	0.016	0.15	1.08		1.23
1300	Second Coat						
1320	Minimum	SF	0.009	0.13	0.60		0.73
1340	Average	"	0.010	0.14	0.68		0.82
1360	Maximum	"	0.011	0.15	0.77		0.92
1500	Ladders						
1520	Brush						
1540	First Coat						
1560	Minimum	LF	0.020	0.19	1.36		1.55
1580	Average	"	0.023	0.21	1.55		1.76
1600	Maximum	"	0.027	0.22	1.81		2.03
1620	Second Coat						
1640	Minimum	LF	0.016	0.19	1.08		1.27

© 2020 BNi Publications Inc. ALL RIGHTS RESERVED

PAINT

ID Code	Description: Component Descriptions	Output: Unit of Meas.	Output: Manhr / Unit	Unit Costs: Material Cost	Unit Costs: Labor Cost	Unit Costs: Equipment Cost	Unit Costs: Total Cost
09 - 91132	**EXT. PAINTING, MISC., Cont'd...**						**09 - 91132**
1660	Average	LF	0.018	0.21	1.20		1.41
1680	Maximum	"	0.020	0.22	1.36		1.58
1700	Spray						
1720	First Coat						
1740	Minimum	LF	0.013	0.13	0.90		1.03
1760	Average	"	0.015	0.14	0.98		1.12
1780	Maximum	"	0.016	0.15	1.08		1.23
1800	Second Coat						
1820	Minimum	LF	0.011	0.13	0.77		0.90
1840	Average	"	0.012	0.14	0.83		0.97
1860	Maximum	"	0.013	0.15	0.90		1.05
3000	Shakes						
3020	Spray						
3040	First Coat						
3060	Minimum	SF	0.003	0.14	0.22		0.36
3080	Average	"	0.004	0.15	0.24		0.39
3100	Maximum	"	0.004	0.17	0.27		0.44
3120	Second Coat						
3140	Minimum	SF	0.003	0.14	0.20		0.34
3160	Average	"	0.003	0.15	0.22		0.37
3180	Maximum	"	0.004	0.17	0.24		0.41
3200	Shingles, Wood						
3220	Roller						
3240	First Coat						
3260	Minimum	SF	0.004	0.15	0.30		0.45
3280	Average	"	0.005	0.17	0.34		0.51
3300	Maximum	"	0.006	0.18	0.38		0.56
3320	Second Coat						
3340	Minimum	SF	0.003	0.15	0.20		0.35
3360	Average	"	0.003	0.17	0.22		0.39
3380	Maximum	"	0.004	0.18	0.24		0.42
3400	Spray						
3420	First Coat						
3440	Minimum	LF	0.003	0.13	0.20		0.33
3460	Average	"	0.003	0.14	0.22		0.36
3480	Maximum	"	0.004	0.15	0.24		0.39
3500	Second Coat						
3520	Minimum	LF	0.002	0.13	0.16		0.29

© 2020 BNi Publications Inc. ALL RIGHTS RESERVED

PAINT

ID Code	Description: Component Descriptions	Output: Unit of Meas.	Output: Manhr / Unit	Unit Costs: Material Cost	Unit Costs: Labor Cost	Unit Costs: Equipment Cost	Unit Costs: Total Cost
09 - 91132	**EXT. PAINTING, MISC., Cont'd...**						**09 - 91132**
3540	Average	LF	0.003	0.14	0.17		0.31
3560	Maximum	"	0.003	0.15	0.18		0.33
4000	Shutters and Louvers						
4020	Brush						
4040	First Coat						
4060	Minimum	EA	0.160	0.19	11.00		11.25
4080	Average	"	0.200	0.21	13.50		13.75
4100	Maximum	"	0.267	0.22	18.25		18.50
4120	Second Coat						
4140	Minimum	EA	0.100	0.19	6.80		6.99
4160	Average	"	0.123	0.21	8.37		8.58
4180	Maximum	"	0.160	0.22	11.00		11.25
4200	Spray						
4220	First Coat						
4240	Minimum	EA	0.053	0.14	3.62		3.76
4260	Average	"	0.064	0.15	4.35		4.50
4280	Maximum	"	0.080	0.17	5.44		5.61
4300	Second Coat						
4320	Minimum	EA	0.040	0.14	2.72		2.86
4340	Average	"	0.053	0.15	3.62		3.77
4360	Maximum	"	0.064	0.17	4.35		4.52
5000	Stairs, metal						
5020	Brush						
5040	First Coat						
5060	Minimum	SF	0.009	0.19	0.60		0.79
5080	Average	"	0.010	0.21	0.68		0.89
5100	Maximum	"	0.011	0.22	0.77		0.99
5120	Second Coat						
5140	Minimum	SF	0.005	0.19	0.34		0.53
5160	Average	"	0.006	0.21	0.38		0.59
5180	Maximum	"	0.007	0.22	0.45		0.67
5200	Spray						
5220	First Coat						
5240	Minimum	SF	0.004	0.14	0.30		0.44
5260	Average	"	0.006	0.15	0.38		0.53
5280	Maximum	"	0.006	0.17	0.41		0.58
5300	Second Coat						
5320	Minimum	SF	0.003	0.14	0.22		0.36

© 2020 BNi Publications Inc. ALL RIGHTS RESERVED

PAINT

ID Code	Component Descriptions	Unit of Meas.	Manhr / Unit	Material Cost	Labor Cost	Equipment Cost	Total Cost
09 - 91132	**EXT. PAINTING, MISC., Cont'd...**						**09 - 91132**
5340	Average	SF	0.004	0.15	0.27		0.42
5360	Maximum	"	0.005	0.17	0.34		0.51
7000	Steel, Structural, Light						
7020	Brush						
7040	First Coat						
7060	Minimum	SF	0.016	0.15	1.08		1.23
7080	Average	"	0.020	0.17	1.36		1.53
7100	Maximum	"	0.027	0.18	1.81		1.99
7120	Second Coat						
7140	Minimum	SF	0.011	0.14	0.77		0.91
7160	Average	"	0.013	0.15	0.90		1.05
7180	Maximum	"	0.016	0.17	1.08		1.25
7200	Roller						
7220	First Coat						
7240	Minimum	SF	0.011	0.15	0.77		0.92
7260	Average	"	0.013	0.17	0.90		1.07
7280	Maximum	"	0.016	0.18	1.08		1.26
7300	Second Coat						
7320	Minimum	SF	0.007	0.14	0.45		0.59
7340	Average	"	0.008	0.15	0.54		0.69
7360	Maximum	"	0.010	0.17	0.68		0.85
7380	Spray						
7400	First Coat						
7420	Minimum	SF	0.007	0.14	0.45		0.59
7440	Average	"	0.008	0.15	0.54		0.69
7460	Maximum	"	0.010	0.17	0.68		0.85
7480	Second Coat						
7500	Minimum	SF	0.006	0.10	0.38		0.48
7520	Average	"	0.007	0.11	0.45		0.56
7540	Maximum	"	0.008	0.13	0.54		0.67
7560	Steel, Medium to Heavy						
7580	Brush						
7600	First Coat						
7620	Minimum	SF	0.008	0.15	0.54		0.69
7640	Average	"	0.009	0.17	0.60		0.77
7660	Maximum	"	0.010	0.18	0.68		0.86
7680	Second Coat						
7700	Minimum	SF	0.007	0.14	0.45		0.59

© 2020 BNi Publications Inc. ALL RIGHTS RESERVED

PAINT

ID Code	Description: Component Descriptions	Output: Unit of Meas.	Output: Manhr / Unit	Unit Costs: Material Cost	Unit Costs: Labor Cost	Unit Costs: Equipment Cost	Unit Costs: Total Cost
09 - 91132	**EXT. PAINTING, MISC., Cont'd...**						**09 - 91132**
7720	Average	SF	0.008	0.15	0.54		0.69
7740	Maximum	"	0.009	0.17	0.64		0.81
7760	Roller						
7780	First Coat						
7800	Minimum	SF	0.007	0.15	0.45		0.60
7820	Average	"	0.008	0.17	0.54		0.71
7840	Maximum	"	0.009	0.18	0.64		0.82
7860	Second Coat						
7880	Minimum	SF	0.005	0.14	0.34		0.48
7900	Average	"	0.006	0.15	0.38		0.53
7920	Maximum	"	0.007	0.17	0.45		0.62
7940	Spray						
7960	First Coat						
7980	Minimum	SF	0.004	0.14	0.30		0.44
8000	Average	"	0.005	0.15	0.34		0.49
8020	Maximum	"	0.006	0.17	0.38		0.55
8040	Second Coat						
8060	Minimum	SF	0.004	0.10	0.24		0.34
8080	Average	"	0.004	0.11	0.27		0.38
8100	Maximum	"	0.004	0.13	0.30		0.43
09 - 91233	**INT. PAINTING, BUILDINGS**						**09 - 91233**
1000	Acoustical Ceiling						
1020	Roller						
1040	First Coat						
1060	Minimum	SF	0.005	0.19	0.34		0.53
1080	Average	"	0.007	0.21	0.45		0.66
1100	Maximum	"	0.010	0.22	0.68		0.90
1120	Second Coat						
1140	Minimum	SF	0.004	0.19	0.27		0.46
1160	Average	"	0.005	0.21	0.34		0.55
1180	Maximum	"	0.007	0.22	0.45		0.67
1200	Spray						
1220	First Coat						
1240	Minimum	SF	0.002	0.15	0.15		0.30
1260	Average	"	0.003	0.17	0.18		0.35
1280	Maximum	"	0.003	0.18	0.22		0.40
1300	Second Coat						

© 2020 BNi Publications Inc. ALL RIGHTS RESERVED

PAINT

ID Code	Description: Component Descriptions	Output: Unit of Meas.	Output: Manhr / Unit	Unit Costs: Material Cost	Unit Costs: Labor Cost	Unit Costs: Equipment Cost	Unit Costs: Total Cost
09 - 91233	**INT. PAINTING, BUILDINGS, Cont'd...**						**09 - 91233**
1320	Minimum	SF	0.002	0.15	0.12		0.27
1340	Average	"	0.002	0.17	0.13		0.30
1360	Maximum	"	0.002	0.18	0.15		0.33
1380	Cabinets and Casework						
1400	Brush						
1420	First Coat						
1440	Minimum	SF	0.008	0.19	0.54		0.73
1460	Average	"	0.009	0.21	0.60		0.81
1480	Maximum	"	0.010	0.22	0.68		0.90
1500	Second Coat						
1520	Minimum	SF	0.007	0.19	0.45		0.64
1540	Average	"	0.007	0.21	0.49		0.70
1560	Maximum	"	0.008	0.22	0.54		0.76
1580	Spray						
1600	First Coat						
1620	Minimum	SF	0.004	0.15	0.27		0.42
1640	Average	"	0.005	0.17	0.32		0.49
1660	Maximum	"	0.006	0.18	0.38		0.56
1680	Second Coat						
1700	Minimum	SF	0.003	0.15	0.21		0.36
1720	Average	"	0.003	0.17	0.23		0.40
1740	Maximum	"	0.004	0.18	0.30		0.48
1760	Ceilings						
1780	Roller						
1800	First Coat						
1820	Minimum	SF	0.003	0.15	0.22		0.37
1840	Average	"	0.004	0.17	0.24		0.41
1860	Maximum	"	0.004	0.18	0.27		0.45
1880	Second Coat						
1900	Minimum	SF	0.003	0.15	0.18		0.33
1920	Average	"	0.003	0.17	0.20		0.37
1940	Maximum	"	0.003	0.18	0.22		0.40
1960	Spray						
1980	First Coat						
2000	Minimum	SF	0.002	0.13	0.13		0.26
2020	Average	"	0.002	0.14	0.15		0.29
2040	Maximum	"	0.003	0.15	0.17		0.32
2060	Second Coat						

© 2020 BNi Publications Inc. ALL RIGHTS RESERVED

PAINT

ID Code	Description: Component Descriptions	Output: Unit of Meas.	Output: Manhr / Unit	Unit Costs: Material Cost	Unit Costs: Labor Cost	Unit Costs: Equipment Cost	Unit Costs: Total Cost
09 - 91233	**INT. PAINTING, BUILDINGS, Cont'd...**						**09 - 91233**
2080	Minimum	SF	0.002	0.13	0.10		0.23
2100	Average	"	0.002	0.14	0.11		0.25
2120	Maximum	"	0.002	0.15	0.13		0.28
2140	Doors, Metal						
2160	Roller						
2180	First Coat						
2200	Minimum	SF	0.005	0.19	0.36		0.55
2220	Average	"	0.006	0.21	0.41		0.62
2240	Maximum	"	0.007	0.22	0.49		0.71
2260	Second Coat						
2280	Minimum	SF	0.004	0.19	0.25		0.44
2300	Average	"	0.004	0.21	0.28		0.49
2320	Maximum	"	0.005	0.22	0.32		0.54
2340	Spray						
2360	First Coat						
2380	Minimum	SF	0.004	0.13	0.30		0.43
2400	Average	"	0.005	0.14	0.34		0.48
2420	Maximum	"	0.006	0.15	0.38		0.53
2440	Second Coat						
2460	Minimum	SF	0.003	0.19	0.22		0.41
2480	Average	"	0.004	0.21	0.24		0.45
2500	Maximum	"	0.004	0.22	0.27		0.49
2520	Doors, Wood						
2540	Brush						
2560	First Coat						
2580	Minimum	SF	0.011	0.19	0.77		0.96
2600	Average	"	0.015	0.21	0.98		1.19
2620	Maximum	"	0.018	0.22	1.20		1.42
2640	Second Coat						
2660	Minimum	SF	0.009	0.14	0.60		0.74
2680	Average	"	0.010	0.15	0.68		0.83
2700	Maximum	"	0.011	0.17	0.77		0.94
2720	Spray						
2740	First Coat						
2760	Minimum	SF	0.002	0.14	0.16		0.30
2780	Average	"	0.003	0.15	0.19		0.34
2800	Maximum	"	0.004	0.17	0.24		0.41
2820	Second Coat						

© 2020 BNi Publications Inc. ALL RIGHTS RESERVED

PAINT

	Description	Output		Unit Costs			
ID Code	Component Descriptions	Unit of Meas.	Manhr / Unit	Material Cost	Labor Cost	Equipment Cost	Total Cost
09 - 91233	**INT. PAINTING, BUILDINGS, Cont'd...**						**09 - 91233**
2840	Minimum	SF	0.002	0.14	0.12		0.26
2860	Average	"	0.002	0.15	0.14		0.29
2880	Maximum	"	0.003	0.17	0.17		0.34
2900	Ductwork						
2920	Brush						
2940	Minimum	LF	0.010	0.15	0.68		0.83
2960	Average	"	0.011	0.17	0.77		0.94
2980	Maximum	"	0.013	0.18	0.90		1.08
3000	Roller						
3020	Minimum	SF	0.007	0.15	0.45		0.60
3040	Average	"	0.007	0.17	0.49		0.66
3060	Maximum	"	0.008	0.18	0.54		0.72
3080	Spray						
3100	Minimum	SF	0.003	0.15	0.19		0.34
3120	Average	"	0.003	0.17	0.20		0.37
3140	Maximum	"	0.003	0.18	0.22		0.40
3160	Floors						
3180	Roller						
3200	First Coat						
3220	Minimum	SF	0.003	0.15	0.17		0.32
3240	Average	"	0.003	0.17	0.20		0.37
3260	Maximum	"	0.003	0.18	0.22		0.40
3280	Second Coat						
3300	Minimum	SF	0.002	0.15	0.13		0.28
3320	Average	"	0.002	0.17	0.15		0.32
3340	Maximum	"	0.002	0.18	0.16		0.34
3360	Spray						
3380	First Coat						
3400	Minimum	SF	0.002	0.14	0.12		0.26
3420	Average	"	0.002	0.15	0.12		0.27
3440	Maximum	"	0.002	0.17	0.14		0.31
3460	Second Coat						
3480	Minimum	SF	0.002	0.14	0.10		0.24
3500	Average	"	0.002	0.15	0.11		0.26
3520	Maximum	"	0.002	0.17	0.12		0.29
3540	Pipes to 6" diameter						
3560	Brush						
3580	Minimum	LF	0.010	0.19	0.68		0.87

© 2020 BNi Publications Inc. ALL RIGHTS RESERVED

PAINT

Description		Output		Unit Costs			
ID Code	Component Descriptions	Unit of Meas.	Manhr / Unit	Material Cost	Labor Cost	Equipment Cost	Total Cost
09 - 91233	**INT. PAINTING, BUILDINGS, Cont'd...**						**09 - 91233**
3600	Average	LF	0.011	0.21	0.77		0.98
3620	Maximum	"	0.013	0.22	0.90		1.12
3640	Spray						
3660	Minimum	LF	0.003	0.15	0.22		0.37
3680	Average	"	0.004	0.17	0.27		0.44
3700	Maximum	"	0.005	0.18	0.36		0.54
3720	Pipes to 12" diameter						
3740	Brush						
3760	Minimum	LF	0.020	0.35	1.36		1.71
3780	Average	"	0.023	0.36	1.55		1.91
3800	Maximum	"	0.027	0.38	1.81		2.19
3820	Spray						
3840	Minimum	LF	0.007	0.31	0.45		0.76
3860	Average	"	0.008	0.33	0.54		0.87
3880	Maximum	"	0.010	0.34	0.68		1.02
3900	Trim						
3920	Brush						
3940	First Coat						
3960	Minimum	LF	0.003	0.19	0.21		0.40
3980	Average	"	0.004	0.21	0.24		0.45
4000	Maximum	"	0.004	0.22	0.30		0.52
4020	Second Coat						
4040	Minimum	LF	0.002	0.19	0.16		0.35
4060	Average	"	0.003	0.21	0.20		0.41
4080	Maximum	"	0.004	0.22	0.30		0.52
4100	Walls						
4120	Roller						
4140	First Coat						
4160	Minimum	SF	0.003	0.15	0.19		0.34
4180	Average	"	0.003	0.17	0.20		0.37
4200	Maximum	"	0.003	0.18	0.22		0.40
4220	Second Coat						
4240	Minimum	SF	0.003	0.15	0.17		0.32
4260	Average	"	0.003	0.17	0.18		0.35
4280	Maximum	"	0.003	0.18	0.20		0.38
4300	Spray						
4320	First Coat						
4340	Minimum	SF	0.001	0.13	0.08		0.21

© 2020 BNi Publications Inc. ALL RIGHTS RESERVED

PAINT

ID Code	Component Descriptions	Unit of Meas.	Manhr / Unit	Material Cost	Labor Cost	Equipment Cost	Total Cost
09 - 91233	**INT. PAINTING, BUILDINGS, Cont'd...**						**09 - 91233**
4360	Average	SF	0.002	0.14	0.10		0.24
4380	Maximum	"	0.002	0.15	0.13		0.28
4400	Second Coat						
4420	Minimum	SF	0.001	0.13	0.07		0.20
4440	Average	"	0.001	0.14	0.09		0.23
4460	Maximum	"	0.002	0.15	0.12		0.27

© 2020 BNi Publications Inc. ALL RIGHTS RESERVED

DIVISION 10
SPECIALTIES

VISUAL DISPLAY UNITS

ID Code	Description: Component Descriptions	Output: Unit of Meas.	Output: Manhr / Unit	Unit Costs: Material Cost	Unit Costs: Labor Cost	Unit Costs: Equipment Cost	Unit Costs: Total Cost
10 - 11130	**CHALKBOARDS**						**10 - 11130**
1020	Chalkboard, metal frame, 1/4" thick						
1040	48"x60"	EA	0.800	490	65.00		550
1060	48"x96"	"	0.889	670	72.00		740
1080	48"x144"	"	1.000	890	82.00		970
1100	48"x192"	"	1.143	1,210	93.00		1,300
1110	Liquid chalkboard						
1120	48"x60"	EA	0.800	650	65.00		720
1140	48"x96"	"	0.889	830	72.00		900
1160	48"x144"	"	1.000	1,240	82.00		1,320
1180	48"x192"	"	1.143	1,420	93.00		1,510
1200	Map rail, deluxe	LF	0.040	8.11	3.26		11.25

DIRECTORIES

ID Code	Component Descriptions	Unit of Meas.	Manhr / Unit	Material Cost	Labor Cost	Equipment Cost	Total Cost
10 - 13001	**IDENTIFYING DEVICES**						**10 - 13001**
1000	Directory and bulletin boards						
1020	Open face boards						
1040	Chrome plated steel frame	SF	0.400	38.25	32.50		71.00
1060	Aluminum framed	"	0.400	66.00	32.50		99.00
1080	Bronze framed	"	0.400	86.00	32.50		120
1100	Stainless steel framed	"	0.400	120	32.50		150
1140	Tack board, aluminum framed	"	0.400	27.00	32.50		60.00
1160	Visual aid board, aluminum framed	"	0.400	27.00	32.50		60.00
1200	Glass encased boards, hinged and keyed						
1210	Aluminum framed	SF	1.000	150	82.00		230
1220	Bronze framed	"	1.000	170	82.00		250
1230	Stainless steel framed	"	1.000	210	82.00		290
1240	Chrome plated steel framed	"	1.000	230	82.00		310
10 - 14004	**SIGNAGE**						**10 - 14004**
2020	Metal plaque						
2040	Cast bronze	SF	0.667	660	54.00		710
2060	Aluminum	"	0.667	380	54.00		430
2080	Metal engraved plaque						
2100	Porcelain steel	SF	0.667	790	54.00		840
2120	Stainless steel	"	0.667	630	54.00		680
2140	Brass	"	0.667	940	54.00		990
2160	Aluminum	"	0.667	580	54.00		630
2180	Metal built-up plaque						

© 2020 BNi Publications Inc. ALL RIGHTS RESERVED

DIRECTORIES

ID Code	Description: Component Descriptions	Output: Unit of Meas.	Output: Manhr / Unit	Unit Costs: Material Cost	Unit Costs: Labor Cost	Unit Costs: Equipment Cost	Unit Costs: Total Cost
10 - 14004	**SIGNAGE, Cont'd...**						**10 - 14004**
2220	Bronze	SF	0.800	710	65.00		780
2240	Copper and bronze	"	0.800	630	65.00		700
2260	Copper and aluminum	"	0.800	700	65.00		770
2280	Metal nameplate plaques						
2300	Cast bronze	SF	0.500	710	40.75		750
2320	Cast aluminum	"	0.500	520	40.75		560
2330	Engraved, 1-1/2" x 6"						
2340	Bronze	EA	0.500	300	40.75		340
2360	Aluminum	"	0.500	230	40.75		270
2400	Letters, on masonry, aluminum, satin finish						
2440	1/2" thick						
2460	2" high	EA	0.320	27.50	26.00		54.00
2480	4" high	"	0.400	41.25	32.50		74.00
2500	6" high	"	0.444	55.00	36.25		91.00
2510	3/4" thick						
2520	8" high	EA	0.500	82.00	40.75		120
2540	10" high	"	0.571	95.00	46.50		140
2550	1" thick						
2560	12" high	EA	0.667	110	54.00		160
2580	14" high	"	0.800	120	65.00		190
2600	16" high	"	1.000	150	82.00		230
2610	For polished aluminum add, 15%						
2620	For clear anodized aluminum add, 15%						
2630	For colored anodic aluminum add, 30%						
2640	For profiled and color enameled letters add, 50%						
2650	Cast bronze, satin finish letters						
2680	3/8" thick						
2720	2" high	EA	0.320	33.25	26.00		59.00
2740	4" high	"	0.400	50.00	32.50		83.00
2760	1/2" thick, 6" high	"	0.444	68.00	36.25		100
2780	5/8" thick, 8" high	"	0.500	100	40.75		140
2790	1" thick						
2800	10" high	EA	0.571	120	46.50		170
2820	12" high	"	0.667	160	54.00		210
2840	14" high	"	0.800	190	65.00		250
2860	16" high	"	1.000	280	82.00		360
2880	Interior door signs, adhesive, flexible						
3060	2" x 8"	EA	0.200	25.75	12.75		38.50

© 2020 BNi Publications Inc. ALL RIGHTS RESERVED

DIRECTORIES

ID Code	Description: Component Descriptions	Output: Unit of Meas.	Output: Manhr / Unit	Unit Costs: Material Cost	Unit Costs: Labor Cost	Unit Costs: Equipment Cost	Unit Costs: Total Cost
10 - 14004	**SIGNAGE, Cont'd...**						**10 - 14004**
3080	4" x 4"	EA	0.200	27.00	12.75		39.75
3100	6" x 7"	"	0.200	34.25	12.75		47.00
3120	6" x 9"	"	0.200	43.75	12.75		57.00
3140	10" x 9"	"	0.200	57.00	12.75		70.00
3160	10" x 12"	"	0.200	74.00	12.75		87.00
3180	Hard plastic type, no frame						
3220	3" x 8"	EA	0.200	57.00	12.75		70.00
3240	4" x 4"	"	0.200	57.00	12.75		70.00
3260	4" x 12"	"	0.200	62.00	12.75		75.00
3280	Hard plastic type, with frame						
3300	3" x 8"	EA	0.200	170	12.75		180
3320	4" x 4"	"	0.200	140	12.75		150
3340	4" x 12"	"	0.200	210	12.75		220
10 - 14530	**SIGNAGE**						**10 - 14530**
0100	Traffic signs						
1000	Reflectorized per OSHA stds., incl. post						
1030	Stop, 24"x24"	EA	0.533	89.00	34.00		120
1050	Yield, 30" triangle	"	0.533	48.50	34.00		83.00
1070	Speed limit, 12"x18"	"	0.533	55.00	34.00		89.00
1090	Directional, 12"x18"	"	0.533	67.00	34.00		100
1100	Exit, 12"x18"	"	0.533	67.00	34.00		100
1120	Entry, 12"x18"	"	0.533	67.00	34.00		100
1140	Warning, 24"x24"	"	0.533	88.00	34.00		120
1160	Informational, 12"x18"	"	0.533	44.75	34.00		79.00
1180	Handicap parking, 12"x18"	"	0.533	46.00	34.00		80.00

TELEPHONE SPECIALTIES

ID Code	Component Descriptions	Unit of Meas.	Manhr / Unit	Material Cost	Labor Cost	Equipment Cost	Total Cost
10 - 17001	**TELEPHONE ENCLOSURES**						**10 - 17001**
1000	Enclosure, wall mounted, shelf, 28" x 30" x 15"	EA	2.000	2,030	160		2,190
1800	Directory shelf, stainless steel, 3 binders	"	1.333	1,740	110		1,850

© 2020 BNi Publications Inc. ALL RIGHTS RESERVED

COMPARTMENTS AND CUBICLES

ID Code	Description Component Descriptions	Output Unit of Meas.	Manhr / Unit	Unit Costs Material Cost	Labor Cost	Equipment Cost	Total Cost
10 - 21130	**TOILET PARTITIONS**						**10 - 21130**
0100	Toilet partition, plastic laminate						
0120	Ceiling mounted	EA	2.667	1,200	220		1,420
0140	Floor mounted	"	2.000	790	160		950
0150	Metal						
0160	Ceiling mounted	EA	2.667	820	220		1,040
0180	Floor mounted	"	2.000	780	160		940
0190	Wheelchair partition, plastic laminate						
0200	Ceiling mounted	EA	2.667	1,800	220		2,020
0210	Floor mounted	"	2.000	1,580	160		1,740
0220	Painted metal						
0240	Ceiling mounted	EA	2.667	1,280	220		1,500
0260	Floor mounted	"	2.000	1,170	160		1,330
0280	Urinal screen, plastic laminate						
2000	Wall hung	EA	1.000	550	82.00		630
2100	Floor mounted	"	1.000	500	82.00		580
2120	Porcelain enameled steel, floor mounted	"	1.000	640	82.00		720
2140	Painted metal, floor mounted	"	1.000	420	82.00		500
2160	Stainless steel, floor mounted	"	1.000	810	82.00		890
2180	Metal toilet partitions						
2200	Front door and side divider, floor mounted						
5040	Porcelain enameled steel	EA	2.000	1,320	160		1,480
5060	Painted steel	"	2.000	780	160		940
5080	Stainless steel	"	2.000	1,930	160		2,090
10 - 21160	**SHOWER STALLS**						**10 - 21160**
1000	Shower receptors						
1010	Precast, terrazzo						
1020	32" x 32"	EA	0.667	700	60.00		760
1040	32" x 48"	"	0.800	740	72.00		810
1050	Concrete						
1060	32" x 32"	EA	0.667	290	60.00		350
1080	48" x 48"	"	0.889	320	79.00		400
1100	Shower door, trim and hardware						
1120	Economy, 24" wide, chrome, tempered glass	EA	0.800	310	72.00		380
1130	Porcelain enameled steel, flush	"	0.800	570	72.00		640
1140	Baked enameled steel, flush	"	0.800	340	72.00		410
1150	Aluminum, tempered glass, 48" wide, sliding	"	1.000	700	89.00		790
1161	Folding	"	1.000	680	89.00		770

© 2020 BNi Publications Inc. ALL RIGHTS RESERVED

COMPARTMENTS AND CUBICLES

ID Code	Description: Component Descriptions	Output: Unit of Meas.	Output: Manhr / Unit	Unit Costs: Material Cost	Unit Costs: Labor Cost	Unit Costs: Equipment Cost	Unit Costs: Total Cost
10 - 21160	**SHOWER STALLS, Cont'd...**						**10 - 21160**
1190	Aluminum and tempered glass, molded plastic						
1200	Complete with receptor and door						
1220	32" x 32"	EA	2.000	860	180		1,040
1230	36" x 36"	"	2.000	970	180		1,150
1240	40" x 40"	"	2.286	1,000	200		1,200
5400	Shower compartment, precast concrete receptor						
5420	Single entry type						
5440	Porcelain enameled steel	EA	8.000	2,420	720		3,140
5460	Baked enameled steel	"	8.000	2,320	720		3,040
5480	Stainless steel	"	8.000	2,230	720		2,950
5500	Double entry type						
5520	Porcelain enameled steel	EA	10.000	4,320	890		5,210
5540	Baked enameled steel	"	10.000	2,950	890		3,840
5560	Stainless steel	"	10.000	4,770	890		5,660
10 - 21230	**CUBICLES**						**10 - 21230**
1020	Hospital track						
1040	Ceiling hung	LF	0.089	7.73	7.24		15.00
1060	Suspended	"	0.114	8.38	9.32		17.75
1080	Hospital metal dividers, galvanized steel						
1100	Baked enamel finish						
1110	54" high						
1120	10" glass light	LF	0.400	130	32.50		160
1140	14" glass light	"	0.400	130	32.50		160
1160	24" glass light	"	0.400	120	32.50		150
1170	60" high						
1180	10" glass light	LF	0.444	140	36.25		180
1190	14" glass light	"	0.444	140	36.25		180
1200	24" glass light	"	0.444	130	36.25		170
1300	Stainless steel						
1310	54" high						
1320	10" glass light	LF	0.444	220	36.25		260
1340	14" glass light	"	0.444	240	36.25		280
1360	24" glass light	"	0.444	290	36.25		330
1370	60" high						
1380	10" glass light	LF	0.500	250	40.75		290
1400	14" glass light	"	0.500	250	40.75		290
1420	24" glass light	"	0.500	310	40.75		350

© 2020 BNi Publications Inc. ALL RIGHTS RESERVED

PARTITIONS

ID Code	Description: Component Descriptions	Output: Unit of Meas.	Output: Manhr / Unit	Unit Costs: Material Cost	Unit Costs: Labor Cost	Unit Costs: Equipment Cost	Unit Costs: Total Cost
10 - 22190	**MOVABLE PARTITIONS**						**10 - 22190**
0100	Partition, movable, 2-1/2" thick, vinyl-gypsum	SF	0.040	24.50	3.26		27.75
0120	Enameled steel frame, with 1/4" thick clear glass	"	0.040	30.00	3.26		33.25
0160	Door frame and hardware for movable partitions	EA	2.667	830	220		1,050
0200	Add for acoustic movable partition	SF					1.82
2900	Accordion partition, 12' high						
3000	Vinyl	SF	0.133	13.25	10.75		24.00
3020	Acoustical	"	0.133	15.75	10.75		26.50
3080	Standard office cubicles, 8' high, steel framed						
3200	Baked enamel finish						
3540	100% flush	LF	0.200	200	16.25		220
3550	75% flush and 25% glass	"	0.222	250	18.00		270
3560	50% flush and 50% glass	"	0.222	350	18.00		370
3570	100% glass	"	0.267	410	21.75		430
3600	Natural hardwood panels						
3620	100% flush	LF	0.267	220	21.75		240
3640	50% flush and 50% glass	"	0.286	290	23.25		310
3650	Plastic laminated panels						
3660	100% flush	LF	0.267	250	21.75		270
3670	75% flush and 25% glass	"	0.286	330	23.25		350
3680	50% flush and 50% glass	"	0.286	360	23.25		380
3700	Vinyl covered panels						
3710	100% flush	LF	0.276	220	22.50		240
3720	75% flush and 25% glass	"	0.296	290	24.25		310
3730	50% and 50% glass	"	0.296	350	24.25		370
3800	Aluminum framed						
3810	Enameled or anodized aluminum panels						
3830	100% flush	LF	0.200	180	16.25		200
3840	75% flush and 25% glass	"	0.222	170	18.00		190
3850	50% flush and 50% glass	"	0.222	150	18.00		170
3860	Vinyl covered panels						
3870	100% flush	LF	0.276	180	22.50		200
3880	75% flush and 25% glass	"	0.296	200	24.25		220
3890	50% flush and 50% glass	"	0.296	230	24.25		250
4000	60" high partitions, steel framed						
4040	Enameled panels	LF	0.178	89.00	14.50		100
4050	Natural hardwood panels, two sides	"	0.186	350	15.25		370
4060	Plastic laminated panels	"	0.186	190	15.25		210
4070	Vinyl covered panels	"	0.178	240	14.50		250

© 2020 BNi Publications Inc. ALL RIGHTS RESERVED

PARTITIONS

ID Code	Description: Component Descriptions	Output: Unit of Meas.	Output: Manhr / Unit	Unit Costs: Material Cost	Unit Costs: Labor Cost	Unit Costs: Equipment Cost	Unit Costs: Total Cost
10 - 22190	**MOVABLE PARTITIONS, Cont'd...**						**10 - 22190**
4100	Aluminum framed						
4120	Anodized or baked enamel panels	LF	0.178	130	14.50		140
4130	Natural hardwood panels	"	0.186	380	15.25		400
4140	Plastic laminated panels	"	0.186	190	15.25		210
4150	Vinyl covered panels	"	0.178	240	14.50		250
4200	Wire mesh partitions						
5000	Wall panels						
5030	4' x 7'	EA	0.500	180	40.75		220
5040	4' x 8'	"	0.533	200	43.50		240
5050	4' x 10'	"	0.615	220	50.00		270
5060	Wall filler panels						
5070	1' x 7'	EA	0.500	98.00	40.75		140
5080	1' x 8'	"	0.533	110	43.50		150
5090	1' x 10'	"	0.571	130	46.50		180
5100	2' x 7'	"	0.500	110	40.75		150
5120	2' x 8'	"	0.533	120	43.50		160
5130	2' x 10'	"	0.571	140	46.50		190
5140	3' x 7'	"	0.500	140	40.75		180
5150	3' x 8'	"	0.533	190	43.50		230
5160	3' x 10'	"	0.571	210	46.50		260
5200	Ceiling panels						
5210	10' x 2'	EA	1.143	140	93.00		230
5220	10' x 4'	"	1.600	210	130		340
5400	Wall panel with service window						
5410	5' wide						
5420	7' high	EA	0.500	570	40.75		610
5430	8' high	"	0.533	590	43.50		630
5440	10' high	"	0.571	570	46.50		620
5500	Doors						
5510	Sliding						
5520	3' x 7'	EA	2.000	380	160		540
5540	3' x 8'	"	2.286	510	190		700
5560	3' x 10'	"	3.200	540	260		800
5580	4' x 7'	"	2.286	480	190		670
5590	4' x 8'	"	3.200	540	260		800
5600	4' x 10'	"	4.000	570	330		900
5620	5' x 7'	"	3.200	530	260		790
5640	5' x 8'	"	4.000	570	330		900

© 2020 BNi Publications Inc. ALL RIGHTS RESERVED

PARTITIONS

	Description	Output		Unit Costs			
ID Code	Component Descriptions	Unit of Meas.	Manhr / Unit	Material Cost	Labor Cost	Equipment Cost	Total Cost
10 - 22190	**MOVABLE PARTITIONS, Cont'd...**						**10 - 22190**
5660	5' x 10'	EA	4.000	650	330		980
5900	Swing door						
6000	3' x 7'	EA	2.000	320	160		480
6020	4' x 7'	"	2.286	350	190		540
6030	Swing door, with 1' transom						
6040	3' x 7'	EA	2.286	470	190		660
6050	4' x 7'	"	3.200	520	260		780
6060	Swing door, with 3' transom						
6070	3' x 7'	EA	3.200	570	260		830
6080	4' x 7'	"	4.000	610	330		940

TOILET, BATH AND LAUNDRY ACCESSORIES

10 - 28160	**BATH ACCESSORIES**						**10 - 28160**
1040	Ash receiver, wall mounted, aluminum	EA	0.400	170	32.50		200
1050	Grab bar, 1-1/2" dia., stainless steel, wall mounted						
1060	24" long	EA	0.400	57.00	32.50		90.00
1080	36" long	"	0.421	64.00	34.25		98.00
1090	42" long	"	0.444	71.00	36.25		110
1100	48" long	"	0.471	78.00	38.25		120
1120	52" long	"	0.500	85.00	40.75		130
1130	1" dia., stainless steel						
1140	12" long	EA	0.348	35.50	28.25		64.00
1160	18" long	"	0.364	42.50	29.75		72.00
1180	24" long	"	0.400	48.25	32.50		81.00
1200	30" long	"	0.421	57.00	34.25		91.00
1220	36" long	"	0.444	64.00	36.25		100
1240	48" long	"	0.471	71.00	38.25		110
1300	Hand dryer, surface mounted, 110 volt	"	1.000	810	82.00		890
1320	Medicine cabinet, 16 x 22, baked enamel, lighted	"	0.320	160	26.00		190
1340	With mirror, lighted	"	0.533	230	43.50		270
1420	Mirror, 1/4" plate glass, up to 10 sf	SF	0.080	12.50	6.52		19.00
1430	Mirror, stainless steel frame						
1440	18"x24"	EA	0.267	95.00	21.75		120
1460	18"x32"	"	0.320	110	26.00		140
1480	18"x36"	"	0.400	110	32.50		140
1500	24"x30"	"	0.400	120	32.50		150
1510	24"x36"	"	0.444	120	36.25		160

© 2020 BNi Publications Inc. ALL RIGHTS RESERVED

TOILET, BATH AND LAUNDRY ACCESSORIES

ID Code	Description: Component Descriptions	Output: Unit of Meas.	Output: Manhr / Unit	Unit Costs: Material Cost	Unit Costs: Labor Cost	Unit Costs: Equipment Cost	Unit Costs: Total Cost
10 - 28160	**BATH ACCESSORIES, Cont'd...**						**10 - 28160**
1520	24"x48"	EA	0.667	170	54.00		220
1530	24"x60"	"	0.800	430	65.00		500
1560	30"x30"	"	0.800	380	65.00		450
1580	30"x72"	"	1.000	680	82.00		760
1600	48"x72"	"	1.333	740	110		850
1640	With shelf, 18"x24"	"	0.320	290	26.00		320
1820	Sanitary napkin dispenser, stainless steel	"	0.533	700	43.50		740
1830	Shower rod, 1" diameter						
1840	Chrome finish over brass	EA	0.400	250	32.50		280
1860	Stainless steel	"	0.400	180	32.50		210
1900	Soap dish, stainless steel, wall mounted	"	0.533	160	43.50		200
1910	Toilet tissue dispenser, stainless, wall mounted						
1920	Single roll	EA	0.200	79.00	16.25		95.00
1940	Double roll	"	0.229	150	18.75		170
1945	Towel dispenser, stainless steel						
1950	Flush mounted	EA	0.444	300	36.25		340
1960	Surface mounted	"	0.400	430	32.50		460
1970	Combination towel and waste receptacle	"	0.533	660	43.50		700
2000	Towel bar, stainless steel						
2020	18" long	EA	0.320	98.00	26.00		120
2040	24" long	"	0.364	130	29.75		160
2060	30" long	"	0.400	140	32.50		170
2070	36" long	"	0.444	150	36.25		190
2080	Toothbrush and tumbler holder	"	0.267	62.00	21.75		84.00
2100	Waste receptacle, stainless steel, wall mounted	"	0.667	500	54.00		550

FIRE PROTECTION SPECIALTIES

ID Code	Component Descriptions	Unit of Meas.	Manhr / Unit	Material Cost	Labor Cost	Equipment Cost	Total Cost
10 - 44001	**FIRE PROTECTION**						**10 - 44001**
1000	Portable fire extinguishers						
1020	Water pump tank type						
1030	2.5 gal.						
1040	Red enameled galvanized	EA	0.533	150	34.00		180
1060	Red enameled copper	"	0.533	220	34.00		250
1080	Polished copper	"	0.533	290	34.00		320
1200	Carbon dioxide type, red enamel steel						
1210	Squeeze grip with hose and horn						
1220	2.5 lb	EA	0.533	230	34.00		260

© 2020 BNi Publications Inc. ALL RIGHTS RESERVED

FIRE PROTECTION SPECIALTIES

ID Code	Description: Component Descriptions	Output: Unit of Meas.	Output: Manhr / Unit	Unit Costs: Material Cost	Unit Costs: Labor Cost	Unit Costs: Equipment Cost	Unit Costs: Total Cost
10 - 44001	**FIRE PROTECTION, Cont'd...**						**10 - 44001**
1240	5 lb	EA	0.615	330	39.25		370
1260	10 lb	"	0.800	340	51.00		390
1280	15 lb	"	1.000	380	64.00		440
1300	20 lb	"	1.000	460	64.00		520
1310	Wheeled type						
1320	125 lb	EA	1.600	4,120	100		4,220
1340	250 lb	"	1.600	5,210	100		5,310
1360	500 lb	"	1.600	6,730	100		6,830
1400	Dry chemical, pressurized type						
1405	Red enameled steel						
1410	2.5 lb	EA	0.533	72.00	34.00		110
1430	5 lb	"	0.615	98.00	39.25		140
1440	10 lb	"	0.800	200	51.00		250
1450	20 lb	"	1.000	260	64.00		320
1460	30 lb	"	1.000	330	64.00		390
1480	Chrome plated steel, 2.5 lb	"	0.533	310	34.00		340
1500	Other type extinguishers						
1510	2.5 gal, stainless steel, pressurized water tanks	EA	0.533	230	34.00		260
1520	Soda and acid type	"	0.533	190	34.00		220
1530	Cartridge operated, water type	"	0.533	170	34.00		200
1540	Loaded stream, water type	"	0.533	210	34.00		240
1550	Foam type	"	0.533	270	34.00		300
1560	40 gal, wheeled foam type	"	1.600	6,160	100		6,260
1600	Fire extinguisher cabinets						
1605	Enameled steel						
1610	8" x 12" x 27"	EA	1.600	160	100		260
1620	8" x 16" x 38"	"	1.600	190	100		290
1625	Aluminum						
1630	8" x 12" x 27"	EA	1.600	240	100		340
1640	8" x 16" x 38"	"	1.600	290	100		390
1655	Stainless steel						
1660	8" x 16" x 38"	EA	1.600	270	100		370

© 2020 BNi Publications Inc. ALL RIGHTS RESERVED

LOCKERS

	Description	Output		Unit Costs			
ID Code	Component Descriptions	Unit of Meas.	Manhr / Unit	Material Cost	Labor Cost	Equipment Cost	Total Cost
10 - 51001	**LOCKERS**						**10 - 51001**
0080	Locker bench, floor mounted, laminated maple						
0100	4'	EA	0.667	390	54.00		440
0120	6'	"	0.667	550	54.00		600
0130	Wardrobe locker, 12" x 60" x 15", baked on enamel						
0140	1-tier	EA	0.400	420	32.50		450
0160	2-tier	"	0.400	450	32.50		480
0180	3-tier	"	0.421	500	34.25		530
0200	4-tier	"	0.421	540	34.25		570
0240	12" x 72" x 15", baked on enamel						
0260	1-tier	EA	0.400	360	32.50		390
0280	2-tier	"	0.400	430	32.50		460
0300	4-tier	"	0.421	530	34.25		560
0320	5-tier	"	0.421	530	34.25		560
1200	15" x 60" x 15", baked on enamel						
1220	1-tier	EA	0.400	470	32.50		500
1240	4-tier	"	0.421	510	34.25		540
2040	Wardrobe locker, single tier type						
2060	12" x 15" x 72"	EA	0.800	320	65.00		390
2080	18" x 15" x 72"	"	0.842	390	69.00		460
2100	12" x 18" x 72"	"	0.889	350	72.00		420
2120	18" x 18" x 72"	"	0.941	450	77.00		530
2140	Double tier type						
2160	12" x 15" x 36"	EA	0.400	250	32.50		280
2180	18" x 15" x 36"	"	0.400	240	32.50		270
2200	12" x 18" x 36"	"	0.400	260	32.50		290
2220	18" x 18" x 36"	"	0.400	280	32.50		310
2240	Two person unit						
2260	18" x 15" x 72"	EA	1.333	670	110		780
2280	18" x 18" x 72"	"	1.600	750	130		880
2300	Duplex unit						
2320	15" x 15" x 72"	EA	0.800	680	65.00		750
2340	15" x 21" x 72"	"	0.800	720	65.00		780
2400	Basket lockers, basket sets with baskets						
2440	24 basket set	SET	4.000	1,680	330		2,010
2460	30 basket set	"	5.000	2,010	410		2,420
2480	36 basket set	"	6.667	2,280	540		2,820
2500	42 basket set	"	8.000	2,550	650		3,200

© 2020 BNi Publications Inc. ALL RIGHTS RESERVED

POSTAL SPECIALTIES

	Description	Output		Unit Costs			
ID Code	Component Descriptions	Unit of Meas.	Manhr / Unit	Material Cost	Labor Cost	Equipment Cost	Total Cost
10 - 55001	**POSTAL SPECIALTIES**						**10 - 55001**
1500	Mail chutes						
1520	Single mail chute						
1530	Finished aluminum	LF	2.000	880	160		1,040
1540	Bronze	"	2.000	1,220	160		1,380
1560	Single mail chute receiving box						
1580	Finished aluminum	EA	4.000	1,300	330		1,630
1600	Bronze	"	4.000	1,560	330		1,890
1620	Twin mail chute, double parallel						
1630	Finished aluminum	FLR	4.000	1,920	330		2,250
1640	Bronze	"	4.000	2,510	330		2,840
1660	Receiving box, 36" x 20" x 12"						
1680	Finished aluminum	EA	6.667	3,000	540		3,540
1690	Bronze	"	6.667	4,040	540		4,580
1700	Locked receiving mail box						
1720	Finished aluminum	EA	4.000	1,320	330		1,650
1730	Bronze	"	4.000	2,570	330		2,900
1750	Commercial postal accessories for mail chutes						
1760	Letter slot, brass	EA	1.333	120	110		230
1770	Bulk mail slot, brass	"	1.333	260	110		370
1780	Mail boxes						
1790	Residential postal accessories						
1800	Letter slot	EA	0.400	99.00	32.50		130
1810	Rural letter box	"	1.000	190	82.00		270
1820	Apartment house, keyed, 3.5" x 4.5" x 16"	"	0.267	180	21.75		200
1830	Ranch style	"	0.400	190	32.50		220
1840	Commercial postal accessories						
1860	Letter box, with combination lock	EA	0.286	140	23.25		160
1880	Key lock	"	0.286	150	23.25		170
1980	Mail box, aluminum w/glass front, 4x5						
2000	Horizontal rear load	EA	0.229	230	18.75		250
2020	Vertical front load	"	0.229	130	18.75		150

© 2020 BNi Publications Inc. ALL RIGHTS RESERVED

WARDROBE AND CLOSET SPECIALTIES

ID Code	Component Descriptions	Unit of Meas.	Manhr / Unit	Material Cost	Labor Cost	Equipment Cost	Total Cost
10 - 57001	**WARDROBE SPECIALTIES**						**10 - 57001**
1000	Hospital wardrobe units, 24" x 24" x 76", with door						
1020	Baked enameled steel	EA	4.444	3,250	360		3,610
1040	Hardwood	"	4.444	1,740	360		2,100
1060	Stainless steel	"	4.444	5,430	360		5,790
1080	Plastic laminated	"	4.444	1,960	360		2,320
2000	Dormitory wardrobe units, 24" x 76", with door						
2020	Hardwood	EA	4.444	1,810	360		2,170
2040	Plastic laminated	"	4.444	1,640	360		2,000
3000	Hat and coat rack						
3020	Single tier						
3040	Baked enameled steel	LF	0.200	210	16.25		230
3060	Stainless steel	"	0.200	230	16.25		250
3080	Aluminum	"	0.200	220	16.25		240
3100	Double tier						
3120	Baked enameled steel	LF	0.229	390	18.75		410
3140	Stainless steel	"	0.229	530	18.75		550
3160	Aluminum	"	0.229	450	18.75		470
10 - 57230	**SHELVING**						**10 - 57230**
0980	Shelving, enamel, closed side and back, 12" x 36"						
1000	5 shelves	EA	1.333	320	110		430
1020	8 shelves	"	1.778	350	140		490
1030	Open						
1040	5 shelves	EA	1.333	160	110		270
1060	8 shelves	"	1.778	180	140		320
2000	Metal storage shelving, baked enamel						
2030	7 shelf unit, 72" or 84" high						
2040	10" shelf	LF	0.800	46.00	65.00		110
2050	12" shelf	"	0.842	51.00	69.00		120
2060	15" shelf	"	0.889	88.00	72.00		160
2070	18" shelf	"	0.941	88.00	77.00		160
2080	24" shelf	"	1.000	110	82.00		190
2090	30" shelf	"	1.067	100	87.00		190
2100	36" shelf	"	1.143	120	93.00		210
2200	4 shelf unit, 40" high						
2230	10" shelf	LF	0.667	63.00	54.00		120
2240	12" shelf	"	0.727	68.00	59.00		130
2250	15" shelf	"	0.800	82.00	65.00		150

© 2020 BNi Publications Inc. ALL RIGHTS RESERVED

WARDROBE AND CLOSET SPECIALTIES

ID Code	Description: Component Descriptions	Output: Unit of Meas.	Output: Manhr / Unit	Unit Costs: Material Cost	Unit Costs: Labor Cost	Unit Costs: Equipment Cost	Unit Costs: Total Cost
10 - 57230	**SHELVING, Cont'd...**						**10 - 57230**
2260	18" shelf	LF	0.842	93.00	69.00		160
2270	24" shelf	"	0.889	140	72.00		210
2300	3 shelf unit, 32" high						
2320	10" shelf	LF	0.400	54.00	32.50		87.00
2340	12" shelf	"	0.421	54.00	34.25		88.00
2350	15" shelf	"	0.444	59.00	36.25		95.00
2360	18" shelf	"	0.471	62.00	38.25		100
2370	24" shelf	"	0.500	68.00	40.75		110
2400	Single shelf unit, attached to masonry						
2410	10" shelf	LF	0.133	19.00	10.75		29.75
2420	12" shelf	"	0.145	21.00	11.75		32.75
2430	15" shelf	"	0.154	21.75	12.50		34.25
2440	18" shelf	"	0.163	24.50	13.25		37.75
2450	24" shelf	"	0.174	29.75	14.25		44.00
2460	For stainless steel, add to material, 120%						
2470	For attachment to gypsum board, add to labor, 50%						
2500	Built-in wood shelves						
2520	Posts and trimmed plywood	LF	0.114	4.94	9.32		14.25
2540	Solid clear pine	"	0.123	7.79	10.00		17.75
2560	Closet shelf, pine with rod	"	0.123	5.28	10.00		15.25
2580	For lumber edge band, add to material	"					2.70
2590	For prefinished shelves, add to material, 225%						

FLAGPOLES

ID Code	Component Descriptions	Unit of Meas.	Manhr / Unit	Material Cost	Labor Cost	Equipment Cost	Total Cost
10 - 75001	**FLAGPOLES**						**10 - 75001**
2020	Installed in concrete base						
2030	Fiberglass						
2040	25' high	EA	5.333	1,670	430		2,100
2080	50' high	"	13.333	4,410	1,090		5,500
2100	Aluminum						
2120	25' high	EA	5.333	1,620	430		2,050
2140	50' high	"	13.333	3,200	1,090		4,290
2160	Bonderized steel						
2180	25' high	EA	6.154	1,810	500		2,310
2200	50' high	"	16.000	3,620	1,300		4,920
2220	Freestanding tapered, fiberglass						
2240	30' high	EA	5.714	1,990	470		2,460

© 2020 BNi Publications Inc. ALL RIGHTS RESERVED

FLAGPOLES

ID Code	Description: Component Descriptions	Output: Unit of Meas.	Output: Manhr / Unit	Unit Costs: Material Cost	Unit Costs: Labor Cost	Unit Costs: Equipment Cost	Unit Costs: Total Cost
10 - 75001	**FLAGPOLES, Cont'd...**						**10 - 75001**
2260	40' high	EA	7.273	2,580	590		3,170
2280	50' high	"	8.000	6,580	650		7,230
2300	60' high	"	9.412	7,030	770		7,800
2400	Wall mounted, with collar, brushed aluminum finish						
2420	15' long	EA	4.000	1,540	330		1,870
2440	18' long	"	4.000	1,740	330		2,070
2460	20' long	"	4.211	1,900	340		2,240
2480	24' long	"	4.706	2,040	380		2,420
2500	Outrigger, wall, including base						
2520	10' long	EA	5.333	1,560	430		1,990
2540	20' long	"	6.667	2,060	540		2,600

PEST CONTROL DEVICES

ID Code	Description	Unit of Meas.	Manhr / Unit	Material Cost	Labor Cost	Equipment Cost	Total Cost
10 - 81001	**PEST CONTROL**						**10 - 81001**
1000	Termite control						
1010	Under slab spraying						
1020	Minimum	SF	0.002	1.24	0.12		1.36
1040	Average	"	0.004	1.24	0.25		1.49
1120	Maximum	"	0.008	1.76	0.51		2.27

© 2020 BNi Publications Inc. ALL RIGHTS RESERVED

DIVISION 11
EQUIPMENT

ARCHITECTURAL EQUIPMENT

ID Code	Description: Component Descriptions	Output: Unit of Meas.	Output: Manhr / Unit	Unit Costs: Material Cost	Unit Costs: Labor Cost	Unit Costs: Equipment Cost	Unit Costs: Total Cost
11 - 11001	**SPECIAL SYSTEMS**						**11 - 11001**
1980	Air compressor, air cooled, two stage						
2000	5.0 cfm, 175 psi	EA	16.000	2,840	1,430		4,270
2020	10 cfm, 175 psi	"	17.778	3,470	1,590		5,060
2030	20 cfm, 175 psi	"	19.048	4,780	1,700		6,480
2040	50 cfm, 125 psi	"	21.053	6,970	1,880		8,850
2050	80 cfm, 125 psi	"	22.857	9,980	2,040		12,020
2055	Single stage, 125 psi						
2060	1.0 cfm	EA	11.429	2,740	1,020		3,760
2080	1.5 cfm	"	11.429	2,790	1,020		3,810
2090	2.0 cfm	"	11.429	2,860	1,020		3,880
8000	Automotive, hose reel, air and water, 50' hose	"	6.667	1,280	600		1,880
8010	Lube equipment, 3 reel, with pumps	"	32.000	6,680	2,860		9,540
8015	Tire changer						
8020	Truck	EA	11.429	15,150	1,020		16,170
8030	Passenger car	"	6.154	3,580	550		4,130
8040	Air hose reel, includes 50' hose	"	6.154	950	550		1,500
8050	Hose reel, 5 reel, motor oil, gear oil, lube, air & water	"	32.000	8,840	2,860		11,700
8100	Water hose reel, 50' hose	"	6.154	950	550		1,500
8120	Pump, for motor or gear oil, fits 55 gal drum	"	0.800	1,240	72.00		1,310
8140	For chassis lube	"	0.800	2,020	72.00		2,090
8490	Fuel dispensing pump, lighted dial, one product						
8500	One hose	EA	6.667	4,410	600		5,010
8520	Two hose	"	6.667	7,780	600		8,380
8530	Two products, two hose	"	6.667	8,200	600		8,800

LOADING DOCK EQUIPMENT

ID Code	Component Descriptions	Unit of Meas.	Manhr / Unit	Material Cost	Labor Cost	Equipment Cost	Total Cost
11 - 13001	**LOADING DOCK EQUIPMENT**						**11 - 13001**
0080	Dock leveler, 10 ton capacity						
0100	6' x 8'	EA	8.000	5,860	650		6,510
0120	7' x 8'	"	8.000	6,730	650		7,380
0160	Bumpers, laminated rubber						
0165	4-1/2" thick						
0170	6" x 14"	EA	0.160	72.00	13.00		85.00
0175	6" x 36"	"	0.178	140	14.50		150
0180	10" x 14"	"	0.200	96.00	16.25		110
0200	10" x 24"	"	0.229	140	18.75		160
0220	10" x 36"	"	0.267	200	21.75		220

© 2020 BNi Publications Inc. ALL RIGHTS RESERVED

LOADING DOCK EQUIPMENT

ID Code	Description: Component Descriptions	Output: Unit of Meas.	Output: Manhr / Unit	Unit Costs: Material Cost	Unit Costs: Labor Cost	Unit Costs: Equipment Cost	Unit Costs: Total Cost
11 - 13001	LOADING DOCK EQUIPMENT, Cont'd...						11 - 13001
0240	12" x 14"	EA	0.211	120	17.25		140
0260	12" x 24"	"	0.250	170	20.50		190
0280	12" x 36"	"	0.296	250	24.25		270
0290	6" thick						
0300	10" x 14"	EA	0.229	110	18.75		130
0320	10" x 24"	"	0.276	120	22.50		140
0340	10" x 36"	"	0.400	260	32.50		290
0350	Extruded rubber bumpers						
0351	T-section, 22" x 22" x 3"	EA	0.160	170	13.00		180
0355	Molded rubber bumpers						
0356	24" x 12" x 3" thick	EA	0.400	95.00	32.50		130
0360	Door seal, 12" x 12", vinyl covered	LF	0.200	58.00	16.25		74.00
1000	Dock boards, heavy duty, 5' x 5'						
1010	5000 lb						
1020	Minimum	EA	6.667	1,380	540		1,920
1040	Maximum	"	6.667	1,520	540		2,060
1050	9000 lb						
1060	Minimum	EA	6.667	1,590	540		2,130
1070	Maximum	"	7.273	1,910	590		2,500
1080	15,000 lb	"	7.273	2,160	590		2,750
1200	Truck shelters						
1220	Minimum	EA	6.154	1,220	500		1,720
1240	Maximum	"	11.429	2,030	930		2,960

SECURITY CONTROL EQUIPMENT

ID Code	Component Descriptions	Unit of Meas.	Manhr / Unit	Material Cost	Labor Cost	Equipment Cost	Total Cost
11 - 15001	SECURITY EQUIPMENT						11 - 15001
1000	Bulletproof teller window						
1020	4' x 4'	EA	13.333	2,660	1,090		3,750
1040	5' x 4'	"	16.000	3,430	1,300		4,730
1045	Bulletproof partitions						
1050	Up to 12' high, 2.5" thick	SF	0.053	230	4.34		230
1060	Counter for banks						
1080	Minimum	LF	1.600	940	130		1,070
1100	Maximum	"	2.667	4,200	220		4,420
1280	Drive-up window						
1300	Minimum	EA	11.429	5,850	930		6,780
1310	Maximum	"	26.667	6,330	2,170		8,500

© 2020 BNi Publications Inc. ALL RIGHTS RESERVED

SECURITY CONTROL EQUIPMENT

	Description	Output		Unit Costs			
ID Code	Component Descriptions	Unit of Meas.	Manhr / Unit	Material Cost	Labor Cost	Equipment Cost	Total Cost
11 - 15001	**SECURITY EQUIPMENT, Cont'd...**						**11 - 15001**
1400	Night depository						
1420	Minimum	EA	11.429	10,960	930		11,890
1440	Maximum	"	26.667	15,580	2,170		17,750
1450	Office safes, 30" x 20" x 20", 1 hr rating	"	2.000	4,260	160		4,420
1460	30" x 16" x 15", 2 hr rating	"	1.600	2,180	130		2,310
1470	30" x 28" x 20", H&G rating	"	1.000	5,050	82.00		5,130
1600	Service windows, pass through painted steel						
1620	24" x 36"	EA	8.000	3,670	650		4,320
1640	48" x 40"	"	10.000	5,720	820		6,540
1660	72" x 40"	"	16.000	7,190	1,300		8,490
1670	Special doors and windows						
1680	3' x 7' bulletproof door with frame	EA	11.429	7,310	930		8,240
1690	12" x 12" vision panel	"	5.714	4,240	470		4,710
1700	Surveillance system						
1720	Minimum	EA	16.000	7,240	1,300		8,540
1740	Maximum	"	80.000	13,090	6,520		19,610
2040	Vault door, 3' wide, 6'6" high						
2060	3-1/2" thick	EA	100.000	4,710	8,150		12,860
2070	7" thick	"	133.333	7,550	10,870		18,420
2080	10" thick	"	160.000	9,430	13,050		22,480
2160	Insulated vault door						
2170	2 hr rating						
2180	32" wide	EA	8.000	4,710	650		5,360
2200	40" wide	"	8.421	5,240	690		5,930
2210	4 hr rating						
2220	32" wide	EA	8.889	5,210	720		5,930
2240	40" wide	"	10.000	6,070	820		6,890
2250	6 hr rating						
2260	32" wide	EA	8.889	6,000	720		6,720
2280	40" wide	"	10.000	7,010	820		7,830
3100	Insulated file room door						
3110	1 hr rating						
3120	32" wide	EA	8.000	4,640	650		5,290
3140	40" wide	"	8.889	5,160	720		5,880

© 2020 BNi Publications Inc. ALL RIGHTS RESERVED

VAULTS

ID Code	Description: Component Descriptions	Output: Unit of Meas.	Output: Manhr / Unit	Unit Costs: Material Cost	Unit Costs: Labor Cost	Unit Costs: Equipment Cost	Unit Costs: Total Cost
11 - 16001	**VAULTS**						**11 - 16001**
1000	Floor safes						
1005	Class C						
1010	1.0 cf	EA	0.667	930	54.00		980
1020	1.3 cf	"	1.000	1,030	82.00		1,110
1040	1.9 cf	"	1.333	1,350	110		1,460
1060	5.2 cf	"	1.333	2,760	110		2,870

RETAIL AND SERVICE EQUIPMENT

ID Code	Description: Component Descriptions	Output: Unit of Meas.	Output: Manhr / Unit	Unit Costs: Material Cost	Unit Costs: Labor Cost	Unit Costs: Equipment Cost	Unit Costs: Total Cost
11 - 21330	**CHECKROOM EQUIPMENT**						**11 - 21330**
1000	Motorized checkroom equipment						
1020	No shelf system, 6'4" height						
1040	7'6" length	EA	8.000	5,510	650		6,160
1060	14'6" length	"	8.000	5,580	650		6,230
1080	28' length	"	8.000	6,720	650		7,370
1100	One shelf, 6'8" height						
1120	7'6" length	EA	8.000	6,760	650		7,410
1140	14'6" length	"	8.000	6,940	650		7,590
1160	28' length	"	8.000	8,340	650		8,990
1180	Two shelves, 7'5" height						
1200	7'6" length	EA	8.000	8,300	650		8,950
1220	14'6" length	"	8.000	10,490	650		11,140
1240	28' length	"	8.000	10,700	650		11,350
1300	Three shelves, 8' height						
1320	7'6" length	EA	16.000	8,490	1,300		9,790
1340	14'6" length	"	16.000	10,690	1,300		11,990
1360	28' length	"	16.000	10,790	1,300		12,090
1400	Four shelves, 8'7" height						
1420	7'6" length	EA	16.000	8,580	1,300		9,880
1440	14'6" length	"	16.000	10,950	1,300		12,250
1460	28' length	"	16.000	11,200	1,300		12,500

© 2020 BNi Publications Inc. ALL RIGHTS RESERVED

LAUNDRY EQUIPMENT

ID Code	Component Descriptions	Unit of Meas.	Manhr / Unit	Material Cost	Labor Cost	Equipment Cost	Total Cost
11 - 23001	**LAUNDRY EQUIPMENT**						**11 - 23001**
1000	High capacity, heavy duty						
1020	Washer extractors						
1030	135 lb						
1040	Standard	EA	6.667	37,840	540		38,380
1060	Pass through	"	6.667	43,480	540		44,020
1070	200 lb						
1080	Standard	EA	6.667	46,470	540		47,010
1100	Pass through	"	6.667	56,430	540		56,970
1120	110 lb dryer	"	6.667	14,230	540		14,770
1140	Hand operated presser	"	8.889	10,210	720		10,930
1160	Mushroom press	"	8.889	6,380	720		7,100
1200	Spreader feeders						
1220	2 station	EA	8.889	72,660	720		73,380
1240	4 station	"	16.000	85,870	1,300		87,170
1300	Delivery carts						
1320	12 bushel	EA	0.100	370	8.15		380
1340	16 bushel	"	0.107	460	8.69		470
1350	18 bushel	"	0.114	570	9.32		580
1360	30 bushel	"	0.133	830	10.75		840
1370	40 bushel	"	0.160	1,000	13.00		1,010
1500	Low capacity						
1520	Pressers						
1530	Air operated	EA	3.200	7,930	260		8,190
1540	Hand operated	"	3.200	6,280	260		6,540
1560	Extractor, low capacity	"	3.200	5,720	260		5,980
1570	Ironer, 48"	"	1.600	4,320	130		4,450
1600	Coin washers						
1610	10 lb capacity	EA	1.600	2,100	130		2,230
1620	20 lb capacity	"	1.600	4,950	130		5,080
1630	Coin dryer	"	1.000	1,000	82.00		1,080
1680	Coin dry cleaner, 20 lb	"	3.200	4,320	260		4,580

© 2020 BNi Publications Inc. ALL RIGHTS RESERVED

MAINTENANCE EQUIPMENT

	Description	Output		Unit Costs			
ID Code	Component Descriptions	Unit of Meas.	Manhr / Unit	Material Cost	Labor Cost	Equipment Cost	Total Cost
11 - 24001	**MAINTENANCE EQUIPMENT**						**11 - 24001**
1000	Vacuum cleaning system						
1010	3 valves						
1020	1.5 hp	EA	8.889	1,010	720		1,730
1030	2.5 hp	"	11.429	1,220	930		2,150
1040	5 valves	"	16.000	1,900	1,300		3,200
1060	7 valves	"	20.000	2,530	1,630		4,160

FOOD SERVICE EQUIPMENT

11 - 26001	**FOOD SERVICE EQUIPMENT**						**11 - 26001**
1000	Unit kitchens						
1020	30" compact kitchen						
1040	Refrigerator, with range, sink	EA	4.000	1,620	330		1,950
1060	Sink only	"	2.667	2,070	220		2,290
1080	Range only	"	2.000	1,680	170		1,850
1100	Cabinet for upper wall section	"	1.143	420	95.00		520
1120	Stainless shield, for rear wall	"	0.320	170	26.50		200
1140	Side wall	"	0.320	120	26.50		150
1200	42" compact kitchen						
1220	Refrigerator with range, sink	EA	4.444	1,980	370		2,350
1240	Sink only	"	4.000	1,080	330		1,410
1260	Cabinet for upper wall section	"	1.333	840	110		950
1280	Stainless shield, for rear wall	"	0.333	670	27.75		700
1290	Side wall	"	0.333	190	27.75		220
1300	54" compact kitchen						
1310	Refrigerator, oven, range, sink	EA	5.714	2,650	470		3,120
1320	Cabinet for upper wall section	"	1.600	1,080	130		1,210
1330	Stainless shield, for						
1340	Rear wall	EA	0.364	670	30.25		700
1350	Side wall	"	0.364	190	30.25		220
1400	60" compact kitchen						
1420	Refrigerator, oven, range, sink	EA	5.714	3,600	470		4,070
1440	Cabinet for upper wall section	"	1.600	250	130		380
1450	Stainless shield, for						
1460	Rear wall	EA	0.364	800	30.25		830
1480	Side wall	"	0.364	200	30.25		230
1490	72" compact kitchen						
1500	Refrigerator, oven, range, sink	EA	6.667	3,810	550		4,360

© 2020 BNi Publications Inc. ALL RIGHTS RESERVED

FOOD SERVICE EQUIPMENT

	Description	Output		Unit Costs			
ID Code	Component Descriptions	Unit of Meas.	Manhr / Unit	Material Cost	Labor Cost	Equipment Cost	Total Cost
11 - 26001	**FOOD SERVICE EQUIPMENT, Cont'd...**						**11 - 26001**
1510	Cabinet for upper wall section	EA	1.600	260	130		390
1520	Stainless shield for						
1540	Rear wall	EA	0.400	870	33.25		900
1550	Side wall	"	0.400	200	33.25		230
1560	Bake oven						
1580	Single deck						
1620	Minimum	EA	1.000	3,830	83.00		3,910
1640	Maximum	"	2.000	7,330	170		7,500
1650	Double deck						
1660	Minimum	EA	1.333	6,830	110		6,940
1670	Maximum	"	2.000	21,340	170		21,510
1680	Triple deck						
1690	Minimum	EA	1.333	24,220	110		24,330
1700	Maximum	"	2.667	43,210	220		43,430
1710	Convection type oven, electric, 40" x 45" x 57"						
1720	Minimum	EA	1.000	3,780	83.00		3,860
1740	Maximum	"	2.000	6,650	170		6,820
1800	Broiler, without oven, 69" x 26" x 39"						
1820	Minimum	EA	1.000	6,020	83.00		6,100
1840	Maximum	"	1.333	9,470	110		9,580
1900	Coffee urns, 10 gallons						
1920	Minimum	EA	2.667	4,580	220		4,800
1940	Maximum	"	4.000	5,180	330		5,510
2000	Fryer, with submerger						
2010	Single						
2020	Minimum	EA	1.600	1,880	130		2,010
2040	Maximum	"	2.667	5,160	220		5,380
2050	Double						
2060	Minimum	EA	2.000	3,110	170		3,280
2080	Maximum	"	2.667	16,700	220		16,920
2100	Griddle, counter						
2110	3' long						
2120	Minimum	EA	1.333	2,570	110		2,680
2140	Maximum	"	1.600	5,300	130		5,430
2150	5' long						
2160	Minimum	EA	2.000	5,580	170		5,750
2180	Maximum	"	2.667	12,850	220		13,070
2200	Kettles, steam, jacketed						

© 2020 BNi Publications Inc. ALL RIGHTS RESERVED

FOOD SERVICE EQUIPMENT

ID Code	Description: Component Descriptions	Output: Unit of Meas.	Output: Manhr / Unit	Unit Costs: Material Cost	Unit Costs: Labor Cost	Unit Costs: Equipment Cost	Unit Costs: Total Cost
11 - 26001	**FOOD SERVICE EQUIPMENT, Cont'd...**						**11 - 26001**
2210	20 gallons						
2220	Minimum	EA	2.000	12,270	170		12,440
2240	Maximum	"	4.000	13,410	330		13,740
2250	40 gallons						
2260	Minimum	EA	2.000	18,820	170		18,990
2270	Maximum	"	4.000	27,180	330		27,510
2280	60 gallons						
2290	Minimum	EA	2.000	20,690	170		20,860
2300	Maximum	"	4.000	29,270	330		29,600
2310	Range						
2320	Heavy duty, single oven, open top						
2330	Minimum	EA	1.000	7,910	83.00		7,990
2340	Maximum	"	2.667	16,650	220		16,870
2350	Fry top						
2360	Minimum	EA	1.000	8,080	83.00		8,160
2380	Maximum	"	2.667	11,320	220		11,540
2390	Steamers, electric						
2400	27 kw						
2420	Minimum	EA	2.000	14,360	170		14,530
2440	Maximum	"	2.667	26,240	220		26,460
2450	18 kw						
2460	Minimum	EA	2.000	7,900	170		8,070
2480	Maximum	"	2.667	18,540	220		18,760
2500	Dishwasher, rack type						
2520	Single tank, 190 racks/hr	EA	4.000	21,050	330		21,380
2530	Double tank						
2540	234 racks/hr	EA	4.444	43,460	370		43,830
2560	265 racks/hr	"	5.333	51,950	440		52,390
2580	Dishwasher, automatic 100 meals/hr	"	2.667	16,900	220		17,120
2590	Disposals						
2620	100 gal/hr	EA	2.667	1,410	220		1,630
2640	120 gal/hr	"	2.759	1,640	230		1,870
2660	250 gal/hr	"	2.857	1,930	240		2,170
2670	Exhaust hood for dishwasher, gutter 4 sides						
2680	4'x4'x2'	EA	2.963	3,420	250		3,670
2690	4'x7'x2'	"	3.200	4,630	270		4,900
2700	Food preparation machines						
2710	Vertical cutter mixers						

© 2020 BNi Publications Inc. ALL RIGHTS RESERVED

FOOD SERVICE EQUIPMENT

ID Code	Description: Component Descriptions	Output: Unit of Meas.	Output: Manhr / Unit	Unit Costs: Material Cost	Unit Costs: Labor Cost	Unit Costs: Equipment Cost	Unit Costs: Total Cost
11 - 26001	**FOOD SERVICE EQUIPMENT, Cont'd...**						**11 - 26001**
2720	25 quart	EA	2.667	13,940	220		14,160
2730	40 quart	"	2.667	18,000	220		18,220
2740	80 quart	"	4.000	23,000	330		23,330
2750	130 quart	"	6.667	30,660	550		31,210
2760	Choppers						
2770	5 lb	EA	2.000	3,890	170		4,060
2780	16 lb	"	2.667	6,130	220		6,350
2790	40 lb	"	4.000	7,710	330		8,040
2800	Mixers, floor models						
2820	20 quart	EA	1.000	4,290	83.00		4,370
2840	60 quart	"	1.000	21,320	83.00		21,400
2860	80 quart	"	1.143	34,840	95.00		34,940
2870	140 quart	"	1.600	41,570	130		41,700
2890	Ice cube maker						
2900	50 lb per day						
2920	Minimum	EA	8.000	2,640	660		3,300
2940	Maximum	"	8.000	3,900	660		4,560
2950	500 lb per day						
2960	Minimum	EA	13.333	6,270	1,110		7,380
2970	Maximum	"	13.333	7,410	1,110		8,520
3000	Ice flakers						
3020	300 lb per day	EA	8.000	4,600	660		5,260
3040	600 lb per day	"	13.333	7,540	1,110		8,650
3050	1000 lb per day	"	17.778	8,600	1,480		10,080
3060	2000 lb per day	"	20.000	16,590	1,660		18,250
3100	Refrigerated cases						
3120	Dairy products						
3140	Multi-deck type	LF	0.533	1,490	44.25		1,530
3160	For rear sliding doors, add	"					280
3180	Delicatessen case, service deli						
3190	Single deck	LF	4.000	1,060	330		1,390
3200	Multi-deck	"	5.000	1,210	420		1,630
3210	Meat case						
3230	Single deck	LF	4.706	910	390		1,300
3240	Multi-deck	"	5.000	1,070	420		1,490
3260	Produce case						
3270	Single deck	LF	4.706	1,060	390		1,450
3280	Multi-deck	"	5.000	1,130	420		1,550

© 2020 BNi Publications Inc. ALL RIGHTS RESERVED

FOOD SERVICE EQUIPMENT

	Description	Output		Unit Costs			
ID Code	Component Descriptions	Unit of Meas.	Manhr / Unit	Material Cost	Labor Cost	Equipment Cost	Total Cost
11 - 26001	**FOOD SERVICE EQUIPMENT, Cont'd...**						**11 - 26001**
3300	Bottle coolers						
3310	6' long						
3320	Minimum	EA	16.000	3,010	1,330		4,340
3330	Maximum	"	16.000	4,460	1,330		5,790
3340	10' long						
3350	Minimum	EA	26.667	3,900	2,210		6,110
3360	Maximum	"	26.667	7,670	2,210		9,880
3420	Frozen food cases						
3440	Chest type	LF	4.706	810	390		1,200
3460	Reach-in, glass door	"	5.000	1,120	420		1,540
3470	Island case, single	"	4.706	1,000	390		1,390
3480	Multi-deck	"	5.000	1,590	420		2,010
3500	Ice storage bins						
3520	500 lb capacity	EA	11.429	1,680	950		2,630
3530	1000 lb capacity	"	22.857	2,510	1,900		4,410

DARKROOM EQUIPMENT

ID Code	Component Descriptions	Unit of Meas.	Manhr / Unit	Material Cost	Labor Cost	Equipment Cost	Total Cost
11 - 27001	**DARKROOM EQUIPMENT**						**11 - 27001**
0600	Dryers						
0620	36" x 25" x 68"	EA	4.000	11,680	360		12,040
0640	48" x 25" x 68"	"	4.000	12,070	360		12,430
0700	Processors, film						
0720	Black and white	EA	4.000	18,620	360		18,980
0740	Color negatives	"	4.000	21,090	360		21,450
0760	Prints	"	4.000	24,180	360		24,540
0780	Transparencies	"	4.000	26,530	360		26,890
1000	Sinks with cabinet and/or stand						
1020	5" sink with stand						
1040	24" x 48"	EA	2.000	1,040	180		1,220
1060	32" x 64"	"	2.667	1,600	240		1,840
1080	38" x 52"	"	2.667	2,250	240		2,490
1100	42" x 132"	"	4.000	3,360	360		3,720
1120	48" x 52"	"	4.000	2,580	360		2,940
1200	5" sink with cabinet						
1220	24" x 48"	EA	2.000	2,150	180		2,330
1240	32" x 64"	"	2.667	2,690	240		2,930
1260	38" x 52"	"	2.667	2,760	240		3,000

© 2020 BNi Publications Inc. ALL RIGHTS RESERVED

DARKROOM EQUIPMENT

ID Code	Component Descriptions	Unit of Meas.	Manhr / Unit	Material Cost	Labor Cost	Equipment Cost	Total Cost
11 - 27001	**DARKROOM EQUIPMENT, Cont'd...**						**11 - 27001**
1280	42" x 132"	EA	4.000	4,490	360		4,850
1290	48" x 52"	"	4.000	3,760	360		4,120
1300	10" sink with stand						
1320	24" x 48"	EA	2.000	1,780	180		1,960
1340	32" x 64"	"	2.667	1,880	240		2,120
1360	38" x 52"	"	2.667	2,550	240		2,790
1400	10" sink with cabinet						
1420	24" x 48"	EA	2.000	1,950	180		2,130
1460	38" x 52"	"	2.667	3,610	240		3,850

RESIDENTIAL EQUIPMENT

ID Code	Component Descriptions	Unit of Meas.	Manhr / Unit	Material Cost	Labor Cost	Equipment Cost	Total Cost
11 - 31001	**RESIDENTIAL EQUIPMENT**						**11 - 31001**
0300	Compactor, 4 to 1 compaction	EA	2.000	1,680	170		1,850
1300	Dishwasher, built-in						
1320	2 cycles	EA	4.000	830	330		1,160
1330	4 or more cycles	"	4.000	2,230	330		2,560
1340	Disposal						
1350	Garbage disposer	EA	2.667	230	220		450
1360	Heaters, electric, built-in						
1362	Ceiling type	EA	2.667	470	220		690
1363	Wall type						
1370	Minimum	EA	2.000	240	170		410
1380	Maximum	"	2.667	820	220		1,040
1390	Hood for range, 2-speed, vented						
1420	30" wide	EA	2.667	650	220		870
1440	42" wide	"	2.667	1,200	220		1,420
1460	Ice maker, automatic						
1480	30 lb per day	EA	1.143	2,200	95.00		2,290
1500	50 lb per day	"	4.000	2,790	330		3,120
1820	Folding access stairs, disappearing metal stair						
1840	8' long	EA	1.143	1,150	95.00		1,240
1850	11' long	"	1.143	1,200	95.00		1,290
1860	12' long	"	1.143	1,280	95.00		1,380
1940	Wood frame, wood stair						
1950	22" x 54" x 8'9" long	EA	0.800	220	66.00		290
1960	25" x 54" x 10' long	"	0.800	270	66.00		340
2020	Ranges, electric						

© 2020 BNi Publications Inc. ALL RIGHTS RESERVED

RESIDENTIAL EQUIPMENT

	Description	Output		Unit Costs			
ID Code	Component Descriptions	Unit of Meas.	Manhr / Unit	Material Cost	Labor Cost	Equipment Cost	Total Cost
11 - 31001	**RESIDENTIAL EQUIPMENT, Cont'd...**						**11 - 31001**
2040	Built-in, 30", 1 oven	EA	2.667	2,410	220		2,630
2050	2 oven	"	2.667	2,790	220		3,010
2060	Countertop, 4 burner, standard	"	2.000	1,390	170		1,560
2070	With grill	"	2.000	3,480	170		3,650
2198	Freestanding, 21", 1 oven	"	2.667	1,250	220		1,470
2200	30", 1 oven	"	1.600	2,440	130		2,570
2220	2 oven	"	1.600	3,970	130		4,100
3600	Water softener						
3620	30 grains per gallon	EA	2.667	1,360	220		1,580
3640	70 grains per gallon	"	4.000	1,720	330		2,050

LABORATORY EQUIPMENT

ID Code	Component Descriptions	Unit of Meas.	Manhr / Unit	Material Cost	Labor Cost	Equipment Cost	Total Cost
11 - 53001	**LABORATORY EQUIPMENT**						**11 - 53001**
1000	Cabinets, base						
1020	Minimum	LF	0.667	510	54.00		560
1040	Maximum	"	0.667	930	54.00		980
1080	Full storage, 7' high						
1100	Minimum	LF	0.667	490	54.00		540
1140	Maximum	"	0.667	930	54.00		980
1150	Wall						
1160	Minimum	LF	0.800	180	65.00		250
1200	Maximum	"	0.800	310	65.00		380
1220	Countertops						
1240	Minimum	SF	0.100	76.00	8.15		84.00
1260	Average	"	0.114	89.00	9.32		98.00
1280	Maximum	"	0.133	100	10.75		110
1300	Tables						
1320	Open underneath	SF	0.400	160	32.50		190
1330	Doors underneath	"	0.500	550	40.75		590
2000	Medical laboratory equipment						
2010	Analyzer						
2020	Chloride	EA	0.400	6,100	33.25		6,130
2060	Blood	"	0.667	33,540	55.00		33,590
2070	Bath, water, utility, countertop unit	"	0.800	1,260	66.00		1,330
2080	Hot plate, lab, countertop	"	0.727	480	60.00		540
2100	Stirrer	"	0.727	580	60.00		640
2120	Incubator, anaerobic, 23x23x36"	"	4.000	10,340	330		10,670

© 2020 BNi Publications Inc. ALL RIGHTS RESERVED

LABORATORY EQUIPMENT

ID Code	Description Component Descriptions	Output Unit of Meas.	Manhr / Unit	Unit Costs Material Cost	Labor Cost	Equipment Cost	Total Cost
11 - 53001	**LABORATORY EQUIPMENT, Cont'd...**						**11 - 53001**
2140	Dry heat bath	EA	1.333	1,120	110		1,230
2160	Incinerator, for sterilizing	"	0.080	770	6.64		780
2170	Meter, serum protein	"	0.100	1,190	8.30		1,200
2180	pH analog, general purpose	"	0.114	1,280	9.48		1,290
2190	Refrigerator, blood bank	"	1.333	9,780	110		9,890
2200	5.4 cf, undercounter type	"	1.333	6,640	110		6,750
2210	Refrigerator/freezer, 4.4 cf, undercounter type	"	1.333	1,260	110		1,370
2220	Sealer, impulse, free standing, 20x12x4"	"	0.267	700	22.25		720
2240	Timer, electric, 1-60 minutes, bench or wall mounted	"	0.444	270	37.00		310
2260	Glassware washer-dryer, undercounter	"	10.000	11,580	830		12,410
2300	Balance, torsion suspension, tabletop, 4.5 lb capacity	"	0.444	1,520	37.00		1,560
2340	Binocular microscope, with in-base illuminator	"	0.308	4,650	25.50		4,680
2400	Centrifuge, table model, 19x16x13"	"	0.320	1,780	26.50		1,810
2420	Clinical model, with four place head	"	0.178	1,930	14.75		1,940

VOCATIONAL SHOP EQUIPMENT

ID Code	Component Descriptions	Unit of Meas.	Manhr / Unit	Material Cost	Labor Cost	Equipment Cost	Total Cost
11 - 57001	**INDUSTRIAL EQUIPMENT**						**11 - 57001**
1000	Vehicular paint spray booth, solid back, 14'4" x 9'6"						
1020	24' deep	EA	8.000	9,610	650		10,260
1040	26'6" deep	"	8.000	10,980	650		11,630
1060	28'6" deep	"	8.000	12,680	650		13,330
1100	Drive through, 14'9" x 9'6"						
1120	24' deep	EA	8.000	10,230	650		10,880
1140	26'6" deep	"	8.000	12,330	650		12,980
1160	28'6" deep	"	8.000	14,070	650		14,720
1180	Water wash, paint spray booth						
1190	5' x 11'2" x 10'8"	EA	8.000	6,940	650		7,590
1200	6' x 11'2" x 10'8"	"	8.000	7,260	650		7,910
1220	8' x 11'2" x 10'8"	"	8.000	7,850	650		8,500
1240	10' x 11'2" x 11'2"	"	8.000	8,900	650		9,550
1260	12' x 12'2" x 11'2"	"	8.000	10,340	650		10,990
1280	14' x 12'2" x 11'2"	"	8.000	11,700	650		12,350
1290	16' x 12'2" x 11'2"	"	8.000	14,250	650		14,900
1300	20' x 12'2" x 11'2"	"	8.000	17,290	650		17,940
1320	Dry type spray booth, with paint arrestors						
1340	5'4" x 7'2" x 6'8"	EA	8.000	4,190	650		4,840
1360	6'4" x 7'2" x 6'8"	"	8.000	5,450	650		6,100

© 2020 BNi Publications Inc. ALL RIGHTS RESERVED

VOCATIONAL SHOP EQUIPMENT

	Description	Output		Unit Costs			
ID Code	Component Descriptions	Unit of Meas.	Manhr / Unit	Material Cost	Labor Cost	Equipment Cost	Total Cost
11 - 57001	**INDUSTRIAL EQUIPMENT, Cont'd...**						**11 - 57001**
1380	8'4" x 7'2" x 9'2"	EA	8.000	6,180	650		6,830
1400	10'4" x 7'2" x 9'2"	"	8.000	7,270	650		7,920
1420	12'4" x 7'6" x 9'2"	"	8.000	7,230	650		7,880
1440	14'4" x 7'6" x 9'8"	"	8.000	9,780	650		10,430
1460	16'4" x 7'7" x 9'8"	"	8.000	11,180	650		11,830
1480	20'4" x 7'7" x 10'8"	"	8.000	12,720	650		13,370
1500	Air compressor, electric						
1510	1 hp						
1520	115 volt	EA	5.333	1,600	430		2,030
1535	7.5 hp						
1540	115 volt	EA	8.000	4,750	650		5,400
1550	230 volt	"	8.000	5,940	650		6,590
1600	Hydraulic lifts						
1620	8,000 lb capacity	EA	20.000	3,280	1,630		4,910
1640	11,000 lb capacity	"	32.000	5,870	2,610		8,480
1660	24,000 lb capacity	"	53.333	10,340	4,350		14,690
1680	Power tools						
1700	Band saws						
1720	10"	EA	0.667	1,400	54.00		1,450
1740	14"	"	0.800	2,100	65.00		2,160
1760	Motorized shaper	"	0.615	1,120	50.00		1,170
1780	Motorized lathe	"	0.667	1,330	54.00		1,380
1800	Bench saws						
1820	9" saw	EA	0.533	3,490	43.50		3,530
1830	10" saw	"	0.571	4,190	46.50		4,240
1840	12" saw	"	0.667	5,160	54.00		5,210
1900	Electric grinders						
1910	1/3 hp	EA	0.320	540	26.00		570
2000	1/2 hp	"	0.348	560	28.25		590
2020	3/4 hp	"	0.348	940	28.25		970

© 2020 BNi Publications Inc. ALL RIGHTS RESERVED

BROADCAST, THEATER, STAGE EQUIPMENT

ID Code	Description: Component Descriptions	Output: Unit of Meas.	Output: Manhr / Unit	Unit Costs: Material Cost	Unit Costs: Labor Cost	Unit Costs: Equipment Cost	Unit Costs: Total Cost
11 - 61001	**THEATER EQUIPMENT**						**11 - 61001**
1000	Roll out stage, steel frame, wood floor						
1020	Manual	SF	0.050	59.00	4.07		63.00
1040	Electric	"	0.080	56.00	6.52		63.00
1100	Portable stages						
1120	8" high	SF	0.040	23.00	3.26		26.25
1140	18" high	"	0.044	26.75	3.62		30.25
1160	36" high	"	0.047	31.25	3.83		35.00
1180	48" high	"	0.050	35.00	4.07		39.00
1300	Band risers						
1320	Minimum	SF	0.040	60.00	3.26		63.00
1340	Maximum	"	0.040	120	3.26		120
1400	Chairs for risers						
1420	Minimum	EA	0.036	780	2.32		780
1440	Maximum	"	0.036	1,250	2.32		1,250

ATHLETIC EQUIPMENT

ID Code	Description: Component Descriptions	Output: Unit of Meas.	Output: Manhr / Unit	Unit Costs: Material Cost	Unit Costs: Labor Cost	Unit Costs: Equipment Cost	Unit Costs: Total Cost
11 - 66001	**ATHLETIC EQUIPMENT**						**11 - 66001**
1000	Basketball backboard						
1020	Fixed	EA	10.000	2,570	820		3,390
1040	Swing-up	"	16.000	4,100	1,300		5,400
1060	Portable, hydraulic	"	4.000	20,150	330		20,480
1080	Suspended type, standard	"	16.000	5,840	1,300		7,140
1200	For glass backboard, add	"					1,700
1220	For electrically operated, add	"					1,980
2000	Bleacher, telescoping, manual						
2020	15 tier, minimum	SEAT	0.160	160	13.00		170
2040	Maximum	"	0.160	380	13.00		390
2060	20 tier, minimum	"	0.178	110	14.50		120
2080	Maximum	"	0.178	330	14.50		340
2100	30 tier, minimum	"	0.267	87.00	21.75		110
2120	Maximum	"	0.267	260	21.75		280
2220	Boxing ring, elevated, complete, 22' x 22'	EA	114.286	10,750	9,320		20,070
2400	Gym divider curtain						
2420	Minimum	SF	0.011	3.49	0.86		4.35
2440	Maximum	"	0.011	5.24	0.86		6.10
2460	Scoreboards, single face						
2480	Minimum	EA	8.000	7,420	650		8,070

© 2020 BNi Publications Inc. ALL RIGHTS RESERVED

ATHLETIC EQUIPMENT

	Description	Output		Unit Costs			
ID Code	Component Descriptions	Unit of Meas.	Manhr / Unit	Material Cost	Labor Cost	Equipment Cost	Total Cost
11 - 66001	**ATHLETIC EQUIPMENT, Cont'd...**						**11 - 66001**
2500	Maximum	EA	40.000	40,300	3,260		43,560
2540	Parallel bars						
2620	Minimum	EA	8.000	1,540	650		2,190
2630	Maximum	"	13.333	8,240	1,090		9,330
11 - 66002	**POLICE EQUIPMENT**						**11 - 66002**
9000	Firing range equipment, rifle						
9040	3 position	EA	26.667	19,950	2,170		22,120
9060	4 position	"	40.000	25,560	3,260		28,820
9080	5 position	"	44.444	31,170	3,620		34,790
9100	6 position	"	47.059	37,400	3,840		41,240

PLAY FIELD EQUIPMENT AND STRUCTURES

11 - 68230	**RECREATIONAL COURTS**						**11 - 68230**
1000	Walls, galvanized steel						
1020	8' high	LF	0.160	16.00	10.25		26.25
1040	10' high	"	0.178	18.75	11.25		30.00
1060	12' high	"	0.211	21.75	13.50		35.25
1200	Vinyl coated						
1220	8' high	LF	0.160	15.25	10.25		25.50
1240	10' high	"	0.178	18.75	11.25		30.00
1260	12' high	"	0.211	20.75	13.50		34.25
2010	Gates, galvanized steel						
2200	Single, 3' transom						
2210	3'x7'	EA	4.000	370	260		630
2220	4'x7'	"	4.571	390	290		680
2230	5'x7'	"	5.333	540	340		880
2240	6'x7'	"	6.400	580	410		990
2245	Double, 3' transom						
2250	10'x7'	EA	16.000	900	1,020		1,920
2260	12'x7'	"	17.778	1,160	1,140		2,300
2270	14'x7'	"	20.000	1,390	1,280		2,670
2275	Double, no transom						
2280	10'x10'	EA	13.333	980	850		1,830
2290	12'x10'	"	16.000	1,170	1,020		2,190
2300	14'x10'	"	17.778	1,340	1,140		2,480
2400	Vinyl coated						
2405	Single, 3' transom						

© 2020 BNi Publications Inc. ALL RIGHTS RESERVED

PLAY FIELD EQUIPMENT AND STRUCTURES

ID Code	Description: Component Descriptions	Output: Unit of Meas.	Output: Manhr / Unit	Unit Costs: Material Cost	Unit Costs: Labor Cost	Unit Costs: Equipment Cost	Unit Costs: Total Cost
11 - 68230	**RECREATIONAL COURTS, Cont'd...**						**11 - 68230**
2410	3'x7'	EA	4.000	730	260		990
2420	4'x7'	"	4.571	790	290		1,080
2430	5'x7'	"	5.333	790	340		1,130
2440	6'x7'	"	6.400	820	410		1,230
2445	Double, 3'						
2450	10'x7'	EA	16.000	2,150	1,020		3,170
2460	12'x7'	"	17.778	2,210	1,140		3,350
2470	14'x7'	"	20.000	2,390	1,280		3,670
2475	Double, no transom						
2480	10'x10'	EA	13.333	2,140	850		2,990
2490	12'x10'	"	16.000	2,180	1,020		3,200
2500	14'x10'	"	17.778	2,390	1,140		3,530
3000	Baseball backstop, regulation, including material						
3020	Galvanized	EA					8,210
3040	Vinyl coated	"					11,430
3100	Softball backstop, regulation, including material						
3110	14' high						
3130	Galvanized	EA					9,350
3140	Vinyl coated	"					14,850
3160	18' high						
3180	Galvanized	EA					10,340
3200	Vinyl coated	"					14,950
3300	20' high						
3320	Galvanized	EA					12,250
3330	Vinyl coated	"					17,640
3340	22' high						
3350	Galvanized	EA					14,160
3360	Vinyl coated	"					20,680
3380	24' high						
3400	Galvanized	EA					14,970
3420	Vinyl coated	"					24,620
4000	Wire and miscellaneous metal fences						
4020	Chicken wire, post 4' o.c.						
4030	2" mesh						
4040	4' high	LF	0.040	1.95	2.55		4.50
4060	6' high	"	0.053	2.20	3.40		5.60
4100	Galvanized steel						
4120	12 gauge, 2" by 4" mesh, posts 5' o.c.						

© 2020 BNi Publications Inc. ALL RIGHTS RESERVED

PLAY FIELD EQUIPMENT AND STRUCTURES

	Description	Output		Unit Costs			
ID Code	Component Descriptions	Unit of Meas.	Manhr / Unit	Material Cost	Labor Cost	Equipment Cost	Total Cost
11 - 68230	**RECREATIONAL COURTS, Cont'd...**						**11 - 68230**
4140	3' high	LF	0.040	3.12	2.55		5.67
4160	5' high	"	0.050	4.45	3.19		7.64
4200	14 gauge, 1" by 2" mesh, posts 5' o.c.						
4210	3' high	LF	0.040	2.61	2.55		5.16
4220	5' high	"	0.050	4.25	3.19		7.44
11 - 68330	**RECREATIONAL FACILITIES**						**11 - 68330**
1000	Bleachers, outdoor, portable, per seat						
1020	10 tiers						
1040	Minimum	EA	0.150	87.00	9.50	8.31	100
1060	Maximum	"	0.200	110	12.75	11.00	130
1100	20 tiers						
1120	Minimum	EA	0.141	92.00	8.94	7.82	110
1140	Maximum	"	0.185	120	11.75	10.25	140
1500	Grandstands, fixed, wood seat, steel frame						
1520	Per seat, 15 tiers						
1540	Minimum	EA	0.240	75.00	15.25	13.25	100
1560	Maximum	"	0.400	130	25.25	22.25	180
1600	30 tiers						
1620	Minimum	EA	0.218	78.00	13.75	12.00	100
1660	Maximum	"	0.343	170	21.75	19.00	210
1700	Seats						
1720	Seat backs only						
1740	Fiberglass	EA	0.080	43.00	5.10		48.00
1760	Steel and wood seat	"	0.080	62.00	5.10		67.00
1800	Seat restoration, fiberglass on wood						
1820	Seats	EA	0.160	31.00	10.25		41.25
1840	Plain bench, no backs	"	0.067	19.00	4.25		23.25
2000	Benches						
2020	Park, precast concrete with backs						
2040	4' long	EA	2.667	1,190	170		1,360
2060	8' long	"	4.000	2,620	260		2,880
2100	Fiberglass, with backs						
2120	4' long	EA	2.000	950	130		1,080
2140	8' long	"	2.667	1,820	170		1,990
2200	Wood, with backs and fiberglass supports						
2220	4' long	EA	2.000	530	130		660
2240	8' long	"	2.667	550	170		720

© 2020 BNi Publications Inc. ALL RIGHTS RESERVED

PLAY FIELD EQUIPMENT AND STRUCTURES

ID Code	Description: Component Descriptions	Output: Unit of Meas.	Output: Manhr / Unit	Unit Costs: Material Cost	Unit Costs: Labor Cost	Unit Costs: Equipment Cost	Unit Costs: Total Cost
11 - 68330	**RECREATIONAL FACILITIES, Cont'd...**						**11 - 68330**
2300	Steel frame, 6' long						
2320	All steel	EA	2.000	450	130		580
2340	Hardwood boards	"	2.000	310	130		440
2360	Players bench, steel frame, fir seat, 10' long	"	2.667	340	170		510
3000	Backstops						
3200	Handball or squash court, outdoor						
3220	Wood	EA					52,860
3240	Masonry	"					39,450
3260	Soccer goal posts	PAIR					3,530
4000	Running track						
4020	Gravel and cinders over stone base	SY	0.060	10.75	3.80	3.32	17.75
4040	Rubber-cork base resilient pavement	"	0.480	15.25	30.50	26.50	72.00
4060	For colored surfaces, add	"	0.048	9.62	3.04	2.66	15.25
4080	Colored rubberized asphalt	"	0.600	20.50	38.00	33.25	92.00
4100	Artificial resilient mat over asphalt	"	1.200	48.25	76.00	67.00	190
4200	Tennis courts						
4240	Bituminous pavement, 2-1/2" thick	SY	0.150	33.25	9.50	8.31	51.00
4300	Colored sealer, acrylic emulsion						
4320	3 coats	SY	0.053	8.10	3.40		11.50
4340	For 2 color seal coating, add	"	0.008	1.48	0.51		1.99
4360	For preparing old courts, add	"	0.005	3.42	0.34		3.76
4400	Net, nylon, 42' long	EA	1.000	470	64.00		530
4520	Paint markings on asphalt, 2 coats	"	8.000	180	510		690
4580	Complete court with fence, etc., bituminous						
4600	Minimum	EA					33,530
4620	Average	"					57,710
4640	Maximum	"					81,900
4680	Clay court						
4700	Minimum	EA					34,900
4720	Average	"					47,570
4740	Maximum	"					72,140
5000	Playground equipment						
5010	Basketball backboard						
5012	Minimum	EA	2.000	1,020	130		1,150
5014	Maximum	"	2.286	1,930	150		2,080
5016	Bike rack, 10' long	"	1.600	650	100		750
5018	Golf shelter, fiberglass	"	2.000	3,240	130		3,370
5020	Ground socket for movable posts						

© 2020 BNi Publications Inc. ALL RIGHTS RESERVED

PLAY FIELD EQUIPMENT AND STRUCTURES

ID Code	Component Descriptions	Unit of Meas.	Manhr / Unit	Material Cost	Labor Cost	Equipment Cost	Total Cost
11 - 68330	**RECREATIONAL FACILITIES, Cont'd...**						**11 - 68330**
5040	Minimum	EA	0.500	150	32.00		180
5060	Maximum	"	0.500	300	32.00		330
5070	Horizontal monkey ladder, 14' long	"	1.333	990	85.00		1,080
5072	Posts, tether ball	"	0.400	480	25.50		510
5074	Multiple purpose, 10' long	"	0.800	490	51.00		540
5080	See-saw, steel						
5100	Minimum	EA	3.200	1,170	200		1,370
5120	Average	"	4.000	2,260	260		2,520
5140	Maximum	"	5.333	3,320	340		3,660
5150	Slide						
5160	Minimum	EA	6.400	2,260	410		2,670
5180	Maximum	"	7.273	5,980	460		6,440
5800	Swings, plain seats						
5810	8' high						
5820	Minimum	EA	5.333	1,080	340		1,420
5840	Maximum	"	6.154	2,050	390		2,440
5850	12' high						
5860	Minimum	EA	6.154	1,650	390		2,040
5880	Maximum	"	8.889	2,980	570		3,550

EXAMINATION AND TREATMENT EQUIPMENT

ID Code	Component Descriptions	Unit of Meas.	Manhr / Unit	Material Cost	Labor Cost	Equipment Cost	Total Cost
11 - 72001	**MEDICAL EQUIPMENT**						**11 - 72001**
1000	Hospital equipment, lights						
1020	Examination, portable	EA	0.667	2,130	55.00		2,190
1200	Meters						
1220	Air flow meter	EA	0.444	110	37.00		150
1240	Oxygen flow meters	"	0.333	140	27.75		170
1300	Racks						
1320	40 chart, revolving open frame; mobile caddy	EA	0.667	1,470	55.00		1,530
1400	Scales						
1420	Clinical, metric with measure rod, 350 lb	EA	0.727	780	60.00		840
1900	Physical therapy						
1930	Chair, hydrotherapy	EA	0.133	820	10.75		830
1940	Diathermy, shortwave, portable, on casters	"	0.320	3,540	26.00		3,570
1950	Exercise bicycle, floor standing, 35" x 15"	"	0.267	3,440	21.75		3,460
1960	Hydrocollator, 4 pack, portable, 129 x 90 x 160"	"	0.114	620	9.32		630
1970	Lamp, infrared, mobile with variable heat control	"	0.615	830	50.00		880

© 2020 BNi Publications Inc. ALL RIGHTS RESERVED

EXAMINATION AND TREATMENT EQUIPMENT

ID Code	Description: Component Descriptions	Output: Unit of Meas.	Output: Manhr / Unit	Unit Costs: Material Cost	Unit Costs: Labor Cost	Unit Costs: Equipment Cost	Unit Costs: Total Cost
11 - 72001	MEDICAL EQUIPMENT, Cont'd...						11 - 72001
1980	Ultraviolet, base mounted	EA	0.615	770	50.00		820
1990	Mirror, posture training, 27" wide and 72" high	"	0.200	850	16.25		870
2000	Parallel bars, adjustable	"	1.000	3,670	82.00		3,750
2020	Platform mat 10'x6', 1" thick	"	0.200	1,180	16.25		1,200
2030	Pulley, duplex, wall mounted	"	2.667	2,010	220		2,230
2040	Rack, crutch, wall mounted, 66 x 16 x 13"	"	0.800	490	65.00		550
2070	Stimulator, galvanic-faradic, handheld	"	0.053	430	4.34		430
2080	Ultrasound stimulator, portable, 13x13x8"	"	0.067	3,680	5.43		3,690
2100	Sandbag set, velcro straps, saddle bag type	"	0.114	190	9.32		200
2120	Whirlpool, 85 gallon	"	4.000	7,070	330		7,400
2141	65 gallon capacity	"	4.000	6,380	330		6,710
2260	Radiology						
2280	Radiographic table, motor driven tilting table	EA	80.000	62,350	6,520		68,870
2290	Fluoroscope image/tv system	"	160.000	103,440	13,050		116,490
2300	Processor for washing and drying radiographs						
2310	Water filter unit, 30" x 48-1/2" x 37-1/2"	EA	13.333	140	1,110		1,250
2340	Base storage cabinets, sectional design						
2350	With backsplash, 24" deep and 35" high	LF	0.667	710	55.00		770
2360	Wall storage cabinets	"	1.000	280	83.00		360
2400	Steam sterilizers						
2410	For heat and moisture stable materials	EA	0.800	6,020	66.00		6,090
2420	For fast drying after sterilization	"	1.000	7,790	83.00		7,870
2430	Compact unit	"	1.000	2,520	83.00		2,600
2440	Semi-automatic	"	4.000	2,980	330		3,310
2450	Floor loading						
2460	Single door	EA	6.667	88,280	550		88,830
2480	Double door	"	8.000	96,710	660		97,370
2490	Utensil washer, sanitizer	"	6.154	19,410	510		19,920
2500	Automatic washer/sterilizer	"	16.000	21,250	1,330		22,580
2510	16 x 16 x 26", including accessories	"	26.667	24,430	2,210		26,640
2520	Steam generator, elec., 10 kw to 180 kw	"	16.000	40,380	1,330		41,710
2550	Surgical scrub						
2560	Minimum	EA	2.667	2,130	220		2,350
2580	Maximum	"	2.667	12,330	220		12,550
2610	Gas sterilizers						
2620	Automatic, freestanding, 21x19x29"	EA	8.000	7,580	660		8,240
2640	Surgical tables						
2660	Minimum	EA	11.429	26,260	950		27,210

© 2020 BNi Publications Inc. ALL RIGHTS RESERVED

EXAMINATION AND TREATMENT EQUIPMENT

	Description	Output		Unit Costs			
ID Code	Component Descriptions	Unit of Meas.	Manhr / Unit	Material Cost	Labor Cost	Equipment Cost	Total Cost
11 - 72001	**MEDICAL EQUIPMENT, Cont'd...**						**11 - 72001**
2680	Maximum	EA	16.000	31,950	1,330		33,280
2720	Surgical lights, ceiling mounted						
2740	Minimum	EA	13.333	10,490	1,110		11,600
2760	Maximum	"	16.000	21,400	1,330		22,730
2880	Water stills						
2900	4 liters/hr	EA	2.667	4,750	220		4,970
2920	8 liters/hr	"	2.667	7,580	220		7,800
2940	19 liters/hr	"	6.667	15,590	550		16,140
3040	X-ray equipment						
3060	Mobile unit						
3080	Minimum	EA	4.000	13,880	330		14,210
3100	Maximum	"	8.000	26,880	660		27,540
3110	Film viewers						
3120	Minimum	EA	1.333	350	110		460
3140	Maximum	"	2.667	1,200	220		1,420
3200	Autopsy table						
3220	Minimum	EA	8.000	19,370	660		20,030
3240	Maximum	"	8.000	27,370	660		28,030
3300	Incubators						
3320	15 cf	EA	4.000	9,670	330		10,000
3330	29 cf	"	6.667	13,110	550		13,660
3340	Infant transport, portable	"	4.211	7,280	350		7,630
3400	Beds						
3420	Stretcher, with pad, 30" x 78"	EA	2.000	5,410	170		5,580
3440	Transfer, for patient transport	"	2.000	6,370	170		6,540
3450	Headwall						
3460	Aluminum, with back frame and console	EA	4.000	5,460	330		5,790
6000	Hospital ground detection system						
6010	Power ground module	EA	2.286	1,710	190		1,900
6020	Ground slave module	"	1.739	720	140		860
6030	Master ground module	"	1.509	640	130		770
6040	Remote indicator	"	1.600	670	130		800
6050	X-ray indicator	"	1.739	1,910	140		2,050
6060	Micro ammeter	"	2.000	2,270	170		2,440
6070	Supervisory module	"	1.739	1,910	140		2,050
6080	Ground cords	"	0.296	180	24.50		200
6100	Hospital isolation monitors, 5 ma						
6110	120v	EA	3.478	3,620	290		3,910

© 2020 BNi Publications Inc. ALL RIGHTS RESERVED

EXAMINATION AND TREATMENT EQUIPMENT

	Description	Output		Unit Costs			
ID Code	Component Descriptions	Unit of Meas.	Manhr / Unit	Material Cost	Labor Cost	Equipment Cost	Total Cost
11 - 72001	**MEDICAL EQUIPMENT, Cont'd...**						**11 - 72001**
6120	208v	EA	3.478	3,620	290		3,910
6130	240v	"	3.478	3,920	290		4,210
6210	Digital clock-timers separate display	"	1.600	1,590	130		1,720
6220	One display	"	1.600	1,020	130		1,150
6230	Remote control	"	1.250	500	100		600
6240	Battery pack	"	1.250	120	100		220
6310	Surgical chronometer clock and 3 timers	"	2.500	2,980	210		3,190
6320	Auxiliary control	"	1.159	810	96.00		910

DENTAL EQUIPMENT

ID Code	Component Descriptions	Unit of Meas.	Manhr / Unit	Material Cost	Labor Cost	Equipment Cost	Total Cost
11 - 74001	**DENTAL EQUIPMENT**						**11 - 74001**
3500	Dental care equipment						
3520	Drill console with accessories	EA	13.333	5,710	1,110		6,820
3540	Amalgamator	"	0.400	610	33.25		640
3560	Lathe	"	0.267	1,410	22.25		1,430
3580	Finish polisher	"	0.533	1,850	44.25		1,890
3590	Model trimmer	"	0.364	1,140	30.25		1,170
3600	Motor, wall mounted	"	0.364	1,280	30.25		1,310
3640	Cleaner, ultrasonic	"	0.800	3,430	66.00		3,500
3660	Curing unit, bench mounted	"	1.333	5,270	110		5,380
3680	Oral evacuation system, dual pump	"	1.000	6,420	83.00		6,500
3700	Sterilizer, table top, self contained	"	0.444	2,710	37.00		2,750
3720	Dental lights						
3740	Light, floor or ceiling mounted	EA	4.000	2,300	330		2,630
3780	X-ray unit						
3790	Portable	EA	2.000	5,710	170		5,880
3820	Wall mounted with remote control	"	6.667	7,990	550		8,540
3830	Illuminator, single panel	"	11.429	820	950		1,770
3840	X-ray film processor	"	6.667	9,970	550		10,520
3850	Shield, portable x-ray, lead lined	"	0.533	1,850	44.25		1,890

© 2020 BNi Publications Inc. ALL RIGHTS RESERVED

RECYCLING SYSTEMS

ID Code	Description: Component Descriptions	Output: Unit of Meas.	Output: Manhr / Unit	Unit Costs: Material Cost	Unit Costs: Labor Cost	Unit Costs: Equipment Cost	Unit Costs: Total Cost
11 - 82001	**WASTE HANDLING**						**11 - 82001**
1000	Incinerator, electric						
1010	100 lb/hr						
1020	Minimum	EA	8.000	16,590	660		17,250
1040	Maximum	"	8.000	28,520	660		29,180
1050	400 lb/hr						
1060	Minimum	EA	16.000	41,790	1,330		43,120
1070	Maximum	"	16.000	52,240	1,330		53,570
1075	1000 lb/hr						
1080	Minimum	EA	24.242	98,330	2,010		100,340
1090	Maximum	"	24.242	147,500	2,010		149,510
1200	Incinerator, medical-waste						
1220	25 lb/hr, 2-7 x 4-0	EA	16.000	14,140	1,330		15,470
1230	50 lb/hr, 2-11 x 4-11	"	16.000	27,410	1,330		28,740
1240	75 lb/hr, 3-8 x 5-0	"	32.000	36,870	2,660		39,530
1250	100 lb/hr, 3-8 x 6-0	"	32.000	55,310	2,660		57,970
1500	Industrial compactor						
1520	1 c.y.	EA	8.889	14,500	740		15,240
1540	3 c.y.	"	11.429	22,620	950		23,570
1560	5 c.y.	"	16.000	43,020	1,330		44,350
2000	Trash chutes, steel, including sprinklers						
2020	18" dia.	LF	4.000	110	330		440
2030	24" dia.	"	4.211	140	340		480
2040	30" dia.	"	4.444	170	360		530
2050	36" dia.	"	4.706	210	380		590
2060	Refuse bottom hopper	EA	4.444	1,840	360		2,200
11 - 82230	**GRAY WATER RECYCLING SYSTEM**						**11 - 82230**
1000	Residential, small commercial, 150 Gallons						
1010	Minimum	EA					4,040
1020	Average	"					4,580
1030	Maximum	"					5,110
1040	250 Gallons						
1050	Minimum	EA					4,710
1060	Average	"					5,250
1070	Maximum	"					5,790
1080	350 Gallons						
1090	Minimum	EA					5,110
1100	Average	"					5,520

© 2020 BNi Publications Inc. ALL RIGHTS RESERVED

RECYCLING SYSTEMS

	Description	Output		Unit Costs			
ID Code	Component Descriptions	Unit of Meas.	Manhr / Unit	Material Cost	Labor Cost	Equipment Cost	Total Cost
11 - 82230	**GRAY WATER RECYCLING SYSTEM, Cont'd...**						**11 - 82230**
1110	Maximum	EA					5,920
1120	450 Gallons						
1130	Minimum	EA					5,380
1140	Average	"					5,790
1150	Maximum	"					6,190
1160	550 Gallons						
1170	Minimum	EA					6,600
1180	Average	"					5,450
1190	Maximum	"					7,000

© 2020 BNi Publications Inc. ALL RIGHTS RESERVED

DIVISION 12
FURNISHINGS

WINDOW BLINDS

ID Code	Description: Component Descriptions	Output: Unit of Meas.	Output: Manhr / Unit	Unit Costs: Material Cost	Unit Costs: Labor Cost	Unit Costs: Equipment Cost	Unit Costs: Total Cost
12 - 21001	**BLINDS**						**12 - 21001**
0990	Venetian blinds						
1000	2" slats	SF	0.020	40.25	1.63		42.00
1020	1" slats	"	0.020	43.00	1.63		44.75

CURTAINS AND DRAPES

ID Code	Component Descriptions	Unit of Meas.	Manhr / Unit	Material Cost	Labor Cost	Equipment Cost	Total Cost
12 - 22001	**WINDOW TREATMENT**						**12 - 22001**
1000	Drapery tracks, wall or ceiling mounted						
1040	Basic traverse rod						
1080	50 to 90"	EA	0.400	56.00	32.50		89.00
1100	84 to 156"	"	0.444	75.00	36.25		110
1120	136 to 250"	"	0.444	110	36.25		150
1140	165 to 312"	"	0.500	170	40.75		210
1160	Traverse rod with stationary curtain rod						
1180	30 to 50"	EA	0.400	84.00	32.50		120
1200	50 to 90"	"	0.400	97.00	32.50		130
1220	84 to 156"	"	0.444	130	36.25		170
1240	136 to 250"	"	0.500	160	40.75		200
1260	Double traverse rod						
1280	30 to 50"	EA	0.400	99.00	32.50		130
1300	50 to 84"	"	0.400	130	32.50		160
1320	84 to 156"	"	0.444	130	36.25		170
1340	136 to 250"	"	0.500	170	40.75		210

MANUFACTURED WOOD CASEWORK

ID Code	Component Descriptions	Unit of Meas.	Manhr / Unit	Material Cost	Labor Cost	Equipment Cost	Total Cost
12 - 32001	**CASEWORK**						**12 - 32001**
0080	Kitchen base cabinet, standard, 24" deep, 35" high						
0100	12" wide	EA	0.800	220	65.00		290
0120	18" wide	"	0.800	250	65.00		310
0140	24" wide	"	0.889	330	72.00		400
0160	27" wide	"	0.889	370	72.00		440
0180	36" wide	"	1.000	440	82.00		520
0200	48" wide	"	1.000	520	82.00		600
0210	Drawer base, 24" deep, 35" high						
0220	15" wide	EA	0.800	280	65.00		350
0230	18" wide	"	0.800	290	65.00		350
0240	24" wide	"	0.889	470	72.00		540

© 2020 BNi Publications Inc. ALL RIGHTS RESERVED

MANUFACTURED WOOD CASEWORK

	Description	Output		Unit Costs			
ID Code	Component Descriptions	Unit of Meas.	Manhr / Unit	Material Cost	Labor Cost	Equipment Cost	Total Cost
12 - 32001	**CASEWORK, Cont'd...**						**12 - 32001**
0250	27" wide	EA	0.889	530	72.00		600
0260	30" wide	"	0.889	630	72.00		700
0270	Sink-ready base cabinet						
0280	30" wide	EA	0.889	290	72.00		360
0290	36" wide	"	0.889	310	72.00		380
0300	42" wide	"	0.889	340	72.00		410
0310	60" wide	"	1.000	400	82.00		480
0320	Corner cabinet, 36" wide	"	1.000	550	82.00		630
4000	Wall cabinet, 12" deep, 12" high						
4020	30" wide	EA	0.800	280	65.00		350
4060	36" wide	"	0.800	290	65.00		350
4070	15" high						
4080	30" wide	EA	0.889	330	72.00		400
4100	36" wide	"	0.889	490	72.00		560
4110	24" high						
4120	30" wide	EA	0.889	360	72.00		430
4140	36" wide	"	0.889	370	72.00		440
4150	30" high						
4160	12" wide	EA	1.000	210	82.00		290
4180	18" wide	"	1.000	240	82.00		320
4200	24" wide	"	1.000	260	82.00		340
4300	27" wide	"	1.000	310	82.00		390
4320	30" wide	"	1.143	340	93.00		430
4340	36" wide	"	1.143	350	93.00		440
4350	Corner cabinet, 30" high						
4360	24" wide	EA	1.333	380	110		490
4380	30" wide	"	1.333	460	110		570
4390	36" wide	"	1.333	500	110		610
5020	Wardrobe	"	2.000	1,020	160		1,180
6980	Vanity with top, laminated plastic						
7000	24" wide	EA	2.000	840	160		1,000
7020	30" wide	"	2.000	940	160		1,100
7040	36" wide	"	2.667	1,080	220		1,300
7060	48" wide	"	3.200	1,210	260		1,470

© 2020 BNi Publications Inc. ALL RIGHTS RESERVED

COUNTERTOPS

ID Code	Description Component Descriptions	Output Unit of Meas.	Manhr / Unit	Unit Costs Material Cost	Labor Cost	Equipment Cost	Total Cost
12 - 36001	**COUNTERTOPS**						**12 - 36001**
1020	Stainless steel, countertop, with backsplash	SF	0.200	250	16.25		270
2000	Acid-proof, kemrock surface	"	0.133	100	10.75		110

RUGS AND MATS

ID Code	Component Descriptions	Unit of Meas.	Manhr / Unit	Material Cost	Labor Cost	Equipment Cost	Total Cost
12 - 48001	**FLOOR MATS**						**12 - 48001**
1020	Recessed entrance mat, 3/8" thick, aluminum link	SF	0.400	59.00	32.50		92.00
1040	Steel, flexible	"	0.400	21.00	32.50		54.00

© 2020 BNi Publications Inc. ALL RIGHTS RESERVED

DIVISION 13
SPECIAL CONSTRUCTION

CONSTRUCTION

ID Code	Description: Component Descriptions	Output: Unit of Meas.	Output: Manhr / Unit	Unit Costs: Material Cost	Unit Costs: Labor Cost	Unit Costs: Equipment Cost	Unit Costs: Total Cost
13 - 11001	**SWIMMING POOL EQUIPMENT**						**13 - 11001**
1100	Diving boards						
1110	14' long						
1120	Aluminum	EA	4.444	4,970	280		5,250
1140	Fiberglass	"	4.444	3,760	280		4,040
1500	Ladders, heavy duty						
1510	2 steps						
1520	Minimum	EA	1.600	1,180	100		1,280
1540	Maximum	"	1.600	1,840	100		1,940
1550	4 steps						
1560	Minimum	EA	2.000	1,260	130		1,390
1580	Maximum	"	2.000	2,020	130		2,150
1600	Lifeguard chair						
1620	Minimum	EA	8.000	3,340	510		3,850
1640	Maximum	"	8.000	5,180	510		5,690
1700	Lights, underwater						
1705	12 volt, with transformer, 100 watt						
1710	Incandescent	EA	2.000	250	130		380
1715	Halogen	"	2.000	210	130		340
1720	LED	"	2.000	670	130		800
1730	110 volt						
1740	Minimum	EA	2.000	1,130	130		1,260
1760	Maximum	"	2.000	2,720	130		2,850
1780	Ground fault interrupter for 110 volt, each light	"	0.667	240	42.50		280
2000	Pool cover						
2020	Reinforced polyethylene	SF	0.062	2.41	3.92		6.33
2030	Vinyl water tube						
2040	Minimum	SF	0.062	1.48	3.92		5.40
2060	Maximum	"	0.062	2.20	3.92		6.12
2100	Slides with water tube						
2120	Minimum	EA	6.667	1,180	430		1,610
2140	Maximum	"	6.667	25,210	430		25,640

© 2020 BNi Publications Inc. ALL RIGHTS RESERVED

SPECIAL ACTIVITY ROOMS

ID Code	Description: Component Descriptions	Output: Unit of Meas.	Output: Manhr / Unit	Unit Costs: Material Cost	Unit Costs: Labor Cost	Unit Costs: Equipment Cost	Unit Costs: Total Cost
13 - 24160	**SAUNAS**						**13 - 24160**
0010	Prefabricated, cedar siding, insulated panels, prehung						
0020	4'x8"x4'-8"x6'-6"	EA					6,500
0030	5'-8"x6'-8"x6'-6"	"					7,870
0040	6'-8"x6'-8"x6'-6"	"					9,160
0050	7'-8"x7'-8"x6'-6"	"					11,060
0060	7'-8"x9'-8"x6'-6"	"					14,570

FABRICATED ENGINEERED STRUCTURES

ID Code	Component Descriptions	Unit of Meas.	Manhr / Unit	Material Cost	Labor Cost	Equipment Cost	Total Cost
13 - 34190	**PRE-ENGINEERED BUILDINGS**						**13 - 34190**
1080	Pre-engineered metal building, 40'x100'						
1100	14' eave height	SF	0.032	9.33	2.60	3.06	15.00
1120	16' eave height	"	0.037	10.50	3.01	3.53	17.00
1140	20' eave height	"	0.048	12.00	3.91	4.60	20.50
1150	60'x100'						
1160	14' eave height	SF	0.032	11.75	2.60	3.06	17.50
1180	16' eave height	"	0.037	13.00	3.01	3.53	19.50
1190	20' eave height	"	0.048	14.50	3.91	4.60	23.00
1195	80'x100'						
1200	14' eave height	SF	0.032	9.03	2.60	3.06	14.75
1210	16' eave height	"	0.037	9.33	3.01	3.53	15.75
1220	20' eave height	"	0.048	10.50	3.91	4.60	19.00
1280	100'x100'						
1300	14' eave height	SF	0.032	8.81	2.60	3.06	14.50
1320	16' eave height	"	0.037	9.18	3.01	3.53	15.75
1340	20' eave height	"	0.048	10.00	3.91	4.60	18.50
1350	100'x150'						
1360	14' eave height	SF	0.032	7.85	2.60	3.06	13.50
1380	16' eave height	"	0.037	8.15	3.01	3.53	14.75
1400	20' eave height	"	0.048	8.74	3.91	4.60	17.25
1410	120'x150'						
1420	14' eave height	SF	0.032	8.30	2.60	3.06	14.00
1440	16' eave height	"	0.037	8.44	3.01	3.53	15.00
1460	20' eave height	"	0.048	8.81	3.91	4.60	17.25
1480	140'x150'						
1500	14' eave height	SF	0.032	7.85	2.60	3.06	13.50
1520	16' eave height	"	0.037	8.05	3.01	3.53	14.50
1540	20' eave height	"	0.048	8.74	3.91	4.60	17.25

© 2020 BNi Publications Inc. ALL RIGHTS RESERVED

FABRICATED ENGINEERED STRUCTURES

	Description	Output		Unit Costs			
ID Code	Component Descriptions	Unit of Meas.	Manhr / Unit	Material Cost	Labor Cost	Equipment Cost	Total Cost
13 - 34190	**PRE-ENGINEERED BUILDINGS, Cont'd...**						**13 - 34190**
1600	160'x200'						
1620	14' eave height	SF	0.032	6.05	2.60	3.06	11.75
1640	16' eave height	"	0.037	6.24	3.01	3.53	12.75
1680	20' eave height	"	0.048	6.60	3.91	4.60	15.00
1690	200'x200'						
1700	14' eave height	SF	0.032	5.21	2.60	3.06	11.00
1720	16' eave height	"	0.037	5.73	3.01	3.53	12.25
1740	20' eave height	"	0.048	6.10	3.91	4.60	14.50
5020	Hollow metal door and frame, 6' x 7'	EA					1,390
5030	Sectional steel overhead door, manually operated						
5040	8' x 8'	EA					2,280
5080	12' x 12'	"					3,040
5100	Roll-up steel door, manually operated						
5120	10' x 10'	EA					1,770
5140	12' x 12'	"					3,180
5160	For gravity ridge ventilator with birdscreen	"					760
5161	9" throat x 10'	"					830
5181	12" throat x 10'	"					1,010
5200	For 20" rotary vent with damper	"					380
5220	For 4' x 3' fixed louver	"					270
5240	For 4' x 3' aluminum sliding window	"					240
5260	For 3' x 9' fiberglass panels	"					180
8020	Liner panel, 26 ga, painted steel	SF	0.020	3.34	1.80		5.14
8040	Wall panel insulated, 26 ga. steel, foam core	"	0.020	10.50	1.80		12.25
8060	Roof panel, 26 ga. painted steel	"	0.011	3.16	1.02		4.18
8080	Plastic (skylight)	"	0.011	7.12	1.02		8.14
9000	Insulation, 3-1/2" thick blanket, R11	"	0.005	2.13	0.48		2.61

© 2020 BNi Publications Inc. ALL RIGHTS RESERVED

DIVISION 14
CONVEYING

CONVEYING EQUIPMENT

ID Code	Description: Component Descriptions	Output: Unit of Meas.	Output: Manhr / Unit	Unit Costs: Material Cost	Unit Costs: Labor Cost	Unit Costs: Equipment Cost	Unit Costs: Total Cost
14 - 11001	**DUMBWAITERS**						**14 - 11001**
0100	28' travel, extruded alum., 4 stops, 100 lbs. capacity	EA					6,760
0120	150 lbs. capacity	"					9,510
0140	200 lbs. capacity	"					13,650

CONVEYING EQUIPMENT

ID Code	Component Descriptions	Unit of Meas.	Manhr / Unit	Material Cost	Labor Cost	Equipment Cost	Total Cost
14 - 21001	**ELEVATORS**						**14 - 21001**
0420	Passenger elevators, electric, geared						
0440	Based on a shaft of 6 stops and 6 openings						
0450	50 fpm, 2000 lb	EA	24.000	145,640	1,520	1,330	148,490
0510	100 fpm, 2000 lb	"	26.667	151,020	1,690	1,480	154,190
0520	150 fpm						
0530	2000 lb	EA	30.000	166,620	1,900	1,660	170,180
0540	3000 lb	"	34.286	209,870	2,170	1,900	213,940
0550	4000 lb	"	40.000	218,380	2,530	2,220	223,130
0560	200 fpm						
0570	2500 lb	EA	34.286	201,370	2,170	1,900	205,440
0580	3000 lb	"	36.923	207,040	2,340	2,050	211,430
0590	4000 lb	"	40.000	218,380	2,530	2,220	223,130
0600	250 fpm						
0610	2500 lb	EA	34.286	208,460	2,170	1,900	212,530
0620	3000 lb	"	36.923	220,010	2,340	2,050	224,400
0630	4000 lb	"	40.000	225,120	2,530	2,220	229,870
0640	300 fpm						
0650	2500 lb	EA	34.286	206,470	2,170	1,900	210,540
0660	3000 lb	"	36.923	218,380	2,340	2,050	222,770
0670	4000 lb	"	24.000	222,210	1,520	1,330	225,060
0680	For each additional; 50 fpm, add per stop, $3000						
0690	500 lb, add per stop, $4000						
0700	Opening, add per stop, $4500						
0710	Stop, add per stop, $4000						
0720	Bonderized steel door, add per opening, $150						
0730	Colored aluminum door, add per opening, $850						
0740	Stainless steel door, add per opening, $600						
0750	Cast bronze door, add per opening, $1100						
0760	Two speed door, add per opening, $360						
0770	Bi-parting door, add per opening, $850						
0780	Custom cab interior add, $4800						

© 2020 BNi Publications Inc. ALL RIGHTS RESERVED

CONVEYING EQUIPMENT

	Description	Output		Unit Costs			
ID Code	Component Descriptions	Unit of Meas.	Manhr / Unit	Material Cost	Labor Cost	Equipment Cost	Total Cost
14 - 21001	ELEVATORS, Cont'd...						14 - 21001
0790	Based on a shaft of 8 stops and 8 openings						
1010	300 fpm						
1020	3000 lb	EA	48.000	270,190	3,040	2,660	275,890
1040	3500 lb	"	48.000	274,450	3,040	2,660	280,150
1060	4000 lb	"	53.333	287,960	3,380	2,960	294,290
1070	5000 lb	"	57.143	319,960	3,620	3,170	326,750
1080	400 fpm						
1090	3000 lb	EA	48.000	283,060	3,040	2,660	288,760
1100	3500 lb	"	48.000	287,960	3,040	2,660	293,660
1120	4000 lb	"	53.333	307,870	3,380	2,960	314,200
1140	5000 lb	"	57.143	350,530	3,620	3,170	357,320
1150	600 fpm						
1160	3000 lb	EA	53.333	398,670	3,380	2,960	405,000
1180	3500 lb	"	57.143	408,130	3,620	3,170	414,920
1190	4000 lb	"	58.537	412,390	3,710	3,240	419,340
1200	5000 lb	"	60.000	423,770	3,800	3,330	430,900
1210	800 fpm						
1220	3000 lb	EA	53.333	472,830	3,380	2,960	479,160
1240	3500 lb	"	57.143	477,810	3,620	3,170	484,600
1260	4000 lb	"	58.537	483,490	3,710	3,240	490,440
1280	5000 lb	"	60.000	486,690	3,800	3,330	493,820
1300	For each additional; 100 fpm add per stop, $13,000						
1310	500 lb, add per stop, $6500						
1320	Opening add per stop, $12,000						
1330	Stop add per stop, $4800						
1340	Bypass floor, add per each, $2000						
1350	Bonderized steel door, add per opening, $150						
1360	Colored aluminum door, add per opening, $900						
1370	Stainless steel door, add per opening, $600						
1380	Cast bronze door, add per opening, $600						
1390	Two speed bi-parting door, add per opening, $1000						
1400	Custom cab interior, add $5000						
1410	Hydraulic, based on a shaft of 3 stops, 3 openings						
1420	50 fpm						
1500	2000 lb	EA	20.000	101,250	1,270	1,110	103,630
1510	2500 lb	"	20.000	108,270	1,270	1,110	110,650
1520	3000 lb	"	20.870	114,260	1,320	1,160	116,740
1530	100 fpm						

© 2020 BNi Publications Inc. ALL RIGHTS RESERVED

CONVEYING EQUIPMENT

	Description	Output		Unit Costs			
ID Code	Component Descriptions	Unit of Meas.	Manhr / Unit	Material Cost	Labor Cost	Equipment Cost	Total Cost
14 - 21001	**ELEVATORS, Cont'd...**						**14 - 21001**
1540	2000 lb	EA	20.000	110,920	1,270	1,110	113,300
1550	2500 lb	"	20.870	116,960	1,320	1,160	119,440
1560	3000 lb	"	21.818	125,140	1,380	1,210	127,730
1570	150 fpm						
1580	2000 lb	EA	20.000	120,160	1,270	1,110	122,540
1590	2500 lb	"	20.870	131,540	1,320	1,160	134,020
1600	3000 lb	"	22.857	140,780	1,450	1,270	143,490
1610	For each additional; 50 fpm add per stop, $3500						
1620	500 lb, add per stop, $3500						
1630	Opening, add, $4200						
1640	Stop, add per stop, $5300						
1650	Bonderized steel door, add per opening, $400						
1660	Colored aluminum door, add per opening, $1500						
1670	Stainless steel door, add per opening, $650						
1680	Cast bronze door, add per opening, $1200						
1690	Two speed door, add per opening, $400						
1700	Bi-parting door, add per opening, $900						
1710	Custom cab interior, add per cab, $5000						
1720	Small elevators, 4 to 6 passenger capacity						
1730	Electric, push						
2010	2 stops	EA	20.000	34,510	1,270	1,110	36,890
2020	3 stops	"	21.818	43,140	1,380	1,210	45,730
2030	4 stops	"	24.000	49,110	1,520	1,330	51,960
2080	Freight elevators, electric						
2090	Based on a shaft of 6 stops and 6 openings						
2100	50 fpm						
2110	3500 lb	EA	26.667	260,160	1,690	1,480	263,330
2120	4000 lb	"	26.667	260,820	1,690	1,480	263,990
2130	5000 lb	"	30.000	265,470	1,900	1,660	269,030
2140	100 fpm						
2150	3500 lb	EA	30.000	272,100	1,900	1,660	275,660
2160	4000 lb	"	30.000	276,080	1,900	1,660	279,640
2170	5000 lb	"	34.286	280,930	2,170	1,900	285,000
2180	200 fpm						
2190	3500 lb	EA	34.286	270,770	2,170	1,900	274,840
2200	4000 lb	"	34.286	272,100	2,170	1,900	276,170
2210	5000 lb	"	40.000	274,760	2,530	2,220	279,510
2220	For elevator with manual door, deduct 15%						

© 2020 BNi Publications Inc. ALL RIGHTS RESERVED

CONVEYING EQUIPMENT

	Description	Output		Unit Costs			
ID Code	Component Descriptions	Unit of Meas.	Manhr / Unit	Material Cost	Labor Cost	Equipment Cost	Total Cost
14 - 21001	**ELEVATORS, Cont'd...**						**14 - 21001**
2230	For variable voltage control, add 20%						
2300	Based on shaft of 8 stops and 8 openings						
2310	100 fpm						
2320	4000 lb	EA	30.000	261,950	1,900	1,660	265,510
2330	6000 lb	"	30.769	264,140	1,950	1,710	267,790
2340	8000 lb	"	32.432	268,780	2,050	1,800	272,630
2350	150 fpm						
2360	4000 lb	EA	34.286	265,470	2,170	1,900	269,540
2370	6000 lb	"	35.294	265,470	2,240	1,960	269,660
2380	8000 lb	"	37.500	273,430	2,380	2,080	277,880
2390	200 fpm						
2400	4000 lb	EA	40.000	266,790	2,530	2,220	271,540
2410	6000 lb	"	41.379	273,430	2,620	2,290	278,340
2420	8000 lb	"	43.636	285,370	2,760	2,420	290,550
2430	For each additional; 50 fpm, add per stop, $2000						
2440	500 lb, add per stop, $600						
2450	Opening, add per stop, $7000						
2460	Stop, add per stop, $5500						
2470	For variable voltage, add 20%						
2480	Hydraulic, based on 3 stops and 3 openings						
2490	50 fpm						
2510	3000 lb	EA	17.143	111,500	1,090	950	113,540
2520	4000 lb	"	17.778	122,110	1,130	990	124,220
2530	6000 lb	"	18.462	142,020	1,170	1,020	144,210
2540	100 fpm						
2550	3000 lb	EA	17.143	126,100	1,090	950	128,140
2560	4000 lb	"	17.778	132,730	1,130	990	134,840
2570	6000 lb	"	18.462	156,620	1,170	1,020	158,810
2580	150 fpm						
2590	3000 lb	EA	17.143	138,040	1,090	950	140,080
2600	4000 lb	"	17.778	148,660	1,130	990	150,770
2610	6000 lb	"	18.462	169,900	1,170	1,020	172,090
2620	For each additional; 50 fpm, add per stop, $2000						
2630	500 lb, add per stop, $600						
2640	Opening, add per stop, $5500						
2650	Stop, add per stop, $5500						
2660	For elevator with manual door deduct from total,						

© 2020 BNi Publications Inc. ALL RIGHTS RESERVED

ESCALATORS

	Description	Output		Unit Costs			
ID Code	Component Descriptions	Unit of Meas.	Manhr / Unit	Material Cost	Labor Cost	Equipment Cost	Total Cost
14 - 31001	**ESCALATORS**						**14 - 31001**
1000	Escalators						
1020	32" wide, floor to floor						
1040	12' high	EA	40.000	180,150	2,530	2,220	184,900
1050	15' high	"	48.000	196,830	3,040	2,660	202,530
1060	18' high	"	60.000	211,990	3,800	3,330	219,120
1070	22' high	"	80.000	209,770	5,070	4,430	219,270
1080	25' high	"	96.000	238,380	6,080	5,320	249,780
1085	48" wide						
1090	12' high	EA	41.379	200,710	2,620	2,290	205,620
1100	15' high	"	50.000	219,050	3,170	2,770	224,990
1120	18' high	"	63.158	235,310	4,000	3,500	242,810
1130	22' high	"	85.714	263,410	5,430	4,750	273,590
1140	25' high	"	96.000	281,260	6,080	5,320	292,660

LIFTS

14 - 41001	**PERSONNEL LIFTS**						**14 - 41001**
1000	Electrically operated, 1 or 2 person lift						
1001	With attached foot platforms						
1020	3 stops	EA					12,390
1040	5 stops	"					19,320
1060	7 stops	"					22,520
2000	For each additional stop, add $1250						
3020	Residential stair climber, per story	EA	6.667	5,750	550		6,300
3030	curved	"	8.000	12,150	660		12,810

WHEELCHAIR LIFTS

14 - 42001	**WHEELCHAIR LIFTS**						**14 - 42001**
1000	600 lb, Residential	EA	8.000	6,830	660		7,490
1001	Commercial	"	8.000	16,150	660		16,810

© 2020 BNi Publications Inc. ALL RIGHTS RESERVED

VEHICLE LIFTS

ID Code	Description: Component Descriptions	Output: Unit of Meas.	Output: Manhr / Unit	Unit Costs: Material Cost	Unit Costs: Labor Cost	Unit Costs: Equipment Cost	Unit Costs: Total Cost
14 - 45001	VEHICLE LIFTS						14 - 45001
1020	Automotive hoist, one post, semi-hydraulic, 8,000 lb	EA	24.000	4,140	1,520	1,330	6,990
1040	Full hydraulic, 8,000 lb	"	24.000	4,270	1,520	1,330	7,120
1060	2 post, semi-hydraulic, 10,000 lb	"	34.286	4,460	2,170	1,900	8,530
1070	Full hydraulic						
1080	10,000 lb	EA	34.286	5,250	2,170	1,900	9,320
1100	13,000 lb	"	60.000	6,570	3,800	3,330	13,700
1120	18,500 lb	"	60.000	10,500	3,800	3,330	17,630
1140	24,000 lb	"	60.000	14,770	3,800	3,330	21,900
1160	26,000 lb	"	60.000	14,380	3,800	3,330	21,510
1170	Pneumatic hoist, fully hydraulic						
1180	11,000 lb	EA	80.000	7,090	5,070	4,430	16,590
1200	24,000 lb	"	80.000	12,800	5,070	4,430	22,300

MATERIAL HANDLING

ID Code	Component Descriptions	Unit of Meas.	Manhr / Unit	Material Cost	Labor Cost	Equipment Cost	Total Cost
14 - 91001	CHUTES						14 - 91001
1020	Linen chutes, stainless steel, with supports						
1030	18" dia.	LF	0.057	170	5.14		180
1040	24" dia.	"	0.062	210	5.53		220
1050	30" dia.	"	0.067	220	6.00		230
1060	Hopper	EA	0.533	2,680	48.00		2,730
1070	Skylight	"	0.800	1,630	72.00		1,700
1080	Sprinkler unit at top	"	0.889	610	80.00		690
1100	For galvanized metal, deduct from material cost, 35%						
1120	For aluminum, deduct from material cost, 25%						

PNEUMATIC TUBE SYSTEMS

ID Code	Component Descriptions	Unit of Meas.	Manhr / Unit	Material Cost	Labor Cost	Equipment Cost	Total Cost
14 - 92001	PNEUMATIC SYSTEMS						14 - 92001
1000	Pneumatic message tube system						
1010	Average, 20 station job						
1020	3" round system	EA	72.727	44,630	6,040		50,670
1040	4" round system	"	80.000	56,380	6,640		63,020
1060	6" round system	"	88.889	96,740	7,380		104,120
1080	4" x 7" oval system	"	160.000	101,900	13,280		115,180
5000	Trash and linen tube system						
5020	10 stations	EA	120.000	30,070	7,600	6,650	44,320
5030	15 stations	"	160.000	37,690	10,140	8,870	56,690

© 2020 BNi Publications Inc. ALL RIGHTS RESERVED

PNEUMATIC TUBE SYSTEMS

	Description	Output		Unit Costs			
ID Code	Component Descriptions	Unit of Meas.	Manhr / Unit	Material Cost	Labor Cost	Equipment Cost	Total Cost
14 - 92001	**PNEUMATIC SYSTEMS, Cont'd...**						**14 - 92001**
5040	20 stations	EA	184.615	50,230	11,700	10,230	72,160
5060	30 stations	"	218.182	65,330	13,820	12,090	91,240

© 2020 BNi Publications Inc. ALL RIGHTS RESERVED

DIVISION 21
FIRE SUPPRESSION

COMPONENTS

ID Code	Description: Component Descriptions	Output: Unit of Meas.	Output: Manhr / Unit	Unit Costs: Material Cost	Unit Costs: Labor Cost	Unit Costs: Equipment Cost	Unit Costs: Total Cost
21 - 11160	**HYDRANTS**						**21 - 11160**
0980	Wall hydrant						
1000	8" thick	EA	1.333	400	120		520
1020	12" thick	"	1.600	470	140		610
1040	18" thick	"	1.778	510	160		670
1060	24" thick	"	2.000	560	180		740
1070	Ground hydrant						
1080	2' deep	EA	1.000	740	89.00		830
1100	4' deep	"	1.143	860	100		960
1120	6' deep	"	1.333	970	120		1,090
1140	8' deep	"	2.000	1,090	180		1,270

FIRE PROTECTION

ID Code	Component Descriptions	Unit of Meas.	Manhr / Unit	Material Cost	Labor Cost	Equipment Cost	Total Cost
21 - 13001	**WET SPRINKLER SYSTEM**						**21 - 13001**
0120	Sprinkler head, 212 deg, brass, exposed piping	EA	0.320	16.00	28.50		44.50
0140	Chrome, concealed piping	"	0.444	19.50	39.75		59.00
0160	Water motor alarm	"	1.333	370	120		490
0180	Fire department inlet connection	"	1.600	270	140		410
0190	Wall plate for fire dept connection	"	0.667	130	60.00		190
0220	Swing check valve flanged iron body, 4"	"	2.667	350	240		590
0240	Check valve, 6"	"	4.000	1,150	360		1,510
0280	Wet pipe valve, flange to groove, 4"	"	0.889	990	79.00		1,070
0290	Flange to flange						
0300	6"	EA	1.333	1,340	120		1,460
0320	8"	"	2.667	2,360	240		2,600
0380	Alarm valve, flange to flange (wet valve)						
0400	4"	EA	0.889	1,530	79.00		1,610
0420	8"	"	6.667	2,420	600		3,020
0800	Inspector's test connection	"	0.667	70.00	60.00		130
1000	Wall hydrant, polished brass, 2-1/2" x 2-1/2", single	"	0.571	470	51.00		520
1020	2-way	"	0.571	1,060	51.00		1,110
1040	3-way	"	0.571	2,170	51.00		2,220
2080	Wet valve trim, includes retard chamber & gauges, 4"-	"	0.667	740	60.00		800
2100	Retard pressure switch for wet systems	"	1.600	1,300	140		1,440
2500	Air maintenance device	"	0.667	390	60.00		450
8000	Wall hydrant non-freeze, 8" thick wall, vacuum	"	0.400	51.00	35.75		87.00
8020	12" thick wall	"	0.400	56.00	35.75		92.00

© 2020 BNi Publications Inc. ALL RIGHTS RESERVED

CARBON-DIOXIDE FIRE EXTINGUISHING SYSTEMS

	Description	Output		Unit Costs			
ID Code	Component Descriptions	Unit of Meas.	Manhr / Unit	Material Cost	Labor Cost	Equipment Cost	Total Cost
21 - 21001	**CO_2 SYSTEM**						**21 - 21001**
0980	CO_2 system, high pressure, 75# cylinder with						
1000	Valve assemblies	EA	1.600	2,660	140		2,800
1020	Storage rack	"	1.143	1,420	100		1,520
1040	Manifold	"	5.714	1,030	510		1,540
1060	Flexible loops	"	0.100	84.00	8.94		93.00
1080	Beam scale for cylinders	"	1.333	700	120		820
1090	Mechanically controlled head	"	0.533	590	47.75		640
1100	Electrically controlled head	"	0.533	590	47.75		640
1120	Stop valves	"	0.800	1,500	72.00		1,570
1140	Check valves	"	1.000	680	89.00		770
1160	Activation station	"	0.800	790	72.00		860
1180	Nozzles	"	0.667	130	60.00		190
1200	Hose reel with 75' of 3/4" hose	"	4.000	4,750	360		5,110
1220	Main/reserve transfer switch	"	1.333	5,530	120		5,650
1240	Pressure switch	"	0.800	490	72.00		560
1260	Heat responsive device	"	1.333	780	120		900
1280	Battery and charger	"	4.000	4,430	360		4,790
2000	Low pressure						
2020	Battery and charger	EA	4.000	4,430	360		4,790
2040	Pressure switch	"	0.889	440	79.00		520
2060	Nozzles	"	0.727	130	65.00		200
2080	Master selector valve	"	1.333	380	120		500
2100	Selector valve	"	1.333	5,530	120		5,650
2120	Low pressure hose reel with 75' of 3/4" hose	"	4.000	6,860	360		7,220
2140	Tank fill lines	"	1.000	1,550	89.00		1,640
2160	Activation stations	"	0.667	780	60.00		840
2180	Electro manual pilot panels	"	1.000	1,550	89.00		1,640

DRY CHEMICAL FIRE EXTINGUISHING SYSTEMS

21 - 24001	**DRY SPRINKLER SYSTEM**						**21 - 24001**
0080	Dry pipe valve, flange to flange						
0100	4"	EA	1.600	2,250	140		2,390
0120	6"	"	2.000	2,820	180		3,000
0140	Trim, 4" and 6", includes gauges	"	0.667	870	60.00		930
0160	Field testing and flushing	"	6.667		600		600
0180	Disinfection	"	6.667		600		600
0200	Pressure switch double circuit, open/close contacts	"	2.000	420	180		600

© 2020 BNi Publications Inc. ALL RIGHTS RESERVED

DRY CHEMICAL FIRE EXTINGUISHING SYSTEMS

	Description	Output		Unit Costs			
ID Code	Component Descriptions	Unit of Meas.	Manhr / Unit	Material Cost	Labor Cost	Equipment Cost	Total Cost
21 - 24001	**DRY SPRINKLER SYSTEM, Cont'd...**						**21 - 24001**
0210	Low air						
0220	Supervisory unit	EA	1.333	1,360	120		1,480
0240	Pressure switch	"	0.667	430	60.00		490

© 2020 BNi Publications Inc. ALL RIGHTS RESERVED

DIVISION 22
PLUMBING

BASIC MATERIALS

	Description	Output		Unit Costs			
ID Code	Component Descriptions	Unit of Meas.	Manhr / Unit	Material Cost	Labor Cost	Equipment Cost	Total Cost
22 - 05236	**VALVES**						**22 - 05236**
0600	Gate valve, 125 lb, bronze, soldered						
0800	1/2"	EA	0.200	37.75	18.00		56.00
1000	3/4"	"	0.200	45.25	18.00		63.00
1010	1"	"	0.267	56.00	23.75		80.00
1030	1-1/2"	"	0.320	97.00	28.50		130
1040	2"	"	0.400	140	35.75		180
1050	2-1/2"	"	0.500	310	44.75		350
1055	Threaded						
1058	1/4", 125 lb	EA	0.320	35.50	28.50		64.00
1059	1/2"						
1060	125 lb	EA	0.320	34.25	28.50		63.00
1075	300 lb	"	0.320	86.00	28.50		110
1078	3/4"						
1083	125 lb	EA	0.320	40.00	28.50		69.00
1088	300 lb	"	0.320	100	28.50		130
1089	1"						
1091	125 lb	EA	0.320	52.00	28.50		81.00
1098	300 lb	"	0.400	150	35.75		190
1099	1-1/2"						
1100	125 lb	EA	0.400	90.00	35.75		130
1115	300 lb	"	0.444	260	39.75		300
1117	2"						
1118	125 lb	EA	0.571	120	51.00		170
1122	300 lb	"	0.667	330	60.00		390
1123	Cast iron, flanged						
1124	2", 150 lb	EA	0.667	490	60.00		550
1125	2-1/2"						
1126	125 lb	EA	0.667	480	60.00		540
1128	250 lb	"	0.667	1,310	60.00		1,370
1130	3"						
1132	125 lb	EA	0.800	570	72.00		640
1134	250 lb	"	0.800	1,200	72.00		1,270
1136	4"						
1138	125 lb	EA	1.143	750	100		850
1140	250 lb	"	1.143	1,610	100		1,710
1144	6"						
1148	125 lb	EA	1.600	1,380	140		1,520
1150	250 lb	"	1.600	3,230	140		3,370

© 2020 BNi Publications Inc. ALL RIGHTS RESERVED

BASIC MATERIALS

ID Code	Description: Component Descriptions	Output: Unit of Meas.	Output: Manhr / Unit	Unit Costs: Material Cost	Unit Costs: Labor Cost	Unit Costs: Equipment Cost	Unit Costs: Total Cost
22 - 05236	VALVES, Cont'd...						22 - 05236
1151	8"						
1152	125 lb	EA	2.000	2,200	180		2,380
1154	250 lb	"	2.000	6,260	180		6,440
1160	OS&Y, flanged						
1165	2"						
1170	125 lb	EA	0.667	450	60.00		510
1180	250 lb	"	0.667	1,190	60.00		1,250
1185	2-1/2"						
1190	125 lb	EA	0.667	470	60.00		530
1200	250 lb	"	0.800	1,470	72.00		1,540
1205	3"						
1210	125 lb	EA	0.800	520	72.00		590
1215	250 lb	"	0.800	1,530	72.00		1,600
1218	4"						
1220	125 lb	EA	1.333	690	120		810
1225	250 lb	"	1.333	2,340	120		2,460
1227	6"						
1228	125 lb	EA	1.600	1,160	140		1,300
1230	250 lb	"	1.600	3,700	140		3,840
3980	Ball valve, bronze, 250 lb, threaded						
4000	1/2"	EA	0.320	20.50	28.50		49.00
4010	3/4"	"	0.320	30.50	28.50		59.00
4020	1"	"	0.400	38.75	35.75		75.00
4030	1-1/4"	"	0.444	57.00	39.75		97.00
4040	1-1/2"	"	0.500	90.00	44.75		130
4050	2"	"	0.571	100	51.00		150
4980	Angle valve, bronze, 150 lb, threaded						
5000	1/2"	EA	0.286	100	25.50		130
5010	3/4"	"	0.320	140	28.50		170
5020	1"	"	0.320	210	28.50		240
5030	1-1/4"	"	0.400	270	35.75		310
5040	1-1/2"	"	0.444	350	39.75		390
5980	Balancing valve, meter connections, circuit setter						
6000	1/2"	EA	0.320	99.00	28.50		130
6010	3/4"	"	0.364	100	32.50		130
6020	1"	"	0.400	130	35.75		170
6030	1-1/4"	"	0.444	190	39.75		230
6040	1-1/2"	"	0.533	230	47.75		280

© 2020 BNi Publications Inc. ALL RIGHTS RESERVED

BASIC MATERIALS

ID Code	Description: Component Descriptions	Output: Unit of Meas.	Output: Manhr / Unit	Unit Costs: Material Cost	Unit Costs: Labor Cost	Unit Costs: Equipment Cost	Unit Costs: Total Cost
22 - 05236	**VALVES, Cont'd...**						**22 - 05236**
6050	2"	EA	0.667	320	60.00		380
6060	2-1/2"	"	0.800	630	72.00		700
6070	3"	"	1.000	920	89.00		1,010
6080	4"	"	1.333	1,290	120		1,410
8100	Pressure reducing valve, bronze, threaded, 250 lb						
8120	1/2"	EA	0.500	200	44.75		240
8140	3/4"	"	0.500	200	44.75		240
8200	1"	"	0.500	310	44.75		350
8210	1-1/4"	"	0.571	450	51.00		500
8220	1-1/2"	"	0.667	520	60.00		580
8225	Pressure regulating valve, bronze, class 300						
8230	1"	EA	0.500	750	44.75		790
8240	1-1/2"	"	0.615	1,000	55.00		1,060
8250	2"	"	0.800	1,130	72.00		1,200
8260	3"	"	1.143	1,280	100		1,380
8270	4"	"	1.600	1,600	140		1,740
8280	5"	"	2.000	2,420	180		2,600
8290	6"	"	2.667	2,460	240		2,700
8480	Solar water temperature regulating valve						
8500	3/4"	EA	0.667	770	60.00		830
8510	1"	"	0.800	790	72.00		860
8520	1-1/4"	"	0.889	850	79.00		930
8530	1-1/2"	"	1.000	950	89.00		1,040
8540	2"	"	1.143	1,170	100		1,270
8550	2-1/2"	"	2.000	2,210	180		2,390
8980	Tempering valve, threaded						
9000	3/4"	EA	0.267	420	23.75		440
9010	1"	"	0.320	530	28.50		560
9020	1-1/4"	"	0.400	790	35.75		830
9030	1-1/2"	"	0.400	900	35.75		940
9040	2"	"	0.500	1,230	44.75		1,270
9050	2-1/2"	"	0.667	2,080	60.00		2,140
9060	3"	"	0.800	2,750	72.00		2,820
9070	4"	"	1.143	5,500	100		5,600
9180	Thermostatic mixing valve, threaded						
9200	1/2"	EA	0.286	130	25.50		160
9210	3/4"	"	0.320	130	28.50		160
9220	1"	"	0.348	480	31.00		510

© 2020 BNi Publications Inc. ALL RIGHTS RESERVED

BASIC MATERIALS

ID Code	Description: Component Descriptions	Output: Unit of Meas.	Output: Manhr / Unit	Unit Costs: Material Cost	Unit Costs: Labor Cost	Unit Costs: Equipment Cost	Unit Costs: Total Cost
22 - 05236	**VALVES, Cont'd...**						**22 - 05236**
9230	1-1/2"	EA	0.400	540	35.75		580
9240	2"	"	0.500	680	44.75		720
9245	Sweat connection						
9250	1/2"	EA	0.286	150	25.50		180
9260	3/4"	"	0.320	180	28.50		210
9265	Mixing valve, sweat connection						
9270	1/2"	EA	0.286	79.00	25.50		100
9280	3/4"	"	0.320	79.00	28.50		110
9480	Liquid level gauge, aluminum body						
9500	3/4"	EA	0.320	400	28.50		430
9505	125 psi, PVC body						
9510	3/4"	EA	0.320	470	28.50		500
9520	150 psi, CRS body						
9530	3/4"	EA	0.320	380	28.50		410
9540	1"	"	0.320	410	28.50		440
9560	175 psi, bronze body, 1/2"	"	0.286	760	25.50		790
22 - 05291	**PIPE HANGERS, HEAVY**						**22 - 05291**
0160	Hangers						
0180	1/2" pipe, clevis pipe hanger						
0200	Black steel	EA	0.267	1.87	23.75		25.50
0210	Galvanized	"	0.267	2.81	23.75		26.50
0230	U bolt	"	0.080	1.74	7.15		8.89
0290	3/4" pipe, clevis pipe hanger						
0300	Black steel	EA	0.267	1.91	23.75		25.75
0310	Galvanized	"	0.267	2.88	23.75		26.75
0390	1" pipe, clevis pipe hanger						
0400	Black steel	EA	0.267	1.96	23.75		25.75
0410	Galvanized	"	0.267	3.09	23.75		26.75
0690	2" pipe, clevis pipe hanger						
0700	Black steel	EA	0.267	2.66	23.75		26.50
0705	Galvanized	"	0.267	4.36	23.75		28.00
0880	3" pipe, clevis pipe hanger						
0900	Black steel	EA	0.267	5.22	23.75		29.00
0910	Galvanized	"	0.267	9.10	23.75		32.75
1080	4" pipe, clevis pipe hanger						
1100	Black steel	EA	0.267	6.39	23.75		30.25
1110	Galvanized	"	0.267	11.50	23.75		35.25

© 2020 BNi Publications Inc. ALL RIGHTS RESERVED

BASIC MATERIALS

	Description	Output		Unit Costs			
ID Code	Component Descriptions	Unit of Meas.	Manhr / Unit	Material Cost	Labor Cost	Equipment Cost	Total Cost
22 - 05291	**PIPE HANGERS, HEAVY, Cont'd...**						**22 - 05291**
1300	6" pipe, clevis pipe hanger						
1320	Black steel	EA	0.320	10.25	28.50		38.75
1330	Galvanized	"	0.320	20.50	28.50		49.00
1560	12" pipe, clevis pipe hanger						
1580	Black steel	EA	0.320	36.00	28.50		65.00
1590	Galvanized	"	0.320	53.00	28.50		82.00
8000	Threaded rod, galvanized, material only						
8010	3/8"	LF					0.68
8020	1/2"	"					1.36
8065	Hex nuts, galvanized						
8070	3/8"	EA					0.23
8080	1/2"	"					0.48
8210	1"	"					3.04
8220	C-clamp, steel, with lock nut						
8230	3/8"	EA	0.100	2.71	8.94		11.75
8240	1/2"	"	0.100	3.04	8.94		12.00
22 - 06291	**PIPE HANGERS, LIGHT**						**22 - 06291**
0010	A band, black iron						
0020	1/2"	EA	0.057	1.03	5.10		6.13
0030	1"	"	0.059	1.11	5.29		6.40
0040	1-1/4"	"	0.062	1.23	5.50		6.73
0050	1-1/2"	"	0.067	1.28	5.96		7.24
0060	2"	"	0.073	1.36	6.50		7.86
0070	2-1/2"	"	0.080	2.03	7.15		9.18
0080	3"	"	0.089	2.48	7.94		10.50
0090	4"	"	0.100	3.26	8.94		12.25
0100	5"	"	0.107	4.13	9.53		13.75
0110	6"	"	0.114	7.14	10.25		17.50
0120	8"	"	0.133	11.50	12.00		23.50
0130	Copper						
0140	1/2"	EA	0.057	1.67	5.10		6.77
0150	3/4"	"	0.059	1.94	5.29		7.23
0160	1"	"	0.059	1.94	5.29		7.23
0170	1-1/4"	"	0.062	2.09	5.50		7.59
0180	1-1/2"	"	0.067	2.24	5.96		8.20
0190	2"	"	0.073	2.37	6.50		8.87
0200	2-1/2"	"	0.080	4.79	7.15		12.00

© 2020 BNi Publications Inc. ALL RIGHTS RESERVED

BASIC MATERIALS

ID Code	Description: Component Descriptions	Output: Unit of Meas.	Output: Manhr / Unit	Unit Costs: Material Cost	Unit Costs: Labor Cost	Unit Costs: Equipment Cost	Unit Costs: Total Cost
22 - 06291	**PIPE HANGERS, LIGHT, Cont'd...**						**22 - 06291**
0210	3"	EA	0.089	4.99	7.94		13.00
0220	4"	"	0.100	5.51	8.94		14.50
0230	Black riser friction hangers						
0240	3/4"	EA	0.067	4.22	5.96		10.25
0250	1"	"	0.070	4.27	6.21		10.50
0260	1-1/4"	"	0.073	5.34	6.50		11.75
0270	1-1/2"	"	0.076	5.80	6.81		12.50
0280	2"	"	0.080	5.91	7.15		13.00
0290	2-1/2"	"	0.089	6.38	7.94		14.25
0300	3"	"	0.100	6.56	8.94		15.50
0310	4"	"	0.114	8.37	10.25		18.50
0360	Short pattern black riser clamps						
0370	1-1/2"	EA	0.073	5.55	6.50		12.00
0380	2"	"	0.076	5.81	6.81		12.50
0390	3"	"	0.080	6.38	7.15		13.50
0400	4"	"	0.089	7.32	7.94		15.25
0410	Copper riser friction hanger						
0420	1/2"	EA	0.062	7.19	5.50		12.75
0430	3/4"	"	0.064	7.40	5.72		13.00
0440	1"	"	0.067	7.54	5.96		13.50
0450	1-1/4"	"	0.070	9.39	6.21		15.50
0460	1-1/2"	"	0.073	10.25	6.50		16.75
0470	2"	"	0.076	10.50	6.81		17.25
0480	2-1/2"	"	0.080	11.25	7.15		18.50
0490	3"	"	0.080	11.50	7.15		18.75
0501	4"	"	0.089	14.75	7.94		22.75
0510	Auto grip hangers, galvanized						
0520	1/2"	EA	0.057	1.03	5.10		6.13
0540	1"	"	0.064	1.22	5.72		6.94
0570	2"	"	0.073	2.03	6.50		8.53
0590	3"	"	0.080	4.02	7.15		11.25
0600	4"	"	0.089	4.98	7.94		13.00
0610	Copper						
0620	1/2"	EA	0.057	1.70	5.10		6.80
0640	1"	"	0.064	1.98	5.72		7.70
0670	2"	"	0.073	3.28	6.50		9.78
0690	3"	"	0.080	6.45	7.15		13.50
0700	4"	"	0.089	7.98	7.94		16.00

© 2020 BNi Publications Inc. ALL RIGHTS RESERVED

BASIC MATERIALS

ID Code	Description: Component Descriptions	Output: Unit of Meas.	Output: Manhr / Unit	Unit Costs: Material Cost	Unit Costs: Labor Cost	Unit Costs: Equipment Cost	Unit Costs: Total Cost
22 - 06291	PIPE HANGERS, LIGHT, Cont'd...						22 - 06291
0710	Split rings (F&M), galvanized						
0730	1/2"	EA	0.062	3.38	5.50		8.88
0750	1"	"	0.067	4.67	5.96		10.75
0780	2"	"	0.076	6.69	6.81		13.50
0800	3"	"	0.084	16.25	7.52		23.75
0810	4"	"	0.089	17.00	7.94		25.00
0820	Copper						
0850	1/2"	EA	0.062	5.11	5.50		10.50
0870	1"	"	0.067	7.06	5.96		13.00
0900	2"	"	0.076	10.00	6.81		16.75
0920	3"	"	0.084	24.50	7.52		32.00
0930	4"	"	0.089	25.50	7.94		33.50
1000	2 hole clips, galvanized						
1030	3/4"	EA	0.053	0.27	4.76		5.03
1040	1"	"	0.055	0.30	4.93		5.23
1050	1-1/4"	"	0.057	0.39	5.10		5.49
1060	1-1/2"	"	0.059	0.48	5.29		5.77
1070	2"	"	0.062	0.63	5.50		6.13
1080	2-1/2"	"	0.064	1.14	5.72		6.86
1090	3"	"	0.067	1.66	5.96		7.62
1110	4"	"	0.073	3.56	6.50		10.00
1120	Perforated strap						
1130	3/4"						
1140	Galvanized, 20 ga.	LF	0.040	0.44	3.57		4.01
1150	Copper, 22 ga.	"	0.040	0.69	3.57		4.26
1160	Threaded rod couplings						
1170	1/4"	EA	0.050	1.57	4.47		6.04
1180	3/4"	"	0.053	1.65	4.76		6.41
1190	1/2"	"	0.057	1.87	5.10		6.97
1200	5/8"	"	0.062	2.88	5.50		8.38
1220	Reducing rod coupling, 1/2" x 3/8"	"	0.057	2.59	5.10		7.69
1230	C-clamps						
1240	3/4"	EA	0.080	2.24	7.15		9.39
1250	Top beam clamp						
1260	3/8"	EA	0.067	3.45	5.96		9.41
1270	1/2"	"	0.073	4.27	6.50		10.75
1280	Side beam connector						
1290	3/8"	EA	0.067	1.49	5.96		7.45

© 2020 BNi Publications Inc. ALL RIGHTS RESERVED

BASIC MATERIALS

ID Code	Description: Component Descriptions	Output: Unit of Meas.	Output: Manhr / Unit	Unit Costs: Material Cost	Unit Costs: Labor Cost	Unit Costs: Equipment Cost	Unit Costs: Total Cost
22 - 06291	**PIPE HANGERS, LIGHT, Cont'd...**						**22 - 06291**
1300	1/2"	EA	0.073	3.31	6.50		9.81
1310	Hex nuts, heavy, material only						
1320	1"	EA					3.80
1330	Heavy washers						
1340	3/8"	EA					0.11
1350	1/2"	"					0.26
1360	5/8"	"					0.53
1370	3/4"	"					1.06
1380	Lag rod, 3/8" x						
1390	4"	EA					0.49
1400	4-1/2"	"					0.49
1410	6"	"					0.51
1420	8"	"					0.92
1430	10"	"					1.08
1440	12"	"					1.34
1450	18"	"					1.81
1740	J-Hooks						
1750	1/2"	EA	0.036	0.79	3.25		4.04
1760	3/4"	"	0.036	0.84	3.25		4.09
1770	1"	"	0.038	0.86	3.40		4.26
1780	1-1/4"	"	0.039	0.91	3.48		4.39
1790	1-1/2"	"	0.040	0.93	3.57		4.50
1800	2"	"	0.040	0.97	3.57		4.54
1810	3"	"	0.042	1.12	3.76		4.88
1820	4"	"	0.042	1.21	3.76		4.97
1830	PVC coated hangers, galvanized, 28 ga.						
1840	1-1/2" x 12"	EA	0.053	1.29	4.76		6.05
1850	2" x 12"	"	0.057	1.41	5.10		6.51
1860	3" x 12"	"	0.062	1.58	5.50		7.08
1870	4" x 12"	"	0.067	1.76	5.96		7.72
1880	Copper, 30 ga.						
1890	1-1/2" x 12"	EA	0.053	1.99	4.76		6.75
1900	2" x 12"	"	0.057	2.36	5.10		7.46
1910	3" x 12"	"	0.062	2.61	5.50		8.11
1920	4" x 12"	"	0.067	2.87	5.96		8.83
2090	Wire hook hangers						
2095	Black wire, 1/2" x						
2100	4"	EA	0.040	0.44	3.57		4.01

© 2020 BNi Publications Inc. ALL RIGHTS RESERVED

BASIC MATERIALS

ID Code	Description: Component Descriptions	Output: Unit of Meas.	Output: Manhr / Unit	Unit Costs: Material Cost	Unit Costs: Labor Cost	Unit Costs: Equipment Cost	Unit Costs: Total Cost
22 - 06291	**PIPE HANGERS, LIGHT, Cont'd...**						**22 - 06291**
2110	6"	EA	0.042	0.50	3.76		4.26
2120	8"	"	0.044	0.55	3.97		4.52
2130	10"	"	0.044	0.70	3.97		4.67
2140	12"	"	0.047	0.83	4.20		5.03
2150	3/4" x						
2160	4"	EA	0.042	0.53	3.76		4.29
2170	6"	"	0.044	0.58	3.97		4.55
2180	8"	"	0.047	0.59	4.20		4.79
2190	10"	"	0.050	0.81	4.47		5.28
2200	12"	"	0.053	0.82	4.76		5.58
2210	1" x						
2220	4"	EA	0.044	0.53	3.97		4.50
2230	6"	"	0.047	0.55	4.20		4.75
2240	8"	"	0.050	0.59	4.47		5.06
2250	10"	"	0.053	0.80	4.76		5.56
2260	12"	"	0.057	0.86	5.10		5.96
4000	Copper wire hooks						
4010	1/2" x						
4020	4"	EA	0.040	0.58	3.57		4.15
4030	6"	"	0.042	0.66	3.76		4.42
4040	8"	"	0.044	0.74	3.97		4.71
4050	10"	"	0.047	0.93	4.20		5.13
4060	12"	"	0.050	1.06	4.47		5.53
4070	3/4" x						
4080	4"	EA	0.042	0.58	3.76		4.34
4090	6"	"	0.044	0.72	3.97		4.69
4100	8"	"	0.047	0.83	4.20		5.03
4110	10"	"	0.050	0.95	4.47		5.42
4120	12"	"	0.053	1.13	4.76		5.89
22 - 06481	**VIBRATION CONTROL**						**22 - 06481**
0120	Vibration isolator, in-line, stainless connector						
0140	1/2"	EA	0.444	100	39.75		140
0180	1"	"	0.500	120	44.75		160
0280	2"	"	0.615	230	55.00		290
0300	3"	"	0.727	400	65.00		470
0340	6"	"	0.889	860	79.00		940

© 2020 BNi Publications Inc. ALL RIGHTS RESERVED

BASIC MATERIALS

	Description	Output		Unit Costs			
ID Code	Component Descriptions	Unit of Meas.	Manhr / Unit	Material Cost	Labor Cost	Equipment Cost	Total Cost
22 - 06931	**SPECIALTIES**						**22 - 06931**
1000	Wall penetration						
1010	Concrete wall, 6" thick						
1020	2" dia.	EA	0.267		17.00		17.00
1040	4" dia.	"	0.400		25.50		25.50
1060	8" dia.	"	0.571		36.50		36.50
1090	12" thick						
1100	2" dia.	EA	0.364		23.25		23.25
1120	4" dia.	"	0.571		36.50		36.50
1140	8" dia.	"	0.889		57.00		57.00
3010	Non-destructive testing, piping systems						
3020	X-ray of welds						
3030	3" dia. pipe	EA	0.800	25.75	72.00		98.00
3040	4" dia. pipe	"	0.800	34.25	72.00		110
3050	6" dia. pipe	"	0.800	34.25	72.00		110
3060	8" dia. pipe	"	1.000	34.25	89.00		120
3070	10" dia. pipe	"	1.000	45.00	89.00		130
3130	Liquid penetration of welds						
3140	2" dia. pipe	EA	0.500	6.44	44.75		51.00
3160	3" dia. pipe	"	0.500	6.44	44.75		51.00
3180	4" dia. pipe	"	0.500	6.44	44.75		51.00
3200	6" dia. pipe	"	0.500	6.44	44.75		51.00
3220	8" dia. pipe	"	0.500	9.66	44.75		54.00
3240	10" dia. pipe	"	0.500	9.66	44.75		54.00

INSULATION

22 - 07161	**EQUIPMENT INSULATION**						**22 - 07161**
0100	Equipment insulation, 2" thick, cellular glass	SF	0.050	3.56	4.47		8.03
0120	Urethane, rigid, jacket, plastered finish	"	0.100	3.81	8.94		12.75
0140	Fiberglass, rigid, with vapor barrier	"	0.044	3.56	3.97		7.53
22 - 07191	**FIBERGLASS PIPE INSULATION**						**22 - 07191**
1030	Fiberglass insulation on 1/2" pipe						
1040	1" thick	LF	0.027	1.25	2.38		3.63
1060	1-1/2" thick	"	0.033	2.64	2.98		5.62
1070	3/4" pipe						
1080	1" thick	LF	0.027	1.52	2.38		3.90
1100	1-1/2" thick	"	0.033	2.78	2.98		5.76
1110	1" pipe						

© 2020 BNi Publications Inc. ALL RIGHTS RESERVED

INSULATION

ID Code	Description: Component Descriptions	Output: Unit of Meas.	Output: Manhr / Unit	Unit Costs: Material Cost	Unit Costs: Labor Cost	Unit Costs: Equipment Cost	Unit Costs: Total Cost
22 - 07191	**FIBERGLASS PIPE INSULATION, Cont'd...**						**22 - 07191**
1120	1" thick	LF	0.027	1.52	2.38		3.90
1140	1-1/2" thick	"	0.033	2.91	2.98		5.89
1310	2" pipe						
1340	1" thick	LF	0.033	2.07	2.98		5.05
1360	1-1/2" thick	"	0.036	3.60	3.25		6.85
1380	2" thick	"	0.040	5.28	3.57		8.85
1430	2-1/2" pipe						
1440	1" thick	LF	0.033	2.21	2.98		5.19
1460	1-1/2" thick	"	0.036	3.88	3.25		7.13
1470	2" thick	"	0.040	5.63	3.57		9.20
1530	3" pipe						
1540	1" thick	LF	0.038	2.49	3.40		5.89
1560	1-1/2" thick	"	0.040	4.02	3.57		7.59
1580	2" thick	"	0.044	6.09	3.97		10.00
1640	4" pipe						
1660	1" thick	LF	0.038	3.19	3.40		6.59
1680	1-1/2" thick	"	0.040	4.57	3.57		8.14
1700	2" thick	"	0.044	7.01	3.97		11.00
1770	5" pipe						
1780	1" thick	LF	0.038	3.67	3.40		7.07
1800	2" thick	"	0.040	7.90	3.57		11.50
1850	6" pipe						
1870	1" thick	LF	0.042	4.15	3.76		7.91
1880	2" thick	"	0.044	8.61	3.97		12.50
1980	8" pipe						
2000	2" thick	LF	0.042	10.75	3.76		14.50
2020	3" thick	"	0.044	17.00	3.97		21.00
2070	10" pipe						
2080	2" thick	LF	0.042	13.25	3.76		17.00
2100	3" thick	"	0.044	19.75	3.97		23.75
2150	12" pipe						
2160	2" thick	LF	0.042	14.75	3.76		18.50
2180	3" thick	"	0.044	22.25	3.97		26.25

© 2020 BNi Publications Inc. ALL RIGHTS RESERVED

INSULATION

ID Code	Component Descriptions	Unit of Meas.	Manhr / Unit	Material Cost	Labor Cost	Equipment Cost	Total Cost
22 - 07193	**EXTERIOR PIPE INSULATION**						**22 - 07193**
0090	Fiberglass insulation, aluminum jacket						
0110	1/2" pipe						
0120	1" thick	LF	0.062	2.16	5.50		7.66
0140	1-1/2" thick	"	0.067	4.08	5.96		10.00
0150	3/4" pipe						
0160	1" thick	LF	0.062	2.55	5.50		8.05
0180	1-1/2" thick	"	0.067	4.33	5.96		10.25
0190	1" pipe						
0200	1" thick	LF	0.062	2.64	5.50		8.14
0220	1-1/2" thick	"	0.067	4.56	5.96		10.50
0250	1-1/4" pipe						
0260	1" thick	LF	0.073	2.95	6.50		9.45
0280	1-1/2" thick	"	0.076	4.94	6.81		11.75
0310	1-1/2" pipe						
0320	1" thick	LF	0.073	3.19	6.50		9.69
0340	1-1/2" thick	"	0.076	5.13	6.81		12.00
0420	2" pipe						
0440	1" thick	LF	0.073	3.60	6.50		10.00
0460	1-1/2" thick	"	0.076	5.36	6.81		12.25
1030	3" pipe						
1040	1" thick	LF	0.080	4.33	7.15		11.50
1060	1-1/2" thick	"	0.084	6.40	7.52		14.00
1750	6" pipe						
1790	1" thick	LF	0.089	7.19	7.94		15.25
1800	2" thick	"	0.094	12.75	8.41		21.25
2250	10" pipe						
2260	2" thick	LF	0.089	18.25	7.94		26.25
2280	3" thick	"	0.094	26.25	8.41		34.75
22 - 07194	**PIPE INSULATION FITTINGS**						**22 - 07194**
0100	Insulation protection saddle						
0200	1" thick covering						
1000	1/2" pipe	EA	0.320	7.75	28.50		36.25
1020	3/4" pipe	"	0.320	7.97	28.50		36.50
1040	1" pipe	"	0.320	8.17	28.50		36.75
1100	2" pipe	"	0.320	9.01	28.50		37.50
1140	3" pipe	"	0.364	10.25	32.50		42.75
1180	6" pipe	"	0.500	10.25	44.75		55.00

© 2020 BNi Publications Inc. ALL RIGHTS RESERVED

INSULATION

ID Code	Description: Component Descriptions	Output: Unit of Meas.	Output: Manhr / Unit	Unit Costs: Material Cost	Unit Costs: Labor Cost	Unit Costs: Equipment Cost	Unit Costs: Total Cost
22 - 07194	**PIPE INSULATION FITTINGS, Cont'd...**						**22 - 07194**
1190	1-1/2" thick covering						
1200	3/4" pipe	EA	0.320	13.00	28.50		41.50
1220	1" pipe	"	0.320	13.25	28.50		41.75
1280	2" pipe	"	0.320	11.50	28.50		40.00
1300	3" pipe	"	0.320	13.00	28.50		41.50
1360	6" pipe	"	0.500	15.25	44.75		60.00
1400	10" pipe	"	0.667	16.00	60.00		76.00

FACILITY WATER DISTRIBUTION

ID Code	Component Descriptions	Unit of Meas.	Manhr / Unit	Material Cost	Labor Cost	Equipment Cost	Total Cost
22 - 11161	**COPPER PIPE**						**22 - 11161**
0600	Type K copper						
0900	1/2"	LF	0.025	3.77	2.23		6.00
1000	3/4"	"	0.027	7.03	2.38		9.41
1020	1"	"	0.029	9.20	2.55		11.75
1100	1-1/4"	"	0.031	11.50	2.75		14.25
1180	1-1/2"	"	0.033	15.00	2.98		18.00
1240	2"	"	0.036	23.00	3.25		26.25
1280	2-1/2"	"	0.040	33.75	3.57		37.25
1300	3"	"	0.042	47.00	3.76		51.00
1340	4"	"	0.044	78.00	3.97		82.00
3000	DWV, copper						
3020	1-1/4"	LF	0.033	10.25	2.98		13.25
3030	1-1/2"	"	0.036	13.00	3.25		16.25
3040	2"	"	0.040	17.00	3.57		20.50
3070	3"	"	0.044	29.00	3.97		33.00
3080	4"	"	0.050	50.00	4.47		54.00
3090	6"	"	0.057	200	5.10		210
4000	Refrigeration tubing, copper, sealed						
4010	1/8"	LF	0.032	0.73	2.86		3.59
4020	3/16"	"	0.033	0.85	2.98		3.83
4030	1/4"	"	0.035	1.02	3.10		4.12
4040	5/16"	"	0.036	1.31	3.25		4.56
4050	3/8"	"	0.038	1.51	3.40		4.91
4060	1/2"	"	0.040	1.98	3.57		5.55
4090	7/8"	"	0.046	4.81	4.08		8.89
4100	1-1/8"	"	0.053	6.90	4.76		11.75
4110	1-3/8"	"	0.062	10.50	5.50		16.00

© 2020 BNi Publications Inc. ALL RIGHTS RESERVED

FACILITY WATER DISTRIBUTION

ID Code	Description: Component Descriptions	Output: Unit of Meas.	Output: Manhr / Unit	Unit Costs: Material Cost	Unit Costs: Labor Cost	Unit Costs: Equipment Cost	Unit Costs: Total Cost
22 - 11161	**COPPER PIPE, Cont'd...**						**22 - 11161**
6000	Type L copper						
6090	1/4"	LF	0.024	1.52	2.10		3.62
6095	3/8"	"	0.024	2.33	2.10		4.43
6100	1/2"	"	0.025	2.71	2.23		4.94
6190	3/4"	"	0.027	4.33	2.38		6.71
6240	1"	"	0.029	6.50	2.55		9.05
6300	1-1/4"	"	0.031	9.31	2.75		12.00
6360	1-1/2"	"	0.033	12.00	2.98		15.00
6400	2"	"	0.036	18.75	3.25		22.00
6460	2-1/2"	"	0.040	27.75	3.57		31.25
6480	3"	"	0.042	37.25	3.76		41.00
6500	3-1/2"	"	0.043	48.75	3.86		53.00
6520	4"	"	0.044	62.00	3.97		66.00
6580	Type M copper						
6600	1/2"	LF	0.025	1.91	2.23		4.14
6620	3/4"	"	0.027	3.12	2.38		5.50
6630	1"	"	0.029	5.06	2.55		7.61
6650	1-1/4"	"	0.031	7.46	2.75		10.25
6660	2"	"	0.036	16.25	3.25		19.50
6670	2-1/2"	"	0.040	23.75	3.57		27.25
6680	3"	"	0.042	31.50	3.76		35.25
6690	4"	"	0.044	55.00	3.97		59.00
22 - 11162	**COPPER FITTINGS**						**22 - 11162**
0460	Coupling, with stop						
0470	1/4"	EA	0.267	0.95	23.75		24.75
0480	3/8"	"	0.320	1.24	28.50		29.75
0485	1/2"	"	0.348	0.99	31.00		32.00
0490	5/8"	"	0.400	2.87	35.75		38.50
0495	3/4"	"	0.444	1.97	39.75		41.75
0498	1"	"	0.471	4.06	42.00		46.00
0499	3"	"	0.800	49.25	72.00		120
0510	4"	"	1.000	110	89.00		200
0520	Reducing coupling						
0530	1/4" x 1/8"	EA	0.320	2.54	28.50		31.00
0540	3/8" x 1/4"	"	0.348	2.79	31.00		33.75
0545	1/2" x						
0550	3/8"	EA	0.400	2.10	35.75		37.75

© 2020 BNi Publications Inc. ALL RIGHTS RESERVED

FACILITY WATER DISTRIBUTION

	Description	Output		Unit Costs			
ID Code	Component Descriptions	Unit of Meas.	Manhr / Unit	Material Cost	Labor Cost	Equipment Cost	Total Cost
22 - 11162	**COPPER FITTINGS, Cont'd...**						**22 - 11162**
0560	1/4"	EA	0.400	2.54	35.75		38.25
0570	1/8"	"	0.400	2.80	35.75		38.50
0575	3/4" x						
0580	3/8"	EA	0.444	4.50	39.75		44.25
0590	1/2"	"	0.444	3.56	39.75		43.25
0595	1" x						
0600	3/8"	EA	0.500	8.08	44.75		53.00
0610	1" x 1/2"	"	0.500	7.82	44.75		53.00
0620	1" x 3/4"	"	0.500	6.59	44.75		51.00
0625	1-1/4" x						
0630	1/2"	EA	0.533	9.82	47.75		58.00
0640	3/4"	"	0.533	9.29	47.75		57.00
0650	1"	"	0.533	9.29	47.75		57.00
0655	1-1/2" x						
0660	1/2"	EA	0.571	16.25	51.00		67.00
0670	3/4"	"	0.571	15.50	51.00		67.00
0680	1"	"	0.571	15.50	51.00		67.00
0690	1-1/4"	"	0.571	15.50	51.00		67.00
0695	2" x						
0700	1/2"	EA	0.667	26.75	60.00		87.00
0710	3/4"	"	0.667	25.50	60.00		86.00
0720	1"	"	0.667	25.00	60.00		85.00
0730	1-1/4"	"	0.667	23.75	60.00		84.00
0740	1-1/2"	"	0.667	23.75	60.00		84.00
0745	2-1/2" x						
0750	1"	EA	0.800	61.00	72.00		130
0760	1-1/4"	"	0.800	61.00	72.00		130
0770	1-1/2"	"	0.800	54.00	72.00		130
0780	2"	"	0.800	53.00	72.00		130
0785	3" x						
0790	1-1/2"	EA	1.000	74.00	89.00		160
0800	2"	"	1.000	66.00	89.00		160
0810	2-1/2"	"	1.000	67.00	89.00		160
0815	4" x						
0820	2"	EA	1.143	150	100		250
0830	2-1/2"	"	1.143	150	100		250
0840	3"	"	1.143	130	100		230
0850	Slip coupling						

© 2020 BNi Publications Inc. ALL RIGHTS RESERVED

FACILITY WATER DISTRIBUTION

	Description	Output		Unit Costs			
ID Code	Component Descriptions	Unit of Meas.	Manhr / Unit	Material Cost	Labor Cost	Equipment Cost	Total Cost
22 - 11162	**COPPER FITTINGS, Cont'd...**						**22 - 11162**
0860	1/4"	EA	0.267	0.78	23.75		24.50
0870	1/2"	"	0.320	1.31	28.50		29.75
0880	3/4"	"	0.400	2.74	35.75		38.50
0890	1"	"	0.444	5.82	39.75		45.50
0900	1-1/4"	"	0.500	8.76	44.75		54.00
1000	1-1/2"	"	0.533	11.75	47.75		60.00
1020	2"	"	0.667	20.00	60.00		80.00
1030	2-1/2"	"	0.667	26.25	60.00		86.00
1040	3"	"	0.800	50.00	72.00		120
1050	4"	"	1.000	93.00	89.00		180
1060	Coupling with drain						
1070	1/2"	EA	0.400	10.00	35.75		45.75
1080	3/4"	"	0.444	14.75	39.75		55.00
1090	1"	"	0.500	18.25	44.75		63.00
1110	Reducer						
1120	3/8" x 1/4"	EA	0.320	2.85	28.50		31.25
1130	1/2" x 3/8"	"	0.320	2.29	28.50		30.75
1135	3/4" x						
1140	1/4"	EA	0.364	4.65	32.50		37.25
1150	3/8"	"	0.364	4.86	32.50		37.25
1160	1/2"	"	0.364	5.06	32.50		37.50
1165	1" x						
1170	1/2"	EA	0.400	6.99	35.75		42.75
1180	3/4"	"	0.400	5.36	35.75		41.00
1185	1-1/4" x						
1190	1/2"	EA	0.444	9.87	39.75		49.50
1200	3/4"	"	0.444	9.87	39.75		49.50
1210	1"	"	0.444	9.87	39.75		49.50
1215	1-1/2" x						
1220	1/2"	EA	0.500	12.75	44.75		58.00
1230	3/4"	"	0.500	12.75	44.75		58.00
1240	1"	"	0.500	12.75	44.75		58.00
1250	1-1/4"	"	0.500	12.75	44.75		58.00
1255	2" x						
1260	1/2"	EA	0.571	25.50	51.00		77.00
1270	3/4"	"	0.571	25.50	51.00		77.00
1280	1"	"	0.571	25.50	51.00		77.00
1290	1-1/4"	"	0.571	24.25	51.00		75.00

© 2020 BNi Publications Inc. ALL RIGHTS RESERVED

FACILITY WATER DISTRIBUTION

	Description	Output		Unit Costs			
ID Code	Component Descriptions	Unit of Meas.	Manhr / Unit	Material Cost	Labor Cost	Equipment Cost	Total Cost
22 - 11162	**COPPER FITTINGS, Cont'd...**						**22 - 11162**
1300	1-1/2"	EA	0.571	24.25	51.00		75.00
1310	2-1/2" x						
1320	1"	EA	0.667	56.00	60.00		120
1330	1-1/4"	"	0.667	50.00	60.00		110
1340	1-1/2"	"	0.667	49.25	60.00		110
1350	2"	"	0.667	48.25	60.00		110
1355	3" x						
1360	1-1/4"	EA	0.800	66.00	72.00		140
1370	1-1/2"	"	0.800	68.00	72.00		140
1380	2"	"	0.800	61.00	72.00		130
1390	2-1/2"	"	0.800	62.00	72.00		130
1395	4" x						
1400	2"	EA	1.000	140	89.00		230
1410	3"	"	1.000	130	89.00		220
1415	Female adapters						
1430	1/4"	EA	0.320	7.40	28.50		36.00
1440	3/8"	"	0.364	7.58	32.50		40.00
1450	1/2"	"	0.400	3.60	35.75		39.25
1460	3/4"	"	0.444	4.94	39.75		44.75
1470	1"	"	0.444	11.50	39.75		51.00
1480	1-1/4"	"	0.500	16.75	44.75		62.00
1490	1-1/2"	"	0.500	26.00	44.75		71.00
1500	2"	"	0.533	35.75	47.75		84.00
1510	2-1/2"	"	0.571	130	51.00		180
1520	3"	"	0.667	200	60.00		260
1530	4"	"	0.800	240	72.00		310
1540	Increasing female adapters						
1545	1/8" x						
1550	3/8"	EA	0.320	7.27	28.50		35.75
1560	1/2"	"	0.320	6.78	28.50		35.25
1570	1/4" x 1/2"	"	0.348	7.10	31.00		38.00
1580	3/8" x 1/2"	"	0.364	7.62	32.50		40.00
1585	1/2" x						
1590	3/4"	EA	0.400	8.08	35.75		43.75
1600	1"	"	0.400	16.25	35.75		52.00
1605	3/4" x						
1610	1"	EA	0.444	17.25	39.75		57.00
1620	1-1/4"	"	0.444	29.25	39.75		69.00

© 2020 BNi Publications Inc. ALL RIGHTS RESERVED

FACILITY WATER DISTRIBUTION

	Description	Output		Unit Costs			
ID Code	Component Descriptions	Unit of Meas.	Manhr / Unit	Material Cost	Labor Cost	Equipment Cost	Total Cost
22 - 11162	**COPPER FITTINGS, Cont'd...**						**22 - 11162**
1625	1" x						
1630	1-1/4"	EA	0.444	31.00	39.75		71.00
1640	1-1/2"	"	0.444	34.00	39.75		74.00
1645	1-1/4" x						
1650	1-1/2"	EA	0.500	37.00	44.75		82.00
1660	2"	"	0.500	47.25	44.75		92.00
1670	1-1/2" x 2"	"	0.533	69.00	47.75		120
1675	Reducing female adapters						
1690	3/8" x 1/4"	EA	0.364	6.55	32.50		39.00
1695	1/2" x						
1700	1/4"	EA	0.400	5.63	35.75		41.50
1710	3/8"	"	0.400	5.63	35.75		41.50
1720	3/4" x 1/2"	"	0.444	7.86	39.75		47.50
1725	1" x						
1730	1/2"	EA	0.444	21.00	39.75		61.00
1740	3/4"	"	0.444	16.75	39.75		57.00
1745	1-1/4" x						
1750	1/2"	EA	0.500	28.50	44.75		73.00
1760	3/4"	"	0.500	35.50	44.75		80.00
1770	1"	"	0.500	35.50	44.75		80.00
1780	1-1/2" x						
1790	1"	EA	0.533	33.25	47.75		81.00
1800	1-1/4"	"	0.533	36.00	47.75		84.00
1805	2" x						
1810	1"	EA	0.571	45.00	51.00		96.00
1820	1-1/4"	"	0.571	67.00	51.00		120
1830	1-1/2"	"	0.571	58.00	51.00		110
1840	Female fitting adapters						
1850	1/2"	FA	0.400	10.00	35.75		45.75
1860	3/4"	"	0.400	13.00	35.75		48.75
1870	3/4" x 1/2"	"	0.421	15.50	37.75		53.00
1880	1"	"	0.444	17.25	39.75		57.00
1890	1-1/4"	"	0.471	27.75	42.00		70.00
1900	1-1/2"	"	0.500	36.50	44.75		81.00
1910	2"	"	0.533	48.50	47.75		96.00
1920	Male adapters						
1930	1/4"	EA	0.364	11.25	32.50		43.75
1940	3/8"	"	0.364	5.63	32.50		38.25

© 2020 BNi Publications Inc. ALL RIGHTS RESERVED

FACILITY WATER DISTRIBUTION

	Description	Output		Unit Costs			
ID Code	Component Descriptions	Unit of Meas.	Manhr / Unit	Material Cost	Labor Cost	Equipment Cost	Total Cost
22 - 11162	**COPPER FITTINGS, Cont'd...**						**22 - 11162**
1950	3"	EA	0.667	140	60.00		200
1960	4"	"	0.800	200	72.00		270
1970	Increasing male adapters						
1980	3/8" x 1/2"	EA	0.364	7.68	32.50		40.25
1985	1/2" x						
1990	3/4"	EA	0.400	6.66	35.75		42.50
2000	1"	"	0.400	15.00	35.75		51.00
2005	3/4" x						
2010	1"	EA	0.421	14.75	37.75		53.00
2020	1-1/4"	"	0.421	18.75	37.75		57.00
2030	1" x 1-1/4"	"	0.444	18.75	39.75		59.00
2035	1-1/2" x						
2040	3/4"	EA	0.471	22.50	42.00		65.00
2050	1"	"	0.471	33.25	42.00		75.00
2060	1-1/4"	"	0.471	36.00	42.00		78.00
2065	2" x						
2070	1"	EA	0.500	85.00	44.75		130
2080	1-1/4"	"	0.500	87.00	44.75		130
2090	1-1/2"	"	0.500	83.00	44.75		130
2100	2" x 2-1/2"	"	0.533	140	47.75		190
8000	Copper pipe fittings						
8010	1/2"						
8020	90 deg ell	EA	0.178	1.62	16.00		17.50
8040	45 deg ell	"	0.178	2.04	16.00		18.00
8060	Tee	"	0.229	2.71	20.50		23.25
8100	Cap	"	0.089	1.10	7.94		9.04
8120	Coupling	"	0.178	1.18	16.00		17.25
8160	Union	"	0.200	8.21	18.00		26.25
8200	3/4"						
8220	90 deg ell	EA	0.200	3.54	18.00		21.50
8240	45 deg ell	"	0.200	4.13	18.00		22.25
8260	Tee	"	0.267	5.92	23.75		29.75
8290	Cap	"	0.094	2.15	8.41		10.50
8300	Coupling	"	0.200	2.40	18.00		20.50
8320	Union	"	0.229	12.00	20.50		32.50
8360	1"						
8380	90 deg ell	EA	0.267	8.21	23.75		32.00
8390	45 deg ell	"	0.267	10.75	23.75		34.50

© 2020 BNi Publications Inc. ALL RIGHTS RESERVED

FACILITY WATER DISTRIBUTION

ID Code	Description: Component Descriptions	Output: Unit of Meas.	Output: Manhr / Unit	Unit Costs: Material Cost	Unit Costs: Labor Cost	Unit Costs: Equipment Cost	Unit Costs: Total Cost
22 - 11162	**COPPER FITTINGS, Cont'd...**						**22 - 11162**
8400	Tee	EA	0.320	13.50	28.50		42.00
8420	Cap	"	0.133	4.00	12.00		16.00
8430	Coupling	"	0.267	5.92	23.75		29.75
8450	Union	"	0.267	15.75	23.75		39.50
8480	1-1/4"						
8500	90 deg ell	EA	0.229	11.25	20.50		31.75
8510	45 deg ell	"	0.229	13.75	20.50		34.25
8520	Tee	"	0.400	18.25	35.75		54.00
8540	Cap	"	0.133	3.19	12.00		15.25
8560	Union	"	0.286	26.25	25.50		52.00
8580	1-1/2"						
8600	90 deg ell	EA	0.286	14.50	25.50		40.00
8610	45 deg ell	"	0.286	17.25	25.50		42.75
8620	Tee	"	0.444	24.00	39.75		64.00
8640	Cap	"	0.133	3.19	12.00		15.25
8660	Coupling	"	0.267	10.75	23.75		34.50
8680	Union	"	0.364	39.75	32.50		72.00
8905	2"						
8910	90 deg ell	EA	0.320	28.50	28.50		57.00
8920	45 deg ell	"	0.500	26.25	44.75		71.00
8930	Tee	"	0.500	41.00	44.75		86.00
8950	Cap	"	0.160	6.61	14.25		20.75
8960	Coupling	"	0.320	17.25	28.50		45.75
8980	Union	"	0.400	43.25	35.75		79.00
9000	2-1/2"						
9020	90 deg ell	EA	0.400	55.00	35.75		91.00
9030	45 deg ell	"	0.400	47.75	35.75		84.00
9040	Tee	"	0.571	55.00	51.00		110
9070	Cap	"	0.200	13.50	18.00		31.50
9080	Coupling	"	0.400	26.25	35.75		62.00
9100	Union	"	0.444	80.00	39.75		120
22 - 11165	**BRASS FITTINGS**						**22 - 11165**
1000	Compression fittings, union						
1020	3/8"	EA	0.133	1.90	12.00		14.00
1030	1/2"	"	0.133	4.11	12.00		16.00
1040	5/8"	"	0.133	5.23	12.00		17.25
1050	Union elbow						

© 2020 BNi Publications Inc. ALL RIGHTS RESERVED

FACILITY WATER DISTRIBUTION

	Description	Output		Unit Costs			
ID Code	Component Descriptions	Unit of Meas.	Manhr / Unit	Material Cost	Labor Cost	Equipment Cost	Total Cost
22 - 11165	**BRASS FITTINGS, Cont'd...**						**22 - 11165**
1060	3/8"	EA	0.133	6.09	12.00		18.00
1070	1/2"	"	0.133	10.25	12.00		22.25
1080	5/8"	"	0.133	13.50	12.00		25.50
1090	Union tee						
1100	3/8"	EA	0.133	5.71	12.00		17.75
1120	1/2"	"	0.133	8.26	12.00		20.25
1130	5/8"	"	0.133	11.50	12.00		23.50
2000	Brass flare fittings, union						
2020	3/8"	EA	0.129	2.47	11.50		14.00
2030	1/2"	"	0.129	3.40	11.50		15.00
2040	5/8"	"	0.129	4.38	11.50		16.00
2050	90 deg elbow union						
2060	3/8"	EA	0.129	5.30	11.50		16.75
2070	1/2"	"	0.129	7.76	11.50		19.25
2080	5/8"	"	0.129	13.50	11.50		25.00
22 - 11167	**CHROME PLATED FITTINGS**						**22 - 11167**
0005	Fittings						
0010	90 ell						
0020	3/8"	EA	0.200	27.50	18.00		45.50
0030	1/2"	"	0.200	35.50	18.00		54.00
0035	45 ell						
0040	3/8"	EA	0.200	35.50	18.00		54.00
0050	1/2"	"	0.200	46.75	18.00		65.00
0055	Tee						
0060	3/8"	EA	0.267	34.00	23.75		58.00
0070	1/2"	"	0.267	40.50	23.75		64.00
0075	Coupling						
0080	3/8"	EA	0.200	21.50	18.00		39.50
0090	1/2"	"	0.200	21.50	18.00		39.50
0095	Union						
0100	3/8"	EA	0.200	35.50	18.00		54.00
0110	1/2"	"	0.200	36.75	18.00		55.00
0115	Tee						
0130	1/2" x 3/8" x 3/8"	EA	0.267	40.50	23.75		64.00
0140	1/2" x 3/8" x 1/2"	"	0.267	41.25	23.75		65.00

© 2020 BNi Publications Inc. ALL RIGHTS RESERVED

FACILITY WATER DISTRIBUTION

ID Code	Description: Component Descriptions	Output: Unit of Meas.	Output: Manhr / Unit	Unit Costs: Material Cost	Unit Costs: Labor Cost	Unit Costs: Equipment Cost	Unit Costs: Total Cost
22 - 11168	**PVC/CPVC PIPE**						**22 - 11168**
0900	PVC schedule 40						
1000	1/2" pipe	LF	0.033	0.50	2.98		3.48
1020	3/4" pipe	"	0.036	0.69	3.25		3.94
1040	1" pipe	"	0.040	0.88	3.57		4.45
1060	1-1/4" pipe	"	0.044	1.13	3.97		5.10
1080	1-1/2" pipe	"	0.050	1.69	4.47		6.16
1100	2" pipe	"	0.057	2.14	5.10		7.24
1110	2-1/2" pipe	"	0.067	3.46	5.96		9.42
1120	3" pipe	"	0.080	4.41	7.15		11.50
1130	4" pipe	"	0.100	6.30	8.94		15.25
1140	6" pipe	"	0.200	11.25	18.00		29.25
1150	8" pipe	"	0.267	16.50	23.75		40.25
2965	PVC schedule 80 pipe						
3070	1-1/2" pipe	LF	0.050	2.15	4.47		6.62
3071	2" pipe	"	0.057	2.91	5.10		8.01
3100	3" pipe	"	0.080	6.00	7.15		13.25
3110	4" pipe	"	0.100	7.84	8.94		16.75
7000	Polypropylene, acid resistant, DWV pipe						
7010	Schedule 40						
7030	1-1/2" pipe	LF	0.057	9.18	5.10		14.25
7040	2" pipe	"	0.067	12.50	5.96		18.50
7050	3" pipe	"	0.080	25.25	7.15		32.50
7060	4" pipe	"	0.100	32.25	8.94		41.25
7070	6" pipe	"	0.200	64.00	18.00		82.00
7800	Polyethylene pipe and fittings						
7900	SDR-21						
8020	3" pipe	LF	0.100	4.12	8.94		13.00
8030	4" pipe	"	0.133	6.51	12.00		18.50
8040	6" pipe	"	0.200	11.25	18.00		29.25
8050	8" pipe	"	0.229	16.50	20.50		37.00
8060	10" pipe	"	0.267	18.50	23.75		42.25
22 - 11169	**STEEL PIPE**						**22 - 11169**
1000	Black steel, extra heavy pipe, threaded						
1030	1/2" pipe	LF	0.032	2.81	2.86		5.67
1100	3/4" pipe	"	0.032	3.64	2.86		6.50
1200	1" pipe	"	0.040	4.68	3.57		8.25
1400	1-1/2" pipe	"	0.044	7.03	3.97		11.00

© 2020 BNi Publications Inc. ALL RIGHTS RESERVED

FACILITY WATER DISTRIBUTION

	Description	Output		Unit Costs			
ID Code	Component Descriptions	Unit of Meas.	Manhr / Unit	Material Cost	Labor Cost	Equipment Cost	Total Cost
22 - 11169	**STEEL PIPE, Cont'd...**						**22 - 11169**
1500	2-1/2" pipe	LF	0.100	14.00	8.94		23.00
1610	3" pipe	"	0.133	18.75	12.00		30.75
1700	4" pipe	"	0.160	28.25	14.25		42.50
1800	5" pipe	"	0.200	38.25	18.00		56.00
1900	6" pipe	"	0.200	47.75	18.00		66.00
2000	8" pipe	"	0.267	70.00	23.75		94.00
2100	10" pipe	"	0.320	110	28.50		140
2200	12" pipe	"	0.400	140	35.75		180
4000	Fittings, malleable iron, threaded, 1/2" pipe						
4010	90 deg ell	EA	0.267	3.37	23.75		27.00
4020	45 deg ell	"	0.267	4.56	23.75		28.25
4030	Tee	"	0.400	3.66	35.75		39.50
4085	3/4" pipe						
4090	90 deg ell	EA	0.267	3.95	23.75		27.75
4100	45 deg ell	"	0.400	6.26	35.75		42.00
4120	Tee	"	0.400	5.31	35.75		41.00
4178	1" pipe						
4180	90 deg ell	EA	0.320	6.12	28.50		34.50
4200	45 deg ell	"	0.320	8.09	28.50		36.50
4210	Tee	"	0.444	9.18	39.75		49.00
4265	1-1/2" pipe						
4270	90 deg ell	EA	0.400	12.50	35.75		48.25
4280	45 deg ell	"	0.400	15.50	35.75		51.00
4300	Tee	"	0.571	18.25	51.00		69.00
4365	2-1/2" pipe						
4370	90 deg ell	EA	1.000	48.25	89.00		140
4380	45 deg ell	"	1.000	68.00	89.00		160
4400	Tee	"	1.333	67.00	120		190
4448	3" pipe						
4450	90 deg ell	EA	1.333	71.00	120		190
4470	45 deg ell	"	1.333	88.00	120		210
4480	Tee	"	2.000	99.00	180		280
4535	4" pipe						
4540	90 deg ell	EA	1.600	150	140		290
4550	45 deg ell	"	1.600	170	140		310
4560	Tee	"	2.667	240	240		480
4710	90 deg ell	"	1.600	420	140		560
4715	6" pipe						

© 2020 BNi Publications Inc. ALL RIGHTS RESERVED

FACILITY WATER DISTRIBUTION

	Description	Output		Unit Costs			
ID Code	Component Descriptions	Unit of Meas.	Manhr / Unit	Material Cost	Labor Cost	Equipment Cost	Total Cost
22 - 11169	**STEEL PIPE, Cont'd...**						**22 - 11169**
4720	45 deg ell	EA	1.600	490	140		630
4730	Tee	"	2.667	750	240		990
4880	8" pipe						
4900	90 deg ell	EA	3.200	840	290		1,130
4905	45 deg ell	"	3.200	950	290		1,240
4910	Tee	"	5.000	590	450		1,040
4980	10" pipe						
4985	90 deg ell	EA	4.000	930	360		1,290
4990	45 deg ell	"	4.000	1,050	360		1,410
5030	Tee	"	5.000	670	450		1,120
5100	12" pipe						
5110	90 deg ell	EA	5.000	1,020	450		1,470
5120	45 deg ell	"	5.000	1,160	450		1,610
5130	Tee	"	6.667	780	600		1,380
22 - 11170	**GALVANIZED STEEL PIPE**						**22 - 11170**
1000	Galvanized pipe						
1020	1/2" pipe	LF	0.080	3.31	7.15		10.50
1040	3/4" pipe	"	0.100	4.31	8.94		13.25
1050	1" pipe	"	0.114	6.61	10.25		16.75
1060	1-1/4" pipe	"	0.133	8.20	12.00		20.25
1070	1-1/2" pipe	"	0.160	9.00	14.25		23.25
1080	2" pipe	"	0.200	13.00	18.00		31.00
1090	2-1/2" pipe	"	0.267	18.75	23.75		42.50
1100	3" pipe	"	0.286	25.75	25.50		51.00
1110	4" pipe	"	0.333	35.25	29.75		65.00
1120	6" pipe	"	0.667	64.00	60.00		120
22 - 11171	**STAINLESS STEEL PIPE**						**22 - 11171**
3900	Stainless steel, schedule 40, threaded						
4000	1/2" pipe	LF	0.114	10.50	10.25		20.75
4005	3/4" pipe	"	0.118	14.75	10.50		25.25
4090	1" pipe	"	0.123	17.25	11.00		28.25
4100	1-1/2" pipe	"	0.133	23.50	12.00		35.50
4200	2" pipe	"	0.145	35.25	13.00		48.25
4220	2-1/2" pipe	"	0.160	49.50	14.25		64.00
4240	3" pipe	"	0.178	69.00	16.00		85.00
4260	4" pipe	"	0.200	89.00	18.00		110

© 2020 BNi Publications Inc. ALL RIGHTS RESERVED

FACILITY WATER DISTRIBUTION

Description		Output		Unit Costs			
ID Code	Component Descriptions	Unit of Meas.	Manhr / Unit	Material Cost	Labor Cost	Equipment Cost	Total Cost
22 - 11191	**BACKFLOW PREVENTERS**						**22 - 11191**
0080	Backflow preventer, flanged, cast iron, with valves						
0100	3" pipe	EA	4.000	4,030	360		4,390
0120	4" pipe	"	4.444	4,700	400		5,100
0130	6" pipe	"	6.667	8,030	600		8,630
0140	8" pipe	"	8.000	12,240	720		12,960
1900	Threaded						
2000	3/4" pipe	EA	0.500	750	44.75		790
2020	2" pipe	"	0.800	1,320	72.00		1,390
3000	Reduced pressure assembly, bronze, threaded						
3100	3/4"	EA	0.500	640	44.75		680
3120	1"	"	0.571	660	51.00		710
3140	1-1/4"	"	0.667	960	60.00		1,020
3160	1-1/2"	"	0.800	1,080	72.00		1,150
22 - 11196	**VACUUM BREAKERS**						**22 - 11196**
1000	Vacuum breaker, atmospheric, threaded connection						
1010	3/4"	EA	0.320	55.00	28.50		84.00
1015	1"	"	0.320	81.00	28.50		110
1018	Anti-siphon, brass						
1020	3/4"	EA	0.320	60.00	28.50		89.00
1030	1"	"	0.320	93.00	28.50		120
1040	1-1/4"	"	0.400	160	35.75		200
1050	1-1/2"	"	0.444	190	39.75		230
1060	2"	"	0.500	300	44.75		340
22 - 11230	**PUMPS**						**22 - 11230**
0900	In-line pump, bronze, centrifugal						
1000	5 gpm, 20' head	EA	0.500	680	44.75		720
1010	20 gpm, 40' head	"	0.500	1,220	44.75		1,260
1015	50 gpm						
1020	50' head	EA	1.000	1,390	89.00		1,480
1065	Cast iron, centrifugal						
1070	50 gpm, 200' head	EA	1.000	1,320	89.00		1,410
1075	100 gpm						
1080	100' head	EA	1.333	2,250	120		2,370
1500	Centrifugal, close coupled, C.I., single stage						
1520	50 gpm, 100' head	EA	1.000	1,580	89.00		1,670
1530	100 gpm, 100' head	"	1.333	1,920	120		2,040
1535	Base mounted						

© 2020 BNi Publications Inc. ALL RIGHTS RESERVED

FACILITY WATER DISTRIBUTION

	Description	Output		Unit Costs			
ID Code	Component Descriptions	Unit of Meas.	Manhr / Unit	Material Cost	Labor Cost	Equipment Cost	Total Cost
22 - 11230	**PUMPS, Cont'd...**						**22 - 11230**
1540	50 gpm, 100' head	EA	1.000	3,220	89.00		3,310
1550	100 gpm, 50' head	"	1.333	3,660	120		3,780
1560	200 gpm, 100' head	"	2.000	4,690	180		4,870
1570	300 gpm, 175' head	"	2.000	6,080	180		6,260
5980	Condensate pump, simplex						
6000	1000 sf EDR, 2 gpm	EA	6.667	1,650	600		2,250
6010	2000 sf EDR, 3 gpm	"	6.667	1,680	600		2,280
6020	4000 sf EDR, 6 gpm	"	7.273	1,690	650		2,340
6030	6000 sf EDR, 9 gpm	"	7.273	1,720	650		2,370
6035	Duplex, bronze						
6040	8000 sf EDR, 12 gpm	EA	7.273	2,360	650		3,010
6050	10,000 sf EDR, 15 gpm	"	10.000	2,450	890		3,340
6060	15,000 sf EDR, 23 gpm	"	11.429	2,940	1,020		3,960
6070	20,000 sf EDR, 30 gpm	"	16.000	3,420	1,430		4,850
6080	25,000 sf EDR, 38 gpm	"	16.000	3,540	1,430		4,970

FACILITY POTABLE-WATER STORAGE TANKS

22 - 12004	**STORAGE TANKS**						**22 - 12004**
0980	Hot water storage tank, cement lined						
1000	10 gallon	EA	2.667	540	240		780
1020	70 gallon	"	4.000	1,700	360		2,060
1040	200 gallon	"	5.714	3,220	510		3,730
1060	900 gallon	"	10.000	13,370	890		14,260
1080	1100 gallon	"	10.000	17,870	890		18,760
1100	2000 gallon	"	10.000	25,520	890		26,410

FACILITY SANITARY SEWERAGE

22 - 13140	**EXTRA HEAVY SOIL PIPE**						**22 - 13140**
0005	Extra heavy soil pipe, single hub						
0010	2" x 5'	EA	0.160	70.00	14.25		84.00
0020	3" x 5'	"	0.170	78.00	15.25		93.00
0030	4" x 5'	"	0.186	90.00	16.75		110
0040	6" x 5'	"	0.200	170	18.00		190
0045	Double hub						
0050	2" x 5'	EA	0.200	78.00	18.00		96.00
0060	4" x 5'	"	0.211	99.00	18.75		120

© 2020 BNi Publications Inc. ALL RIGHTS RESERVED

FACILITY SANITARY SEWERAGE

	Description	Output		Unit Costs			
ID Code	Component Descriptions	Unit of Meas.	Manhr / Unit	Material Cost	Labor Cost	Equipment Cost	Total Cost
22 - 13140	**EXTRA HEAVY SOIL PIPE, Cont'd...**						**22 - 13140**
0070	5" x 5'	EA	0.222	150	19.75		170
0080	6" x 5'	"	0.229	180	20.50		200
0085	Single hub						
0090	3" x 10'	EA	0.133	160	12.00		170
0100	4" x 10'	"	0.138	200	12.25		210
0110	6" x 10'	"	0.145	310	13.00		320
22 - 13150	**SERVICE WEIGHT PIPE**						**22 - 13150**
0005	Service weight pipe, single hub						
0020	3" x 5'	EA	0.170	52.00	15.25		67.00
0030	4" x 5'	"	0.178	60.00	16.00		76.00
0050	6" x 5'	"	0.200	110	18.00		130
0085	Double hub						
0100	3" x 5'	EA	0.216	58.00	19.25		77.00
0110	4" x 5'	"	0.229	66.00	20.50		87.00
0130	6" x 5'	"	0.267	120	23.75		140
0155	Single hub						
0170	3" x 10'	EA	0.216	68.00	19.25		87.00
0180	4" x 10'	"	0.229	89.00	20.50		110
0200	6" x 10'	"	0.267	150	23.75		170
0395	1/8 bend						
0410	3"	EA	0.320	11.00	28.50		39.50
0420	4"	"	0.364	16.00	32.50		48.50
0440	6"	"	0.400	27.00	35.75		63.00
0525	1/4 bend						
0540	3"	EA	0.320	13.00	28.50		41.50
0550	4"	"	0.364	20.25	32.50		53.00
0570	6"	"	0.400	35.50	35.75		71.00
0635	Sweep						
0650	3"	EA	0.320	21.00	28.50		49.50
0660	4"	"	0.364	31.00	32.50		64.00
0680	6"	"	0.400	62.00	35.75		98.00
0765	Sanitary T						
0790	3"	EA	0.571	22.00	51.00		73.00
0820	4"	"	0.667	27.00	60.00		87.00
0840	6"	"	0.727	61.00	65.00		130
0845	Wye						
0870	3"	EA	0.444	23.00	39.75		63.00

© 2020 BNi Publications Inc. ALL RIGHTS RESERVED

FACILITY SANITARY SEWERAGE

ID Code	Description: Component Descriptions	Output: Unit of Meas.	Output: Manhr / Unit	Unit Costs: Material Cost	Unit Costs: Labor Cost	Unit Costs: Equipment Cost	Unit Costs: Total Cost
22 - 13150	**SERVICE WEIGHT PIPE, Cont'd...**						**22 - 13150**
0900	4"	EA	0.471	30.75	42.00		73.00
0990	6"	"	0.571	71.00	51.00		120
22 - 13160	**C.I. PIPE, ABOVE GROUND**						**22 - 13160**
0980	No hub pipe						
1000	1-1/2" pipe	LF	0.057	12.25	5.10		17.25
1010	2" pipe	"	0.067	10.75	5.96		16.75
1100	3" pipe	"	0.080	15.00	7.15		22.25
1200	4" pipe	"	0.133	19.50	12.00		31.50
1300	6" pipe	"	0.160	34.50	14.25		48.75
1400	8" pipe	"	0.267	55.00	23.75		79.00
1500	10" pipe	"	0.320	87.00	28.50		120
4980	No hub fittings, 1-1/2" pipe						
5000	1/4 bend	EA	0.267	11.25	23.75		35.00
5060	1/8 bend	"	0.267	9.48	23.75		33.25
5100	Sanitary tee	"	0.400	15.75	35.75		52.00
5180	Coupling	"					22.50
5200	Wye	"	0.400	19.75	35.75		56.00
5370	2" pipe						
5380	1/4 bend	EA	0.320	13.00	28.50		41.50
5440	1/8 bend	"	0.320	10.50	28.50		39.00
5480	Sanitary tee	"	0.533	18.00	47.75		66.00
5560	Coupling	"					19.75
5600	Wye	"	0.667	16.75	60.00		77.00
5980	3" pipe						
6000	1/4 bend	EA	0.400	18.00	35.75		54.00
6080	1/8 bend	"	0.400	15.00	35.75		51.00
6120	Sanitary tee	"	0.500	22.00	44.75		67.00
6260	Coupling	"					22.50
6280	Wye	"	0.667	23.75	60.00		84.00
6810	4" pipe						
6820	1/4 bend	EA	0.400	26.00	35.75		62.00
6900	1/8 bend	"	0.400	19.00	35.75		55.00
6940	Sanitary tee	"	0.667	34.00	60.00		94.00
7100	Coupling	"					22.00
7120	Wye	"	0.667	38.75	60.00		99.00
7595	6" pipe						
7600	1/4 bend	EA	0.667	65.00	60.00		130

© 2020 BNi Publications Inc. ALL RIGHTS RESERVED

FACILITY SANITARY SEWERAGE

ID Code	Description: Component Descriptions	Output: Unit of Meas.	Output: Manhr / Unit	Unit Costs: Material Cost	Unit Costs: Labor Cost	Unit Costs: Equipment Cost	Unit Costs: Total Cost
22 - 13160	**C.I. PIPE, ABOVE GROUND, Cont'd...**						**22 - 13160**
7640	1/8 bend	EA	0.667	44.00	60.00		100
7660	Sanitary tee	"	0.800	99.00	72.00		170
7700	Coupling	"					56.00
7710	Wye	"	0.800	100	72.00		170
7980	8" pipe						
7990	1/4 bend	EA	0.667	110	60.00		170
8030	1/8 bend	"	0.667	80.00	60.00		140
8050	Sanitary tee	"	1.000	250	89.00		340
8090	Coupling	"					110
8100	Wye	"	0.800	150	72.00		220
8355	10" pipe						
8360	1/4 bend	EA	0.667	230	60.00		290
8380	1/8 bend	"	0.667	150	60.00		210
8420	Coupling	"					140
8450	Wye	"	1.333	330	120		450
8490	10x6" wye	"	1.333	240	120		360
22 - 13161	**C.I. PIPE, BELOW GROUND**						**22 - 13161**
1010	No hub pipe						
1020	1-1/2" pipe	LF	0.040	9.96	3.57		13.50
1030	2" pipe	"	0.044	10.25	3.97		14.25
1120	3" pipe	"	0.050	14.25	4.47		18.75
1220	4" pipe	"	0.067	18.25	5.96		24.25
1320	6" pipe	"	0.073	31.50	6.50		38.00
1420	8" pipe	"	0.089	49.00	7.94		57.00
1520	10" pipe	"	0.100	82.00	8.94		91.00
5000	Fittings, 1-1/2"						
5010	1/4 bend	EA	0.229	12.00	20.50		32.50
5080	1/8 bend	"	0.229	9.93	20.50		30.50
5220	Wye	"	0.320	16.75	28.50		45.25
5370	2"						
5390	1/4 bend	EA	0.267	13.00	23.75		36.75
5460	1/8 bend	"	0.267	11.00	23.75		34.75
6000	3"						
6020	1/4 bend	EA	0.320	18.00	28.50		46.50
6100	1/8 bend	"	0.320	15.00	28.50		43.50
6300	Wye	"	0.500	23.75	44.75		69.00
6820	4"						

© 2020 BNi Publications Inc. ALL RIGHTS RESERVED

FACILITY SANITARY SEWERAGE

ID Code	Description: Component Descriptions	Output: Unit of Meas.	Output: Manhr / Unit	Unit Costs: Material Cost	Unit Costs: Labor Cost	Unit Costs: Equipment Cost	Unit Costs: Total Cost
22 - 13161	**C.I. PIPE, BELOW GROUND, Cont'd...**						**22 - 13161**
6840	1/4 bend	EA	0.320	26.00	28.50		55.00
6920	1/8 bend	"	0.320	19.00	28.50		47.50
7140	Wye	"	0.500	38.75	44.75		84.00
7780	6"						
7800	1/4 bend	EA	0.500	41.25	44.75		86.00
7820	1/8 bend	"	0.500	27.75	44.75		73.00
7990	8"						
8000	1/4 bend	EA	0.500	110	44.75		150
8040	1/8 bend	"	0.500	80.00	44.75		120
8120	Wye	"	0.667	150	60.00		210
8350	10"						
8370	1/4 bend	EA	0.500	230	44.75		270
8390	1/8 bend	"	0.500	150	44.75		190
8410	Plug	"					70.00
8460	Wye	"	1.000	330	89.00		420
22 - 13163	**ABS DWV PIPE**						**22 - 13163**
1480	Schedule 40 ABS						
1500	1-1/2" pipe	LF	0.040	1.58	3.57		5.15
1520	2" pipe	"	0.044	2.11	3.97		6.08
1530	3" pipe	"	0.057	4.32	5.10		9.42
1540	4" pipe	"	0.080	6.12	7.15		13.25
1550	6" pipe	"	0.100	12.50	8.94		21.50
22 - 13165	**PLASTIC PIPE**						**22 - 13165**
1000	Fiberglass reinforced pipe						
1010	2" pipe	LF	0.062	4.38	5.50		9.88
1020	3" pipe	"	0.067	6.23	5.96		12.25
1030	4" pipe	"	0.073	8.14	6.50		14.75
1040	6" pipe	"	0.080	15.50	7.15		22.75
1050	8" pipe	"	0.133	22.75	12.00		34.75
1060	10" pipe	"	0.160	34.00	14.25		48.25
1070	12" pipe	"	0.200	44.25	18.00		62.00

© 2020 BNi Publications Inc. ALL RIGHTS RESERVED

FACILITY SANITARY SEWERAGE

	Description	Output		Unit Costs			
ID Code	Component Descriptions	Unit of Meas.	Manhr / Unit	Material Cost	Labor Cost	Equipment Cost	Total Cost
22 - 13167	**DRAINS, ROOF & FLOOR**						**22 - 13167**
1020	Floor drain, cast iron, with cast iron top						
1030	2"	EA	0.667	180	60.00		240
1040	3"	"	0.667	180	60.00		240
1050	4"	"	0.667	390	60.00		450
1060	6"	"	0.800	500	72.00		570
1090	Roof drain, cast iron						
1100	2"	EA	0.667	280	60.00		340
1110	3"	"	0.667	290	60.00		350
1120	4"	"	0.667	370	60.00		430
1130	5"	"	0.800	540	72.00		610
1140	6"	"	0.800	550	72.00		620
22 - 13168	**TRAPS**						**22 - 13168**
0980	Bucket trap, threaded						
1000	3/4"	EA	0.500	230	44.75		270
1010	1"	"	0.533	650	47.75		700
1020	1-1/4"	"	0.615	770	55.00		830
1030	1-1/2"	"	0.727	1,160	65.00		1,230
1080	Inverted bucket steam trap, threaded						
1100	3/4"	EA	0.500	280	44.75		320
1110	1"	"	0.500	550	44.75		590
1120	1-1/4"	"	0.444	830	39.75		870
1140	1-1/2"	"	0.667	880	60.00		940
1480	Float trap, 15 psi						
1500	3/4"	EA	0.500	200	44.75		240
1510	1"	"	0.533	310	47.75		360
1520	1-1/4"	"	0.571	400	51.00		450
1530	1-1/2"	"	0.667	510	60.00		570
1540	2"	"	0.800	880	72.00		950
1980	Float and thermostatic trap, 15 psi						
2000	3/4"	EA	0.500	210	44.75		250
2010	1"	"	0.533	240	47.75		290
2020	1-1/4"	"	0.571	370	51.00		420
2030	1-1/2"	"	0.667	480	60.00		540
2040	2"	"	0.800	880	72.00		950
2135	Steam trap, cast iron body, threaded, 125 psi						
2140	3/4"	EA	0.500	250	44.75		290
2150	1"	"	0.533	290	47.75		340

© 2020 BNi Publications Inc. ALL RIGHTS RESERVED

FACILITY SANITARY SEWERAGE

ID Code	Description: Component Descriptions	Output: Unit of Meas.	Output: Manhr / Unit	Unit Costs: Material Cost	Unit Costs: Labor Cost	Unit Costs: Equipment Cost	Unit Costs: Total Cost
22 - 13168	**TRAPS, Cont'd...**						**22 - 13168**
2160	1-1/4"	EA	0.571	430	51.00		480
2170	1-1/2"	"	0.667	690	60.00		750
22 - 13192	**CLEANOUTS**						**22 - 13192**
0980	Cleanout, wall						
1000	2"	EA	0.533	240	47.75		290
1020	3"	"	0.533	340	47.75		390
1040	4"	"	0.667	340	60.00		400
1042	6"	"	0.800	560	72.00		630
1046	8"	"	1.000	780	89.00		870
1050	Floor						
1060	2"	EA	0.667	220	60.00		280
1080	3"	"	0.667	290	60.00		350
1100	4"	"	0.800	300	72.00		370
1120	6"	"	1.000	410	89.00		500
1140	8"	"	1.143	770	100		870
22 - 13193	**GREASE TRAPS**						**22 - 13193**
1000	Grease traps, cast iron, 3" pipe						
1020	35 gpm, 70 lb capacity	EA	8.000	5,320	720		6,040
1040	50 gpm, 100 lb capacity	"	10.000	6,780	890		7,670

PLUMBING EQUIPMENT

ID Code	Component Descriptions	Unit of Meas.	Manhr / Unit	Material Cost	Labor Cost	Equipment Cost	Total Cost
22 - 33001	**DOMESTIC WATER HEATERS**						**22 - 33001**
0900	Water heater, electric						
1000	6 gal	EA	1.333	450	120		570
1020	10 gal	"	1.333	460	120		580
1030	15 gal	"	1.333	450	120		570
1040	20 gal	"	1.600	630	140		770
1050	30 gal	"	1.600	650	140		790
1060	40 gal	"	1.600	710	140		850
1070	52 gal	"	2.000	800	180		980
1080	66 gal	"	2.000	970	180		1,150
1090	80 gal	"	2.000	1,050	180		1,230
1100	100 gal	"	2.667	1,300	240		1,540
1120	120 gal	"	2.667	1,670	240		1,910
2980	Oil fired						
3000	20 gal	EA	4.000	1,430	360		1,790

© 2020 BNi Publications Inc. ALL RIGHTS RESERVED

PLUMBING EQUIPMENT

ID Code	Description: Component Descriptions	Output: Unit of Meas.	Output: Manhr / Unit	Unit Costs: Material Cost	Unit Costs: Labor Cost	Unit Costs: Equipment Cost	Unit Costs: Total Cost
22 - 33001	**DOMESTIC WATER HEATERS, Cont'd...**						**22 - 33001**
3020	50 gal	EA	5.714	2,230	510		2,740
5000	Tankless water heater, natural gas						
5010	Minimum	EA	5.333	770	480		1,250
5020	Average	"	8.000	880	720		1,600
5030	Maximum	"	16.000	990	1,430		2,420
5040	Propane						
5050	Minimum	EA	5.333	660	480		1,140
5060	Average	"	8.000	770	720		1,490
5070	Maximum	"	16.000	880	1,430		2,310
8000	For trim and rough-in						
8100	Minimum	EA	2.667	210	240		450
8150	Average	"	4.000	300	360		660
8200	Maximum	"	8.000	860	720		1,580
22 - 33002	**SOLAR WATER HEATERS**						**22 - 33002**
1000	Hydronic system, 100-120 Gallons including material						
1010	Minimum	EA					14,360
1020	Average	"					15,120
1030	Maximum	"					15,880
1040	Direct-Solar, 100-120 Gallons						
1050	Minimum	EA					9,530
1060	Average	"					9,970
1070	Maximum	"					10,420
1080	Indirect-Solar tank, 50-80 Gallons						
1090	Minimum	EA					1,910
1100	Average	"					2,350
1110	Maximum	"					2,800
1120	100-120 Gallons						
1130	Minimum	EA					3,180
1140	Average	"					3,620
1150	Maximum	"					4,070
1160	Solar water collector panel, 3 x 8						
1170	Minimum	EA	1.000	1,020	89.00		1,110
1180	Average	"	1.143	1,050	100		1,150
1190	Maximum	"	1.333	1,080	120		1,200
1200	4 x 7						
1210	Minimum	EA	1.000	1,140	89.00		1,230
1220	Average	"	1.143	1,180	100		1,280

© 2020 BNi Publications Inc. ALL RIGHTS RESERVED

PLUMBING EQUIPMENT

ID Code	Component Descriptions	Unit of Meas.	Manhr / Unit	Material Cost	Labor Cost	Equipment Cost	Total Cost
22 - 33002	**SOLAR WATER HEATERS, Cont'd...**						**22 - 33002**
1230	Maximum	EA	1.333	1,210	120		1,330
1240	4 x 8						
1250	Minimum	EA	1.000	1,210	89.00		1,300
1260	Average	"	1.143	1,240	100		1,340
1270	Maximum	"	1.333	1,270	120		1,390
1280	4 x 10						
1290	Minimum	EA	1.143	1,400	100		1,500
1300	Average	"	1.333	1,520	120		1,640
1310	Maximum	"	1.600	1,650	140		1,790
1320	Passive tube tank system, 12 Tube						
1330	Minimum	EA	1.000	760	89.00		850
1340	Average	"	1.143	950	100		1,050
1350	Maximum	"	1.333	1,140	120		1,260
1360	24 Tube						
1370	Minimum	EA	1.000	1,140	89.00		1,230
1380	Average	"	1.143	1,330	100		1,430
1390	Maximum	"	1.333	1,520	120		1,640
1400	27 Tube						
1410	Minimum	EA	1.000	1,270	89.00		1,360
1420	Average	"	1.143	1,590	100		1,690
1430	Maximum	"	1.333	1,910	120		2,030

PLUMBING FIXTURES

ID Code	Component Descriptions	Unit of Meas.	Manhr / Unit	Material Cost	Labor Cost	Equipment Cost	Total Cost
22 - 42009	**FIXTURE CARRIERS**						**22 - 42009**
0980	Water fountain, wall carrier						
1000	Minimum	EA	0.800	70.00	72.00		140
1020	Average	"	1.000	94.00	89.00		180
1040	Maximum	"	1.333	120	120		240
1110	Lavatory, wall carrier						
1120	Minimum	EA	0.800	160	72.00		230
1140	Average	"	1.000	230	89.00		320
1160	Maximum	"	1.333	290	120		410
1380	Sink, industrial, wall carrier						
1400	Minimum	EA	0.800	210	72.00		280
1420	Average	"	1.000	240	89.00		330
1440	Maximum	"	1.333	300	120		420
1510	Toilets, water closets, wall carrier						

© 2020 BNi Publications Inc. ALL RIGHTS RESERVED

PLUMBING FIXTURES

ID Code	Component Descriptions	Unit of Meas.	Manhr / Unit	Material Cost	Labor Cost	Equipment Cost	Total Cost
22 - 42009	**FIXTURE CARRIERS, Cont'd...**						**22 - 42009**
1520	Minimum	EA	0.800	310	72.00		380
1540	Average	"	1.000	360	89.00		450
1560	Maximum	"	1.333	470	120		590
1570	Floor support						
1580	Minimum	EA	0.667	150	60.00		210
1600	Average	"	0.800	180	72.00		250
1620	Maximum	"	1.000	200	89.00		290
1630	Urinals, wall carrier						
1640	Minimum	EA	0.800	160	72.00		230
1660	Average	"	1.000	210	89.00		300
1680	Maximum	"	1.333	250	120		370
1690	Floor support						
1700	Minimum	EA	0.667	130	60.00		190
1720	Average	"	0.800	190	72.00		260
1740	Maximum	"	1.000	220	89.00		310
22 - 42135	**URINALS**						**22 - 42135**
0980	Urinal, flush valve, floor mounted						
1000	Minimum	EA	2.000	600	180		780
1010	Average	"	2.667	710	240		950
1020	Maximum	"	4.000	830	360		1,190
1040	Wall mounted						
1060	Minimum	EA	2.000	490	180		670
1080	Average	"	2.667	680	240		920
1100	Maximum	"	4.000	880	360		1,240
8980	For trim and rough-in						
9000	Minimum	EA	2.000	220	180		400
9020	Average	"	4.000	320	360		680
9040	Maximum	"	5.333	430	480		910
22 - 42136	**WATER CLOSETS**						**22 - 42136**
0980	Water closet flush tank, floor mounted						
1000	Minimum	EA	2.000	400	180		580
1010	Average	"	2.667	790	240		1,030
1020	Maximum	"	4.000	1,240	360		1,600
1030	Handicapped						
1040	Minimum	EA	2.667	540	240		780
1050	Average	"	4.000	970	360		1,330
1060	Maximum	"	8.000	1,850	720		2,570

© 2020 BNi Publications Inc. ALL RIGHTS RESERVED

PLUMBING FIXTURES

	Description	Output		Unit Costs			
ID Code	Component Descriptions	Unit of Meas.	Manhr / Unit	Material Cost	Labor Cost	Equipment Cost	Total Cost
22 - 42136	**WATER CLOSETS, Cont'd...**						**22 - 42136**
1180	Bowl, with flush valve, floor mounted						
1200	Minimum	EA	2.000	560	180		740
1220	Average	"	2.667	610	240		850
1240	Maximum	"	4.000	1,200	360		1,560
1250	Wall mounted						
1260	Minimum	EA	2.000	560	180		740
1280	Average	"	2.667	650	240		890
1300	Maximum	"	4.000	1,250	360		1,610
8980	For trim and rough-in						
9000	Minimum	EA	2.000	250	180		430
9020	Average	"	2.667	300	240		540
9040	Maximum	"	4.000	400	360		760
22 - 42162	**LAVATORIES**						**22 - 42162**
1980	Lavatory, countertop, porcelain enamel on cast iron						
2000	Minimum	EA	1.600	230	140		370
2010	Average	"	2.000	350	180		530
2020	Maximum	"	2.667	630	240		870
2080	Wall hung, china						
2100	Minimum	EA	1.600	320	140		460
2110	Average	"	2.000	370	180		550
2120	Maximum	"	2.667	930	240		1,170
2280	Handicapped						
2300	Minimum	EA	2.000	520	180		700
2310	Average	"	2.667	600	240		840
2320	Maximum	"	4.000	1,000	360		1,360
8980	For trim and rough-in						
9000	Minimum	EA	2.000	270	180		450
9020	Average	"	2.667	450	240		690
9040	Maximum	"	4.000	560	360		920
22 - 42164	**SINKS**						**22 - 42164**
0980	Service sink, 24"x29"						
1000	Minimum	EA	2.000	770	180		950
1020	Average	"	2.667	960	240		1,200
1040	Maximum	"	4.000	1,410	360		1,770
2000	Kitchen sink, single, stainless steel, single bowl						
2020	Minimum	EA	1.600	340	140		480
2040	Average	"	2.000	390	180		570

© 2020 BNi Publications Inc. ALL RIGHTS RESERVED

PLUMBING FIXTURES

ID Code	Description: Component Descriptions	Output: Unit of Meas.	Output: Manhr / Unit	Unit Costs: Material Cost	Unit Costs: Labor Cost	Unit Costs: Equipment Cost	Unit Costs: Total Cost
22 - 42164	**SINKS, Cont'd...**						**22 - 42164**
2060	Maximum	EA	2.667	710	240		950
2070	Double bowl						
2080	Minimum	EA	2.000	390	180		570
2100	Average	"	2.667	430	240		670
2120	Maximum	"	4.000	750	360		1,110
2190	Porcelain enamel, cast iron, single bowl						
2200	Minimum	EA	1.600	240	140		380
2220	Average	"	2.000	320	180		500
2240	Maximum	"	2.667	490	240		730
2250	Double bowl						
2260	Minimum	EA	2.000	330	180		510
2280	Average	"	2.667	470	240		710
2300	Maximum	"	4.000	670	360		1,030
2980	Mop sink, 24"x36"x10"						
3000	Minimum	EA	1.600	590	140		730
3020	Average	"	2.000	710	180		890
3040	Maximum	"	2.667	950	240		1,190
5980	Washing machine box						
6000	Minimum	EA	2.000	210	180		390
6040	Average	"	2.667	310	240		550
6060	Maximum	"	4.000	370	360		730
8980	For trim and rough-in						
9000	Minimum	EA	2.667	350	240		590
9020	Average	"	4.000	530	360		890
9040	Maximum	"	5.333	680	480		1,160
22 - 42190	**BATHS**						**22 - 42190**
0980	Bathtub, 5' long						
1000	Minimum	EA	2.667	640	240		880
1020	Average	"	4.000	1,400	360		1,760
1040	Maximum	"	8.000	3,190	720		3,910
1050	6' long						
1060	Minimum	EA	2.667	720	240		960
1080	Average	"	4.000	1,460	360		1,820
1100	Maximum	"	8.000	4,140	720		4,860
1110	Square tub, whirlpool, 4'x4'						
1120	Minimum	EA	4.000	2,200	360		2,560
1140	Average	"	8.000	3,110	720		3,830

© 2020 BNi Publications Inc. ALL RIGHTS RESERVED

PLUMBING FIXTURES

ID Code	Description: Component Descriptions	Output: Unit of Meas.	Output: Manhr / Unit	Unit Costs: Material Cost	Unit Costs: Labor Cost	Unit Costs: Equipment Cost	Unit Costs: Total Cost
22 - 42190	**BATHS, Cont'd...**						**22 - 42190**
1160	Maximum	EA	10.000	9,500	890		10,390
1170	5'x5'						
1180	Minimum	EA	4.000	2,200	360		2,560
1200	Average	"	8.000	3,110	720		3,830
1220	Maximum	"	10.000	9,680	890		10,570
1230	6'x6'						
1240	Minimum	EA	4.000	2,680	360		3,040
1260	Average	"	8.000	3,910	720		4,630
1280	Maximum	"	10.000	11,220	890		12,110
8980	For trim and rough-in						
9000	Minimum	EA	2.667	230	240		470
9020	Average	"	4.000	330	360		690
9040	Maximum	"	8.000	950	720		1,670
22 - 42230	**SHOWERS**						**22 - 42230**
0980	Shower, fiberglass, 36"x34"x84"						
1000	Minimum	EA	5.714	690	510		1,200
1020	Average	"	8.000	970	720		1,690
1040	Maximum	"	8.000	1,400	720		2,120
2980	Steel, 1 piece, 36"x36"						
3000	Minimum	EA	5.714	640	510		1,150
3020	Average	"	8.000	970	720		1,690
3040	Maximum	"	8.000	1,140	720		1,860
3980	Receptor, molded stone, 36"x36"						
4000	Minimum	EA	2.667	270	240		510
4020	Average	"	4.000	450	360		810
4040	Maximum	"	6.667	690	600		1,290
8980	For trim and rough-in						
9000	Minimum	EA	3.636	270	330		600
9020	Average	"	4.444	450	400		850
9040	Maximum	"	8.000	560	720		1,280
22 - 42260	**DISPOSALS & ACCESSORIES**						**22 - 42260**
0040	Disposal, continuous feed						
0050	Minimum	EA	1.600	83.00	140		220
0070	Maximum	"	2.667	440	240		680
0200	Batch feed, 1/2 hp						
0220	Minimum	EA	1.600	320	140		460
0240	Maximum	"	2.667	1,090	240		1,330

© 2020 BNi Publications Inc. ALL RIGHTS RESERVED

PLUMBING FIXTURES

ID Code	Component Descriptions	Unit of Meas.	Manhr / Unit	Material Cost	Labor Cost	Equipment Cost	Total Cost
22 - 42260	**DISPOSALS & ACCESSORIES, Cont'd...**						**22 - 42260**
1100	Hot water dispenser						
1110	Minimum	EA	1.600	230	140		370
1130	Maximum	"	2.667	580	240		820
22 - 42390	**FAUCETS**						**22 - 42390**
0980	Kitchen						
1000	Minimum	EA	1.333	95.00	120		210
1020	Average	"	1.600	270	140		410
1040	Maximum	"	2.000	330	180		510
1050	Bath						
1060	Minimum	EA	1.333	95.00	120		210
1080	Average	"	1.600	280	140		420
1100	Maximum	"	2.000	430	180		610
1110	Lavatory, domestic						
1120	Minimum	EA	1.333	100	120		220
1140	Average	"	1.600	320	140		460
1160	Maximum	"	2.000	530	180		710
1170	Hospital, patient rooms						
1180	Minimum	EA	2.000	140	180		320
1200	Average	"	2.667	440	240		680
1220	Maximum	"	4.000	780	360		1,140
1230	Operating room						
1240	Minimum	EA	2.000	290	180		470
1260	Average	"	2.667	640	240		880
1280	Maximum	"	4.000	930	360		1,290
1290	Washroom						
1300	Minimum	EA	1.333	130	120		250
1320	Average	"	1.600	320	140		460
1340	Maximum	"	2.000	580	180		760
1350	Handicapped						
1360	Minimum	EA	1.600	140	140		280
1380	Average	"	2.000	420	180		600
1400	Maximum	"	2.667	650	240		890
1410	Shower						
1420	Minimum	EA	1.333	130	120		250
1440	Average	"	1.600	370	140		510
1460	Maximum	"	2.000	580	180		760
1480	For trim and rough-in						

© 2020 BNi Publications Inc. ALL RIGHTS RESERVED

PLUMBING FIXTURES

ID Code	Component Descriptions	Unit of Meas.	Manhr / Unit	Material Cost	Labor Cost	Equipment Cost	Total Cost
22 - 42390	**FAUCETS, Cont'd...**						**22 - 42390**
1500	Minimum	EA	1.600	89.00	140		230
1520	Average	"	2.000	140	180		320
1540	Maximum	"	4.000	230	360		590
22 - 42398	**HOSE BIBBS**						**22 - 42398**
0005	Hose bibb						
0010	1/2"	EA	0.267	10.50	23.75		34.25
0200	3/4"	"	0.267	11.00	23.75		34.75

DRINKING FOUNTAINS AND WATER COOLERS

ID Code	Component Descriptions	Unit of Meas.	Manhr / Unit	Material Cost	Labor Cost	Equipment Cost	Total Cost
22 - 47001	**MISCELLANEOUS FIXTURES**						**22 - 47001**
0900	Electric water cooler						
1000	Floor mounted	EA	2.667	1,110	240		1,350
1020	Wall mounted	"	2.667	1,040	240		1,280
1980	Wash fountain						
2000	Wall mounted	EA	4.000	2,660	360		3,020
2020	Circular, floor supported	"	8.000	4,660	720		5,380
4000	Deluge shower and eye wash	"	4.000	1,110	360		1,470

POOLS AND FOUNTAIN PLUMBING SYSTEMS

ID Code	Component Descriptions	Unit of Meas.	Manhr / Unit	Material Cost	Labor Cost	Equipment Cost	Total Cost
22 - 51007	**SOLAR WATER HEATERS, POOLS**						**22 - 51007**
1000	Solar Water Heater, 1000 BTU/SF panel, 4x8	EA	1.000	140	89.00		230
1100	4 x 10	"	1.143	160	100		260
1200	4 x 12	"	1.333	180	120		300
2200	Panel Mounting Kit, 4x8	"	0.400	38.00	35.75		74.00
2300	4 x 10	"	0.444	57.00	39.75		97.00
2400	4 x 12	"	0.500	70.00	44.75		110

GAS AND VACUUM SYSTEMS

ID Code	Component Descriptions	Unit of Meas.	Manhr / Unit	Material Cost	Labor Cost	Equipment Cost	Total Cost
22 - 66001	**GLASS PIPE**						**22 - 66001**
0980	Glass pipe						
1000	1-1/2" dia.	LF	0.160	15.50	14.25		29.75
1020	2" dia.	"	0.178	21.00	16.00		37.00
1040	3" dia.	"	0.200	28.00	18.00		46.00
1060	4" dia.	"	0.229	51.00	20.50		72.00
1080	6" dia.	"	0.267	94.00	23.75		120

© 2020 BNi Publications Inc. ALL RIGHTS RESERVED

DIVISION 23
HVAC

INSULATION

ID Code	Description: Component Descriptions	Output: Unit of Meas.	Output: Manhr / Unit	Unit Costs: Material Cost	Unit Costs: Labor Cost	Unit Costs: Equipment Cost	Unit Costs: Total Cost
23 - 07131	**DUCTWORK INSULATION**						**23 - 07131**
0980	Fiberglass duct insulation, plain blanket						
1000	1-1/2" thick	SF	0.010	0.22	0.89		1.11
1060	2" thick	"	0.013	0.30	1.19		1.49
1500	With vapor barrier						
1520	1-1/2" thick	SF	0.010	0.26	0.89		1.15
1540	2" thick	"	0.013	0.33	1.19		1.52
2000	Rigid with vapor barrier						
2020	2" thick	SF	0.027	1.45	2.38		3.83
2040	3" thick	"	0.032	1.99	2.86		4.85
2060	4" thick	"	0.040	2.54	3.57		6.11
2080	6" thick	"	0.053	3.99	4.76		8.75
3180	Weatherproof, poly, 3" thick, w/vapor barrier	"	0.080	3.08	7.15		10.25
3200	Urethane board with vapor barrier	"	0.100	4.35	8.94		13.25

CONTROLS

ID Code	Component Descriptions	Unit of Meas.	Manhr / Unit	Material Cost	Labor Cost	Equipment Cost	Total Cost
23 - 09131	**HVAC CONTROLS**						**23 - 09131**
1000	Pressure gauge, direct reading gauge cock and siphon	EA	0.500	130	44.75		170
1210	Control valve, 1", modulating						
1220	2-way	EA	0.667	1,010	60.00		1,070
1240	3-way	"	1.000	1,140	89.00		1,230
1260	Self contained control valve w/ sensing elmnt, 3/4"	"	0.500	190	44.75		230
1920	Inst air syst 2-1/2 hp comp, rcvr refrg dryer	"					8,490
2020	Thermostat primary control device	"					190
2040	Humidistat primary control device	"					150
2060	Timers primary control device, indoor/outdoor, 24 hour	"					300
2080	Thermometer, dir. reading, 3 dial	"					150
4380	Control dampers, round						
4400	6" dia.	EA	0.320	130	28.50		160
4401	8" dia	"	0.320	180	28.50		210
4402	10" dia	"	0.320	240	28.50		270
4403	12" dia	"	0.320	320	28.50		350
4404	16" dia	"	0.400	470	35.75		510
4405	18" dia	"	0.400	500	35.75		540
4406	20" dia	"	0.400	670	35.75		710
4407	Rectangular, parallel blade standard leakage						
4480	12" x 12"	EA	0.400	95.00	35.75		130
4525	16" x 16"	"	0.400	140	35.75		180

© 2020 BNi Publications Inc. ALL RIGHTS RESERVED

CONTROLS

	Description	Output		Unit Costs			
ID Code	Component Descriptions	Unit of Meas.	Manhr / Unit	Material Cost	Labor Cost	Equipment Cost	Total Cost
23 - 09131	**HVAC CONTROLS, Cont'd...**						**23 - 09131**
4530	20" x 20"	EA	0.400	170	35.75		210
4600	48" x 48"	"	1.143	500	100		600
4680	48" x 60"	"	1.333	640	120		760
4700	48" x 72"	"	1.333	780	120		900
4980	Low leakage						
5000	12" x 12"	EA	0.400	180	35.75		220
5040	16" x 16"	"	0.400	230	35.75		270
5140	36" x 36"	"	0.667	590	60.00		650
5220	48" x 48"	"	1.143	1,070	100		1,170
5320	48" x 72"	"	1.333	1,660	120		1,780
5980	Rectangular, opposed horizontal blade						
6000	12" x 12"	EA	0.400	120	35.75		160
6040	16" x 16"	"	0.400	170	35.75		210
6080	24" x 24"	"	0.400	230	35.75		270
6140	36" x 36"	"	0.667	380	60.00		440
6320	48" x 72"	"	1.333	1,070	120		1,190

HYDRONIC PIPING AND PUMPS

23 - 21136	**EXPANSION TANKS**						**23 - 21136**
0980	Expansion tank, 125 psi, steel						
1000	20 gallon	EA	1.000	820	89.00		910
1060	80 gallon	"	2.286	1,250	200		1,450
23 - 21137	**STRAINERS**						**23 - 21137**
0980	Strainer, Y pattern, 125 psi, cast iron body, threaded						
1000	3/4"	EA	0.286	13.75	25.50		39.25
1010	1"	"	0.320	17.75	28.50		46.25
1020	1-1/4"	"	0.400	22.25	35.75		58.00
1030	1-1/2"	"	0.400	28.25	35.75		64.00
1040	2"	"	0.500	42.00	44.75		87.00
1980	250 psi, brass body, threaded						
2000	3/4"	EA	0.320	36.00	28.50		65.00
2010	1"	"	0.320	50.00	28.50		79.00
2040	1-1/4"	"	0.400	63.00	35.75		99.00
2100	1-1/2"	"	0.400	88.00	35.75		120
2120	2"	"	0.500	150	44.75		190
2130	Cast iron body, threaded						
2140	3/4"	EA	0.320	21.00	28.50		49.50

© 2020 BNi Publications Inc. ALL RIGHTS RESERVED

HYDRONIC PIPING AND PUMPS

	Description	Output		Unit Costs			
ID Code	Component Descriptions	Unit of Meas.	Manhr / Unit	Material Cost	Labor Cost	Equipment Cost	Total Cost
23 - 21137	**STRAINERS, Cont'd...**						**23 - 21137**
2160	1"	EA	0.320	26.75	28.50		55.00
2180	1-1/4"	"	0.400	35.50	35.75		71.00
2200	1-1/2"	"	0.400	47.00	35.75		83.00
2220	2"	"	0.500	60.00	44.75		100

AIR DISTRIBUTION

23 - 31130	**METAL DUCTWORK**						**23 - 31130**
0090	Rectangular duct						
0100	Galvanized steel						
1000	Minimum	LB	0.073	0.92	6.50		7.42
1010	Average	"	0.089	1.15	7.94		9.09
1020	Maximum	"	0.133	1.76	12.00		13.75
1080	Aluminum						
1100	Minimum	LB	0.160	2.41	14.25		16.75
1120	Average	"	0.200	3.21	18.00		21.25
1140	Maximum	"	0.267	3.98	23.75		27.75
1160	Fittings						
1180	Minimum	EA	0.267	7.62	23.75		31.25
1200	Average	"	0.400	11.50	35.75		47.25
1220	Maximum	"	0.800	16.75	72.00		89.00
1230	For work						
1240	10-20' high, add per pound, $.30						
1260	30-50', add per pound, $.50						

AIR DUCT ACCESSORIES

23 - 33130	**DAMPERS**						**23 - 33130**
0980	Horizontal parallel aluminum backdraft damper						
1000	12" x 12"	EA	0.200	58.00	18.00		76.00
1010	16" x 16"	"	0.229	60.00	20.50		81.00
1030	24" x 24"	"	0.400	92.00	35.75		130
1060	36" x 36"	"	0.571	210	51.00		260
1100	48" x 48"	"	0.800	380	72.00		450
2000	"Up", parallel dampers						
2010	12" x 12"	EA	0.200	94.00	18.00		110
2020	16" x 16"	"	0.229	130	20.50		150
2040	24" x 24"	"	0.400	160	35.75		200

© 2020 BNi Publications Inc. ALL RIGHTS RESERVED

AIR DUCT ACCESSORIES

	Description	Output		Unit Costs			
ID Code	Component Descriptions	Unit of Meas.	Manhr / Unit	Material Cost	Labor Cost	Equipment Cost	Total Cost
23 - 33130	**DAMPERS, Cont'd...**						**23 - 33130**
2070	36" x 36"	EA	0.571	280	51.00		330
2100	48" x 48"	"	0.800	540	72.00		610
3000	"Down", parallel dampers						
3010	12" x 12"	EA	0.200	94.00	18.00		110
3020	16" x 16"	"	0.229	130	20.50		150
3040	24" x 24"	"	0.400	160	35.75		200
3070	36" x 36"	"	0.571	280	51.00		330
3100	48" x 48"	"	0.800	540	72.00		610
3980	Fire damper, 1.5 hr rating						
4000	12" x 12"	EA	0.400	38.25	35.75		74.00
4010	16" x 16"	"	0.400	61.00	35.75		97.00
4030	24" x 24"	"	0.400	77.00	35.75		110
4060	36" x 36"	"	0.800	130	72.00		200
4090	48" x 48"	"	1.143	250	100		350
23 - 33460	**FLEXIBLE DUCTWORK**						**23 - 33460**
1010	Flexible duct, 1.25" fiberglass						
1020	5" dia.	LF	0.040	3.47	3.57		7.04
1040	6" dia.	"	0.044	3.86	3.97		7.83
1060	7" dia.	"	0.047	4.77	4.20		8.97
1080	8" dia.	"	0.050	5.00	4.47		9.47
1100	10" dia.	"	0.057	6.66	5.10		11.75
1120	12" dia.	"	0.062	7.27	5.50		12.75
1140	14" dia.	"	0.067	9.12	5.96		15.00
1160	16" dia.	"	0.073	13.75	6.50		20.25
9000	Flexible duct connector, 3" wide fabric	"	0.133	2.42	12.00		14.50

HVAC FANS

23 - 34001	**EXHAUST FANS**						**23 - 34001**
0160	Belt drive roof exhaust fans						
1020	640 cfm, 2618 fpm	EA	1.000	1,140	89.00		1,230
1030	940 cfm, 2604 fpm	"	1.000	1,480	89.00		1,570
1040	1050 cfm, 3325 fpm	"	1.000	1,320	89.00		1,410
1050	1170 cfm, 2373 fpm	"	1.000	1,920	89.00		2,010
1110	2440 cfm, 4501 fpm	"	1.000	1,500	89.00		1,590
1120	2760 cfm, 4950 fpm	"	1.000	1,660	89.00		1,750
1140	3890 cfm, 6769 fpm	"	1.000	1,890	89.00		1,980
1160	2380 cfm, 3382 fpm	"	1.000	2,100	89.00		2,190

© 2020 BNi Publications Inc. ALL RIGHTS RESERVED

HVAC FANS

ID Code	Description: Component Descriptions	Output: Unit of Meas.	Output: Manhr / Unit	Unit Costs: Material Cost	Unit Costs: Labor Cost	Unit Costs: Equipment Cost	Unit Costs: Total Cost
23 - 34001	**EXHAUST FANS, Cont'd...**						**23 - 34001**
1180	2880 cfm, 3859 fpm	EA	1.000	2,200	89.00		2,290
1200	3200 cfm, 4173 fpm	"	1.333	2,220	120		2,340
1260	3660 cfm, 3437 fpm	"	1.333	2,260	120		2,380
3020	Direct drive fans						
3040	60 to 390 cfm	EA	1.000	930	89.00		1,020
3060	145 to 590 cfm	"	1.000	1,130	89.00		1,220
3080	295 to 860 cfm	"	1.000	1,370	89.00		1,460
3100	235 to 1300 cfm	"	1.000	1,470	89.00		1,560
3120	415 to 1630 cfm	"	1.000	1,660	89.00		1,750
3160	590 to 2045 cfm	"	1.000	1,920	89.00		2,010

AIR OUTLETS AND INLETS

ID Code	Component Descriptions	Unit of Meas.	Manhr / Unit	Material Cost	Labor Cost	Equipment Cost	Total Cost
23 - 37131	**DIFFUSERS**						**23 - 37131**
1980	Ceiling diffusers, round, baked enamel finish						
2000	6" dia.	EA	0.267	40.25	23.75		64.00
2060	12" dia.	"	0.333	69.00	29.75		99.00
2100	16" dia.	"	0.364	100	32.50		130
2140	20" dia.	"	0.400	140	35.75		180
2480	Rectangular						
2500	6x6"	EA	0.267	43.00	23.75		67.00
2540	12x12"	"	0.400	76.00	35.75		110
2580	18x18"	"	0.400	120	35.75		160
2620	24x24"	"	0.500	170	44.75		210
3000	Lay in, flush mounted, perforated face, with grid						
3010	6x6/24x24	EA	0.320	62.00	28.50		91.00
3080	12x12/24x24	"	0.320	72.00	28.50		100
3140	18x18/24x24	"	0.320	110	28.50		140
5000	Two-way slot diffuser with balancing damper, 4'	"	0.800	69.00	72.00		140
23 - 37134	**REGISTERS AND GRILLES**						**23 - 37134**
0980	Lay in flush mounted, perforated face, return						
1000	6x6/24x24	EA	0.320	54.00	28.50		83.00
1020	8x8/24x24	"	0.320	54.00	28.50		83.00
1040	9x9/24x24	"	0.320	58.00	28.50		87.00
1060	10x10/24x24	"	0.320	63.00	28.50		92.00
1080	12x12/24x24	"	0.320	63.00	28.50		92.00
3040	Rectangular, ceiling return, single deflection						
3060	10x10	EA	0.400	32.25	35.75		68.00

© 2020 BNi Publications Inc. ALL RIGHTS RESERVED

AIR OUTLETS AND INLETS

	Description	Output		Unit Costs			
ID Code	Component Descriptions	Unit of Meas.	Manhr / Unit	Material Cost	Labor Cost	Equipment Cost	Total Cost
23 - 37134	**REGISTERS AND GRILLES, Cont'd...**						**23 - 37134**
3080	12x12	EA	0.400	37.50	35.75		73.00
3100	14x14	"	0.400	45.75	35.75		82.00
3120	16x8	"	0.400	37.50	35.75		73.00
3140	16x16	"	0.400	37.50	35.75		73.00
3220	24x12	"	0.400	100	35.75		140
3240	24x18	"	0.400	130	35.75		170
3260	36x24	"	0.444	250	39.75		290
3280	36x30	"	0.444	370	39.75		410
4980	Wall, return air register						
5000	12x12	EA	0.200	53.00	18.00		71.00
5020	16x16	"	0.200	79.00	18.00		97.00
5040	18x18	"	0.200	93.00	18.00		110
5060	20x20	"	0.200	110	18.00		130
5080	24x24	"	0.200	150	18.00		170
5980	Ceiling, return air grille						
6000	6x6	EA	0.267	31.00	23.75		55.00
6020	8x8	"	0.320	38.50	28.50		67.00
6040	10x10	"	0.320	47.75	28.50		76.00
6980	Ceiling, exhaust grille, aluminum egg crate						
7000	6x6	EA	0.267	21.25	23.75		45.00
7020	8x8	"	0.320	21.25	28.50		49.75
7040	10x10	"	0.320	23.50	28.50		52.00
7060	12x12	"	0.400	29.00	35.75		65.00
7080	14x14	"	0.400	38.00	35.75		74.00
7100	16x16	"	0.400	44.75	35.75		81.00
7120	18x18	"	0.400	54.00	35.75		90.00
23 - 37232	**RELIEF VENTILATORS**						**23 - 37232**
0980	Intake ventilator, aluminum, with screen, no curbs						
1000	12" x 12"	EA	0.667	230	60.00		290
1020	16" x 16"	"	0.800	310	72.00		380
1100	36" x 36"	"	1.333	1,210	120		1,330
1140	48" x 48"	"	1.600	1,990	140		2,130

© 2020 BNi Publications Inc. ALL RIGHTS RESERVED

AIR OUTLETS AND INLETS

ID Code	Description: Component Descriptions	Output: Unit of Meas.	Output: Manhr / Unit	Unit Costs: Material Cost	Unit Costs: Labor Cost	Unit Costs: Equipment Cost	Unit Costs: Total Cost
23 - 37238	**PENTHOUSE LOUVERS**						**23 - 37238**
0100	Penthouse louvers						
1000	12" high, extruded aluminum, 4" louver						
1020	6' perimeter	EA	2.000	520	180		700
1080	12' perimeter	"	2.000	1,270	180		1,450
1160	20' perimeter	"	5.333	2,620	480		3,100
2000	16" high x 4' perimeter	"	2.000	430	180		610
2020	6' perimeter	"	2.000	610	180		790
2080	12' perimeter	"	2.000	1,420	180		1,600
2160	20' perimeter	"	5.333	3,060	480		3,540
3000	20" high x 4' perimeter	"	2.000	610	180		790
3020	6' perimeter	"	2.000	650	180		830
3080	12' perimeter	"	2.000	1,600	180		1,780
3160	20' perimeter	"	5.333	3,280	480		3,760
4000	24" high x 4' perimeter	"	2.000	610	180		790
4020	6' perimeter	"	2.000	740	180		920
4080	12' perimeter	"	2.000	1,780	180		1,960
4160	20' perimeter	"	5.333	3,700	480		4,180

CENTRAL HEATING EQUIPMENT

ID Code	Component Descriptions	Unit of Meas.	Manhr / Unit	Material Cost	Labor Cost	Equipment Cost	Total Cost
23 - 52230	**BOILERS**						**23 - 52230**
0900	Cast iron, gas fired, hot water						
1000	115 mbh	EA	20.000	3,150	1,270	1,110	5,530
1020	175 mbh	"	21.818	3,750	1,380	1,210	6,340
1040	235 mbh	"	24.000	4,800	1,520	1,330	7,650
1060	940 mbh	"	48.000	16,530	3,040	2,660	22,230
1080	1600 mbh	"	60.000	22,860	3,800	3,330	29,990
1100	3000 mbh	"	80.000	36,980	5,070	4,430	46,480
1120	6000 mbh	"	120.000	74,000	7,600	6,650	88,250
1130	Steam						
1140	115 mbh	EA	20.000	3,470	1,270	1,110	5,850
1160	175 mbh	"	21.818	4,180	1,380	1,210	6,770
1180	235 mbh	"	24.000	4,950	1,520	1,330	7,800
1200	940 mbh	"	48.000	17,280	3,040	2,660	22,980
1220	1600 mbh	"	60.000	22,360	3,800	3,330	29,490
1240	3000 mbh	"	80.000	33,300	5,070	4,430	42,800
1260	6000 mbh	"	120.000	70,280	7,600	6,650	84,530
1980	Electric, hot water						

© 2020 BNi Publications Inc. ALL RIGHTS RESERVED

CENTRAL HEATING EQUIPMENT

	Description	Output		Unit Costs			
ID Code	Component Descriptions	Unit of Meas.	Manhr / Unit	Material Cost	Labor Cost	Equipment Cost	Total Cost
23 - 52230	BOILERS, Cont'd...						23 - 52230
2000	115 mbh	EA	12.000	5,630	760	670	7,060
2020	175 mbh	"	12.000	6,230	760	670	7,660
2040	235 mbh	"	12.000	7,110	760	670	8,540
2060	940 mbh	"	24.000	17,290	1,520	1,330	20,140
2080	1600 mbh	"	48.000	24,490	3,040	2,660	30,190
2100	3000 mbh	"	60.000	36,580	3,800	3,330	43,710
2120	6000 mbh	"	80.000	42,430	5,070	4,430	51,930
2130	Steam						
2140	115 mbh	EA	12.000	7,110	760	670	8,540
2160	175 mbh	"	12.000	8,700	760	670	10,130
2180	235 mbh	"	12.000	9,500	760	670	10,930
2190	940 mbh	"	24.000	18,910	1,520	1,330	21,760
2200	1600 mbh	"	48.000	31,700	3,040	2,660	37,400
2220	3000 mbh	"	60.000	45,040	3,800	3,330	52,170
2240	6000 mbh	"	80.000	46,800	5,070	4,430	56,300
3980	Oil fired, hot water						
4000	115 mbh	EA	16.000	4,150	1,010	890	6,050
4010	175 mbh	"	18.462	5,270	1,170	1,020	7,460
4020	235 mbh	"	21.818	7,280	1,380	1,210	9,870
4030	940 mbh	"	40.000	13,820	2,530	2,220	18,570
4040	1600 mbh	"	48.000	21,900	3,040	2,660	27,600
4060	3000 mbh	"	60.000	31,750	3,800	3,330	38,880
4080	6000 mbh	"	120.000	73,440	7,600	6,650	87,690
4190	Steam						
4200	115 mbh	EA	16.000	4,150	1,010	890	6,050
4220	175 mbh	"	18.462	5,270	1,170	1,020	7,460
4240	235 mbh	"	21.818	6,720	1,380	1,210	9,310
4260	940 mbh	"	40.000	13,420	2,530	2,220	18,170
4280	1600 mbh	"	48.000	21,900	3,040	2,660	27,600
4300	3000 mbh	"	60.000	29,550	3,800	3,330	36,680
4320	6000 mbh	"	120.000	73,440	7,600	6,650	87,690

© 2020 BNi Publications Inc. ALL RIGHTS RESERVED

FURNACES

ID Code	Description: Component Descriptions	Output: Unit of Meas.	Output: Manhr / Unit	Unit Costs: Material Cost	Unit Costs: Labor Cost	Unit Costs: Equipment Cost	Unit Costs: Total Cost
23 - 54130	**FURNACES**						**23 - 54130**
0980	Electric, hot air						
1000	40 mbh	EA	4.000	850	360		1,210
1020	60 mbh	"	4.211	920	380		1,300
1040	80 mbh	"	4.444	1,000	400		1,400
1060	100 mbh	"	4.706	1,130	420		1,550
1080	125 mbh	"	4.848	1,380	430		1,810
1100	160 mbh	"	5.000	1,900	450		2,350
1120	200 mbh	"	5.161	2,760	460		3,220
1140	400 mbh	"	5.333	4,890	480		5,370
1980	Gas fired hot air						
2000	40 mbh	EA	4.000	850	360		1,210
2020	60 mbh	"	4.211	910	380		1,290
2040	80 mbh	"	4.444	1,050	400		1,450
2060	100 mbh	"	4.706	1,090	420		1,510
2080	125 mbh	"	4.848	1,200	430		1,630
2100	160 mbh	"	5.000	1,430	450		1,880
2120	200 mbh	"	5.161	2,540	460		3,000
2140	400 mbh	"	5.333	4,540	480		5,020
2980	Oil fired hot air						
3000	40 mbh	EA	4.000	1,140	360		1,500
3020	60 mbh	"	4.211	1,890	380		2,270
3040	80 mbh	"	4.444	1,900	400		2,300
3060	100 mbh	"	4.706	1,930	420		2,350
3080	125 mbh	"	4.848	2,000	430		2,430
3100	160 mbh	"	5.000	2,300	450		2,750
3120	200 mbh	"	5.161	2,700	460		3,160
3140	400 mbh	"	5.333	4,480	480		4,960

CENTRAL COOLING EQUIPMENT

ID Code	Component Descriptions	Unit of Meas.	Manhr / Unit	Material Cost	Labor Cost	Equipment Cost	Total Cost
23 - 63001	**CONDENSING UNITS**						**23 - 63001**
0980	Air cooled condenser, single circuit						
1000	3 ton	EA	1.333	1,810	120		1,930
1030	5 ton	"	1.333	2,720	120		2,840
1040	7.5 ton	"	3.810	4,450	340		4,790
1050	20 ton	"	4.000	13,220	360		13,580
1060	25 ton	"	4.000	19,920	360		20,280
1070	30 ton	"	4.000	22,710	360		23,070

© 2020 BNi Publications Inc. ALL RIGHTS RESERVED

CENTRAL COOLING EQUIPMENT

ID Code	Description Component Descriptions	Output Unit of Meas.	Output Manhr / Unit	Unit Costs Material Cost	Unit Costs Labor Cost	Unit Costs Equipment Cost	Unit Costs Total Cost
23 - 63001	**CONDENSING UNITS, Cont'd...**						**23 - 63001**
1080	40 ton	EA	5.714	29,370	510		20,880
1090	50 ton	"	5.714	35,720	510		36,230
1100	60 ton	"	5.000	41,120	450		41,570
1480	With low ambient dampers						
1500	3 ton	EA	2.000	1,980	180		2,160
1530	5 ton	"	2.000	3,120	180		3,300
1550	7.5 ton	"	4.000	4,790	360		5,150
1570	20 ton	"	5.333	13,260	480		13,740
1590	25 ton	"	5.333	20,190	480		20,670
1610	30 ton	"	5.333	23,190	480		23,670
1630	40 ton	"	6.667	30,780	600		31,380
1650	50 ton	"	7.273	37,120	650		37,770
1670	60 ton	"	7.273	42,570	650		43,220
2980	Dual circuit						
3000	10 ton	EA	4.000	4,240	360		4,600
3010	15 ton	"	5.714	6,200	510		6,710
3030	20 ton	"	5.714	12,680	510		13,190
3040	25 ton	"	5.714	20,200	510		20,710
3050	30 ton	"	5.714	23,500	510		24,010
3060	40 ton	"	6.667	33,600	600		34,200
3070	50 ton	"	6.667	37,120	600		37,720
3080	60 ton	"	6.667	38,530	600		39,130
3100	80 ton	"	8.889	48,180	790		48,970
3120	100 ton	"	8.889	56,400	790		57,190
3130	120 ton	"	8.889	66,970	790		67,760
4030	With low ambient dampers						
4050	15 ton	EA	5.714	6,910	510		7,420
4080	20 ton	"	5.714	13,480	510		13,990
4100	25 ton	"	5.714	21,380	510		21,890
4120	30 ton	"	5.714	24,110	510		24,620
4140	40 ton	"	6.667	34,770	600		35,370
4160	50 ton	"	6.667	38,310	600		38,910
4180	60 ton	"	6.667	39,710	600		40,310
4190	80 ton	"	8.889	50,510	790		51,300
4200	100 ton	"	8.889	58,750	790		59,540
4210	120 ton	"	8.889	70,030	790		70,820

© 2020 BNi Publications Inc. ALL RIGHTS RESERVED

PACKAGED WATER CHILLERS

	Description	Output		Unit Costs			
ID Code	Component Descriptions	Unit of Meas.	Manhr / Unit	Material Cost	Labor Cost	Equipment Cost	Total Cost
23 - 64001	**CHILLERS**						**23 - 64001**
0980	Chiller, reciprocal						
1000	Air cooled, remote condenser, starter						
1020	20 ton	EA	8.000	31,460	510	440	32,410
1030	25 ton	"	8.000	35,500	510	440	36,450
1040	30 ton	"	8.000	37,540	510	440	38,490
1050	40 ton	"	12.000	55,510	760	670	56,940
1060	50 ton	"	13.333	61,620	840	740	63,200
1120	100 ton	"	24.000	101,140	1,520	1,330	103,990
1160	200 ton	"	40.000	184,300	2,530	2,220	189,050
2000	Water cooled, with starter						
2020	20 ton	EA	8.000	26,970	510	440	27,920
2030	25 ton	"	8.000	29,890	510	440	30,840
2040	30 ton	"	12.000	35,960	760	670	37,390
2050	40 ton	"	12.000	49,450	760	670	50,880
2060	50 ton	"	13.333	53,940	840	740	55,520
2120	100 ton	"	24.000	78,670	1,520	1,330	81,520
2160	200 ton	"	40.000	125,860	2,530	2,220	130,610
2980	Packaged, air cooled, with starter						
3000	20 ton	EA	6.000	30,120	380	330	30,830
3010	25 ton	"	6.000	32,590	380	330	33,300
3020	30 ton	"	6.000	37,990	380	330	38,700
3030	40 ton	"	6.000	43,600	380	330	44,310
3980	Heat recovery, air cooled, with starter						
4010	50 ton	EA	12.000	58,430	760	670	59,860
4040	100 ton	"	24.000	83,610	1,520	1,330	86,460
4045	Water cooled, with starter						
4050	40 ton	EA	12.000	49,900	760	670	51,330
4090	100 ton	"	26.667	90,120	1,690	1,480	93,290
4980	Centrifugal, single bundle condenser, with starter						
5000	80 ton	EA	34.286	117,990	2,170	1,900	122,060
5040	230 ton	"	53.333	135,930	3,380	2,960	142,260
5070	460 ton	"	80.000	205,960	5,070	4,430	215,460
5090	670 ton	"	96.000	274,660	6,080	5,320	286,060

© 2020 BNi Publications Inc. ALL RIGHTS RESERVED

COOLING TOWERS

ID Code	Component Descriptions	Unit of Meas.	Manhr / Unit	Material Cost	Labor Cost	Equipment Cost	Total Cost
23 - 65001	**COOLING TOWERS**						**23 - 65001**
5980	Cooling tower, propeller type						
6000	100 ton	EA	8.000	14,750	510	440	15,700
6030	400 ton	"	24.000	49,260	1,520	1,330	52,110
6060	1000 ton	"	60.000	118,280	3,800	3,330	125,410
6065	Centrifugal						
6070	100 ton	EA	8.000	20,460	510	440	21,410
6110	400 ton	"	24.000	59,000	1,520	1,330	61,850
6140	1000 ton	"	60.000	138,040	3,800	3,330	145,170

AIR HANDLING

ID Code	Component Descriptions	Unit of Meas.	Manhr / Unit	Material Cost	Labor Cost	Equipment Cost	Total Cost
23 - 74001	**AIR HANDLING UNITS**						**23 - 74001**
0980	Air handling unit, medium pressure, single zone						
1000	1500 cfm	EA	5.000	4,840	450		5,290
1060	3000 cfm	"	8.889	6,360	790		7,150
1180	4000 cfm	"	10.000	8,150	890		9,040
2000	5000 cfm	"	10.667	10,270	950		11,220
2120	6000 cfm	"	11.429	13,200	1,020		14,220
2240	7000 cfm	"	12.308	15,150	1,100		16,250
3000	8500 cfm	"	13.333	18,630	1,190		19,820
3120	10,500 cfm	"	16.000	20,460	1,430		21,890
3240	12,500 cfm	"	17.778	23,550	1,590		25,140
8980	Rooftop air handling units						
9000	4950 cfm	EA	8.889	13,910	790		14,700
9060	7370 cfm	"	11.429	17,640	1,020		18,660
9080	9790 cfm	"	13.333	18,770	1,190		19,960
9100	14,300 cfm	"	11.429	26,510	1,020		27,530
9120	21,725 cfm	"	11.429	37,550	1,020		38,570
9140	33,000 cfm	"	13.333	53,020	1,190		54,210
23 - 74009	**ROOF CURBS**						**23 - 74009**
0980	8" high, insulated, with liner and raised can						
1000	15" x 15"	EA	0.400	120	35.75		160
1060	21" x 21"	"	0.400	150	35.75		190
1160	36" x 36"	"	0.571	210	51.00		260
1200	48" x 48"	"	0.615	600	55.00		650
1260	60" x 60"	"	0.800	1,100	72.00		1,170
1320	72" x 72"	"	1.000	1,760	89.00		1,850

© 2020 BNi Publications Inc. ALL RIGHTS RESERVED

HVAC EQUIPMENT

ID Code	Description: Component Descriptions	Output: Unit of Meas.	Output: Manhr / Unit	Unit Costs: Material Cost	Unit Costs: Labor Cost	Unit Costs: Equipment Cost	Unit Costs: Total Cost
23 - 81132	**ROOFTOP UNITS**						**23 - 81132**
0980	Packaged, single zone rooftop unit, with roof curb						
1000	2 ton	EA	8.000	4,130	720		4,850
1020	3 ton	"	8.000	4,340	720		5,060
1040	4 ton	"	10.000	4,740	890		5,630
1060	5 ton	"	13.333	5,140	1,190		6,330
1070	7.5 ton	"	16.000	7,470	1,430		8,900
23 - 81230	**COMPUTER ROOM A/C**						**23 - 81230**
1010	Air cooled, alarm, high efficiency filter, elec. heat						
1020	3 ton	EA	6.154	18,640	550		19,190
1040	5 ton	"	6.667	19,900	600		20,500
1060	7.5 ton	"	8.000	36,060	720		36,780
1070	10 ton	"	10.000	37,680	890		38,570
1080	15 ton	"	11.429	41,410	1,020		42,430
1090	Steam heat						
1100	3 ton	EA	6.154	18,130	550		18,680
1120	5 ton	"	6.667	19,290	600		19,890
1140	7.5 ton	"	8.000	30,720	720		31,440
1160	10 ton	"	10.000	31,650	890		32,540
1180	15 ton	"	11.429	35,000	1,020		36,020
1190	Hot water heat						
1200	3 ton	EA	6.154	18,130	550		18,680
1220	5 ton	"	6.667	19,290	600		19,890
1240	7.5 ton	"	8.000	30,720	720		31,440
1260	10 ton	"	10.000	31,650	890		32,540
1300	15 ton	"	11.429	35,110	1,020		36,130
1310	Air cooled condenser, low ambient damper						
1320	3 ton	EA	1.600	1,810	140		1,950
1340	5 ton	"	2.000	2,850	180		3,030
1360	7.5 ton	"	4.000	4,380	360		4,740
1400	10 ton	"	5.714	6,400	510		6,910
1420	15 ton	"	4.706	7,070	420		7,490
3010	Water cooled, high efficiency filter, alarm, elec. heat						
3020	3 ton	EA	5.714	18,800	510		19,310
3040	5 ton	"	6.667	20,250	600		20,850
3060	7.5 ton	"	10.000	32,260	890		33,150
3080	10 ton	"	11.429	33,470	1,020		34,490
3100	15 ton	"	13.333	39,170	1,190		40,360

© 2020 BNi Publications Inc. ALL RIGHTS RESERVED

HVAC EQUIPMENT

	Description	Output		Unit Costs			
ID Code	Component Descriptions	Unit of Meas.	Manhr / Unit	Material Cost	Labor Cost	Equipment Cost	Total Cost
23 - 81230	**COMPUTER ROOM A/C, Cont'd...**						**23 - 81230**
3110	Steam heat						
3120	3 ton	EA	5.714	21,470	510		21,980
3140	5 ton	"	6.667	24,500	600		25,100
3160	7.5 ton	"	10.000	34,560	890		35,450
3180	10 ton	"	11.429	35,780	1,020		36,800
3200	15 ton	"	13.333	41,600	1,190		42,790
3210	Hot water heat						
3220	3 ton	EA	5.714	21,470	510		21,980
3240	5 ton	"	6.667	22,920	600		23,520
3260	7.5 ton	"	10.000	34,560	890		35,450
3280	10 ton	"	11.429	35,780	1,020		36,800
3300	15 ton	"	13.333	41,600	1,190		42,790

CONVECTION HEATING AND COOLING UNITS

23 - 82190	**FAN COIL UNITS**						**23 - 82190**
0980	Fan coil unit, 2 pipe, complete						
1000	200 cfm ceiling hung	EA	2.667	1,220	240		1,460
1020	Floor mounted	"	2.000	1,150	180		1,330
1100	300 cfm, ceiling hung	"	3.200	1,290	290		1,580
1130	Floor mounted	"	2.667	1,230	240		1,470
1200	400 cfm, ceiling hung	"	3.810	1,360	340		1,700
1220	Floor mounted	"	2.667	1,310	240		1,550
1300	500 cfm, ceiling hung	"	4.000	1,580	360		1,940
1310	Floor mounted	"	3.077	1,520	280		1,800
1400	600 cfm, ceiling hung	"	4.420	2,000	400		2,400
1420	Floor mounted	"	3.636	1,860	330		2,190
23 - 82390	**UNIT HEATERS**						**23 - 82390**
0980	Steam unit heater, horizontal						
1000	12,500 btuh, 200 cfm	EA	1.333	560	120		680
1010	17,000 btuh, 300 cfm	"	1.333	740	120		860
1020	40,000 btuh, 500 cfm	"	1.333	900	120		1,020
1030	60,000 btuh, 700 cfm	"	1.333	940	120		1,060
1040	70,000 btuh, 1000 cfm	"	2.000	980	180		1,160
1045	Vertical						
1050	12,500 btuh, 200 cfm	EA	1.333	560	120		680
1060	17,000 btuh, 300 cfm	"	1.333	930	120		1,050
1070	40,000 btuh, 500 cfm	"	1.333	900	120		1,020

© 2020 BNi Publications Inc. ALL RIGHTS RESERVED

CONVECTION HEATING AND COOLING UNITS

	Description	Output		Unit Costs			
ID Code	Component Descriptions	Unit of Meas.	Manhr / Unit	Material Cost	Labor Cost	Equipment Cost	Total Cost
23 - 82390	**UNIT HEATERS, Cont'd...**						**23 - 82390**
1080	60,000 btuh, 700 cfm	EA	1.333	940	120		1,060
1090	70,000 btuh, 1000 cfm	"	1.333	980	120		1,100
1980	Gas unit heater, horizontal						
2000	27,400 btuh	EA	3.200	860	290		1,150
2010	38,000 btuh	"	3.200	900	290		1,190
2020	56,000 btuh	"	3.200	940	290		1,230
2030	82,200 btuh	"	3.200	980	290		1,270
2040	103,900 btuh	"	5.000	1,090	450		1,540
2060	125,700 btuh	"	5.000	1,280	450		1,730
2080	133,200 btuh	"	5.000	1,380	450		1,830
2090	149,000 btuh	"	5.000	1,630	450		2,080
2100	172,000 btuh	"	5.000	1,760	450		2,210
2120	190,000 btuh	"	5.000	1,850	450		2,300
2130	225,000 btuh	"	5.000	2,030	450		2,480
3980	Hot water unit heater, horizontal						
4000	12,500 btuh, 200 cfm	EA	1.333	450	120		570
4010	17,000 btuh, 300 cfm	"	1.333	500	120		620
4020	25,000 btuh, 500 cfm	"	1.333	580	120		700
4030	30,000 btuh, 700 cfm	"	1.333	680	120		800
4040	50,000 btuh, 1000 cfm	"	2.000	740	180		920
4050	60,000 btuh, 1300 cfm	"	2.000	780	180		960
4055	Vertical						
4060	12,500 btuh, 200 cfm	EA	1.333	660	120		780
4070	17,000 btuh, 300 cfm	"	1.333	660	120		780
4080	25,000 btuh, 500 cfm	"	1.333	660	120		780
4090	30,000 btuh, 700 cfm	"	1.333	660	120		780
4100	50,000 btuh, 1000 cfm	"	1.333	690	120		810
4120	60,000 btuh, 1300 cfm	"	1.333	850	120		970
5000	Cabinet unit heaters, ceiling, exposed, hot water						
5010	200 cfm	EA	2.667	1,290	240		1,530
5030	300 cfm	"	3.200	1,380	290		1,670
5050	400 cfm	"	3.810	1,440	340		1,780
5070	600 cfm	"	4.211	1,480	380		1,860
5090	800 cfm	"	5.000	1,850	450		2,300
5120	1000 cfm	"	5.714	2,410	510		2,920
5140	1200 cfm	"	6.667	2,590	600		3,190
5160	2000 cfm	"	8.889	4,040	790		4,830

© 2020 BNi Publications Inc. ALL RIGHTS RESERVED

RESISTANCE HEATING

ID Code	Description: Component Descriptions	Output: Unit of Meas.	Output: Manhr / Unit	Unit Costs: Material Cost	Unit Costs: Labor Cost	Unit Costs: Equipment Cost	Unit Costs: Total Cost
23 - 83330	**ELECTRIC HEATING**						**23 - 83330**
1000	Baseboard heater						
1020	2', 375w	EA	1.000	41.75	83.00		120
1040	3', 500w	"	1.000	49.50	83.00		130
1060	4', 750w	"	1.143	55.00	95.00		150
1100	5', 935w	"	1.333	78.00	110		190
1120	6', 1125w	"	1.600	92.00	130		220
1140	7', 1310w	"	1.818	100	150		250
1160	8', 1500w	"	2.000	120	170		290
1180	9', 1680w	"	2.222	130	180		310
1200	10', 1875w	"	2.286	180	190		370
1210	Unit heater, wall mounted						
1220	1500w	EA	1.667	220	140		360
1240	2500w	"	1.818	240	150		390
1260	4000w	"	2.286	320	190		510
1270	Thermostat						
1280	Integral	EA	0.500	37.50	41.50		79.00
1300	Line voltage	"	0.500	38.50	41.50		80.00
1320	Electric heater connection	"	0.250	1.65	20.75		22.50

HUMIDITY CONTROL EQUIPMENT

ID Code	Component Descriptions	Unit of Meas.	Manhr / Unit	Material Cost	Labor Cost	Equipment Cost	Total Cost
23 - 84160	**DEHUMIDIFIERS**						**23 - 84160**
1000	Desiccant dehumidifier, 1125 cfm	EA					35,230

© 2020 BNi Publications Inc. ALL RIGHTS RESERVED

DIVISION 26
ELECTRICAL

CONDUCTORS, CONDUIT AND RACEWAYS

ID Code	Description: Component Descriptions	Output: Unit of Meas.	Output: Manhr / Unit	Unit Costs: Material Cost	Unit Costs: Labor Cost	Unit Costs: Equipment Cost	Unit Costs: Total Cost
26 - 05134	COPPER CONDUCTORS						26 - 05134
0980	Copper conductors, type THW, solid						
1000	#14	LF	0.004	0.12	0.33		0.45
1040	#12	"	0.005	0.18	0.41		0.59
1060	#10	"	0.006	0.28	0.50		0.78
1070	Stranded						
1080	#14	LF	0.004	0.13	0.33		0.46
1100	#12	"	0.005	0.16	0.41		0.57
1120	#10	"	0.006	0.25	0.50		0.75
1140	#8	"	0.008	0.41	0.66		1.07
1160	#6	"	0.009	0.67	0.74		1.41
1180	#4	"	0.010	1.04	0.83		1.87
1200	#3	"	0.010	1.32	0.83		2.15
1220	#2	"	0.012	1.66	0.99		2.65
1240	#1	"	0.014	2.10	1.16		3.26
1260	1/0	"	0.016	2.50	1.32		3.82
1280	2/0	"	0.020	3.14	1.66		4.80
1300	3/0	"	0.025	3.96	2.07		6.03
1520	4/0	"	0.028	4.95	2.32		7.27
1540	250 MCM	"	0.030	6.10	2.50		8.60
1560	300 MCM	"	0.033	7.18	2.76		9.94
1580	350 MCM	"	0.040	8.40	3.32		11.75
1600	400 MCM	"	0.044	9.57	3.68		13.25
1620	500 MCM	"	0.052	11.75	4.28		16.00
1640	600 MCM	"	0.059	15.75	4.91		20.75
1660	750 MCM	"	0.067	19.75	5.53		25.25
1680	1000 MCM	"	0.076	24.75	6.32		31.00
2010	THHN-THWN, solid						
2020	#14	LF	0.004	0.12	0.33		0.45
2040	#12	"	0.005	0.18	0.41		0.59
2060	#10	"	0.006	0.28	0.50		0.78
2070	Stranded						
2080	#14	LF	0.004	0.12	0.33		0.45
2100	#12	"	0.005	0.18	0.41		0.59
2120	#10	"	0.006	0.28	0.50		0.78
2140	#8	"	0.008	0.49	0.66		1.15
2160	#6	"	0.009	0.77	0.74		1.51
2180	#4	"	0.010	1.22	0.83		2.05
2200	#2	"	0.012	1.70	0.99		2.69

© 2020 BNi Publications Inc. ALL RIGHTS RESERVED

CONDUCTORS, CONDUIT AND RACEWAYS

	Description	Output		Unit Costs			
ID Code	Component Descriptions	Unit of Meas.	Manhr / Unit	Material Cost	Labor Cost	Equipment Cost	Total Cost
26 - 05134	**COPPER CONDUCTORS, Cont'd...**						**26 - 05134**
2220	#1	LF	0.014	2.15	1.16		3.31
2240	1/0	"	0.016	2.65	1.32		3.97
2260	2/0	"	0.020	3.27	1.66		4.93
2280	3/0	"	0.025	4.11	2.07		6.18
2300	4/0	"	0.028	5.14	2.32		7.46
2320	250 MCM	"	0.030	6.29	2.50		8.79
2340	350 MCM	"	0.040	7.50	3.32		10.75
26 - 05135	**SHEATHED CABLE**						**26 - 05135**
6700	Non-metallic sheathed cable						
6705	Type NM cable with ground						
6710	#14/2	LF	0.015	0.28	1.24		1.52
6720	#12/2	"	0.016	0.44	1.32		1.76
6730	#10/2	"	0.018	0.69	1.47		2.16
6740	#8/2	"	0.020	1.13	1.66		2.79
6750	#6/2	"	0.025	1.78	2.07		3.85
6760	#14/3	"	0.026	0.39	2.14		2.53
6770	#12/3	"	0.027	0.62	2.21		2.83
6780	#10/3	"	0.027	0.99	2.25		3.24
6790	#8/3	"	0.028	1.66	2.28		3.94
6800	#6/3	"	0.028	2.68	2.32		5.00
6810	#4/3	"	0.032	5.55	2.65		8.20
6820	#2/3	"	0.035	8.34	2.88		11.25
6825	Type UF cable with ground						
6830	#14/2	LF	0.016	0.33	1.32		1.65
6840	#12/2	"	0.019	0.49	1.58		2.07
6850	#10/2	"	0.020	0.79	1.66		2.45
6860	#8/2	"	0.023	1.36	1.89		3.25
6870	#6/2	"	0.027	2.12	2.25		4.37
6880	#14/3	"	0.020	0.46	1.66		2.12
6890	#12/3	"	0.022	0.70	1.81		2.51
6900	#10/3	"	0.025	1.08	2.07		3.15
6910	#8/3	"	0.028	2.05	2.32		4.37
6920	#6/3	"	0.032	3.33	2.65		5.98

© 2020 BNi Publications Inc. ALL RIGHTS RESERVED

CONDUCTORS, CONDUIT AND RACEWAYS

ID Code	Description: Component Descriptions	Output: Unit of Meas.	Output: Manhr / Unit	Unit Costs: Material Cost	Unit Costs: Labor Cost	Unit Costs: Equipment Cost	Unit Costs: Total Cost
26 - 05261	**GROUNDING**						**26 - 05261**
0400	Ground rods, copper clad, 1/2" x						
0510	6'	EA	0.667	17.00	55.00		72.00
0520	8'	"	0.727	23.50	60.00		84.00
0530	10'	"	1.000	29.25	83.00		110
0535	5/8" x						
0540	5'	EA	0.615	21.00	51.00		72.00
0550	6'	"	0.727	22.75	60.00		83.00
0560	8'	"	1.000	29.25	83.00		110
0570	10'	"	1.250	36.25	100		140
0580	3/4" x						
0590	8'	EA	0.727	52.00	60.00		110
0600	10'	"	0.800	57.00	66.00		120
1060	Ground rod clamp						
1080	5/8"	EA	0.123	6.94	10.25		17.25
1100	3/4"	"	0.123	9.82	10.25		20.00
26 - 05292	**CONDUIT SPECIALTIES**						**26 - 05292**
8005	Rod beam clamp, 1/2"	EA	0.050	6.94	4.15		11.00
8007	Hanger rod						
8010	3/8"	LF	0.040	1.46	3.32		4.78
8020	1/2"	"	0.050	3.65	4.15		7.80
8030	All thread rod						
8040	1/4"	LF	0.030	0.47	2.50		2.97
8060	3/8"	"	0.040	0.53	3.32		3.85
8080	1/2"	"	0.050	1.00	4.15		5.15
8100	5/8"	"	0.080	1.76	6.64		8.40
8120	Hanger channel, 1-1/2"						
8140	No holes	EA	0.030	4.62	2.50		7.12
8160	Holes	"	0.030	5.70	2.50		8.20
8170	Channel strap						
8180	1/2"	EA	0.050	1.44	4.15		5.59
8200	3/4"	"	0.050	1.93	4.15		6.08
8220	1"	"	0.050	2.47	4.15		6.62
8240	1-1/4"	"	0.080	1.99	6.64		8.63
8260	1-1/2"	"	0.080	2.39	6.64		9.03
8280	2"	"	0.080	2.59	6.64		9.23
8290	2-1/2"	"	0.123	4.97	10.25		15.25
8300	3"	"	0.123	5.41	10.25		15.75

© 2020 BNi Publications Inc. ALL RIGHTS RESERVED

CONDUCTORS, CONDUIT AND RACEWAYS

	Description	Output		Unit Costs			
ID Code	Component Descriptions	Unit of Meas.	Manhr / Unit	Material Cost	Labor Cost	Equipment Cost	Total Cost
26 - 05292	**CONDUIT SPECIALTIES, Cont'd...**						**26 - 05292**
8310	3-1/2"	EA	0.123	6.74	10.25		17.00
8320	4"	"	0.145	7.62	12.00		19.50
8340	5"	"	0.145	12.25	12.00		24.25
8360	6"	"	0.145	14.00	12.00		26.00
8410	Conduit penetrations, roof and wall, 8" thick						
8420	1/2"	EA	0.615		51.00		51.00
8460	3/4"	"	0.615		51.00		51.00
8480	1"	"	0.800		66.00		66.00
8500	1-1/4"	"	0.800		66.00		66.00
8520	1-1/2"	"	0.800		66.00		66.00
8540	2"	"	1.600		130		130
8560	2-1/2"	"	1.600		130		130
8580	3"	"	1.600		130		130
8590	3-1/2"	"	2.000		170		170
8600	4"	"	2.000		170		170
9505	Fireproofing, for conduit penetrations						
9510	1/2"	EA	0.500	3.57	41.50		45.00
9520	3/4"	"	0.500	3.70	41.50		45.25
9530	1"	"	0.500	3.78	41.50		45.25
9540	1-1/4"	"	0.727	8.89	65.00		74.00
9550	1-1/2"	"	0.727	5.25	60.00		65.00
9560	2"	"	0.727	5.39	60.00		65.00
9570	2-1/2"	"	0.899	10.25	75.00		85.00
9580	3"	"	0.899	10.75	80.00		91.00
9590	3-1/2"	"	1.250	12.50	100		110
9600	4"	"	1.509	15.00	130		140
26 - 05334	**SURFACE MOUNTED RACEWAY**						**26 - 05334**
0980	Single Raceway						
1000	3/4" x 17/32" Conduit	LF	0.040	1.93	3.32		5.25
1020	Mounting Strap	EA	0.053	0.52	4.42		4.94
1040	Connector	"	0.053	0.69	4.42		5.11
1060	Elbow						
2000	45 degree	EA	0.050	8.80	4.15		13.00
2020	90 degree	"	0.050	2.80	4.15		6.95
2040	internal	"	0.050	3.53	4.15		7.68
2050	external	"	0.050	3.26	4.15		7.41
2060	Switch	"	0.400	22.75	33.25		56.00

© 2020 BNi Publications Inc. ALL RIGHTS RESERVED

CONDUCTORS, CONDUIT AND RACEWAYS

ID Code	Component Descriptions	Unit of Meas.	Manhr / Unit	Material Cost	Labor Cost	Equipment Cost	Total Cost
26 - 05334	**SURFACE MOUNTED RACEWAY, Cont'd...**						**26 - 05334**
2100	Utility Box	EA	0.400	15.25	33.25		48.50
2110	Receptacle	"	0.400	27.00	33.25		60.00
2140	3/4" x 21/32" Conduit	LF	0.040	2.19	3.32		5.51
2160	Mounting Strap	EA	0.053	0.81	4.42		5.23
2180	Connector	"	0.053	0.83	4.42		5.25
2200	Elbow						
2210	45 degree	EA	0.050	10.75	4.15		15.00
2220	90 degree	"	0.050	2.99	4.15		7.14
2240	internal	"	0.050	4.06	4.15		8.21
2260	external	"	0.050	4.06	4.15		8.21
3000	Switch	"	0.400	22.75	33.25		56.00
3010	Utility Box	"	0.400	15.25	33.25		48.50
3020	Receptacle	"	0.400	27.00	33.25		60.00
26 - 05335	**PULL BOXES AND CABINETS**						**26 - 05335**
2500	Galvanized pull boxes, screw cover						
2520	4x4x4	EA	0.190	7.83	15.75		23.50
2540	4x6x4	"	0.190	9.33	15.75		25.00
2560	6x6x4	"	0.190	11.75	15.75		27.50
2580	6x8x4	"	0.190	14.00	15.75		29.75
2620	8x8x4	"	0.250	17.50	20.75		38.25
3020	8x10x4	"	0.242	20.25	20.00		40.25
3040	8x12x4	"	0.250	22.25	20.75		43.00
3050	Screw cover						
3060	10x10x4	EA	0.308	22.25	25.50		47.75
3200	12x12x6	"	0.444	32.75	37.00		70.00
3220	12x15x6	"	0.444	39.00	37.00		76.00
3240	12x18x6	"	0.500	43.25	41.50		85.00
3250	15x18x6	"	0.571	48.25	47.50		96.00
3260	18x24x6	"	0.615	90.00	51.00		140
3270	18x30x6	"	0.727	100	60.00		160
3280	24x36x6	"	0.727	150	60.00		210
26 - 05337	**WIREWAYS**						**26 - 05337**
0960	Wireway, hinge cover type						
0980	2-1/2" x 2-1/2"						
1000	1' section	EA	0.154	17.50	12.75		30.25
1040	2'	"	0.190	25.00	15.75		40.75
1060	3'	"	0.250	33.75	20.75		55.00

© 2020 BNi Publications Inc. ALL RIGHTS RESERVED

CONDUCTORS, CONDUIT AND RACEWAYS

	Description	Output		Unit Costs			
ID Code	Component Descriptions	Unit of Meas.	Manhr / Unit	Material Cost	Labor Cost	Equipment Cost	Total Cost
26 - 05337	**WIREWAYS, Cont'd...**						**26 - 05337**
1080	5'	EA	0.381	57.00	31.50		89.00
1100	10'	"	0.667	120	55.00		180
1110	4" x 4"						
1120	1'	EA	0.250	19.25	20.75		40.00
1140	2'	"	0.250	28.00	20.75		48.75
1160	3'	"	0.308	41.50	25.50		67.00
1180	4'	"	0.308	57.00	25.50		83.00
1200	10'	"	0.800	170	66.00		240
26 - 05339	**PULL AND JUNCTION BOXES**						**26 - 05339**
1050	4"						
1060	Octagon box	EA	0.114	3.96	9.48		13.50
1070	Box extension	"	0.059	6.67	4.91		11.50
1080	Plaster ring	"	0.059	3.66	4.91		8.57
1100	Cover blank	"	0.059	1.61	4.91		6.52
1120	Square box	"	0.114	5.70	9.48		15.25
1140	Box extension	"	0.059	5.59	4.91		10.50
1160	Plaster ring	"	0.059	3.06	4.91		7.97
1180	Cover blank	"	0.059	1.57	4.91		6.48
1190	4-11/16"						
1200	Square box	EA	0.114	11.50	9.48		21.00
1240	Box extension	"	0.059	12.50	4.91		17.50
1260	Plaster ring	"	0.059	7.61	4.91		12.50
1280	Cover blank	"	0.059	2.82	4.91		7.73
1300	Switch and device boxes						
1320	2 gang	EA	0.114	17.25	9.48		26.75
1340	3 gang	"	0.114	30.25	9.48		39.75
1360	4 gang	"	0.160	40.50	13.25		54.00
2000	Device covers						
2020	2 gang	EA	0.059	13.75	4.91		18.75
2040	3 gang	"	0.059	14.25	4.91		19.25
2060	4 gang	"	0.059	19.25	4.91		24.25
2100	Handy box	"	0.114	4.25	9.48		13.75
2120	Extension	"	0.059	4.00	4.91		8.91
2140	Switch cover	"	0.059	2.12	4.91		7.03
2160	Switch box with knockout	"	0.145	6.38	12.00		18.50
2200	Weatherproof cover, spring type	"	0.080	11.75	6.64		18.50
2220	Cover plate, dryer receptacle 1 gang plastic	"	0.100	1.81	8.30		10.00

© 2020 BNi Publications Inc. ALL RIGHTS RESERVED

CONDUCTORS, CONDUIT AND RACEWAYS

ID Code	Description: Component Descriptions	Output: Unit of Meas.	Output: Manhr / Unit	Unit Costs: Material Cost	Unit Costs: Labor Cost	Unit Costs: Equipment Cost	Unit Costs: Total Cost
26 - 05339	**PULL AND JUNCTION BOXES, Cont'd...**						**26 - 05339**
2240	For 4" receptacle, 2 gang	EA	0.100	3.23	8.30		11.50
2260	Duplex receptacle cover plate, plastic	"	0.059	0.79	4.91		5.70
3005	4", vertical bracket box, 1-1/2" with						
3010	RMX clamps	EA	0.145	8.22	12.00		20.25
3020	BX clamps	"	0.145	8.82	12.00		20.75
3025	4", octagon device cover						
3030	1 switch	EA	0.059	4.82	4.91		9.73
3040	1 duplex recept	"	0.059	4.82	4.91		9.73
3105	4", square face bracket boxes, 1-1/2"						
3110	RMX	EA	0.145	9.81	12.00		21.75
3120	BX	"	0.145	10.75	12.00		22.75
26 - 05341	**ALUMINUM CONDUIT**						**26 - 05341**
1010	Aluminum conduit						
1020	1/2"	LF	0.030	2.21	2.50		4.71
1040	3/4"	"	0.040	2.84	3.32		6.16
1060	1"	"	0.050	3.99	4.15		8.14
1080	1-1/4"	"	0.059	5.33	4.91		10.25
1100	1-1/2"	"	0.080	6.62	6.64		13.25
1120	2"	"	0.089	8.83	7.37		16.25
1140	2-1/2"	"	0.100	14.00	8.30		22.25
1160	3"	"	0.107	18.25	8.85		27.00
1180	3-1/2"	"	0.123	22.00	10.25		32.25
1200	4"	"	0.145	26.00	12.00		38.00
1220	5"	"	0.182	37.25	15.00		52.00
1240	6"	"	0.200	49.25	16.50		66.00
26 - 05342	**EMT CONDUIT**						**26 - 05342**
0080	EMT conduit						
0100	1/2"	LF	0.030	0.60	2.50		3.10
1020	3/4"	"	0.040	1.09	3.32		4.41
1030	1"	"	0.050	1.82	4.15		5.97
1040	1-1/4"	"	0.059	2.92	4.91		7.83
1060	1-1/2"	"	0.080	3.69	6.64		10.25
1080	2"	"	0.089	4.62	7.37		12.00
1100	2-1/2"	"	0.100	9.21	8.30		17.50
1120	3"	"	0.123	10.25	10.25		20.50
1140	3-1/2"	"	0.145	14.50	12.00		26.50
1160	4"	"	0.182	14.00	15.00		29.00

© 2020 BNi Publications Inc. ALL RIGHTS RESERVED

CONDUCTORS, CONDUIT AND RACEWAYS

	Description	Output		Unit Costs			
ID Code	Component Descriptions	Unit of Meas.	Manhr / Unit	Material Cost	Labor Cost	Equipment Cost	Total Cost
26 - 05343	**FLEXIBLE CONDUIT**						**26 - 05343**
0080	Flexible conduit, steel						
0100	3/8"	LF	0.030	0.75	2.50		3.25
1020	1/2	"	0.030	0.85	2.50		3.35
1040	3/4"	"	0.040	1.17	3.32		4.49
1060	1"	"	0.040	2.23	3.32		5.55
1080	1-1/4"	"	0.050	2.79	4.15		6.94
1100	1-1/2"	"	0.059	4.63	4.91		9.54
1120	2"	"	0.080	5.69	6.64		12.25
1140	2-1/2"	"	0.089	6.93	7.37		14.25
1160	3"	"	0.107	12.00	8.85		20.75
26 - 05344	**GALVANIZED CONDUIT**						**26 - 05344**
1980	Galvanized rigid steel conduit						
2000	1/2"	LF	0.040	2.97	3.32		6.29
2040	3/4"	"	0.050	3.28	4.15		7.43
2060	1"	"	0.059	4.74	4.91		9.65
2080	1-1/4"	"	0.080	6.55	6.64		13.25
2100	1-1/2"	"	0.089	7.71	7.37		15.00
2120	2"	"	0.100	9.81	8.30		18.00
2140	2-1/2"	"	0.145	18.00	12.00		30.00
2160	3"	"	0.182	18.75	15.00		33.75
2180	3-1/2"	"	0.190	27.00	15.75		42.75
2200	4"	"	0.211	30.75	17.50		48.25
2220	5"	"	0.286	57.00	23.75		81.00
2240	6"	"	0.381	82.00	31.50		110
26 - 05345	**PLASTIC COATED CONDUIT**						**26 - 05345**
0980	Rigid steel conduit, plastic coated						
1000	1/2"	LF	0.050	7.06	4.15		11.25
1040	3/4"	"	0.059	8.20	4.91		13.00
1060	1"	"	0.080	10.50	6.64		17.25
1080	1-1/4"	"	0.100	13.50	8.30		21.75
1100	1-1/2"	"	0.123	16.25	10.25		26.50
1120	2"	"	0.145	21.25	12.00		33.25
1140	2-1/2"	"	0.190	32.25	15.75		48.00
1160	3"	"	0.222	40.75	18.50		59.00
1180	3-1/2"	"	0.250	49.75	20.75		71.00
1200	4"	"	0.308	60.00	25.50		86.00
1220	5"	"	0.381	100	31.50		130

© 2020 BNi Publications Inc. ALL RIGHTS RESERVED

CONDUCTORS, CONDUIT AND RACEWAYS

	Description	Output		Unit Costs			
ID Code	Component Descriptions	Unit of Meas.	Manhr / Unit	Material Cost	Labor Cost	Equipment Cost	Total Cost
26 - 05347	**STEEL CONDUIT**						**26 - 05347**
7980	Intermediate metal conduit (IMC)						
8000	1/2"	LF	0.030	2.02	2.50		4.52
8040	3/4"	"	0.040	2.48	3.32		5.80
8060	1"	"	0.050	3.76	4.15		7.91
8080	1-1/4"	"	0.059	4.81	4.91		9.72
8100	1-1/2"	"	0.080	6.02	6.64		12.75
8120	2"	"	0.089	7.86	7.37		15.25
8140	2-1/2"	"	0.119	15.50	9.91		25.50
8160	3"	"	0.145	20.00	12.00		32.00
8180	3-1/2"	"	0.182	23.50	15.00		38.50
8200	4"	"	0.190	26.00	15.75		41.75
26 - 05361	**CABLE TRAY**						**26 - 05361**
1010	Cable tray, 6"	LF	0.059	22.00	4.91		27.00
1020	Ventilated cover	"	0.030	8.89	2.50		11.50
1030	Solid cover	"	0.030	6.93	2.50		9.43

TRANSFORMERS

26 - 12001	**MEDIUM-VOLTAGE TRANSFORMERS**						**26 - 12001**
0080	Floor mtd, one phase, int. dry, 480v-120/240v						
0100	3 kva	EA	1.818	690	150		840
1080	5 kva	"	3.077	920	260		1,180
1100	7.5 kva	"	3.478	1,240	290		1,530
1120	10 kva	"	3.810	1,550	320		1,870
1140	15 kva	"	4.301	2,070	360		2,430
1240	100 kva	"	11.594	8,340	960		9,300
1980	Three phase, 480v-120/208v						
2000	15 kva	EA	6.015	2,270	500		2,770
2040	30 kva	"	9.412	2,720	780		3,500
2060	45 kva	"	10.811	3,620	900		4,520
2140	225 kva	"	15.385	13,010	1,280		14,290

© 2020 BNi Publications Inc. ALL RIGHTS RESERVED

SERVICE AND DISTRIBUTION

	Description	Output		Unit Costs			
ID Code	Component Descriptions	Unit of Meas.	Manhr / Unit	Material Cost	Labor Cost	Equipment Cost	Total Cost
26 - 24130	**SWITCHBOARDS**						**26 - 24130**
0580	Switchboard, 90" high, no main disconnect, 208/120v						
0581	400a	EA	7.921	3,040	660		3,700
1000	600a	"	8.000	4,710	660		5,370
1020	1000a	"	8.000	5,930	660		6,590
1040	1200a	"	10.000	6,280	830		7,110
1060	1600a	"	11.940	6,900	990		7,890
1080	2000a	"	14.035	7,410	1,160		8,570
1100	2500a	"	16.000	7,500	1,330		8,830
1520	277/480v						
1540	600a	EA	8.163	5,410	680		6,090
1560	800a	"	8.163	5,930	680		6,610
1580	1600a	"	11.940	7,470	990		8,460
1600	2000a	"	14.035	7,990	1,160		9,150
1620	2500a	"	16.000	8,510	1,330		9,840
1640	3000a	"	27.586	9,800	2,290		12,090
1660	4000a	"	29.630	11,870	2,460		14,330
26 - 24160	**PANELBOARDS**						**26 - 24160**
1500	3 phase, 480/277v, main lugs only, 120a, 30 circuits	EA	3.478	1,670	290		1,960
1510	277/480v, 4 wire, flush surface						
1520	225a, 30 circuits	EA	4.000	2,720	330		3,050
1540	400a, 30 circuits	"	5.000	3,690	420		4,110
1560	600a, 42 circuits	"	6.015	7,120	500		7,620
2000	208/120v, main circuit breaker, 3 phase, 4 wire						
2030	100a						
2035	12 circuits	EA	5.096	1,440	420		1,860
2040	20 circuits	"	6.299	1,790	520		2,310
2060	30 circuits	"	7.018	2,640	580		3,220
2110	400a						
2120	30 circuits	EA	14.815	5,550	1,230		6,780
2140	42 circuits	"	16.000	6,660	1,330		7,990
2180	600a, 42 circuits	"	18.182	12,950	1,510		14,460
2510	120/208v, flush, 3 ph., 4 wire, main only						
2515	100a						
2520	12 circuits	EA	5.096	1,020	420		1,440
2540	20 circuits	"	6.299	1,410	520		1,930
2560	30 circuits	"	7.018	2,090	580		2,670
2610	400a						

© 2020 BNi Publications Inc. ALL RIGHTS RESERVED

SERVICE AND DISTRIBUTION

ID Code	Component Descriptions	Unit of Meas.	Manhr / Unit	Material Cost	Labor Cost	Equipment Cost	Total Cost
26 - 24160	**PANELBOARDS, Cont'd...**						**26 - 24160**
2620	30 circuits	EA	14.815	4,080	1,230		5,310
2640	42 circuits	"	16.000	5,940	1,330		7,270
2680	600a, 42 circuits	"	18.182	9,260	1,510		10,770
26 - 24190	**MOTOR CONTROLS**						**26 - 24190**
0080	Motor generator set, 3 phase, 480/277v, w/controls						
0100	10kw	EA	27.586	13,490	2,290		15,780
1020	15kw	"	30.769	17,600	2,550		20,150
1040	20kw	"	32.000	19,530	2,660		22,190
1100	40kw	"	38.095	27,490	3,160		30,650
1180	100kw	"	61.538	40,960	5,110		46,070
1240	200kw	"	72.727	93,620	6,040		99,660
1280	300kw	"	80.000	117,020	6,640		123,660
2010	2 pole, 230 volt starter, w/NEMA-1						
2020	1 hp, 9a, size 00	EA	1.000	150	83.00		230
2040	2 hp, 18a, size 0	"	1.000	170	83.00		250
2060	3 hp, 27a, size 1	"	1.000	240	83.00		320
2080	5 hp, 45a, size 1p	"	1.000	240	83.00		320
2100	7-1/2 hp, 45a, size 2	"	1.000	580	83.00		660
2120	15 hp, 90a, size 3	"	1.000	880	83.00		960

BASIC MATERIALS

ID Code	Component Descriptions	Unit of Meas.	Manhr / Unit	Material Cost	Labor Cost	Equipment Cost	Total Cost
26 - 27268	**RECEPTACLES**						**26 - 27268**
0490	Contractor grade duplex receptacles, 15a 120v						
0510	Duplex	EA	0.200	1.60	16.50		18.00
1000	125 volt, 20a, duplex, standard grade	"	0.200	12.00	16.50		28.50
1040	Ground fault interrupter type	"	0.296	38.75	24.50		63.00
1520	250 volt, 20a, 2 pole, single, grounding type	"	0.200	20.00	16.50		36.50
1540	120/208v, 4 pole, single receptacle, twist lock						
1560	20a	EA	0.348	23.75	28.75		53.00
1580	50a	"	0.348	45.25	28.75		74.00
1590	125/250v, 3 pole, flush receptacle						
1600	30a	EA	0.296	24.00	24.50		48.50
1620	50a	"	0.296	29.75	24.50		54.00
1640	60a	"	0.348	77.00	28.75		110
1660	277v, 20a, 2 pole, grounding type, twist lock	"	0.200	13.00	16.50		29.50
2020	Dryer receptacle, 250v, 30a/50a, 3 wire	"	0.296	18.00	24.50		42.50
2040	Clock receptacle, 2 pole, grounding type	"	0.200	12.00	16.50		28.50

© 2020 BNi Publications Inc. ALL RIGHTS RESERVED

BASIC MATERIALS

ID Code	Component Descriptions	Unit of Meas.	Manhr / Unit	Material Cost	Labor Cost	Equipment Cost	Total Cost
26 - 27268	**RECEPTACLES, Cont'd...**						**26 - 27268**
3000	125v, 20a single recept. grounding type						
3010	Standard grade	EA	0.200	13.00	16.50		29.50
3020	Specification	"	0.200	15.50	16.50		32.00
3030	Hospital	"	0.200	16.25	16.50		32.75
3040	Isolated ground orange	"	0.250	55.00	20.75		76.00
3045	Duplex						
3050	Specification grade	EA	0.200	13.00	16.50		29.50
3060	Hospital	"	0.200	26.50	16.50		43.00
3070	Isolated ground orange	"	0.250	55.00	20.75		76.00
3100	GFI hospital grade recepts, 20a, 125v, duplex	"	0.296	58.00	24.50		83.00

LOW-VOLTAGE CIRCUIT PROTECTIVE DEVICES

ID Code	Component Descriptions	Unit of Meas.	Manhr / Unit	Material Cost	Labor Cost	Equipment Cost	Total Cost
26 - 28130	**FUSES**						**26 - 28130**
1000	Fuse, one-time, 250v						
1010	30a	EA	0.050	2.77	4.15		6.92
1020	60a	"	0.050	4.68	4.15		8.83
1040	100a	"	0.050	19.50	4.15		23.75
1060	200a	"	0.050	47.25	4.15		51.00
1080	400a	"	0.050	110	4.15		110
1100	600a	"	0.050	180	4.15		180
1120	600v						
1140	30a	EA	0.050	14.00	4.15		18.25
1160	60a	"	0.050	22.25	4.15		26.50
1180	100a	"	0.050	42.00	4.15		46.25
1200	200a	"	0.050	110	4.15		110
1220	400a	"	0.050	230	4.15		230
26 - 28161	**CIRCUIT BREAKERS**						**26 - 28161**
0950	Molded case, 240v, 15-60a, bolt-on						
1000	1 pole	EA	0.250	20.00	20.75		40.75
1060	2 pole	"	0.348	42.50	28.75		71.00
1080	70-100a, 2 pole	"	0.533	120	44.25		160
1100	15-60a, 3 pole	"	0.400	150	33.25		180
1120	70-100a, 3 pole	"	0.615	240	51.00		290
1980	480v, 2 pole						
2000	15-60a	EA	0.296	310	24.50		330
2080	70-100a	"	0.400	390	33.25		420
2090	3 pole						

© 2020 BNi Publications Inc. ALL RIGHTS RESERVED

LOW-VOLTAGE CIRCUIT PROTECTIVE DEVICES

ID Code	Description: Component Descriptions	Output: Unit of Meas.	Output: Manhr / Unit	Unit Costs: Material Cost	Unit Costs: Labor Cost	Unit Costs: Equipment Cost	Unit Costs: Total Cost
26 - 28161	**CIRCUIT BREAKERS, Cont'd...**						**26 - 28161**
2100	15-60a	EA	0.400	390	33.25		420
2120	70-100a	"	0.444	470	37.00		510
2140	70-225a	"	0.615	960	51.00		1,010
5000	Load center circuit breakers, 240v						
5010	1 pole, 10-60a	EA	0.250	21.25	20.75		42.00
5015	2 pole						
5020	10-60a	EA	0.400	43.25	33.25		77.00
5030	70-100a	"	0.667	130	55.00		190
5040	110-150a	"	0.727	280	60.00		340
5045	3 pole						
5050	10-60a	EA	0.500	120	41.50		160
5060	70-100a	"	0.727	180	60.00		240
5065	Load center, GFI breakers, 240v						
5070	1 pole, 15-30a	EA	0.296	160	24.50		180
5080	2 pole, 15-30a	"	0.400	280	33.25		310
5090	Key operated breakers, 240v, 1 pole, 10-30a	"	0.296	100	24.50		120
5095	Tandem breakers, 240v						
5100	1 pole, 15-30a	EA	0.400	35.00	33.25		68.00
5110	2 pole, 15-30a	"	0.533	64.00	44.25		110
5120	Bolt-on, GFI breakers, 240v, 1 pole, 15-30a	"	0.348	150	28.75		180
26 - 28164	**SWITCHES**						**26 - 28164**
1080	Fused interrupter load, 35kv, 20a						
1100	1 pole	EA	16.000	26,000	1,330		27,330
1120	2 pole	"	17.021	28,170	1,410		29,580
1250	Weatherproof switch, including box & cover, 20a						
1260	1 pole	EA	16.000	28,160	1,330		29,490
1280	2 pole	"	17.021	30,330	1,410		31,740
5175	Specification grade toggle switches, 20a, 120-277v						
5180	Single pole	EA	0.200	3.93	16.50		20.50
5190	Double pole	"	0.296	9.43	24.50		34.00
5200	3 way	"	0.250	10.25	20.75		31.00
5210	4 way	"	0.296	31.00	24.50		56.00
5495	Switch plates, plastic ivory						
5510	1 gang	EA	0.080	0.43	6.64		7.07
5520	2 gang	"	0.100	1.01	8.30		9.31
5530	3 gang	"	0.119	1.58	9.91		11.50
5540	4 gang	"	0.145	4.04	12.00		16.00

© 2020 BNi Publications Inc. ALL RIGHTS RESERVED

LOW-VOLTAGE CIRCUIT PROTECTIVE DEVICES

ID Code	Description: Component Descriptions	Output: Unit of Meas.	Output: Manhr / Unit	Unit Costs: Material Cost	Unit Costs: Labor Cost	Unit Costs: Equipment Cost	Unit Costs: Total Cost
26 - 28164	**SWITCHES, Cont'd...**						**26 - 28164**
5550	5 gang	EA	0.160	4.23	13.25		17.50
5560	6 gang	"	0.182	4.99	15.00		20.00
5565	Stainless steel						
5570	1 gang	EA	0.080	3.64	6.64		10.25
5580	2 gang	"	0.100	5.06	8.30		13.25
5590	3 gang	"	0.123	7.76	10.25		18.00
5600	4 gang	"	0.145	13.25	12.00		25.25
5610	5 gang	"	0.160	15.50	13.25		28.75
5620	6 gang	"	0.182	19.50	15.00		34.50
5625	Brass						
5630	1 gang	EA	0.080	6.79	6.64		13.50
5640	2 gang	"	0.100	14.50	8.30		22.75
5650	3 gang	"	0.123	22.50	10.25		32.75
5660	4 gang	"	0.145	26.00	12.00		38.00
5670	5 gang	"	0.160	32.25	13.25		45.50
5680	6 gang	"	0.182	38.75	15.00		54.00
26 - 28166	**SAFETY SWITCHES**						**26 - 28166**
0080	Fused, 3 phase, 30 amp, 600v, heavy duty						
1010	NEMA 1	EA	1.143	240	95.00		340
1020	NEMA 3r	"	1.143	540	95.00		630
1040	NEMA 4	"	1.600	1,510	130		1,640
1060	NEMA 12	"	1.739	480	140		620
1070	60a						
1080	NEMA 1	EA	1.143	330	95.00		430
1100	NEMA 3r	"	1.143	640	95.00		730
1120	NEMA 4	"	1.600	1,670	130		1,800
1140	NEMA 12	"	1.739	570	140		710
1150	100a						
1160	NEMA 1	EA	1.739	570	140		710
1200	NEMA 3r	"	1.739	990	140		1,130
1220	NEMA 4	"	2.000	3,550	170		3,720
1240	NEMA 12	"	2.500	860	210		1,070
1250	200a						
1260	NEMA 1	EA	2.500	840	210		1,050
1280	NEMA 3r	"	2.500	1,370	210		1,580
1300	NEMA 4	"	2.759	4,660	230		4,890
1320	NEMA 12	"	3.478	1,270	290		1,560

© 2020 BNi Publications Inc. ALL RIGHTS RESERVED

LOW-VOLTAGE CIRCUIT PROTECTIVE DEVICES

	Description	Output		Unit Costs			
ID Code	Component Descriptions	Unit of Meas.	Manhr / Unit	Material Cost	Labor Cost	Equipment Cost	Total Cost
26 - 28166	**SAFETY SWITCHES, Cont'd...**						**26 - 28166**
2000	Non-fused, 240-600v, heavy duty, 3 phase, 30 amp						
2020	NEMA 1	EA	1.143	170	95.00		260
2040	NEMA 3r	"	1.143	270	95.00		360
2060	NEMA 4	"	1.739	1,070	140		1,210
2080	NEMA 12	"	1.739	320	140		460
2090	60a						
2100	NEMA1	EA	1.143	220	95.00		310
2120	NEMA 3r	"	1.143	410	95.00		500
2140	NEMA 4	"	1.739	1,160	140		1,300
2160	NEMA 12	"	1.739	390	140		530
2170	100a						
2180	NEMA 1	EA	1.739	360	140		500
2200	NEMA 3r	"	1.739	570	140		710
2220	NEMA 4	"	2.500	2,340	210		2,550
2240	NEMA 12	"	2.500	550	210		760
2260	200a, NEMA 1	"	2.500	550	210		760
2680	600a, NEMA 12	"	12.308	3,080	1,020		4,100
26 - 28168	**SAFETY SWITCHES, HEAVY DUTY**						**26 - 28168**
0980	Safety switch, 600v, 3 pole, heavy duty, NEMA-1						
1000	30a	EA	1.000	210	83.00		290
1020	60a	"	1.143	280	95.00		380
1040	100a	"	1.600	530	130		660
1100	200a	"	2.500	830	210		1,040
1120	400a	"	5.517	2,140	460		2,600
1140	600a	"	8.000	3,800	660		4,460
1160	800a	"	10.526	8,600	870		9,470
1200	1200a	"	14.286	10,680	1,190		11,870
26 - 28169	**TRANSFER SWITCHES**						**26 - 28169**
0980	Automatic transfer switch 600v, 3 pole						
1000	30a	EA	3.478	2,680	290		2,970
1040	100a	"	4.762	3,550	400		3,950
1140	400a	"	10.000	8,010	830		8,840
1180	800a	"	18.182	14,740	1,510		16,250
1220	1200a	"	22.857	24,300	1,900		26,200
1280	2600a	"	42.105	63,670	3,490		67,160

© 2020 BNi Publications Inc. ALL RIGHTS RESERVED

POWER GENERATION

ID Code	Description Component Descriptions	Output Unit of Meas.	Manhr / Unit	Unit Costs Material Cost	Labor Cost	Equipment Cost	Total Cost
26 - 32001	**GENERATORS**						**26 - 32001**
0980	Diesel generator, with auto transfer switch						
1040	50kw	EA	30.769	32,530	2,550		35,080
1220	125kw	"	50.000	48,980	4,150		53,130
1320	300kw	"	100.000	80,230	8,300		88,530
1500	750kw	"	200.000	233,920	16,600		250,520

BATTERY EQUIPMENT

ID Code	Component Descriptions	Unit of Meas.	Manhr / Unit	Material Cost	Labor Cost	Equipment Cost	Total Cost
26 - 33530	**UNINTERRUPTIBLE POWER**						**26 - 33530**
1010	Uninterruptible power systems, (UPS), 3 kva	EA	8.000	7,750	660		8,410
1020	5 kva	"	11.004	8,690	910		9,600
1030	7.5 kva	"	16.000	10,430	1,330		11,760
1040	10 kva	"	21.978	13,040	1,820		14,860
1050	15 kva	"	22.857	15,650	1,900		17,550
1060	20 kva	"	24.024	21,730	1,990		23,720
1070	25 kva	"	25.000	27,820	2,080		29,900
1080	30 kva	"	25.974	28,680	2,160		30,840
1090	35 kva	"	27.027	30,420	2,240		32,660
1100	40 kva	"	27.972	33,030	2,320		35,350
1110	45 kva	"	28.986	34,770	2,410		37,180
1120	50 kva	"	29.963	37,380	2,490		39,870
1130	62.5 kva	"	32.000	44,330	2,660		46,990
1140	75 kva	"	34.934	51,280	2,900		54,180
1150	100 kva	"	36.036	68,670	2,990		71,660
1160	150 kva	"	50.000	104,310	4,150		108,460
1170	200 kva	"	55.172	139,080	4,580		143,660
1180	300 kva	"	74.766	208,620	6,210		214,830
1190	400 kva	"	89.888	312,100	7,460		319,560
1200	500 kva	"	109.589	390,120	9,100		399,220

FACILITY LIGHTNING PROTECTION

ID Code	Component Descriptions	Unit of Meas.	Manhr / Unit	Material Cost	Labor Cost	Equipment Cost	Total Cost
26 - 41130	**LIGHTNING PROTECTION**						**26 - 41130**
0100	Lightning protection						
0980	Copper point, nickel plated, 12'						
1000	1/2" dia.	EA	1.000	46.25	83.00		130
1020	5/8" dia.	"	1.000	52.00	83.00		140

© 2020 BNi Publications Inc. ALL RIGHTS RESERVED

LIGHTING

	Description	Output		Unit Costs			
ID Code	Component Descriptions	Unit of Meas.	Manhr / Unit	Material Cost	Labor Cost	Equipment Cost	Total Cost
26 - 51100	**INTERIOR LIGHTING**						**26 - 51100**
0010	Recessed fluorescent fixtures, 2'x2'						
0015	2 lamp	EA	0.727	72.00	60.00		130
0020	4 lamp	"	0.727	98.00	60.00		160
0030	2 lamp w/flange	"	1.000	91.00	83.00		170
0040	4 lamp w/flange	"	1.000	110	83.00		190
0045	1'x4'						
0050	2 lamp	EA	0.667	74.00	55.00		130
0060	3 lamp	"	0.667	100	55.00		160
0070	2 lamp w/flange	"	0.727	91.00	60.00		150
0080	3 lamp w/flange	"	0.727	120	60.00		180
0085	2'x4'						
0090	2 lamp	EA	0.727	91.00	60.00		150
0100	3 lamp	"	0.727	110	60.00		170
0110	4 lamp	"	0.727	100	60.00		160
0120	2 lamp w/flange	"	1.000	110	83.00		190
0130	3 lamp w/flange	"	1.000	120	83.00		200
0140	4 lamp w/flange	"	1.000	120	83.00		200
0145	4'x4'						
0150	4 lamp	EA	1.000	370	83.00		450
0160	6 lamp	"	1.000	430	83.00		510
0170	8 lamp	"	1.000	460	83.00		540
0180	4 lamp w/flange	"	1.509	450	130		580
0190	6 lamp w/flange	"	1.509	560	130		690
0200	8 lamp, w/flange	"	1.509	620	130		750
0205	Surface mounted incandescent fixtures						
0210	40w	EA	0.667	110	55.00		160
0220	75w	"	0.667	110	55.00		160
0230	100w	"	0.667	120	55.00		180
0240	150w	"	0.667	160	55.00		210
0245	Pendant						
0250	40w	EA	0.800	90.00	66.00		160
0260	75w	"	0.800	99.00	66.00		160
0270	100w	"	0.800	110	66.00		180
0280	150w	"	0.800	130	66.00		200
0287	Recessed incandescent fixtures						
0290	40w	EA	1.509	150	130		280
0300	75w	"	1.509	160	130		290
0310	100w	"	1.509	180	130		310

© 2020 BNi Publications Inc. ALL RIGHTS RESERVED

LIGHTING

ID Code	Description / Component Descriptions	Output: Unit of Meas.	Output: Manhr / Unit	Unit Costs: Material Cost	Unit Costs: Labor Cost	Unit Costs: Equipment Cost	Unit Costs: Total Cost
26 - 51100	**INTERIOR LIGHTING, Cont'd...**						**26 - 51100**
0320	150w	EA	1.509	190	130		320
0325	Exit lights, 120v						
0330	Recessed	EA	1.250	51.00	100		150
0340	Back mount	"	0.727	82.00	60.00		140
0350	Universal mount	"	0.727	86.00	60.00		150
0360	Emergency battery units, 6v-120v, 50 unit	"	1.509	180	130		310
0370	With 1 head	"	1.509	200	130		330
0380	With 2 heads	"	1.509	230	130		360
0395	Light track single circuit						
0400	2'	EA	0.500	44.25	41.50		86.00
0410	4'	"	0.500	52.00	41.50		94.00
0420	8'	"	1.000	72.00	83.00		160
0430	12'	"	1.509	100	130		230
0595	Fixtures, square						
0600	R-20	EA	0.145	45.00	12.00		57.00
0610	R-30	"	0.145	70.00	12.00		82.00
0660	Mini spot	"	0.145	42.75	12.00		55.00
26 - 51401	**INDUSTRIAL LIGHTING**						**26 - 51401**
0100	Surface mounted fluorescent, wrap around lens						
0110	1 lamp	EA	0.800	90.00	66.00		160
0120	2 lamps	"	0.889	130	74.00		200
0140	4 lamps	"	1.000	140	83.00		220
0250	Wall mounted fluorescent						
0300	2-20w lamps	EA	0.500	93.00	41.50		130
0320	2-30w lamps	"	0.500	110	41.50		150
0340	2-40w lamps	"	0.667	110	55.00		160
0350	Indirect, with wood shielding, 2049w lamps						
0360	4'	EA	1.000	110	83.00		190
0380	8'	"	1.600	140	130		270
0390	Industrial fluorescent, 2 lamp						
0400	4'	EA	0.727	72.00	60.00		130
0420	8'	"	1.333	110	110		220
0490	Strip fluorescent						
0510	4'						
0520	1 lamp	EA	0.667	45.50	55.00		100
0540	2 lamps	"	0.667	55.00	55.00		110
0550	8'						

© 2020 BNi Publications Inc. ALL RIGHTS RESERVED

LIGHTING

	Description	Output		Unit Costs			
ID Code	Component Descriptions	Unit of Meas.	Manhr / Unit	Material Cost	Labor Cost	Equipment Cost	Total Cost
26 - 51401	**INDUSTRIAL LIGHTING, Cont'd...**						**26 - 51401**
0560	1 lamp	EA	0.727	66.00	60.00		130
0580	2 lamps	"	0.889	99.00	74.00		170
0640	Wire guard for strip fixture, 4' long	"	0.348	10.25	28.75		39.00
0660	Strip fluorescent, 8' long, two 4' lamps	"	1.333	140	110		250
0680	With four 4' lamps	"	1.600	170	130		300
0690	Wet location fluorescent, plastic housing						
0695	4' long						
0700	1 lamp	EA	1.000	110	83.00		190
0720	2 lamps	"	1.333	160	110		270
0730	8' long						
0740	2 lamps	EA	1.600	280	130		410
0760	4 lamps	"	1.739	370	140		510
1000	Parabolic troffer, 2'x2'						
1020	With 2 U lamps	EA	1.000	130	83.00		210
1060	With 3 U lamps	"	1.143	150	95.00		250
1080	2'x4'						
1100	With 2 40w lamps	EA	1.143	150	95.00		250
1120	With 3 40w lamps	"	1.333	150	110		260
1140	With 4 40w lamps	"	1.333	160	110		270
1180	1'x4'						
1220	With 1 T-12 lamp, 9 cell	EA	0.727	78.00	60.00		140
1240	With 2 T-12 lamps	"	0.889	88.00	74.00		160
1260	With 1 T-12 lamp, 20 cell	"	0.727	88.00	60.00		150
1280	With 2 T-12 lamps	"	0.889	98.00	74.00		170
1480	Steel sided surface fluorescent, 2'x4'						
1500	3 lamps	EA	1.333	150	110		260
1520	4 lamps	"	1.333	170	110		280
2100	Outdoor sign fluor., 1 lamp, remote ballast						
2120	4' long	EA	6.015	3,160	500		3,660
2140	6' long	"	8.000	3,800	660		4,460
2620	Recess mounted, commercial, 2'x2', 13" high						
2640	100w	EA	4.000	1,050	330		1,380
2660	250w	"	4.494	1,160	370		1,530
3120	High pressure sodium, hi-bay open						
3140	400w	EA	1.739	460	140		600
3160	1000w	"	2.424	790	200		990
3170	Enclosed						
3180	400w	EA	2.424	740	200		940

© 2020 BNi Publications Inc. ALL RIGHTS RESERVED

LIGHTING

ID Code	Description: Component Descriptions	Output: Unit of Meas.	Output: Manhr / Unit	Unit Costs: Material Cost	Unit Costs: Labor Cost	Unit Costs: Equipment Cost	Unit Costs: Total Cost
26 - 51401	**INDUSTRIAL LIGHTING, Cont'd...**						**26 - 51401**
3200	1000w	EA	2.963	1,030	250		1,280
3210	Metal halide hi-bay, open						
3220	400w	EA	1.739	280	140		420
3240	1000w	"	2.424	580	200		780
3250	Enclosed						
3260	400w	EA	2.424	640	200		840
3280	1000w	"	2.963	610	250		860
3500	High pressure sodium, low-bay surface mounted						
3520	100w	EA	1.000	230	83.00		310
3540	150w	"	1.143	260	95.00		350
3560	250w	"	1.333	290	110		400
3580	400w	"	1.600	360	130		490
3590	Metal halide, low-bay, pendant mounted						
3600	175w	EA	1.333	370	110		480
3620	250w	"	1.600	510	130		640
3660	400w	"	2.222	550	180		730
4000	Indirect luminaire, square, metal halide, freestanding						
4020	175w	EA	1.000	460	83.00		540
4040	250w	"	1.000	500	83.00		580
4060	400w	"	1.000	510	83.00		590
4070	High pressure sodium						
4080	150w	EA	1.000	830	83.00		910
4100	250w	"	1.000	890	83.00		970
4120	400w	"	1.000	980	83.00		1,060
4125	Round, metal halide						
4140	175w	EA	1.000	960	83.00		1,040
4160	250w	"	1.000	990	83.00		1,070
4180	400w	"	1.000	1,040	83.00		1,120
4190	High pressure sodium						
4200	150w	EA	1.000	910	83.00		990
4220	250w	"	1.000	1,060	83.00		1,140
4240	400w	"	1.000	1,110	83.00		1,190
4250	Wall mounted, metal halide						
4260	175w	EA	2.500	420	210		630
4280	250w	"	2.500	410	210		620
4300	400w	"	3.200	450	270		720
4310	High pressure sodium						
4320	150w	EA	2.500	390	210		600

© 2020 BNi Publications Inc. ALL RIGHTS RESERVED

LIGHTING

	Description	Output		Unit Costs			
ID Code	Component Descriptions	Unit of Meas.	Manhr / Unit	Material Cost	Labor Cost	Equipment Cost	Total Cost
26 - 51401	**INDUSTRIAL LIGHTING, Cont'd...**						**26 - 51401**
4340	250w	EA	2.500	410	210		620
4360	400w	"	3.200	420	270		690
4480	Wall pack lithonia, high pressure sodium						
4500	35w	EA	0.889	60.00	74.00		130
4520	55w	"	1.000	76.00	83.00		160
4540	150w	"	1.600	190	130		320
4560	250w	"	1.739	200	140		340
4570	Low pressure sodium						
4580	35w	EA	1.739	320	140		460
4600	55w	"	2.000	440	170		610
4610	Wall pack hubbell, high pressure sodium						
4620	35w	EA	0.889	270	74.00		340
4640	150w	"	1.600	330	130		460
4660	250w	"	1.739	430	140		570
4700	Compact fluorescent						
4720	2-7w	EA	1.000	160	83.00		240
4740	2-13w	"	1.333	180	110		290
4760	1-18w	"	1.333	220	110		330
6000	Handball & racquet ball court, 2'x2', metal halide						
6020	250w	EA	2.500	580	210		790
6040	400w	"	2.759	690	230		920
6060	High pressure sodium						
6080	250w	EA	2.500	640	210		850
6100	400w	"	2.759	690	230		920
6120	Bollard light, 42" w/found., high pressure sodium						
6160	70w	EA	2.581	970	210		1,180
6180	100w	"	2.581	990	210		1,200
6200	150w	"	2.581	1,000	210		1,210
8000	Light fixture lamps						
8010	Lamp						
8020	20w med. bi-pin base, cool white, 24"	EA	0.145	7.78	12.00		19.75
8040	30w cool white, rapid start, 36"	"	0.145	9.86	12.00		21.75
8060	40w cool white U, 3"	"	0.145	21.50	12.00		33.50
8080	40w cool white, rapid start, 48"	"	0.145	9.18	12.00		21.25
8100	70w high pressure sodium, mogul base	"	0.200	71.00	16.50		88.00
8120	75w slimline, 96"	"	0.200	21.25	16.50		37.75
8130	100w						
8140	Incandescent, 100a, inside frost	EA	0.100	3.69	8.30		12.00

© 2020 BNi Publications Inc. ALL RIGHTS RESERVED

LIGHTING

	Description	Output		Unit Costs			
ID Code	Component Descriptions	Unit of Meas.	Manhr / Unit	Material Cost	Labor Cost	Equipment Cost	Total Cost
26 - 51401	**INDUSTRIAL LIGHTING, Cont'd...**						**26 - 51401**
8160	Mercury vapor, clear, mogul base	EA	0.200	64.00	16.50		81.00
8180	High pressure sodium, mogul base	"	0.200	98.00	16.50		110
8190	150w						
8200	Par 38 flood or spot, incandescent	EA	0.100	21.75	8.30		30.00
8220	High pressure sodium, 1/2 mogul base	"	0.200	87.00	16.50		100
8230	175w						
8240	Mercury vapor, clear, mogul base	EA	0.200	39.00	16.50		56.00
8260	Metal halide, clear, mogul base	"	0.200	78.00	16.50		95.00
8270	High pressure sodium, mogul base	"	0.200	87.00	16.50		100
8530	250w						
8540	Mercury vapor, clear, mogul base	EA	0.200	54.00	16.50		71.00
8560	Metal halide, clear, mogul base	"	0.200	78.00	16.50		95.00
8580	High pressure sodium, mogul base	"	0.200	92.00	16.50		110
8590	400w						
8600	Mercury vapor, clear, mogul base	EA	0.200	59.00	16.50		76.00
8620	Metal halide, clear, mogul base	"	0.200	78.00	16.50		95.00
8640	High pressure sodium, mogul base	"	0.200	95.00	16.50		110
8650	1000w						
8660	Mercury vapor, clear, mogul base	EA	0.250	140	20.75		160
8680	High pressure sodium, mogul base	"	0.250	250	20.75		270

EXTERIOR LIGHTING

26 - 56003	**EXTERIOR LIGHTING**						**26 - 56003**
1200	Exterior light fixtures						
1210	Rectangle, high pressure sodium						
1220	70w	EA	2.500	330	210		540
1240	100w	"	2.581	340	210		550
1260	150w	"	2.581	360	210		570
1280	250w	"	2.759	490	230		720
1300	400w	"	3.478	550	290		840
1310	Flood, rectangular, high pressure sodium						
1320	70w	EA	2.500	240	210		450
1340	100w	"	2.581	280	210		490
1360	150w	"	2.581	260	210		470
1400	400w	"	3.478	320	290		610
1420	1000w	"	4.494	540	370		910
1430	Round						

© 2020 BNi Publications Inc. ALL RIGHTS RESERVED

EXTERIOR LIGHTING

	Description	Output		Unit Costs			
ID Code	Component Descriptions	Unit of Meas.	Manhr / Unit	Material Cost	Labor Cost	Equipment Cost	Total Cost
26 - 56003	**EXTERIOR LIGHTING, Cont'd...**						**26 - 56003**
1440	400w	EA	3.478	600	290		890
1460	1000w	"	4.494	940	370		1,310
1470	Round, metal halide						
1480	400w	EA	3.478	670	290		960
1500	1000w	"	4.494	990	370		1,360
1980	Light fixture arms, cobra head, 6', high press. sodium						
2000	100w	EA	2.000	360	170		530
2021	150w	"	2.500	570	210		780
2060	250w	"	2.500	600	210		810
2080	400w	"	2.963	620	250		870
2090	Flood, metal halide						
2100	400w	EA	3.478	600	290		890
2120	1000w	"	4.494	820	370		1,190
6260	1500w	"	6.015	1,030	500		1,530
6270	Mercury vapor						
6280	250w	EA	2.759	390	230		620
6300	400w	"	3.478	440	290		730
6360	Incandescent						
6380	300w	EA	1.739	91.00	140		230
6410	500w	"	2.000	160	170		330
6420	1000w	"	3.200	180	270		450
26 - 56004	**ENERGY EFFICIENT EXTERIOR LIGHTING**						**26 - 56004**
1000	Solar Powered, LED area light, 100 Watt, Zone 4						
1010	Minimum	EA	1.333	1,190	110		1,300
1020	Average	"	1.600	1,330	130		1,460
1030	Maximum	"	2.000	1,450	170		1,620
1040	Zone 2						
1050	Minimum	EA	1.333	1,570	110		1,680
1060	Average	"	1.600	1,630	130		1,760
1070	Maximum	"	2.000	1,690	170		1,860
1080	Zone 4DD						
1090	Minimum	EA	1.333	1,800	110		1,910
1100	Average	"	1.600	1,930	130		2,060
1110	Maximum	"	2.000	2,060	170		2,230
1120	Zone 2DD						
1130	Minimum	EA	1.333	1,990	110		2,100
1140	Average	"	1.600	2,150	130		2,280

© 2020 BNi Publications Inc. ALL RIGHTS RESERVED

EXTERIOR LIGHTING

	Description	Output		Unit Costs			
ID Code	Component Descriptions	Unit of Meas.	Manhr / Unit	Material Cost	Labor Cost	Equipment Cost	Total Cost
26 - 56004	**ENERGY EFFICIENT EXTERIOR LIGHTING, Cont'd...**						**26 - 56004**
1150	Maximum	EA	2.000	2,300	170		2,470

© 2020 BNi Publications Inc. ALL RIGHTS RESERVED

DIVISION 27
COMMUNICATIONS

COMMUNICATIONS

	Description	Output		Unit Costs			
ID Code	Component Descriptions	Unit of Meas.	Manhr / Unit	Material Cost	Labor Cost	Equipment Cost	Total Cost
27 - 32001	**TELEPHONE SYSTEMS**						**27 - 32001**
0480	Communication cable						
0490	25 pair	LF	0.026	0.98	2.14		3.12
0520	100 pair	"	0.029	4.67	2.37		7.04
0560	400 pair	"	0.044	16.50	3.68		20.25
0700	Cable tap in manhole or junction box						
0800	25 pair cable	EA	3.810	6.81	320		330
1020	100 pair cable	"	15.094	27.50	1,250		1,280
1100	400 pair cable	"	61.538	110	5,110		5,220
2000	Cable terminations, manhole or junction box						
2020	25 pair cable	EA	3.756	6.81	310		320
2060	100 pair cable	"	15.094	27.50	1,250		1,280
2140	400 pair cable	"	61.538	85.00	5,110		5,200

AUDIO-VIDEO SYSTEMS

27 - 41005	**TELEVISION SYSTEMS**						**27 - 41005**
0100	TV outlet, self terminating, w/cover plate	EA	0.308	5.95	25.50		31.50
0120	Thru splitter	"	1.600	13.00	130		140
0140	End of line	"	1.333	10.75	110		120
0480	In line splitter multitap						
0490	4 way	EA	1.818	21.75	150		170
0520	2 way	"	1.702	16.25	140		160
1000	Equipment cabinet	"	1.600	54.00	130		180
1010	Antenna						
1020	Broad band UHF	EA	3.478	110	290		400
1040	Lightning arrester	"	0.727	33.00	60.00		93.00
1060	TV cable	LF	0.005	0.49	0.41		0.90

© 2020 BNi Publications Inc. ALL RIGHTS RESERVED

DIVISION 28
SAFETY & SECURITY

ELECTRONIC

ID Code	Description Component Descriptions	Output Unit of Meas.	Manhr / Unit	Unit Costs Material Cost	Labor Cost	Equipment Cost	Total Cost
28 - 16005	**SECURITY SYSTEMS**						**28 - 16005**
1000	Sensors						
1020	Balanced magnetic door switch, surface mounted	EA	0.500	180	41.50		220
1040	With remote test	"	1.000	230	83.00		310
1060	Flush mounted	"	1.860	170	150		320
1080	Mounted bracket	"	0.348	12.75	28.75		41.50
1100	Mounted bracket spacer	"	0.348	11.25	28.75		40.00
1120	Photoelectric sensor, for fence						
1140	6 beam	EA	2.759	18,550	230		18,780
1160	9 beam	"	4.255	22,680	350		23,030
1170	Photoelectric sensor, 12 volt dc						
1180	500' range	EA	1.600	550	130		680
1190	800' range	"	2.000	610	170		780
1195	Capacitance wire grid kit						
1200	Surface	EA	1.000	150	83.00		230
1220	Duct	"	1.600	110	130		240
1240	Tube grid kit	"	0.500	190	41.50		230
1260	Vibration sensor, 30 max per zone	"	0.500	230	41.50		270
1280	Audio sensor, 30 max per zone	"	0.500	240	41.50		280
1290	Inertia sensor						
1300	Outdoor	EA	0.727	180	60.00		240
1320	Indoor	"	0.500	110	41.50		150
2000	Ultrasonic transmitter, 20 max per zone						
2020	Omni-directional	EA	1.600	130	130		260
2040	Directional	"	1.333	140	110		250
2050	Transceiver						
2060	Omni-directional	EA	1.000	140	83.00		220
2080	Directional	"	1.000	150	83.00		230
2560	Passive infrared sensor, 20 max per zone	"	1.600	980	130		1,110
3020	Access/secure unit, balanced magnetic switch	"	1.600	610	130		740
3040	Photoelectric sensor	"	1.600	1,000	130		1,130
3060	Photoelectric fence sensor	"	1.600	1,030	130		1,160
3080	Capacitance sensor	"	1.739	1,190	140		1,330
3100	Audio and vibration sensor	"	1.600	1,050	130		1,180
3120	Inertia sensor	"	1.600	1,410	130		1,540
3160	Ultrasonic sensor	"	1.739	1,610	140		1,750
3200	infrared sensor	"	2.000	1,000	170		1,170
4020	Monitor panel, with access/secure tone, standard	"	1.739	690	140		830
4040	High security	"	2.000	1,000	170		1,170

© 2020 BNi Publications Inc. ALL RIGHTS RESERVED

ELECTRONIC

	Description	Output		Unit Costs			
ID Code	Component Descriptions	Unit of Meas.	Manhr / Unit	Material Cost	Labor Cost	Equipment Cost	Total Cost
28 - 16005	**SECURITY SYSTEMS, Cont'd...**						**28 - 16005**
4060	Emergency power indicator	EA	0.500	410	41.50		450
4070	Monitor rack with 115v power supply						
4080	1 zone	EA	1.000	570	83.00		650
4100	10 zone	"	2.500	2,910	210		3,120
4120	Monitor cabinet, wall mounted						
4140	1 zone	EA	1.000	880	83.00		960
4160	5 zone	"	1.600	1,450	130		1,580
4180	10 zone	"	1.739	3,160	140		3,300
4200	20 zone	"	2.000	4,410	170		4,580
4240	Floor mounted, 50 zone	"	4.000	4,590	330		4,920
5000	Security system accessories						
5020	Tamper assembly for monitor cabinet	EA	0.444	110	37.00		150
5040	Monitor panel blank	"	0.348	15.25	28.75		44.00
5060	Audible alarm	"	0.500	130	41.50		170
5080	Audible alarm control	"	0.348	530	28.75		560
5090	Termination screw, terminal cabinet						
5100	25 pair	EA	1.600	380	130		510
5120	50 pair	"	2.500	610	210		820
5140	150 pair	"	5.000	990	420		1,410
5150	Universal termination, cabinets & panel						
5160	Remote test	EA	1.739	89.00	140		230
5180	No remote test	"	0.727	63.00	60.00		120
5220	High security line supervision termination	"	1.000	440	83.00		520
5240	Door cord for capacitance sensor, 12"	"	0.500	15.25	41.50		57.00
5260	Insulation block kit for capacitance sensor	"	0.348	71.00	28.75		100
5280	Termination block for capacitance sensor	"	0.348	15.50	28.75		44.25
5360	Guard alert display	"	0.615	1,610	51.00		1,660
5380	Uninterrupted power supply	"	8.000	1,640	660		2,300
5390	Plug-in 40kva transformer						
5400	12 volt	EA	0.348	65.00	28.75		94.00
5420	18 volt	"	0.348	43.00	28.75		72.00
5440	24 volt	"	0.348	30.50	28.75		59.00
5520	Test relay	"	0.348	98.00	28.75		130
5580	Coaxial cable, 50 ohm	LF	0.006	0.44	0.50		0.94
6010	Door openers	EA	0.500	110	41.50		150
6015	Push buttons						
6020	Standard	EA	0.348	25.25	28.75		54.00
6030	Weatherproof	"	0.444	38.00	37.00		75.00

© 2020 BNi Publications Inc. ALL RIGHTS RESERVED

ELECTRONIC

ID Code	Description: Component Descriptions	Output: Unit of Meas.	Output: Manhr / Unit	Unit Costs: Material Cost	Unit Costs: Labor Cost	Unit Costs: Equipment Cost	Unit Costs: Total Cost
28 - 16005	**SECURITY SYSTEMS, Cont'd...**						**28 - 16005**
6040	Bells	EA	0.727	97.00	60.00		160
6045	Horns						
6050	Standard	EA	1.000	100	83.00		180
6060	Weatherproof	"	1.250	190	100		290
6070	Chimes	"	0.667	150	55.00		200
6080	Flasher	"	0.615	110	51.00		160
6090	Motion detectors	"	1.509	440	130		570
6100	Intercom units	"	0.727	100	60.00		160
6110	Remote annunciator	"	5.000	4,560	420		4,980

FIRE SAFETY

ID Code	Component Descriptions	Unit of Meas.	Manhr / Unit	Material Cost	Labor Cost	Equipment Cost	Total Cost
28 - 31001	**FIRE ALARM SYSTEMS**						**28 - 31001**
1000	Master fire alarm box, pedestal mounted	EA	16.000	8,210	1,330		9,540
1020	Master fire alarm box	"	6.015	4,220	500		4,720
1040	Box light	"	0.500	150	41.50		190
1060	Ground assembly for box	"	0.667	120	55.00		180
1080	Bracket for pole type box	"	0.727	150	60.00		210
1090	Pull station						
1100	Waterproof	EA	0.500	75.00	41.50		120
1110	Manual	"	0.400	56.00	33.25		89.00
1120	Horn, waterproof	"	1.000	110	83.00		190
1140	Interior alarm	"	0.727	71.00	60.00		130
1160	Coded transmitter, automatic	"	2.000	1,080	170		1,250
1180	Control panel, 8 zone	"	8.000	2,480	660		3,140
1200	Battery charger and cabinet	"	2.000	830	170		1,000
1240	Batteries, nickel cadmium or lead calcium	"	5.000	620	420		1,040
2500	CO_2 pressure switch connection	"	0.727	120	60.00		180
3000	Annunciator panels						
3020	Fire detection annunciator, remote type, 8 zone	EA	1.818	420	150		570
3100	12 zone	"	2.000	530	170		700
3120	16 zone	"	2.500	670	210		880
4000	Fire alarm systems						
4010	Bell	EA	0.615	130	51.00		180
4020	Weatherproof bell	"	0.667	82.00	55.00		140
4030	Horn	"	0.727	75.00	60.00		140
4040	Siren	"	2.000	760	170		930
4050	Chime	"	0.615	94.00	51.00		140

© 2020 BNi Publications Inc. ALL RIGHTS RESERVED

FIRE SAFETY

ID Code	Description: Component Descriptions	Output: Unit of Meas.	Output: Manhr / Unit	Unit Costs: Material Cost	Unit Costs: Labor Cost	Unit Costs: Equipment Cost	Unit Costs: Total Cost
28 - 31001	FIRE ALARM SYSTEMS, Cont'd...						28 - 31001
4060	Audio/visual	EA	0.727	140	60.00		200
4070	Strobe light	"	0.727	130	60.00		190
4080	Smoke detector	"	0.667	200	55.00		250
4090	Heat detector	"	0.500	35.25	41.50		77.00
4100	Thermal detector	"	0.500	32.75	41.50		74.00
4110	Ionization detector	"	0.533	170	44.25		210
4120	Duct detector	"	2.759	550	230		780
4130	Test switch	"	0.500	94.00	41.50		140
4140	Remote indicator	"	0.571	58.00	47.50		110
4150	Door holder	"	0.727	200	60.00		260
4160	Telephone jack	"	0.296	3.81	24.50		28.25
4170	Fireman phone	"	1.000	500	83.00		580
4180	Speaker	"	0.800	100	66.00		170
4185	Remote fire alarm annunciator panel						
4190	24 zone	EA	6.667	2,580	550		3,130
4200	48 zone	"	13.008	5,160	1,080		6,240
4205	Control panel						
4210	12 zone	EA	2.963	1,750	250		2,000
4220	16 zone	"	4.444	2,300	370		2,670
4230	24 zone	"	6.667	3,520	550		4,070
4240	48 zone	"	16.000	6,570	1,330		7,900
4250	Power supply	"	1.509	410	130		540
4260	Status command	"	5.000	11,130	420		11,550
4270	Printer	"	1.509	3,550	130		3,680
4280	Transponder	"	0.899	180	75.00		250
4290	Transformer	"	0.667	250	55.00		300
4300	Transceiver	"	0.727	360	60.00		420
4310	Relays	"	0.500	140	41.50		180
4320	Flow switch	"	2.000	460	170		630
4330	Tamper switch	"	2.963	280	250		530
4340	End of line resistor	"	0.348	20.00	28.75		48.75
4350	Printed circuit card	"	0.500	180	41.50		220
4360	Central processing unit	"	6.154	12,170	510		12,680
4370	UPS backup to CPU	"	8.999	22,280	750		23,030
8020	Smoke detector, fixed temp. & rate of rise comb.	"	1.600	360	130		490

© 2020 BNi Publications Inc. ALL RIGHTS RESERVED

DIVISION 31
EARTHWORK

SITE CLEARING

ID Code	Description: Component Descriptions	Output: Unit of Meas.	Output: Manhr / Unit	Unit Costs: Material Cost	Unit Costs: Labor Cost	Unit Costs: Equipment Cost	Unit Costs: Total Cost
31 - 11001	**CLEAR WOODED AREAS**						**31 - 11001**
0980	Clear wooded area						
1000	Light density	ACRE	60.000		3,800	3,330	7,130
1500	Medium density	"	80.000		5,070	4,430	9,500
1800	Heavy density	"	96.000		6,080	5,320	11,400

SELECTIVE TREE AND SHRUB REMOVAL AND TRIMMING

ID Code	Description: Component Descriptions	Output: Unit of Meas.	Output: Manhr / Unit	Unit Costs: Material Cost	Unit Costs: Labor Cost	Unit Costs: Equipment Cost	Unit Costs: Total Cost
31 - 13005	**TREE CUTTING & CLEARING**						**31 - 13005**
0980	Cut trees and clear out stumps						
1000	9" to 12" dia.	EA	4.800		300	270	570
1400	To 24" dia.	"	6.000		380	330	710
1600	24" dia. and up	"	8.000		510	440	950
5000	Loading and trucking						
5010	For machine load, per load, round trip						
5020	1 mile	EA	0.960		61.00	53.00	110
5025	3 mile	"	1.091		69.00	60.00	130
5030	5 mile	"	1.200		76.00	67.00	140
5035	10 mile	"	1.600		100	89.00	190
5040	20 mile	"	2.400		150	130	290
5050	Hand loaded, round trip						
5060	1 mile	EA	2.000		130	150	280
5065	3 mile	"	2.286		140	170	320
5070	5 mile	"	2.667		170	200	370
5080	10 mile	"	3.200		200	240	440
5100	20 mile	"	4.000		250	300	550
6000	Tree trimming for pole line construction						
6020	Light cutting	LF	0.012		0.76	0.66	1.42
6040	Medium cutting	"	0.016		1.01	0.88	1.90
6060	Heavy cutting	"	0.024		1.52	1.33	2.85

EARTHWORK, EXCAVATION & FILL

ID Code	Description: Component Descriptions	Output: Unit of Meas.	Output: Manhr / Unit	Unit Costs: Material Cost	Unit Costs: Labor Cost	Unit Costs: Equipment Cost	Unit Costs: Total Cost
31 - 22130	**ROUGH GRADING**						**31 - 22130**
1000	Site grading, cut & fill, sandy clay, 200' haul, 75 hp	CY	0.032		2.02	2.40	4.42
1100	Spread topsoil by equipment on site	"	0.036		2.25	2.66	4.91
1200	Site grading (cut and fill to 6") less than 1 acre						
1300	75 hp dozer	CY	0.053		3.37	4.00	7.37
1400	1.5 c.y. backhoe/loader	"	0.080		5.06	6.00	11.00

© 2020 BNi Publications Inc. ALL RIGHTS RESERVED

EARTHWORK, EXCAVATION & FILL

	Description	Output		Unit Costs			
ID Code	Component Descriptions	Unit of Meas.	Manhr / Unit	Material Cost	Labor Cost	Equipment Cost	Total Cost
31 - 23131	**BASE COURSE**						**31 - 23131**
1019	Base course, crushed stone						
1020	3" thick	SY	0.004	3.34	0.33	0.44	4.11
1030	4" thick	"	0.004	4.50	0.36	0.47	5.34
1040	6" thick	"	0.005	6.75	0.39	0.51	7.66
1050	8" thick	"	0.005	9.00	0.44	0.58	10.00
1060	10" thick	"	0.006	11.25	0.48	0.62	12.25
1070	12" thick	"	0.007	13.50	0.56	0.73	14.75
2500	Base course, bank run gravel						
3020	4" deep	SY	0.004	3.17	0.35	0.46	3.98
3040	6" deep	"	0.005	4.85	0.38	0.50	5.73
3060	8" deep	"	0.005	6.41	0.42	0.55	7.38
3070	10" deep	"	0.005	8.02	0.44	0.58	9.05
3080	12" deep	"	0.006	9.58	0.51	0.67	10.75
4000	Prepare and roll sub-base						
4020	Minimum	SY	0.004		0.33	0.44	0.77
4030	Average	"	0.005		0.42	0.55	0.97
4040	Maximum	"	0.007		0.56	0.73	1.29
31 - 23132	**BORROW**						**31 - 23132**
1000	Borrow fill, FOB at pit						
1005	Sand, haul to site, round trip						
1010	10 mile	CY	0.080	23.75	6.74	8.80	39.25
1020	20 mile	"	0.133	23.75	11.25	14.75	49.75
1030	30 mile	"	0.200	23.75	16.75	22.00	63.00
3980	Place borrow fill and compact						
4000	Less than 1 in 4 slope	CY	0.040	23.75	3.37	4.40	31.50
4100	Greater than 1 in 4 slope	"	0.053	23.75	4.49	5.86	34.00
31 - 23137	**GRAVEL AND STONE**						**31 - 23137**
0120	FOB at plant, material only						
1000	No. 21 crusher run stone	CY					33.00
1100	No. 26 crusher run stone	"					33.00
1140	No. 57 stone	"					33.00
1150	No. 67 gravel	"					33.00
1180	No. 68 stone	"					33.00
1220	No. 78 stone	"					33.00
1235	No. 78 gravel, (pea gravel)	"					33.00
1250	No. 357 or B-3 stone	"					33.00
1260	Structural & foundation backfill						

© 2020 BNi Publications Inc. ALL RIGHTS RESERVED

EARTHWORK, EXCAVATION & FILL

	Description	Output		Unit Costs			
ID Code	Component Descriptions	Unit of Meas.	Manhr / Unit	Material Cost	Labor Cost	Equipment Cost	Total Cost
31 - 23137	**GRAVEL AND STONE, Cont'd...**						**31 - 23137**
1400	No. 21 crusher run stone	TON					24.50
1500	No. 26 crusher run stone	"					24.50
1600	No. 57 stone	"					24.50
2160	No. 67 gravel	"					24.50
2210	No. 68 stone	"					24.50
2220	No. 78 stone	"					24.50
2280	No. 78 gravel (pea gravel)	"					24.50
3240	No. 357 or B-3 stone	"					24.50
31 - 23163	**BULK EXCAVATION**						**31 - 23163**
1000	Excavation, by small dozer						
1020	Large areas	CY	0.016		1.01	1.20	2.21
1040	Small areas	"	0.027		1.68	2.00	3.68
1060	Trim banks	"	0.040		2.53	3.00	5.53
1200	Drag line						
1220	1-1/2 c.y. bucket						
1240	Sand or gravel	CY	0.040		2.53	2.21	4.75
1260	Light clay	"	0.053		3.37	2.95	6.33
1280	Heavy clay	"	0.060		3.80	3.32	7.12
1300	Unclassified	"	0.064		4.05	3.54	7.60
1400	2 c.y. bucket						
1420	Sand or gravel	CY	0.037		2.33	2.04	4.38
1440	Light clay	"	0.048		3.04	2.66	5.70
1460	Heavy clay	"	0.053		3.37	2.95	6.33
1480	Unclassified	"	0.056		3.57	3.12	6.70
1500	2-1/2 c.y. bucket						
1520	Sand or gravel	CY	0.034		2.17	1.90	4.07
1540	Light clay	"	0.044		2.76	2.41	5.18
1560	Heavy clay	"	0.048		3.04	2.66	5.70
1580	Unclassified	"	0.051		3.20	2.80	6.00
1600	3 c.y. bucket						
1620	Sand or gravel	CY	0.030		1.90	1.66	3.56
1640	Light clay	"	0.040		2.53	2.21	4.75
1660	Heavy clay	"	0.044		2.76	2.41	5.18
1680	Unclassified	"	0.046		2.89	2.53	5.42
1700	Hydraulic excavator						
1720	1 c.y. capacity						
1740	Light material	CY	0.040		2.53	2.21	4.75

© 2020 BNi Publications Inc. ALL RIGHTS RESERVED

EARTHWORK, EXCAVATION & FILL

	Description	Output		Unit Costs			
ID Code	Component Descriptions	Unit of Meas.	Manhr / Unit	Material Cost	Labor Cost	Equipment Cost	Total Cost
31 - 23163	**BULK EXCAVATION, Cont'd...**						**31 - 23163**
1760	Medium material	CY	0.048		3.04	2.66	5.70
1780	Wet material	"	0.060		3.80	3.32	7.12
1790	Blasted rock	"	0.069		4.34	3.80	8.14
1800	1-1/2 c.y. capacity						
1820	Light material	CY	0.010		0.84	1.10	1.94
1840	Medium material	"	0.013		1.12	1.46	2.59
1860	Wet material	"	0.016		1.34	1.76	3.10
1880	Blasted rock	"	0.020		1.68	2.20	3.88
1900	2 c.y. capacity						
1920	Light material	CY	0.009		0.74	0.97	1.72
1940	Medium material	"	0.011		0.96	1.25	2.22
1960	Wet material	"	0.013		1.12	1.46	2.59
1980	Blasted rock	"	0.016		1.34	1.76	3.10
2000	Wheel mounted front-end loader						
2020	7/8 c.y. capacity						
2040	Light material	CY	0.020		1.68	2.20	3.88
2060	Medium material	"	0.023		1.92	2.51	4.44
2080	Wet material	"	0.027		2.24	2.93	5.18
2100	Blasted rock	"	0.032		2.69	3.52	6.21
2200	1-1/2 c.y. capacity						
2220	Light material	CY	0.011		0.96	1.25	2.22
2240	Medium material	"	0.012		1.03	1.35	2.39
2260	Wet material	"	0.013		1.12	1.46	2.59
2280	Blasted rock	"	0.015		1.22	1.60	2.82
2300	2-1/2 c.y. capacity						
2320	Light material	CY	0.009		0.79	1.03	1.82
2340	Medium material	"	0.010		0.84	1.10	1.94
2360	Wet material	"	0.011		0.89	1.17	2.07
2380	Blasted rock	"	0.011		0.96	1.25	2.22
2400	3-1/2 c.y. capacity						
2420	Light material	CY	0.009		0.74	0.97	1.72
2440	Medium material	"	0.009		0.79	1.03	1.82
2460	Wet material	"	0.010		0.84	1.10	1.94
2480	Blasted rock	"	0.011		0.89	1.17	2.07
2500	6 c.y. capacity						
2520	Light material	CY	0.005		0.44	0.58	1.03
2540	Medium material	"	0.006		0.48	0.62	1.11
2560	Wet material	"	0.006		0.51	0.67	1.19

© 2020 BNi Publications Inc. ALL RIGHTS RESERVED

EARTHWORK, EXCAVATION & FILL

	Description	Output		Unit Costs			
ID Code	Component Descriptions	Unit of Meas.	Manhr / Unit	Material Cost	Labor Cost	Equipment Cost	Total Cost
31 - 23163	**BULK EXCAVATION, Cont'd...**						**31 - 23163**
2580	Blasted rock	CY	0.007		0.56	0.73	1.29
2600	Track mounted front-end loader						
2620	1-1/2 c.y. capacity						
2640	Light material	CY	0.013		1.12	1.46	2.59
2660	Medium material	"	0.015		1.22	1.60	2.82
2680	Wet material	"	0.016		1.34	1.76	3.10
2700	Blasted rock	"	0.018		1.49	1.95	3.45
2720	2-3/4 c.y. capacity						
2740	Light material	CY	0.008		0.67	0.88	1.55
2760	Medium material	"	0.009		0.74	0.97	1.72
2780	Wet material	"	0.010		0.84	1.10	1.94
2790	Blasted rock	"	0.011		0.96	1.25	2.22
31 - 23164	**BUILDING EXCAVATION**						**31 - 23164**
0090	Structural excavation, unclassified earth						
0100	3/8 c.y. backhoe	CY	0.107		8.99	11.75	20.75
0110	3/4 c.y. backhoe	"	0.080		6.74	8.80	15.50
0120	1 c.y. backhoe	"	0.067		5.62	7.33	13.00
0600	Foundation backfill and compaction by machine	"	0.160		13.50	17.50	31.00
31 - 23165	**HAND EXCAVATION**						**31 - 23165**
0980	Excavation						
1000	To 2' deep						
1020	Normal soil	CY	0.889		57.00		57.00
1040	Sand and gravel	"	0.800		51.00		51.00
1060	Medium clay	"	1.000		64.00		64.00
1080	Heavy clay	"	1.143		73.00		73.00
1100	Loose rock	"	1.333		85.00		85.00
1200	To 6' deep						
1220	Normal soil	CY	1.143		73.00		73.00
1240	Sand and gravel	"	1.000		64.00		64.00
1260	Medium clay	"	1.333		85.00		85.00
1280	Heavy clay	"	1.600		100		100
1300	Loose rock	"	2.000		130		130
2020	Backfilling foundation without compaction, 6" lifts	"	0.500		32.00		32.00
2200	Compaction of backfill around structures or in trench						
2220	By hand with air tamper	CY	0.571		36.50		36.50
2240	By hand with vibrating plate tamper	"	0.533		34.00		34.00
2250	1 ton roller	"	0.400		25.25	30.00	55.00

© 2020 BNi Publications Inc. ALL RIGHTS RESERVED

EARTHWORK, EXCAVATION & FILL

	Description	Output		Unit Costs			
ID Code	Component Descriptions	Unit of Meas.	Manhr / Unit	Material Cost	Labor Cost	Equipment Cost	Total Cost
31 - 23165	**HAND EXCAVATION, Cont'd...**						**31 - 23165**
5400	Miscellaneous hand labor						
5440	Trim slopes, sides of excavation	SF	0.001		0.08		0.08
5450	Trim bottom of excavation	"	0.002		0.10		0.10
5460	Excavation around obstructions and services	CY	2.667		170		170
31 - 23167	**UTILITY EXCAVATION**						**31 - 23167**
2080	Trencher, sandy clay, 8" wide trench						
2100	18" deep	LF	0.018		1.12	1.33	2.45
2200	24" deep	"	0.020		1.26	1.50	2.76
2300	36" deep	"	0.023		1.44	1.71	3.16
6080	Trench backfill, 95% compaction						
7000	Tamp by hand	CY	0.500		32.00		32.00
7050	Vibratory compaction	"	0.400		25.50		25.50
7060	Trench backfilling, with borrow sand, place & compact	"	0.400	22.75	25.50		48.25
31 - 23168	**ROADWAY EXCAVATION**						**31 - 23168**
0100	Roadway excavation						
0110	1/4 mile haul	CY	0.016		1.34	1.76	3.10
0120	2 mile haul	"	0.027		2.24	2.93	5.18
0130	5 mile haul	"	0.040		3.37	4.40	7.77
0150	Excavation of open ditches	"	0.011		0.96	1.25	2.22
0160	Trim banks, swales or ditches	SY	0.013		1.12	1.46	2.59
0165	Bulk swale excavation by dragline						
0170	Small jobs	CY	0.060		3.80	3.32	7.12
0180	Large jobs	"	0.034		2.17	1.90	4.07
3000	Spread base course	"	0.020		1.68	2.20	3.88
3100	Roll and compact	"	0.027		2.24	2.93	5.18
31 - 23169	**HAULING MATERIAL**						**31 - 23169**
0090	Haul material by 10 c.y. dump truck, round trip distance						
0100	1 mile	CY	0.044		2.81	3.33	6.14
0110	2 mile	"	0.053		3.37	4.00	7.37
0120	5 mile	"	0.073		4.60	5.45	10.00
0130	10 mile	"	0.080		5.06	6.00	11.00
0140	20 mile	"	0.089		5.63	6.66	12.25
0150	30 mile	"	0.107		6.75	8.00	14.75

© 2020 BNi Publications Inc. ALL RIGHTS RESERVED

EARTHWORK, EXCAVATION & FILL

	Description	Output		Unit Costs			
ID Code	Component Descriptions	Unit of Meas.	Manhr / Unit	Material Cost	Labor Cost	Equipment Cost	Total Cost
31 - 23195	**WELLPOINT SYSTEMS**						**31 - 23195**
0980	Pumping, gas driven, 50' hose						
1000	3" header pipe	DAY	8.000		510	600	1,110
1010	6" header pipe	"	10.000		630	750	1,380
1080	Wellpoint system per job; 150' length of PVC header						
1100	6" header pipe, 2" wellpoints, 5' centers	LF	0.032	60.00	2.02	2.40	64.00
1110	8" header pipe	"	0.040	72.00	2.53	3.00	78.00
1120	10" header pipe	"	0.053	110	3.37	4.00	120
1200	Jetting wellpoint system						
1220	14' long	EA	0.533	82.00	33.75	40.00	160
1230	18' long	"	0.667	94.00	42.25	50.00	190
1240	Sand filter for wellpoints	LF	0.013	3.93	0.84	1.00	5.77
2000	Replacement of wellpoint components	EA	0.160		10.25	12.00	22.25
31 - 23336	**TRENCHING**						**31 - 23336**
0100	Trenching and continuous footing excavation						
0980	By gradall						
1000	1 c.y. capacity						
1020	Light soil	CY	0.023		1.92	2.51	4.44
1040	Medium soil	"	0.025		2.07	2.70	4.78
1060	Heavy/wet soil	"	0.027		2.24	2.93	5.18
1080	Loose rock	"	0.029		2.45	3.20	5.65
1090	Blasted rock	"	0.031		2.59	3.38	5.98
1095	By hydraulic excavator						
1100	1/2 c.y. capacity						
1120	Light soil	CY	0.027		2.24	2.93	5.18
1140	Medium soil	"	0.029		2.45	3.20	5.65
1160	Heavy/wet soil	"	0.032		2.69	3.52	6.21
1180	Loose rock	"	0.036		2.99	3.91	6.91
1190	Blasted rock	"	0.040		3.37	4.40	7.77
1200	1 c.y. capacity						
1220	Light soil	CY	0.019		1.58	2.07	3.65
1240	Medium soil	"	0.020		1.68	2.20	3.88
1260	Heavy/wet soil	"	0.021		1.79	2.34	4.14
1280	Loose rock	"	0.023		1.92	2.51	4.44
1300	Blasted rock	"	0.025		2.07	2.70	4.78
1400	1-1/2 c.y. capacity						
1420	Light soil	CY	0.017		1.42	1.85	3.27
1440	Medium soil	"	0.018		1.49	1.95	3.45

© 2020 BNi Publications Inc. ALL RIGHTS RESERVED

EARTHWORK, EXCAVATION & FILL

ID Code	Description: Component Descriptions	Output: Unit of Meas.	Output: Manhr / Unit	Unit Costs: Material Cost	Unit Costs: Labor Cost	Unit Costs: Equipment Cost	Unit Costs: Total Cost
31 - 23336	**TRENCHING, Cont'd...**						**31 - 23336**
1460	Heavy/wet soil	CY	0.019		1.58	2.07	3.65
1480	Loose rock	"	0.020		1.68	2.20	3.88
1500	Blasted rock	"	0.021		1.79	2.34	4.14
1600	2 c.y. capacity						
1620	Light soil	CY	0.016		1.34	1.76	3.10
1640	Medium soil	"	0.017		1.42	1.85	3.27
1660	Heavy/wet soil	"	0.018		1.49	1.95	3.45
1680	Loose rock	"	0.019		1.58	2.07	3.65
1690	Blasted rock	"	0.020		1.68	2.20	3.88
1700	2-1/2 c.y. capacity						
1720	Light soil	CY	0.015		1.22	1.60	2.82
1740	Medium soil	"	0.015		1.28	1.67	2.96
1760	Heavy/wet soil	"	0.016		1.34	1.76	3.10
1780	Loose rock	"	0.017		1.42	1.85	3.27
1790	Blasted rock	"	0.018		1.49	1.95	3.45
1800	Trencher, chain, 1' wide to 4' deep						
1940	Light soil	CY	0.020		1.26	1.50	2.76
1960	Medium soil	"	0.023		1.44	1.71	3.16
1980	Heavy soil	"	0.027		1.68	2.00	3.68
3000	Hand excavation						
3100	Bulk, wheeled 100'						
3120	Normal soil	CY	0.889		57.00		57.00
3140	Sand or gravel	"	0.800		51.00		51.00
3160	Medium clay	"	1.143		73.00		73.00
3180	Heavy clay	"	1.600		100		100
3200	Loose rock	"	2.000		130		130
3300	Trenches, up to 2' deep						
3320	Normal soil	CY	1.000		64.00		64.00
3340	Sand or gravel	"	0.889		57.00		57.00
3360	Medium clay	"	1.333		85.00		85.00
3380	Heavy clay	"	2.000		130		130
3390	Loose rock	"	2.667		170		170
3400	Trenches, to 6' deep						
3420	Normal soil	CY	1.143		73.00		73.00
3440	Sand or gravel	"	1.000		64.00		64.00
3460	Medium clay	"	1.600		100		100
3480	Heavy clay	"	2.667		170		170
3500	Loose rock	"	4.000		260		260

© 2020 BNi Publications Inc. ALL RIGHTS RESERVED

EARTHWORK, EXCAVATION & FILL

	Description	Output		Unit Costs			
ID Code	Component Descriptions	Unit of Meas.	Manhr / Unit	Material Cost	Labor Cost	Equipment Cost	Total Cost
31 - 23336	**TRENCHING, Cont'd...**						**31 - 23336**
3590	Backfill trenches						
3600	With compaction						
3620	By hand	CY	0.667		42.50		42.50
3640	By 60 hp tracked dozer	"	0.020		1.26	1.50	2.76
3650	By 200 hp tracked dozer	"	0.009		0.74	0.97	1.72
3660	By small front-end loader	"	0.023		1.44	1.71	3.16
3700	Spread dumped fill or gravel, no compaction						
3740	6" layers	SY	0.013		0.84	1.00	1.84
3760	12" layers	"	0.016		1.01	1.20	2.21
3800	Compaction in 6" layers						
3820	By hand with air tamper	SY	0.016		1.02	0.34	1.36
3890	Backfill trenches, sand bedding, no compaction						
3900	By hand	CY	0.667	23.00	42.50		66.00
3940	By small front-end loader	"	0.023	23.00	1.92	2.51	27.50

SOIL STABILIZATION & TREATMENT

31 - 31160	**SOIL TREATMENT**						**31 - 31160**
1100	Soil treatment, termite control pretreatment						
1120	Under slabs	SF	0.004	0.38	0.28		0.66
1140	By walls	"	0.005	0.38	0.34		0.72

SOIL STABILIZATION

31 - 32003	**GEOTEXTILE**						**31 - 32003**
0060	Filter cloth, light reinforcement						
1180	Woven						
1200	12'-6" wide x 50' long	SF	0.001	0.34	0.07		0.41
1300	Various lengths	"	0.001	0.51	0.07		0.58
1380	Non-woven						
1390	14'-8" wide x 430' long	SF	0.001	0.18	0.07		0.25
1400	Various lengths	"	0.001	0.26	0.07		0.33

© 2020 BNi Publications Inc. ALL RIGHTS RESERVED

SOIL STABILIZATION

ID Code	Description / Component Descriptions	Output: Unit of Meas.	Output: Manhr / Unit	Unit Costs: Material Cost	Unit Costs: Labor Cost	Unit Costs: Equipment Cost	Unit Costs: Total Cost
31 - 32005	**SOIL STABILIZATION**						**31 - 32005**
0100	Straw bale secured with rebar	LF	0.027	7.54	1.70		9.24
0120	Filter barrier, 18" high filter fabric	"	0.080	1.82	5.10		6.92
0130	Sediment fence, 36" fabric with 6" mesh	"	0.100	4.32	6.38		10.75
1000	Soil stabilization with tar paper, burlap, straw and	SF	0.001	0.36	0.07		0.43

GABIONS

ID Code	Component Descriptions	Unit of Meas.	Manhr / Unit	Material Cost	Labor Cost	Equipment Cost	Total Cost
31 - 36001	**SLOPE PROTECTION**						**31 - 36001**
2060	Gabions, stone filled						
2080	6" deep	SY	0.200	30.25	12.75	15.00	58.00
2090	9" deep	"	0.229	37.00	14.50	17.25	69.00
2100	12" deep	"	0.267	49.00	17.00	20.00	86.00
2120	18" deep	"	0.320	63.00	20.25	24.00	110
2140	36" deep	"	0.533	110	33.75	40.00	180

RIPRAP

ID Code	Component Descriptions	Unit of Meas.	Manhr / Unit	Material Cost	Labor Cost	Equipment Cost	Total Cost
31 - 37001	**RIPRAP**						**31 - 37001**
0100	Riprap						
0110	Crushed stone blanket, max size 2-1/2"	TON	0.533	35.25	34.00	48.25	120
0120	Stone, quarry run, 300 lb. stones	"	0.492	44.25	31.50	44.50	120
0130	400 lb. stones	"	0.457	46.00	29.25	41.50	120
0140	500 lb. stones	"	0.427	48.00	27.25	38.75	110
0150	750 lb. stones	"	0.400	49.75	25.50	36.25	110
0160	Dry concrete riprap in bags 3" thick, 80 lb. per bag	BAG	0.027	5.96	1.70	2.41	10.00

SHORING AND UNDERPINNING

ID Code	Component Descriptions	Unit of Meas.	Manhr / Unit	Material Cost	Labor Cost	Equipment Cost	Total Cost
31 - 41160	**STEEL SHEET PILING**						**31 - 41160**
1000	Steel sheet piling,12" wide						
1100	20' long	SF	0.096	21.00	7.82	9.20	38.00
1200	35' long	"	0.069	21.00	5.59	6.57	33.25
1300	50' long	"	0.048	21.00	3.91	4.60	29.50
1400	Over 50' long	"	0.044	21.00	3.55	4.18	28.75

© 2020 BNi Publications Inc. ALL RIGHTS RESERVED

SHORING AND UNDERPINNING

ID Code	Component Descriptions	Unit of Meas.	Manhr / Unit	Material Cost	Labor Cost	Equipment Cost	Total Cost
31 - 41330	**TRENCH SHEETING**						**31 - 41330**
0980	Closed timber, including pull and salvage, excavation						
1000	8' deep	SF	0.064	2.87	4.08	5.80	12.75
1200	10' deep	"	0.067	2.93	4.30	6.10	13.50
1300	12' deep	"	0.071	3.01	4.54	6.44	14.00
1400	14' deep	"	0.075	3.14	4.80	6.82	15.00
1600	16' deep	"	0.080	3.16	5.10	7.25	15.50
1800	18' deep	"	0.091	3.26	5.83	8.28	17.25
2000	20' deep	"	0.098	4.12	6.28	8.92	19.25

EXCAVATION SUPPORT AND PROTECTION

ID Code	Component Descriptions	Unit of Meas.	Manhr / Unit	Material Cost	Labor Cost	Equipment Cost	Total Cost
31 - 50001	**TRENCH BOX**						**31 - 50001**
0100	10' x 10' Steel, 4" Wall, Rental Rate						
0120	4 Hour	EA					160
0140	Day	"					160
0160	Week	"					400
0180	Month	"					1,210
0200	10 x 24 Steel 6" Wall, Rental Rate						
0220	4 Hour	EA					310
0240	Day	"					310
0260	Week	"					780
0280	Month	"					2,340

COFFERDAMS

ID Code	Component Descriptions	Unit of Meas.	Manhr / Unit	Material Cost	Labor Cost	Equipment Cost	Total Cost
31 - 52001	**COFFERDAMS**						**31 - 52001**
0980	Cofferdam, steel, driven from shore						
1000	15' deep	SF	0.137	20.25	11.25	13.25	44.50
1020	20' deep	"	0.128	20.25	10.50	12.25	43.00
1040	25' deep	"	0.120	20.25	9.78	11.50	41.50
1060	30' deep	"	0.113	20.25	9.21	10.75	40.25
1080	40' deep	"	0.107	20.25	8.69	10.25	39.25
1090	Driven from barge						
1100	20' deep	SF	0.148	20.25	12.00	14.25	46.50
1120	30' deep	"	0.137	20.25	11.25	13.25	44.50
1140	40' deep	"	0.128	20.25	10.50	12.25	43.00
1160	50' deep	"	0.120	20.25	9.78	11.50	41.50

© 2020 BNi Publications Inc. ALL RIGHTS RESERVED

PILES AND CAISSONS

ID Code	Description Component Descriptions	Output Unit of Meas.	Manhr / Unit	Unit Costs Material Cost	Labor Cost	Equipment Cost	Total Cost
31 - 62001	**PILE TESTING**						**31 - 62001**
1000	Pile test						
1020	50 ton to 100 ton	EA					20,670
1030	To 200 ton	"					29,160
1060	To 300 ton	"					33,730
1080	To 400 ton	"					39,500
1100	To 600 ton	"					49,380
31 - 62165	**STEEL PILES**						**31 - 62165**
1000	H-section piles						
1010	8x8						
1020	36 lb/ft						
1021	30' long	LF	0.080	21.75	6.52	7.66	36.00
1022	40' long	"	0.064	21.75	5.21	6.13	33.00
1023	50' long	"	0.053	21.75	4.34	5.11	31.25
1030	10x10						
1040	42 lb/ft						
1041	30' long	LF	0.080	25.50	6.52	7.66	39.75
1042	40' long	"	0.064	25.50	5.21	6.13	36.75
1043	50' long	"	0.053	25.50	4.34	5.11	35.00
1060	57 lb/ft						
1061	30' long	LF	0.080	34.50	6.52	7.66	48.75
1063	40' long	"	0.064	34.50	5.21	6.13	45.75
1065	50' long	"	0.053	34.50	4.34	5.11	44.00
1070	12x12						
1080	53 lb/ft						
1081	30' long	LF	0.087	32.00	7.11	8.36	47.50
1083	40' long	"	0.069	32.00	5.59	6.57	44.25
1085	50' long	"	0.053	32.00	4.34	5.11	41.50
1100	74 lb/ft						
1101	30' long	LF	0.087	44.75	7.11	8.36	60.00
1103	40' long	"	0.069	44.75	5.59	6.57	57.00
1105	50' long	"	0.053	44.75	4.34	5.11	54.00
1110	14x14						
1120	73 lb/ft						
1121	40' long	LF	0.087	44.25	7.11	8.36	60.00
1123	50' long	"	0.069	44.25	5.59	6.57	57.00
1125	60' long	"	0.053	44.25	4.34	5.11	54.00
1140	89 lb/ft						

© 2020 BNi Publications Inc. ALL RIGHTS RESERVED

PILES AND CAISSONS

ID Code	Description Component Descriptions	Output Unit of Meas.	Manhr / Unit	Unit Costs Material Cost	Labor Cost	Equipment Cost	Total Cost
31 - 62165	**STEEL PILES, Cont'd...**						**31 - 62165**
1142	40' long	LF	0.087	54.00	7.11	8.36	70.00
1144	50' long	"	0.069	54.00	5.59	6.57	66.00
1146	60' long	"	0.053	54.00	4.34	5.11	63.00
1160	102 lb/ft						
1161	40' long	LF	0.087	62.00	7.11	8.36	78.00
1163	50' long	"	0.069	62.00	5.59	6.57	74.00
1165	60' long	"	0.053	62.00	4.34	5.11	71.00
1180	117 lb/ft						
1182	40' long	LF	0.091	71.00	7.45	8.76	87.00
1184	50' long	"	0.071	71.00	5.79	6.81	84.00
1186	60' long	"	0.055	71.00	4.47	5.25	81.00
4010	Splice						
4020	8"	EA	1.333	110	85.00		200
4060	10"	"	1.600	120	100		220
4080	12"	"	1.600	160	100		260
4100	14"	"	2.000	200	130		330
4110	Driving cap						
4120	8"	EA	0.800	51.00	51.00		100
4140	10"	"	1.000	51.00	64.00		110
4160	12"	"	1.000	51.00	64.00		110
4200	14"	"	1.143	51.00	73.00		120
4210	Standard point						
4220	8"	EA	0.800	80.00	51.00		130
4240	10"	"	1.000	94.00	64.00		160
4260	12"	"	1.143	110	73.00		180
4280	14"	"	1.333	130	85.00		210
4290	Heavy duty point						
4300	8"	EA	0.889	67.00	57.00		120
4305	10"	"	1.143	79.00	73.00		150
4310	12"	"	1.333	100	85.00		190
4320	14"	"	1.600	130	100		230
5000	Tapered friction piles, fluted casing, up to 50'						
5002	With 4000 psi concrete no reinforcing						
5040	12" dia.	LF	0.048	21.50	3.91	4.60	30.00
5060	14" dia.	"	0.049	24.75	4.01	4.71	33.50
5080	16" dia.	"	0.051	29.75	4.12	4.84	38.75
5100	18" dia.	"	0.056	33.50	4.60	5.41	43.50

© 2020 BNi Publications Inc. ALL RIGHTS RESERVED

PILES AND CAISSONS

ID Code	Description: Component Descriptions	Output: Unit of Meas.	Output: Manhr / Unit	Unit Costs: Material Cost	Unit Costs: Labor Cost	Unit Costs: Equipment Cost	Unit Costs: Total Cost
31 - 62166	**STEEL PIPE PILES**						**31 - 62166**
1000	Concrete filled, 3000# concrete, up to 40'						
1100	8" dia.	LF	0.069	24.00	5.59	6.57	36.25
1120	10" dia.	"	0.071	31.00	5.79	6.81	43.50
1140	12" dia.	"	0.074	36.00	6.02	7.07	49.00
1160	14" dia.	"	0.077	39.50	6.26	7.36	53.00
1180	16" dia.	"	0.080	45.00	6.52	7.66	59.00
1200	18" dia.	"	0.083	62.00	6.80	8.00	77.00
2000	Pipe piles, non-filled						
2020	8" dia.	LF	0.053	21.75	4.34	5.11	31.25
2040	10" dia.	"	0.055	27.25	4.47	5.25	37.00
2060	12" dia.	"	0.056	33.25	4.60	5.41	43.25
2080	14" dia.	"	0.060	35.00	4.89	5.75	45.75
2100	16" dia.	"	0.062	40.00	5.05	5.93	51.00
2120	18" dia.	"	0.064	52.00	5.21	6.13	63.00
2520	Splice						
2540	8" dia.	EA	1.600	97.00	100		200
2560	10" dia.	"	1.600	110	100		210
2580	12" dia.	"	2.000	120	130		250
2600	14" dia.	"	2.000	130	130		260
2620	16" dia.	"	2.667	160	170		330
2640	18" dia.	"	2.667	210	170		380
2680	Standard point						
2700	8" dia.	EA	1.600	130	100		230
2740	10" dia.	"	1.600	170	100		270
2760	12" dia.	"	2.000	180	130		310
2780	14" dia.	"	2.000	200	130		330
2800	16" dia.	"	2.667	250	170		420
2820	18" dia.	"	2.667	360	170		530
2880	Heavy duty point						
2900	8" dia	EA	2.000	230	130		360
2920	10" dia.	"	2.000	320	130		450
2940	12" dia.	"	2.667	340	170		510
2960	14" dia.	"	2.667	470	170		640
2980	16" dia.	"	3.200	470	200		670
3000	18" dia.	"	3.200	520	200		720

© 2020 BNi Publications Inc. ALL RIGHTS RESERVED

PILES AND CAISSONS

ID Code	Description: Component Descriptions	Output: Unit of Meas.	Output: Manhr / Unit	Unit Costs: Material Cost	Unit Costs: Labor Cost	Unit Costs: Equipment Cost	Unit Costs: Total Cost
31 - 62190	**WOOD AND TIMBER PILES**						**31 - 62190**
0080	Treated wood piles, 12" butt, 8" tip						
0100	25' long	LF	0.096	18.00	7.82	9.20	35.00
0110	30' long	"	0.080	19.25	6.52	7.66	33.50
0120	35' long	"	0.069	19.25	5.59	6.57	31.50
0125	40' long	"	0.060	19.25	4.89	5.75	30.00
0128	12" butt, 7" tip						
0130	40' long	LF	0.060	21.75	4.89	5.75	32.50
0132	45' long	"	0.053	21.75	4.34	5.11	31.25
0134	50' long	"	0.048	24.75	3.91	4.60	33.25
0150	55' long	"	0.044	24.75	3.55	4.18	32.50
0160	60' long	"	0.040	24.75	3.26	3.83	31.75

BORED PILES

ID Code	Component Descriptions	Unit of Meas.	Manhr / Unit	Material Cost	Labor Cost	Equipment Cost	Total Cost
31 - 63135	**PRESTRESSED PILING**						**31 - 63135**
0980	Prestressed concrete piling, less than 60' long						
1000	10" sq.	LF	0.040	20.75	3.26	3.83	27.75
1002	12" sq.	"	0.042	29.00	3.40	4.00	36.50
1004	14" sq.	"	0.043	30.25	3.47	4.08	37.75
1006	16" sq.	"	0.044	37.25	3.55	4.18	45.00
1008	18" sq.	"	0.047	51.00	3.81	4.48	59.00
1010	20" sq.	"	0.048	70.00	3.91	4.60	79.00
1012	24" sq.	"	0.049	89.00	4.01	4.71	98.00
1100	More than 60' long						
1120	12" sq.	LF	0.034	29.75	2.79	3.28	35.75
1140	14" sq.	"	0.035	32.75	2.84	3.34	39.00
1160	16" sq.	"	0.036	39.25	2.89	3.40	45.50
1180	18" sq.	"	0.036	52.00	2.95	3.47	58.00
1200	20" sq.	"	0.037	70.00	3.01	3.53	77.00
1220	24" sq.	"	0.038	83.00	3.07	3.60	90.00
1480	Straight cylinder, less than 60' long						
1500	12" dia.	LF	0.044	27.00	3.55	4.18	34.75
1540	14" dia.	"	0.045	36.50	3.64	4.27	44.50
1560	16" dia.	"	0.046	44.50	3.72	4.38	53.00
1580	18" dia.	"	0.047	56.00	3.81	4.48	64.00
1600	20" dia.	"	0.048	67.00	3.91	4.60	76.00
1620	24" dia.	"	0.049	83.00	4.01	4.71	92.00
1680	More than 60' long						

© 2020 BNi Publications Inc. ALL RIGHTS RESERVED

BORED PILES

ID Code	Description Component Descriptions	Output Unit of Meas.	Manhr / Unit	Unit Costs Material Cost	Labor Cost	Equipment Cost	Total Cost
31 - 63135	**PRESTRESSED PILING, Cont'd...**						**31 - 63135**
1700	12" dia.	LF	0.035	27.00	2.84	3.34	33.25
1720	14" dia.	"	0.036	36.50	2.89	3.40	42.75
1740	16" dia.	"	0.036	44.50	2.95	3.47	51.00
1760	18" dia.	"	0.037	56.00	3.01	3.53	63.00
1780	20" dia.	"	0.038	67.00	3.07	3.60	74.00
1800	24" dia.	"	0.038	84.00	3.13	3.68	91.00
3000	Concrete sheet piling						
3100	12" thick x 20' long	SF	0.096	31.00	7.82	9.20	48.00
3120	25' long	"	0.087	31.00	7.11	8.36	46.50
3130	30' long	"	0.080	31.00	6.52	7.66	45.25
3140	35' long	"	0.074	31.00	6.02	7.07	44.00
3150	40' long	"	0.069	31.00	5.59	6.57	43.25
3200	16" thick x 40' long	"	0.053	42.25	4.34	5.11	52.00
3220	45' long	"	0.051	42.25	4.12	4.84	51.00
3240	50' long	"	0.048	42.25	3.91	4.60	51.00
3260	55' long	"	0.046	42.25	3.72	4.38	50.00
3280	60' long	"	0.044	42.25	3.55	4.18	50.00

CAISSONS

ID Code	Component Descriptions	Unit of Meas.	Manhr / Unit	Material Cost	Labor Cost	Equipment Cost	Total Cost
31 - 64001	**CAISSONS, INCLUDES CASING**						**31 - 64001**
1000	Caisson, 3000# conc., 60# reinf./CY, stable ground						
1020	18" dia., 0.065 CY/ LF	LF	0.192	17.75	15.75	18.50	52.00
1040	24" dia., 0.116 CY/ LF	"	0.200	28.50	16.25	19.25	64.00
1060	30" dia., 0.182 CY/ LF	"	0.240	43.50	19.50	23.00	86.00
1080	36" dia., 0.262 CY/ LF	"	0.274	61.00	22.25	26.25	110
1100	48" dia., 0.465 CY/ LF	"	0.320	110	26.00	30.75	170
1120	60" dia., 0.727 CY/ LF	"	0.436	180	35.50	41.75	260
1140	72" dia., 1.05 CY/ LF	"	0.533	270	43.50	51.00	360
1160	84" dia., 1.43 CY/ LF	"	0.686	320	56.00	66.00	440
1500	Wet ground, casing required but pulled						
1520	18" dia.	LF	0.240	19.50	19.50	23.00	62.00
1540	24" dia.	"	0.267	31.00	21.75	25.50	78.00
1560	30" dia.	"	0.300	47.50	24.50	28.75	100
1580	36" dia.	"	0.320	66.00	26.00	30.75	120
1600	48" dia.	"	0.400	120	32.50	38.25	190
1620	60" dia.	"	0.533	200	43.50	51.00	300
1640	72" dia.	"	0.800	290	65.00	77.00	430

© 2020 BNi Publications Inc. ALL RIGHTS RESERVED

CAISSONS

	Description	Output		Unit Costs			
ID Code	Component Descriptions	Unit of Meas.	Manhr / Unit	Material Cost	Labor Cost	Equipment Cost	Total Cost
31 - 64001	**CAISSONS, INCLUDES CASING, Cont'd...**						**31 - 64001**
1660	84" dia.	LF	1.200	350	98.00	110	560
2000	Soft rock						
2020	18" dia.	LF	0.686	19.50	56.00	66.00	140
2040	24" dia.	"	1.200	31.00	98.00	110	240
2060	30" dia.	"	1.600	47.50	130	150	330
2080	36" dia.	"	2.400	66.00	200	230	500
2100	48" dia.	"	3.200	120	260	310	690
2120	60" dia.	"	4.800	200	390	460	1,050
2140	72" dia.	"	5.333	290	430	510	1,240
2160	84" dia.	"	6.000	350	490	580	1,410

© 2020 BNi Publications Inc. ALL RIGHTS RESERVED

DIVISION 32
EXTERIOR IMPROVEMENTS

PAVING

ID Code	Description: Component Descriptions	Output: Unit of Meas.	Output: Manhr / Unit	Unit Costs: Material Cost	Unit Costs: Labor Cost	Unit Costs: Equipment Cost	Unit Costs: Total Cost
32 - 11171	**ASPHALT REPAIR**						**32 - 11171**
0010	Coal tar seal coat, rubber add., fuel resist.	SY	0.011	2.68	0.72		3.40
0020	Bituminous surface treatment, single	"	0.008	2.46	0.51		2.97
0030	Double	"	0.001	3.27	0.05		3.32
0040	Bituminous prime coat	"	0.001	1.65	0.06		1.71
0050	Tack coat	"	0.001	0.80	0.05		0.85
0910	Crack sealing, concrete paving	LF	0.005	1.37	0.34		1.71
4000	Bituminous paving for pipe trench, 4" thick	SY	0.160	15.75	10.25	8.86	34.75
6010	Polypropylene, nonwoven paving fabric	"	0.004	2.02	0.25		2.27
6020	Rubberized asphalt	"	0.073	3.27	4.64		7.91
6040	Asphalt slurry seal	"	0.047	8.08	3.00		11.00

FLEXIBLE PAVEMENT

ID Code	Component Descriptions	Unit of Meas.	Manhr / Unit	Material Cost	Labor Cost	Equipment Cost	Total Cost
32 - 12160	**ASPHALT SURFACES**						**32 - 12160**
0050	Asphalt wearing surface, flexible pavement						
0100	1" thick	SY	0.016	4.52	1.30	1.53	7.35
0120	1-1/2" thick	"	0.019	6.82	1.56	1.84	10.25
0130	2" thick	"	0.024	9.09	1.95	2.30	13.25
0140	3" thick	"	0.032	13.75	2.60	3.06	19.50
1000	Binder course						
1010	1-1/2" thick	SY	0.018	6.45	1.44	1.70	9.60
1030	2" thick	"	0.022	8.58	1.77	2.09	12.50
1040	3" thick	"	0.029	12.75	2.37	2.78	18.00
1050	4" thick	"	0.032	17.00	2.60	3.06	22.75
1060	5" thick	"	0.036	21.25	2.89	3.40	27.50
1070	6" thick	"	0.040	25.75	3.26	3.83	32.75
2000	Bituminous sidewalk, no base						
2020	2" thick	SY	0.028	9.84	1.78	1.56	13.25
2040	3" thick	"	0.030	14.75	1.90	1.66	18.25

RIGID PAVING

ID Code	Component Descriptions	Unit of Meas.	Manhr / Unit	Material Cost	Labor Cost	Equipment Cost	Total Cost
32 - 13130	**CONCRETE PAVING**						**32 - 13130**
1080	Concrete paving, reinforced, 5000 psi concrete						
2000	6" thick	SY	0.150	28.75	12.25	14.25	55.00
2005	7" thick	"	0.160	33.50	13.00	15.25	62.00
2010	8" thick	"	0.171	38.25	14.00	16.50	69.00
2015	9" thick	"	0.185	43.00	15.00	17.75	76.00

© 2020 BNi Publications Inc. ALL RIGHTS RESERVED

RIGID PAVING

ID Code	Description: Component Descriptions	Output: Unit of Meas.	Output: Manhr / Unit	Unit Costs: Material Cost	Unit Costs: Labor Cost	Unit Costs: Equipment Cost	Unit Costs: Total Cost
32 - 13130	**CONCRETE PAVING, Cont'd...**						**32 - 13130**
2020	10" thick	SY	0.200	47.75	16.25	19.25	83.00
2030	11" thick	"	0.218	53.00	17.75	21.00	92.00
2040	12" thick	"	0.240	57.00	19.50	23.00	100
2045	15" thick	"	0.300	72.00	24.50	28.75	130
2050	Concrete paving, for pipe trench, reinforced						
2051	7" thick	SY	0.240	57.00	15.25	13.25	86.00
2052	8" thick	"	0.267	62.00	17.00	14.75	94.00
2053	9" thick	"	0.300	66.00	19.00	16.50	100
2054	10" thick	"	0.343	71.00	21.75	19.00	110
2060	Fibrous concrete						
2070	5" thick	SY	0.185	29.50	15.00	17.75	62.00
2080	8" thick	"	0.200	37.50	16.25	19.25	73.00
4000	Roller compacted concrete (RCC)						
4040	8" thick	SY	0.240	36.25	19.50	23.00	79.00
4060	12" thick	"	0.300	55.00	24.50	28.75	110
8980	Steel edge forms up to						
9000	12" deep	LF	0.027	1.08	1.70		2.78
9020	15" deep	"	0.032	1.39	2.04		3.43
9030	Paving finishes						
9040	Belt dragged	SY	0.040		2.55		2.55
9060	Curing	"	0.008	0.41	0.51		0.92
32 - 13131	**SIDEWALKS**						**32 - 13131**
6000	Walks, with wire mesh, base not incl.						
6010	4" thick	SF	0.027	1.99	1.70		3.69
6020	5" thick	"	0.032	2.70	2.04		4.74
6030	6" thick	"	0.040	3.32	2.55		5.87

UNIT PAVING

ID Code	Component Descriptions	Unit of Meas.	Manhr / Unit	Material Cost	Labor Cost	Equipment Cost	Total Cost
32 - 14160	**PAVERS, MASONRY**						**32 - 14160**
4010	Brick walk laid on sand, sand joints						
4020	Laid flat (4.5 per sf)	SF	0.089	4.16	6.91		11.00
4040	Laid on edge (7.2 per sf)	"	0.133	6.66	10.25		17.00
4080	Precast concrete patio blocks						
4100	2" thick						
5010	Natural	SF	0.027	3.58	2.07		5.65
5020	Colors	"	0.027	4.53	2.07		6.60
5080	Exposed aggregates, local aggregate						

© 2020 BNi Publications Inc. ALL RIGHTS RESERVED

UNIT PAVING

ID Code	Description: Component Descriptions	Output: Unit of Meas.	Output: Manhr / Unit	Unit Costs: Material Cost	Unit Costs: Labor Cost	Unit Costs: Equipment Cost	Unit Costs: Total Cost
32 - 14160	**PAVERS, MASONRY, Cont'd...**						**32 - 14160**
5100	Natural	SF	0.027	10.00	2.07		12.00
5120	Colors	"	0.027	10.00	2.07		12.00
5130	Granite or limestone aggregate	"	0.027	10.00	2.07		12.00
5140	White tumblestone aggregate	"	0.027	10.75	2.07		12.75
5960	Stone pavers, set in mortar						
5990	Bluestone						
6000	1" thick						
6010	Irregular	SF	0.200	10.00	15.50		25.50
6020	Snapped rectangular	"	0.160	14.75	12.50		27.25
6060	1-1/2" thick, random rectangular	"	0.200	18.00	15.50		33.50
6070	2" thick, random rectangular	"	0.229	20.50	17.75		38.25
6080	Slate						
6090	Natural cleft						
6100	Irregular, 3/4" thick	SF	0.229	10.75	17.75		28.50
6110	Random rectangular						
6120	1-1/4" thick	SF	0.200	23.25	15.50		38.75
6130	1-1/2" thick	"	0.222	26.25	17.25		43.50
6140	Granite blocks						
6150	3" thick, 3" to 6" wide						
7020	4" to 12" long	SF	0.267	13.25	20.75		34.00
7030	6" to 15" long	"	0.229	8.67	17.75		26.50
9800	Crushed stone, white marble, 3" thick	"	0.016	1.88	1.02		2.90

PAVING SPECIALTIES

ID Code	Component Descriptions	Unit of Meas.	Manhr / Unit	Material Cost	Labor Cost	Equipment Cost	Total Cost
32 - 17003	**GUARDRAILS**						**32 - 17003**
1000	Pipe bollard, steel pipe, concrete filled, painted						
1020	6" dia.	EA	0.667	280	42.50		320
1040	8" dia.	"	1.000	370	64.00		430
1060	12" dia.	"	2.667	580	170		750
2000	Corrugated steel, guardrail, galvanized	LF	0.040	35.00	2.53	2.21	39.75
2020	End section, wrap around or flared	EA	0.800	94.00	51.00		140
2080	Timber guardrail, 4" x 8"	LF	0.030	40.75	1.90	1.66	44.25
3000	Guardrail, 3 cables, 3/4" dia.						
3020	Steel posts	LF	0.120	20.00	7.60	6.65	34.25
3040	Wood posts	"	0.096	20.00	6.08	5.32	31.50
3050	Steel box beam						
3060	6" x 6"	LF	0.133	82.00	8.44	7.38	98.00

© 2020 BNi Publications Inc. ALL RIGHTS RESERVED

PAVING SPECIALTIES

	Description	Output		Unit Costs			
ID Code	Component Descriptions	Unit of Meas.	Manhr / Unit	Material Cost	Labor Cost	Equipment Cost	Total Cost
32 - 17003	**GUARDRAILS, Cont'd...**						**32 - 17003**
4000	6" x 8"	LF	0.150	87.00	9.50	8.31	100
4010	Concrete posts	EA	0.400	54.00	25.50		80.00
4100	Barrel type impact barrier	"	0.800	680	51.00		730
4200	Light shield, 6' high	LF	0.160	44.75	10.25		55.00
32 - 17004	**PARKING BARRIERS**						**32 - 17004**
1000	Timber, treated, 4' long						
1020	4" x 4"	EA	0.667	12.50	42.50		55.00
1200	6" x 6"	"	0.800	23.50	51.00		75.00
1780	Precast concrete, 6' long, with dowels						
1800	12" x 6"	EA	0.400	65.00	25.50		91.00
1900	12" x 8"	"	0.444	77.00	28.25		110
32 - 17230	**PAVEMENT MARKINGS**						**32 - 17230**
0080	Pavement line marking, paint						
0100	4" wide	LF	0.002	0.29	0.12		0.41
0120	6" wide	"	0.004	0.29	0.28		0.57
0125	8" wide	"	0.007	0.44	0.42		0.86
0140	Reflective paint, 4" wide	"	0.007	0.60	0.42		1.02
0145	Airfield markings, retro-reflective						
0150	White	LF	0.007	0.99	0.42		1.41
0160	Yellow	"	0.007	1.06	0.42		1.48
0165	Preformed tape, 4" wide						
0170	Inlaid reflective	LF	0.001	2.67	0.07		2.74
0190	Reflective paint	"	0.002	1.82	0.12		1.94
0195	Thermoplastic						
0200	White	LF	0.004	1.12	0.25		1.37
0210	Yellow	"	0.004	1.12	0.25		1.37
0220	12" wide, thermoplastic, white	"	0.011	3.18	0.72		3.90
0230	Directional arrows, reflective preformed tape	EA	0.800	160	51.00		210
0240	Messages, reflective preformed tape (per letter)	"	0.400	80.00	25.50		110
0250	Handicap symbol, preformed tape	"	0.800	33.00	51.00		84.00
2000	Parking stall painting	"	0.160	7.62	10.25		17.75

© 2020 BNi Publications Inc. ALL RIGHTS RESERVED

SITE IMPROVEMENTS

	Description	Output		Unit Costs			
ID Code	Component Descriptions	Unit of Meas.	Manhr / Unit	Material Cost	Labor Cost	Equipment Cost	Total Cost
32 - 31130	**CHAIN LINK FENCE**						**32 - 31130**
0230	Chain link fence, 9 ga., galvanized, with posts 10' o.c.						
0250	4' high	LF	0.057	7.38	3.64		11.00
0260	5' high	"	0.073	9.87	4.64		14.50
0270	6' high	"	0.100	11.25	6.38		17.75
0280	7' high	"	0.123	12.75	7.85		20.50
1000	8' high	"	0.160	14.75	10.25		25.00
1040	For barbed wire with hangers, add						
1050	3 strand	LF	0.040	2.69	2.55		5.24
1060	6 strand	"	0.067	4.56	4.25		8.81
1070	Corner or gate post, 3" post						
1080	4' high	EA	0.267	86.00	17.00		100
1084	5' high	"	0.296	95.00	19.00		110
1085	6' high	"	0.348	110	22.25		130
1086	7' high	"	0.400	130	25.50		160
1087	8' high	"	0.444	130	28.25		160
1089	4" post						
1090	4' high	EA	0.296	150	19.00		170
1091	5' high	"	0.348	170	22.25		190
1092	6' high	"	0.400	190	25.50		220
1093	7' high	"	0.444	210	28.25		240
1094	8' high	"	0.500	230	32.00		260
1100	Gate with gate posts, galvanized, 3' wide						
1102	4' high	EA	2.000	95.00	130		230
1104	5' high	"	2.667	120	170		290
1106	6' high	"	2.667	150	170		320
1108	7' high	"	4.000	180	260		440
1109	8' high	"	4.000	190	260		450
1161	Fabric, galvanized chain link, 2" mesh, 9 ga.						
1163	4' high	LF	0.027	4.01	1.70		5.71
1164	5' high	"	0.032	4.91	2.04		6.95
1165	6' high	"	0.040	6.87	2.55		9.42
1166	8' high	"	0.053	11.50	3.40		15.00
1400	Line post, no rail fitting, galvanized, 2-1/2" dia.						
1410	4' high	EA	0.229	28.00	14.50		42.50
1420	5' high	"	0.250	30.50	16.00		46.50
1430	6' high	"	0.267	33.25	17.00		50.00
1440	7' high	"	0.320	37.75	20.50		58.00
1450	8' high	"	0.400	42.25	25.50		68.00

© 2020 BNi Publications Inc. ALL RIGHTS RESERVED

SITE IMPROVEMENTS

ID Code	Description: Component Descriptions	Output: Unit of Meas.	Output: Manhr / Unit	Unit Costs: Material Cost	Unit Costs: Labor Cost	Unit Costs: Equipment Cost	Unit Costs: Total Cost
32 - 31130	**CHAIN LINK FENCE, Cont'd...**						**32 - 31130**
1460	1-7/8" H beam						
1470	4' high	EA	0.229	34.75	14.50		49.25
1480	5' high	"	0.250	39.00	16.00		55.00
1490	6' high	"	0.267	46.50	17.00		64.00
1500	7' high	"	0.320	53.00	20.50		74.00
1510	8' high	"	0.400	57.00	25.50		83.00
1550	2-1/4" H beam						
1560	4' high	EA	0.229	25.50	14.50		40.00
1570	5' high	"	0.250	31.75	16.00		47.75
1580	6' high	"	0.267	36.25	17.00		53.00
1590	7' high	"	0.320	42.25	20.50		63.00
1600	8' high	"	0.400	49.00	25.50		75.00
1980	Vinyl coated, 9 ga., with posts 10' o.c.						
2000	4' high	LF	0.057	7.98	3.64		11.50
2010	5' high	"	0.073	9.50	4.64		14.25
2020	6' high	"	0.100	11.25	6.38		17.75
2030	7' high	"	0.123	12.50	7.85		20.25
2040	8' high	"	0.160	14.25	10.25		24.50
2045	For barbed wire w/hangers, add						
2050	3 strand	LF	0.040	2.90	2.55		5.45
2060	6 Strand	"	0.067	4.68	4.25		8.93
2080	Corner, or gate post, 4' high						
2100	3" dia.	EA	0.267	99.00	17.00		120
2110	4" dia.	"	0.267	150	17.00		170
2120	6" dia.	"	0.320	180	20.50		200
2160	Gate, with posts, 3' wide						
2180	4' high	EA	2.000	110	130		240
2190	5' high	"	2.667	130	170		300
2200	6' high	"	2.667	150	170		320
2210	7' high	"	4.000	170	260		430
2220	8' high	"	4.000	190	260		450
2500	Line post, no rail fitting, 2-1/2" dia.						
2510	4' high	EA	0.229	26.25	14.50		40.75
2520	5' high	"	0.250	28.50	16.00		44.50
2530	6' high	"	0.267	31.25	17.00		48.25
2540	7' high	"	0.320	35.50	20.50		56.00
2550	8' high	"	0.400	39.50	25.50		65.00
2600	Corner post, no top rail fitting, 4" dia.						

© 2020 BNi Publications Inc. ALL RIGHTS RESERVED

SITE IMPROVEMENTS

Description		Output		Unit Costs			
ID Code	Component Descriptions	Unit of Meas.	Manhr / Unit	Material Cost	Labor Cost	Equipment Cost	Total Cost
32 - 31130	**CHAIN LINK FENCE, Cont'd...**						**32 - 31130**
2610	4' high	EA	0.267	140	17.00		160
2620	5' high	"	0.296	160	19.00		180
2630	6' high	"	0.348	180	22.25		200
2640	7' high	"	0.400	200	25.50		230
2650	8' high	"	0.444	210	28.25		240
3000	Fabric, vinyl, chain link, 2" mesh, 9 ga.						
3010	4' high	LF	0.027	3.76	1.70		5.46
3020	5' high	"	0.032	4.59	2.04		6.63
3030	6' high	"	0.040	6.43	2.55		8.98
3040	8' high	"	0.053	10.75	3.40		14.25
8000	Swing gates, galvanized, 4' high						
8010	Single gate						
8020	3' wide	EA	2.000	200	130		330
8025	4' wide	"	2.000	220	130		350
8028	Double gate						
8030	10' wide	EA	3.200	530	200		730
8035	12' wide	"	3.200	570	200		770
8040	14' wide	"	3.200	590	200		790
8045	16' wide	"	3.200	660	200		860
8050	18' wide	"	4.571	710	290		1,000
8055	20' wide	"	4.571	750	290		1,040
8060	22' wide	"	4.571	830	290		1,120
8065	24' wide	"	5.333	850	340		1,190
8070	26' wide	"	5.333	890	340		1,230
8075	28' wide	"	6.400	950	410		1,360
8080	30' wide	"	6.400	1,010	410		1,420
8085	5' high						
8088	Single gate						
8090	3' wide	EA	2.667	210	170		380
8095	4' wide	"	2.667	250	170		420
8098	Double gate						
8100	10' wide	EA	4.000	570	260		830
8105	12' wide	"	4.000	610	260		870
8110	14' wide	"	4.000	1,050	260		1,310
8115	16' wide	"	4.000	690	260		950
8120	18' wide	"	4.571	710	290		1,000
8125	20' wide	"	4.571	800	290		1,090
8130	22' wide	"	4.571	840	290		1,130

© 2020 BNi Publications Inc. ALL RIGHTS RESERVED

SITE IMPROVEMENTS

	Description	Output		Unit Costs			
ID Code	Component Descriptions	Unit of Meas.	Manhr / Unit	Material Cost	Labor Cost	Equipment Cost	Total Cost
32 - 31130	**CHAIN LINK FENCE, Cont'd...**						**32 - 31130**
8135	24' wide	EA	5.333	880	340		1,220
8140	26' wide	"	5.333	920	340		1,260
8145	28' wide	"	6.400	1,020	410		1,430
8150	30' wide	"	6.400	1,060	410		1,470
8155	6' high						
8158	Single gate						
8160	3' wide	EA	2.667	270	170		440
8165	4' wide	"	2.667	290	170		460
8168	Double gate						
8170	10' wide	EA	4.000	650	260		910
8175	12' wide	"	4.000	740	260		1,000
8180	14' wide	"	4.000	780	260		1,040
8185	16' wide	"	4.000	840	260		1,100
8190	18' wide	"	4.571	900	290		1,190
8195	20' wide	"	4.571	930	290		1,220
8200	22' wide	"	4.571	1,000	290		1,290
8205	24' wide	"	5.333	1,070	340		1,410
8210	26' wide	"	5.333	1,110	340		1,450
8215	28' wide	"	6.400	1,200	410		1,610
8220	30' wide	"	6.400	1,260	410		1,670
8225	7' high						
8228	Single gate						
8230	3' wide	EA	4.000	330	260		590
8235	4' wide	"	4.000	370	260		630
8238	Double gate						
8240	10' wide	EA	5.333	850	340		1,190
8245	12' wide	"	5.333	930	340		1,270
8250	14' wide	"	5.333	1,000	340		1,340
8255	16' wide	"	5.333	1,070	340		1,410
8260	18' wide	"	6.400	1,140	410		1,550
8265	20' wide	"	6.400	1,210	410		1,620
8270	22' wide	"	6.400	1,270	410		1,680
8275	24' wide	"	8.000	1,370	510		1,880
8280	26' wide	"	8.000	1,460	510		1,970
8285	28' wide	"	10.000	1,540	640		2,180
8290	30' wide	"	10.000	1,760	640		2,400
8295	8' high						
8298	Single gate						

© 2020 BNi Publications Inc. ALL RIGHTS RESERVED

SITE IMPROVEMENTS

ID Code	Description: Component Descriptions	Output: Unit of Meas.	Output: Manhr / Unit	Unit Costs: Material Cost	Unit Costs: Labor Cost	Unit Costs: Equipment Cost	Unit Costs: Total Cost
32 - 31130	**CHAIN LINK FENCE, Cont'd...**						**32 - 31130**
8300	3' wide	EA	4.000	370	260		630
8305	4' wide	"	4.000	400	260		660
8308	Double gate						
8310	10' wide	EA	5.333	970	340		1,310
8315	12' wide	"	5.333	1,030	340		1,370
8320	14' wide	"	5.333	1,100	340		1,440
8325	16' wide	"	5.333	1,210	340		1,550
8330	18' wide	"	6.400	1,260	410		1,670
8335	20' wide	"	6.400	1,300	410		1,710
8340	22' wide	"	6.400	1,380	410		1,790
8345	24' wide	"	8.000	1,530	510		2,040
8350	26' wide	"	8.000	1,580	510		2,090
8355	28' wide	"	10.000	1,680	640		2,320
8360	30' wide	"	10.000	1,850	640		2,490
8505	Vinyl coated swing gates, 4' high						
8508	Single gate						
8510	3' wide	EA	2.000	310	130		440
8515	4' wide	"	2.000	340	130		470
8518	Double gate						
8520	10' wide	EA	3.200	800	200		1,000
8525	12' wide	"	3.200	860	200		1,060
8530	14' wide	"	3.200	880	200		1,080
8535	16' wide	"	3.200	990	200		1,190
8540	18' wide	"	4.571	1,060	290		1,350
8545	20' wide	"	4.571	1,120	290		1,410
8550	22' wide	"	4.571	1,250	290		1,540
8555	24' wide	"	5.333	1,280	340		1,620
8560	26' wide	"	5.333	1,340	340		1,680
8565	28' wide	"	6.400	1,420	410		1,830
8570	30' wide	"	6.400	1,510	410		1,920
8575	5' high						
8578	Single gate						
8580	3' wide	EA	2.667	320	170		490
8585	4' wide	"	2.667	370	170		540
8588	Double gate						
8590	10' wide	EA	4.000	860	260		1,120
8595	12' wide	"	4.000	920	260		1,180
8600	14' wide	"	4.000	990	260		1,250

© 2020 BNi Publications Inc. ALL RIGHTS RESERVED

SITE IMPROVEMENTS

ID Code	Description: Component Descriptions	Output: Unit of Meas.	Output: Manhr / Unit	Unit Costs: Material Cost	Unit Costs: Labor Cost	Unit Costs: Equipment Cost	Unit Costs: Total Cost
32 - 31130	**CHAIN LINK FENCE, Cont'd...**						**32 - 31130**
8605	16' wide	EA	4.000	1,030	260		1,290
8610	18' wide	"	4.571	1,060	290		1,350
8615	20' wide	"	4.571	1,200	290		1,490
8620	22' wide	"	4.571	1,260	290		1,550
8625	24' wide	"	5.333	1,320	340		1,660
8630	26' wide	"	5.333	1,380	340		1,720
8635	28' wide	"	6.400	1,540	410		1,950
8640	30' wide	"	6.400	1,590	410		2,000
8645	6' high						
8648	Single gate						
8650	3' wide	EA	2.667	400	170		570
8655	4' wide	"	2.667	440	170		610
8658	Double gate						
8660	10' wide	EA	4.000	980	260		1,240
8665	12' wide	"	4.000	1,100	260		1,360
8670	14' wide	"	4.000	1,170	260		1,430
8675	16' wide	"	4.000	1,260	260		1,520
8680	18' wide	"	4.571	1,350	290		1,640
8685	20' wide	"	4.571	1,400	290		1,690
8690	22' wide	"	4.571	1,500	290		1,790
8695	24' wide	"	5.333	1,610	340		1,950
8698	26' wide	"	5.333	1,660	340		2,000
8700	28' wide	"	6.400	1,800	410		2,210
8705	30' wide	"	6.400	1,900	410		2,310
8710	7' high						
8713	Single gate						
8715	3' wide	EA	4.000	490	260		750
8720	4' wide	"	4.000	550	260		810
8723	Double gate						
8725	10' wide	FA	5.333	1,270	340		1,610
8730	12' wide	"	5.333	1,400	340		1,740
8735	14' wide	"	5.333	1,500	340		1,840
8740	16' wide	"	5.333	1,600	340		1,940
8745	18' wide	"	6.400	1,700	410		2,110
8750	20' wide	"	6.400	1,810	410		2,220
8755	22' wide	"	6.400	1,910	410		2,320
8760	24' wide	"	8.000	2,050	510		2,560
8765	26' wide	"	8.000	2,190	510		2,700

© 2020 BNi Publications Inc. ALL RIGHTS RESERVED

SITE IMPROVEMENTS

	Description	Output		Unit Costs			
ID Code	Component Descriptions	Unit of Meas.	Manhr / Unit	Material Cost	Labor Cost	Equipment Cost	Total Cost
32 - 31130	**CHAIN LINK FENCE, Cont'd...**						**32 - 31130**
8770	28' wide	EA	10.000	2,300	640		2,940
8775	30' wide	"	10.000	2,640	640		3,280
8780	8' high						
8783	Single gate						
8785	3' wide	EA	4.000	550	260		810
8790	4' wide	"	4.000	590	260		850
8793	Double gate						
8795	10' wide	EA	5.333	1,450	340		1,790
8800	12' wide	"	5.333	1,540	340		1,880
8805	14' wide	"	5.333	1,660	340		2,000
8810	16' wide	"	5.333	1,810	340		2,150
8815	18' wide	"	6.400	1,890	410		2,300
8820	20' wide	"	6.400	1,960	410		2,370
8825	22' wide	"	6.400	2,060	410		2,470
8830	24' wide	"	8.000	2,290	510		2,800
8840	28' wide	"	8.000	2,370	510		2,880
8845	30' wide	"	10.000	2,520	640		3,160
8900	Motor operator for gates, no wiring	"					5,830
9000	Drilling fence post holes						
9010	In soil						
9020	By hand	EA	0.400		25.50		25.50
9030	By machine auger	"	0.200		12.75	4.25	17.00
9050	In rock						
9060	By jackhammer	EA	2.667		170	57.00	230
9070	By rock drill	"	0.800		51.00	17.00	68.00
9100	Aluminum privacy slats, installed vertically	SF	0.020	0.99	1.27		2.26
9120	Post hole, dig by hand	EA	0.533		34.00		34.00
9130	Set fence post in concrete	"	0.400	9.68	25.50		35.25
32 - 31901	**SHRUB & TREE MAINTENANCE**						**32 - 31901**
1000	Moving shrubs on site						
1020	12" ball	EA	1.000		64.00		64.00
1040	24" ball	"	1.333		85.00		85.00
1220	3' high	"	0.800		51.00		51.00
1240	4' high	"	0.889		57.00		57.00
1260	5' high	"	1.000		64.00		64.00
1280	18" spread	"	1.143		73.00		73.00
1300	30" spread	"	1.333		85.00		85.00

© 2020 BNi Publications Inc. ALL RIGHTS RESERVED

SITE IMPROVEMENTS

	Description	Output		Unit Costs			
ID Code	Component Descriptions	Unit of Meas.	Manhr / Unit	Material Cost	Labor Cost	Equipment Cost	Total Cost
32 - 31901	**SHRUB & TREE MAINTENANCE, Cont'd...**						**32 - 31901**
2000	Moving trees on site						
2020	24" ball	EA	1.200		76.00	67.00	140
2040	48" ball	"	1.600		100	89.00	190
3020	Trees						
3040	3' high	EA	0.480		30.50	26.50	57.00
3060	6' high	"	0.533		33.75	29.50	63.00
3080	8' high	"	0.600		38.00	33.25	71.00
3100	10' high	"	0.800		51.00	44.25	95.00
3110	Palm trees						
3120	7' high	EA	0.600		38.00	33.25	71.00
3140	10' high	"	0.800		51.00	44.25	95.00
3142	20' high	"	2.400		150	130	290
3144	40' high	"	4.800		300	270	570
3148	Guying trees						
3150	4" dia.	EA	0.400	13.00	25.50		38.50
3160	8" dia.	"	0.500	13.00	32.00		45.00
32 - 31902	**FERTILIZING**						**32 - 31902**
0080	Fertilizing (23#/1000 sf)						
0100	By square yard	SY	0.002	0.03	0.12	0.04	0.20
0120	By acre	ACRE	10.000	190	640	210	1,040
2980	Liming (70#/1000 sf)						
3000	By square yard	SY	0.003	0.03	0.17	0.05	0.25
3020	By acre	ACRE	13.333	190	850	280	1,320
32 - 31903	**WEED CONTROL**						**32 - 31903**
1000	Weed control, bromicil, 15 lb./acre, wettable powder	ACRE	4.000	330	260		590
1100	Vegetation control, by application of plant killer	SY	0.003	0.02	0.20		0.22
1200	Weed killer, lawns and fields	"	0.002	0.27	0.10		0.37

PLANTING IRRIGATION

32 - 84004	**LAWN IRRIGATION**						**32 - 84004**
0480	Residential system, complete						
0490	Minimum	ACRE					19,250
0520	Maximum	"					36,630
0580	Commercial system, complete						
0600	Minimum	ACRE					29,220
0620	Maximum	"					46,190

© 2020 BNi Publications Inc. ALL RIGHTS RESERVED

PLANTING IRRIGATION

	Description	Output		Unit Costs			
ID Code	Component Descriptions	Unit of Meas.	Manhr / Unit	Material Cost	Labor Cost	Equipment Cost	Total Cost
32 - 84004	LAWN IRRIGATION, Cont'd...						32 - 84004
1000	Components						
1200	Pipe						
1400	Schedule 40, PVC						
1410	1/2"	LF	0.042	0.43	2.68		3.11
1420	3/4"	"	0.044	0.61	2.83		3.44
1430	1"	"	0.046	0.93	2.91		3.84
1450	1-1/4"	"	0.046	1.23	2.91		4.14
1460	1-1/2"	"	0.047	1.46	3.00		4.46
1470	2"	"	0.050	2.16	3.19		5.35
1480	2-1/2"	"	0.053	3.27	3.40		6.67
1490	3"	"	0.057	4.47	3.64		8.11
1500	4"	"	0.067	6.32	4.25		10.50
1510	6"	"	0.080	11.00	5.10		16.00
3000	Fittings						
3200	Tee						
3210	1/2"	EA	0.133	0.84	8.51		9.35
3220	3/4"	"	0.133	0.97	8.51		9.48
3230	1"	"	0.133	1.75	8.51		10.25
3340	1-1/4"	"	0.145	3.17	9.28		12.50
3350	1-1/2"	"	0.160	3.61	10.25		13.75
3360	2"	"	0.178	5.11	11.25		16.25
3370	2-1/2"	"	0.200	19.25	12.75		32.00
3380	3"	"	0.229	24.75	14.50		39.25
3390	4"	"	0.267	37.50	17.00		55.00
3400	6"	"	0.320	160	20.50		180
3450	Ell						
3460	1/2"	EA	0.123	0.72	7.85		8.57
3470	3/4"	"	0.133	0.80	8.51		9.31
3480	1"	"	0.133	1.41	8.51		9.92
3490	1-1/4"	"	0.133	2.47	8.51		11.00
3500	1-1/2"	"	0.133	2.82	8.51		11.25
3510	2"	"	0.145	4.23	9.28		13.50
3520	2-1/2"	"	0.160	14.00	10.25		24.25
3530	3"	"	0.178	17.50	11.25		28.75
3540	4"	"	0.200	24.75	12.75		37.50
3550	6"	"	0.267	110	17.00		130
3600	Coupling						
3610	1/2"	EA	0.123	0.53	7.85		8.38

© 2020 BNi Publications Inc. ALL RIGHTS RESERVED

PLANTING IRRIGATION

	Description	Output		Unit Costs			
ID Code	Component Descriptions	Unit of Meas.	Manhr / Unit	Material Cost	Labor Cost	Equipment Cost	Total Cost
32 - 84004	**LAWN IRRIGATION, Cont'd...**						**32 - 84004**
3620	3/4"	EA	0.133	0.72	8.51		9.23
3630	1"	"	0.133	1.25	8.51		9.76
3640	1-1/4"	"	0.133	1.58	8.51		10.00
3650	1-1/2"	"	0.133	1.66	8.51		10.25
3660	2"	"	0.145	2.64	9.28		12.00
3670	2-1/2"	"	0.160	5.99	10.25		16.25
3680	3"	"	0.178	10.50	11.25		21.75
3690	4"	"	0.200	11.25	12.75		24.00
3700	6"	"	0.267	47.75	17.00		65.00
3800	45 Ell						
3810	1/2"	EA	0.123	1.06	7.85		8.91
3820	3/4"	"	0.133	1.75	8.51		10.25
3830	1"	"	0.133	2.47	8.51		11.00
3840	1-1/4"	"	0.133	3.35	8.51		11.75
3850	1-1/2"	"	0.133	4.04	8.51		12.50
3860	2"	"	0.145	5.27	9.28		14.50
3870	2-1/2"	"	0.160	15.75	10.25		26.00
3880	3"	"	0.178	22.75	11.25		34.00
3890	4"	"	0.200	36.00	12.75		48.75
3900	6"	"	0.267	120	17.00		140
4000	Riser, 1/2" diameter						
4010	2" (close)	EA	0.200	0.60	12.75		13.25
4020	3"	"	0.200	0.72	12.75		13.50
4030	4"	"	0.229	0.88	14.50		15.50
4040	5"	"	0.229	0.97	14.50		15.50
4050	6"	"	0.229	1.25	14.50		15.75
4100	3/4" diameter						
4110	2" (close)	EA	0.200	0.60	12.75		13.25
4120	3"	"	0.200	0.72	12.75		13.50
4130	4"	"	0.229	0.80	14.50		15.25
4140	5"	"	0.229	1.06	14.50		15.50
4150	6"	"	0.229	1.13	14.50		15.75
4200	1" diameter						
4210	2" (close)	EA	0.200	0.88	12.75		13.75
4220	3"	"	0.200	1.13	12.75		14.00
4230	4"	"	0.229	1.32	14.50		15.75
4240	5"	"	0.229	1.66	14.50		16.25
4250	6"	"	0.229	1.73	14.50		16.25

© 2020 BNi Publications Inc. ALL RIGHTS RESERVED

PLANTING IRRIGATION

ID Code	Description: Component Descriptions	Output: Unit of Meas.	Output: Manhr / Unit	Unit Costs: Material Cost	Unit Costs: Labor Cost	Unit Costs: Equipment Cost	Unit Costs: Total Cost
32 - 84004	**LAWN IRRIGATION, Cont'd...**						**32 - 84004**
5000	Valve Box						
5010	Concrete, Square						
5020	12" x 22"	EA	1.000	86.00	64.00		150
5030	18" x 20"	"	1.143	110	73.00		180
5040	24" x 13"	"	1.333	96.00	85.00		180
5050	Round						
5060	12"	EA	0.800	38.00	51.00		89.00
5400	Plastic						
5410	Square						
5420	12"	EA	1.000	40.00	64.00		100
5430	18"	"					150
5500	Round						
5510	6"	EA	1.000	9.99	64.00		74.00
5520	10"	"	1.000	22.00	64.00		86.00
5530	12"	"	1.143	32.00	73.00		110
6000	Sprinkler, Pop-Up						
6100	Spray						
6110	2" high	EA	1.333	5.93	85.00		91.00
6120	3" high	"	1.333	6.50	85.00		92.00
6130	4" high	"	1.600	7.06	100		110
6140	6" high	"	1.600	14.75	100		110
6150	12" high	"	1.600	19.00	100		120
6200	Rotor						
6210	4" high	EA	1.333	27.75	85.00		110
6220	6" high	"	1.600	37.00	100		140
6300	Impact						
6310	Brass	EA	1.333	37.00	85.00		120
6320	Plastic	"	1.600	18.50	100		120
6400	Shrub Head						
6410	Spray	EA	1.333	11.25	85.00		96.00
6420	Rotor	"	1.600	27.75	100		130
6900	Time Clocks						
6910	Minimum	EA	2.000	180	130		310
6920	Average	"	2.667	350	170		520
6930	Maximum	"	4.000	3,180	260		3,440
7000	Valves						
7100	Anti-siphon						
7110	Brass						

© 2020 BNi Publications Inc. ALL RIGHTS RESERVED

PLANTING IRRIGATION

	Description	Output		Unit Costs			
ID Code	Component Descriptions	Unit of Meas.	Manhr / Unit	Material Cost	Labor Cost	Equipment Cost	Total Cost
32 - 84004	**LAWN IRRIGATION, Cont'd...**						**32 - 84004**
7120	3/4"	EA	1.333	76.00	85.00		160
7130	1"	"	1.333	94.00	85.00		180
7140	Plastic						
7150	3/4"	EA	1.333	55.00	85.00		140
7160	1"	"	1.333	66.00	85.00		150
7200	Ball Valve						
7210	Plastic						
7220	1/2"	EA	1.333	7.23	85.00		92.00
7230	3/4"	"	1.333	7.89	85.00		93.00
7240	1"	"	1.333	10.50	85.00		96.00
7250	1-1/2"	"	1.600	21.00	100		120
7260	2"	"	1.600	27.50	100		130
7270	Brass						
7280	1/2"	EA	1.333	13.50	85.00		99.00
7290	3/4"	"	1.333	21.50	85.00		110
7300	1"	"	1.333	33.50	85.00		120
7310	1-1/2"	"	1.600	64.00	100		160
7320	2"	"	1.600	97.00	100		200
7400	Gate valves, Brass						
7410	1/2"	EA	1.333	15.75	85.00		100
7420	3/4"	"	1.333	26.25	85.00		110
7430	1"	"	1.333	36.75	85.00		120
7450	1-1/2"	"	1.600	63.00	100		160
7460	2"	"	1.600	79.00	100		180
8000	Vacuum Breakers						
8010	Brass						
8020	3/4"	EA	2.667	54.00	170		220
8030	1"	"	2.667	74.00	170		240
8040	1-1/2"	"	2.667	150	170		320
8050	2"	"	2.667	210	170		380
8060	Plastic						
8070	3/4"	EA	2.667	66.00	170		240
8080	1"	"	2.667	77.00	170		250
8090	1-1/2"	"	2.667	120	170		290
8100	2"	"	2.667	140	170		310
8400	Backflow Preventors, Brass						
8410	3/4"	EA	26.667	320	1,700		2,020
8420	1"	"	26.667	360	1,700		2,060

© 2020 BNi Publications Inc. ALL RIGHTS RESERVED

PLANTING IRRIGATION

ID Code	Description: Component Descriptions	Output: Unit of Meas.	Output: Manhr / Unit	Unit Costs: Material Cost	Unit Costs: Labor Cost	Unit Costs: Equipment Cost	Unit Costs: Total Cost
32 - 84004	**LAWN IRRIGATION, Cont'd...**						**32 - 84004**
8430	1-1/2"	EA	26.667	780	1,700		2,480
8440	2"	"	32.000	970	2,040		3,010
8600	Pressure Regulators, Brass						
8610	3/4"	EA	0.800	140	51.00		190
8620	1"	"	0.800	190	51.00		240
8630	1-1/2"	"	0.889	570	57.00		630
8640	2"	"	1.000	680	64.00		740
8800	Quick Coupler Valve						
8810	3/4"	EA	1.333	100	85.00		190
8820	1"	"	1.333	150	85.00		240

PLANTING

ID Code	Description	Unit of Meas.	Manhr / Unit	Material Cost	Labor Cost	Equipment Cost	Total Cost
32 - 91191	**TOPSOIL**						**32 - 91191**
0005	Spread topsoil, with equipment						
0010	Minimum	CY	0.080		6.74	8.80	15.50
0020	Maximum	"	0.100		8.43	11.00	19.50
0080	By hand						
0100	Minimum	CY	0.800		51.00		51.00
0110	Maximum	"	1.000		64.00		64.00
0980	Area prep. seeding (grade, rake and clean)						
1000	Square yard	SY	0.006		0.40		0.40
1020	By acre	ACRE	32.000		2,040		2,040
2000	Remove topsoil and stockpile on site						
2020	4" deep	CY	0.067		5.62	7.33	13.00
2040	6" deep	"	0.062		5.19	6.76	12.00
2200	Spreading topsoil from stock pile						
2220	By loader	CY	0.073		6.13	8.00	14.25
2240	By hand	"	0.800		67.00	88.00	160
2260	Top dress by hand	SY	0.008		0.67	0.88	1.55
2280	Place imported top soil						
2300	By loader						
2320	4" deep	SY	0.008		0.67	0.88	1.55
2340	6" deep	"	0.009		0.74	0.97	1.72
2360	By hand						
2370	4" deep	SY	0.089		5.67		5.67
2380	6" deep	"	0.100		6.38		6.38
5980	Plant bed preparation, 18" deep						

© 2020 BNi Publications Inc. ALL RIGHTS RESERVED

PLANTING

ID Code	Description: Component Descriptions	Output: Unit of Meas.	Output: Manhr / Unit	Unit Costs: Material Cost	Unit Costs: Labor Cost	Unit Costs: Equipment Cost	Unit Costs: Total Cost
32 - 91191	**TOPSOIL, Cont'd...**						**32 - 91191**
6000	With backhoe/loader	SY	0.020		1.68	2.20	3.88
6010	By hand	"	0.133		8.51		8.51

TURF AND GRASSES

ID Code	Component Descriptions	Unit of Meas.	Manhr / Unit	Material Cost	Labor Cost	Equipment Cost	Total Cost
32 - 92190	**SEEDING**						**32 - 92190**
0980	Mechanical seeding, 175 lb/acre						
1000	By square yard	SY	0.002	0.23	0.10	0.03	0.36
1020	By acre	ACRE	8.000	930	510	170	1,610
2040	450 lb/acre						
2060	By square yard	SY	0.002	0.59	0.12	0.04	0.76
2080	By acre	ACRE	10.000	2,310	640	210	3,160
5980	Seeding by hand, 10 lb per 100 SY						
6000	By square yard	SY	0.003	0.66	0.17		0.83
6010	By acre	ACRE	13.333	2,580	850		3,430
8010	Reseed disturbed areas	SF	0.004	0.06	0.25		0.31

PLANTS

ID Code	Component Descriptions	Unit of Meas.	Manhr / Unit	Material Cost	Labor Cost	Equipment Cost	Total Cost
32 - 93230	**PLANTS**						**32 - 93230**
0100	Euonymus coloratus, 18" (Purple Wintercreeper)	EA	0.133	3.01	8.51		11.50
0150	Hedera Helix, 2-1/4" pot (English ivy)	"	0.133	1.25	8.51		9.76
0200	Liriope muscari, 2" clumps	"	0.080	5.26	5.10		10.25
0250	Santolina, 12"	"	0.080	6.02	5.10		11.00
0280	Vinca major or minor, 3" pot	"	0.080	0.98	5.10		6.08
0300	Cortaderia argentia, 2 gallon (Pampas Grass)	"	0.080	19.00	5.10		24.00
0350	Ophiopogan japonicus, 1 quart (4" pot)	"	0.080	5.26	5.10		10.25
0400	Ajuga reptans, 2-3/4" pot (carpet bugle)	"	0.080	0.98	5.10		6.08
0450	Pachysandra terminalis, 2-3/4" pot (Japanese Spurge)	"	0.080	1.33	5.10		6.43
32 - 93330	**SHRUBS**						**32 - 93330**
0100	Juniperus conferia litoralis, 18"-24" (Shore Juniper)	EA	0.320	44.00	20.50		65.00
0150	Horizontalis plumosa, 18"-24" (Andorra Juniper)	"	0.320	46.75	20.50		67.00
0200	Sabina tamar-iscfolia-tamarix juniper, 18"-24"	"	0.320	46.75	20.50		67.00
0250	Chin San Jose, 18"-24" (San Jose Juniper)	"	0.320	46.75	20.50		67.00
0300	Sargenti, 18"-24" (Sargent's Juniper)	"	0.320	44.00	20.50		65.00
0350	Nandina domestica, 18"-24" (Heavenly Bamboo)	"	0.320	29.50	20.50		50.00
0400	Raphiolepis Indica Springtime, 18"-24"	"	0.320	31.75	20.50		52.00
0450	Osmanthus Heterophyllus Gulftide, 18"-24"	"	0.320	34.00	20.50		55.00

© 2020 BNi Publications Inc. ALL RIGHTS RESERVED

PLANTS

ID Code	Component Descriptions	Unit of Meas.	Manhr / Unit	Material Cost	Labor Cost	Equipment Cost	Total Cost
32 - 93330	**SHRUBS, Cont'd...**						**32 - 93330**
0460	Ilex Cornuta Burfordi Nana, 18"-24"	EA	0.320	38.75	20.50		59.00
0550	Glabra, 18"-24" (Inkberry Holly)	"	0.320	36.50	20.50		57.00
0600	Azalea, Indica types, 18"-24"	"	0.320	41.25	20.50		62.00
0650	Kurume types, 18"-24"	"	0.320	46.00	20.50		67.00
0700	Berberis Julianae, 18"-24" (Wintergreen Barberry)	"	0.320	27.00	20.50		47.50
0800	Pieris Japonica Japanese, 18"-24"	"	0.320	27.00	20.50		47.50
0900	Ilex Cornuta Rotunda, 18"-24"	"	0.320	32.00	20.50		53.00
1000	Juniperus Horiz. Plumosa, 24"-30"	"	0.400	29.50	25.50		55.00
1200	Rhodopendrow Hybrids, 24"-30"	"	0.400	79.00	25.50		100
1400	Aucuba Japonica Varigata, 24"-30"	"	0.400	26.75	25.50		52.00
1600	Ilex Crenata Willow Leaf, 24"-30"	"	0.400	29.50	25.50		55.00
1620	Cleyera Japonica, 30"-36"	"	0.500	34.50	32.00		67.00
1700	Pittosporum Tobira, 30"-36"	"	0.500	40.25	32.00		72.00
1800	Prumus Laurocerasus, 30"-36"	"	0.500	74.00	32.00		110
1900	Ilex Cornuta Burfordi, 30"-36" (Burford Holly)	"	0.500	39.25	32.00		71.00
2000	Abelia Grandiflora, 24"-36" (Yew Podocarpus)	"	0.400	27.00	25.50		53.00
2100	Podocarpos Macrophylla, 24"-36"	"	0.400	44.00	25.50		70.00
2500	Pyracantha Coccinea Lalandi, 3'-4' (Firethorn)	"	0.500	25.25	32.00		57.00
2520	Photinia Frazieri, 3'-4' (Red Photinia)	"	0.500	40.00	32.00		72.00
2600	Forsythia Suspensa, 3'-4' (Weeping Forsythia)	"	0.500	25.25	32.00		57.00
2700	Camellia Japonica, 3'-4' (Common Camellia)	"	0.500	44.50	32.00		77.00
2800	Juniperus Chin Torulosa, 3'-4' (Hollywood Juniper)	"	0.500	47.25	32.00		79.00
2900	Cupressocyparis Leylandii, 3'-4'	"	0.500	39.75	32.00		72.00
3000	Ilex Opaca Fosteri, 5'-6' (Foster's Holly)	"	0.667	160	42.50		200
3200	Opaca, 5'-6' (American Holly)	"	0.667	230	42.50		270
3300	Nyrica Cerifera, 4'-5' (Southern Wax Myrtles)	"	0.571	50.00	36.50		87.00
3400	Ligustrum Japonicum, 4'-5' (Japanese Privet)	"	0.571	39.25	36.50		76.00
32 - 93430	**TREES**						**32 - 93430**
0100	Cornus Florida, 5'-6' (White flowering Dogwood)	EA	0.667	120	42.50		160
0120	Prunus Serrulata Kwanzan, 6'-8' (Kwanzan Cherry)	"	0.800	130	51.00		180
0130	Caroliniana, 6'-8' (Carolina Cherry Laurel)	"	0.800	150	51.00		200
0140	Cercis Canadensis, 6'-8' (Eastern Redbud)	"	0.800	100	51.00		150
0200	Koelreuteria Paniculata, 8'-10' (Goldenrain Tree)	"	1.000	180	64.00		240
0250	Acer Platanoides, 1-3/4"-2" (11'-13')	"	1.333	240	85.00		330
0300	Rubrum, 1-3/4"-2" (11'-13') (Red Maple)	"	1.333	180	85.00		260
0350	Saccharum, 1-3/4"-2" (Sugar Maple)	"	1.333	320	85.00		400
0400	Fraxinus Pennsylvanica, 1-3/4"-2"	"	1.333	150	85.00		240

© 2020 BNi Publications Inc. ALL RIGHTS RESERVED

PLANTS

ID Code	Description: Component Descriptions	Output: Unit of Meas.	Output: Manhr / Unit	Unit Costs: Material Cost	Unit Costs: Labor Cost	Unit Costs: Equipment Cost	Unit Costs: Total Cost
32 - 93430	TREES, Cont'd...						32 - 93430
0450	Celtis Occidentalis, 1-3/4"-2"	EA	1.333	230	85.00		310
0460	Glenditsia Triacantos Inermis, 2"	"	1.333	210	85.00		300
1000	Prunus Cerasifera "Thundercloud", 6'-8'	"	0.800	120	51.00		170
1200	Yeodensis, 6'-8' (Yoshino Cherry)	"	0.800	130	51.00		180
1400	Lagerstroemia Indica, 8'-10' (Crapemyrtle)	"	1.000	210	64.00		270
1600	Crataegus Phaenopyrum, 8'-10'	"	1.000	320	64.00		380
1800	Quercus Borealis, 1-3/4"-2" (Northern Red Oak)	"	1.333	190	85.00		280
2000	Quercus Acutissima, 1-3/4"-2" (8'-10')	"	1.333	180	85.00		260
2100	Saliz Babylonica, 1-3/4"-2" (Weeping Willow)	"	1.333	90.00	85.00		180
2200	Tilia Cordata Greenspire, 1-3/4"-2" (10'-12')	"	1.333	400	85.00		480
2300	Malus, 2"-2-1/2" (8'-10') (Flowering Crabapple)	"	1.333	190	85.00		280
2400	Platanus Occidentalis, (12'-14')	"	1.600	300	100		400
2500	Pyrus Calleryana Bradford, 2"-2-1/2"	"	1.333	230	85.00		310
2600	Quercus Palustris, 2"-2-1/2" (12'-14') (Pin Oak)	"	1.333	260	85.00		350
2700	Phellos, 2-1/2"-3" (Willow Oak)	"	1.600	280	100		380
2800	Nigra, 2"-2-1/2" (Water Oak)	"	1.333	240	85.00		330
3000	Magnolia Soulangeana, 4'-5' (Saucer Magnolia)	"	0.667	140	42.50		180
3100	Grandiflora, 6'-8' (Southern Magnolia)	"	0.800	190	51.00		240
3200	Cedrus Deodara, 10'-12' (Deodare Cedar)	"	1.333	320	85.00		400
3300	Gingko Biloba, 10'-12' (2"-2-1/2")	"	1.333	300	85.00		390
3400	Pinus Thunbergi, 5'-6' (Japanese Black Pine)	"	0.667	120	42.50		160
3500	Strobus, 6'-8' (White Pine)	"	0.800	130	51.00		180
3600	Taeda, 6'-8' (Loblolly Pine)	"	0.800	110	51.00		160
3700	Quercus Virginiana, 2"-2-1/2" (Live Oak)	"	1.600	280	100		380

PLANT ACCESSORIES

ID Code	Component Descriptions	Unit of Meas.	Manhr / Unit	Material Cost	Labor Cost	Equipment Cost	Total Cost
32 - 94002	LANDSCAPE ACCESSORIES						32 - 94002
0100	Steel edging, 3/16" x 4"	LF	0.010	1.29	0.63		1.92
0200	Landscaping stepping stones, 15"x15", white	EA	0.040	5.83	2.55		8.38
6000	Wood chip mulch	CY	0.533	40.75	34.00		75.00
6010	2" thick	SY	0.016	2.49	1.02		3.51
6020	4" thick	"	0.023	4.70	1.45		6.15
6030	6" thick	"	0.029	7.04	1.85		8.89
6200	Gravel mulch, 3/4" stone	CY	0.800	32.25	51.00		83.00
6300	White marble chips, 1" deep	SF	0.008	0.63	0.51		1.14
6980	Peat moss						
7000	2" thick	SY	0.018	3.46	1.13		4.59

© 2020 BNi Publications Inc. ALL RIGHTS RESERVED

PLANT ACCESSORIES

ID Code	Description: Component Descriptions	Output: Unit of Meas.	Output: Manhr / Unit	Unit Costs: Material Cost	Unit Costs: Labor Cost	Unit Costs: Equipment Cost	Unit Costs: Total Cost
32 - 94002	**LANDSCAPE ACCESSORIES, Cont'd...**						**32 - 94002**
7020	4" thick	SY	0.027	6.66	1.70		8.36
7030	6" thick	"	0.033	10.25	2.12		12.25
7980	Landscaping timbers, treated lumber						
8000	4" x 4"	LF	0.027	3.58	1.70		5.28
8020	6" x 6"	"	0.029	8.32	1.82		10.25
8040	8" x 8"	"	0.033	10.00	2.12		12.00
32 - 94330	**PREFABRICATED PLANTERS**						**32 - 94330**
1000	Concrete precast, circular						
1020	24" dia., 18" high	EA	0.800	420	51.00		470
1040	42" dia., 30" high	"	1.000	560	64.00		620
2000	Fiberglass, circular						
2040	36" dia., 27" high	EA	0.400	710	25.50		740
2060	60" dia., 39" high	"	0.444	1,640	28.25		1,670
2100	Tapered, circular						
2120	24" dia., 36" high	EA	0.364	560	23.25		580
2140	40" dia., 36" high	"	0.400	930	25.50		960
2200	Square						
2220	2' by 2', 17" high	EA	0.364	480	23.25		500
2240	4' by 4', 39" high	"	0.444	1,640	28.25		1,670
2300	Rectangular						
2320	4' by 1', 18" high	EA	0.400	530	25.50		560

© 2020 BNi Publications Inc. ALL RIGHTS RESERVED

DIVISION 33
UTILITIES

SITE RESTORATION

ID Code	Description: Component Descriptions	Output: Unit of Meas.	Output: Manhr / Unit	Unit Costs: Material Cost	Unit Costs: Labor Cost	Unit Costs: Equipment Cost	Unit Costs: Total Cost
33 - 01101	**PIPELINE RESTORATION**						**33 - 01101**
0980	Relining existing water main						
1000	6" dia.	LF	0.240	8.62	19.50	23.00	51.00
1020	8" dia.	"	0.253	9.71	20.50	24.25	54.00
1040	10" dia.	"	0.267	10.75	21.75	25.50	58.00
1060	12" dia.	"	0.282	11.75	23.00	27.00	62.00
1080	14" dia.	"	0.300	12.75	24.50	28.75	66.00
1100	16" dia.	"	0.320	13.75	26.00	30.75	71.00
1120	18" dia.	"	0.343	15.00	28.00	32.75	76.00
1140	20" dia.	"	0.369	16.50	30.00	35.50	82.00
1160	24" dia.	"	0.400	17.50	32.50	38.25	89.00
1180	36" dia.	"	0.480	19.00	39.25	46.00	100
1200	48" dia.	"	0.533	21.25	43.50	51.00	120
1220	72" dia.	"	0.600	27.00	49.00	58.00	140
1980	Replacing in-line gate valves						
2000	6" valve	EA	3.200	970	260	310	1,540
2020	8" valve	"	4.000	1,520	330	380	2,230
2040	10" valve	"	4.800	2,290	390	460	3,140
2060	12" valve	"	6.000	3,970	490	580	5,030
2080	16" valve	"	6.857	9,010	560	660	10,230
2090	18" valve	"	8.000	13,630	650	770	15,050
2100	20" valve	"	9.600	18,740	780	920	20,440
2120	24" valve	"	12.000	26,780	980	1,150	28,910
2140	36" valve	"	16.000	73,340	1,300	1,530	76,180

TUNNELING, BORING & JACKING

ID Code	Component Descriptions	Unit of Meas.	Manhr / Unit	Material Cost	Labor Cost	Equipment Cost	Total Cost
33 - 05231	**PIPE JACKING**						**33 - 05231**
1080	Pipe casing, horizontal jacking						
1100	18" dia.	LF	0.711	110	45.50	64.00	220
1200	21" dia.	"	0.762	130	48.75	69.00	250
1300	24" dia.	"	0.800	140	51.00	73.00	260
1400	27" dia.	"	0.800	150	51.00	73.00	270
1500	30" dia.	"	0.842	170	54.00	76.00	300
1600	36" dia.	"	0.914	190	58.00	83.00	330
1700	42" dia.	"	1.000	220	64.00	91.00	370
1800	48" dia.	"	1.067	270	68.00	97.00	430

© 2020 BNi Publications Inc. ALL RIGHTS RESERVED

DISTRIBUTION PIPING

ID Code	Description Component Descriptions	Output Unit of Meas.	Manhr / Unit	Unit Costs Material Cost	Labor Cost	Equipment Cost	Total Cost
33 - 11003	**CHILLED WATER SYSTEMS**						**33 - 11003**
0100	Chilled water pipe, 2" thick insulation, w/casing						
1020	Align and tack weld on sleepers						
1030	1-1/2" dia.	LF	0.022	26.75	1.38	1.20	29.25
1040	3" dia.	"	0.034	43.00	2.17	1.90	47.00
1050	4" dia.	"	0.048	49.50	3.04	2.66	55.00
1060	6" dia.	"	0.060	56.00	3.80	3.32	63.00
1070	8" dia.	"	0.069	79.00	4.34	3.80	87.00
1080	10" dia.	"	0.080	100	5.06	4.43	110
1090	12" dia.	"	0.096	120	6.08	5.32	130
1100	14" dia.	"	0.104	160	6.61	5.78	170
1120	16" dia.	"	0.120	210	7.60	6.65	220
1200	Align and tack weld on trench bottom						
1210	18" dia.	LF	0.133	210	8.44	7.38	230
1220	20" dia.	"	0.150	280	9.50	8.31	300
2000	Preinsulated fittings						
2050	Align and tack weld on sleepers						
2100	Elbows						
2110	1-1/2"	EA	0.500	640	44.75		680
2120	3"	"	0.800	810	72.00		880
2140	4"	"	1.000	1,050	89.00		1,140
2150	6"	"	1.333	1,450	120		1,570
2160	8"	"	1.600	2,060	140		2,200
2200	Tees						
2210	1-1/2"	EA	0.533	980	47.75		1,030
2220	3"	"	0.889	1,380	79.00		1,460
2510	4"	"	1.143	1,050	100		1,150
2520	6"	"	1.600	2,260	140		2,400
2530	8"	"	2.000	3,010	180		3,190
2540	Reducers						
2550	3"	EA	0.667	780	60.00		840
2560	4"	"	0.800	1,230	72.00		1,300
2570	6"	"	1.000	1,660	89.00		1,750
2580	8"	"	1.333	1,800	120		1,920
2590	Anchors, not including concrete						
2600	4"	EA	1.000	410	89.00		500
2610	6"	"	1.000	600	89.00		690
2902	Align and tack weld on trench bottom						
2910	Elbows						

© 2020 BNi Publications Inc. ALL RIGHTS RESERVED

DISTRIBUTION PIPING

	Description	Output		Unit Costs			
ID Code	Component Descriptions	Unit of Meas.	Manhr / Unit	Material Cost	Labor Cost	Equipment Cost	Total Cost
33 - 11003	**CHILLED WATER SYSTEMS, Cont'd...**						**33 - 11003**
2920	10"	EA	1.500	2,360	95.00	83.00	2,540
2930	12"	"	1.714	2,820	110	95.00	3,020
2940	14"	"	1.846	3,550	120	100	3,770
2950	16"	"	2.000	3,820	130	110	4,060
2960	18"	"	2.182	4,260	140	120	4,520
2970	20"	"	2.400	5,140	150	130	5,430
3030	Tees						
3035	10"	EA	1.500	3,820	95.00	83.00	4,000
3040	12"	"	1.714	4,850	110	95.00	5,050
3050	14"	"	1.846	5,140	120	100	5,360
3060	16"	"	2.000	5,440	130	110	5,680
3070	18"	"	2.182	6,170	140	120	6,430
3080	20"	"	2.400	7,060	150	130	7,350
3210	Reducers						
3215	10"	EA	1.000	2,910	63.00	55.00	3,030
3220	12"	"	1.091	3,530	69.00	60.00	3,660
3230	14"	"	1.200	3,820	76.00	67.00	3,960
3240	16"	"	1.333	4,700	84.00	74.00	4,860
3250	18"	"	1.500	5,140	95.00	83.00	5,320
3260	20"	"	1.714	5,440	110	95.00	5,640
3320	Anchors, not including concrete						
3340	10"	EA	1.000	780	63.00	55.00	900
3350	12"	"	1.091	860	69.00	60.00	990
3360	14"	"	1.200	990	76.00	67.00	1,130
3370	16"	"	1.333	1,390	84.00	74.00	1,550
3380	18"	"	1.500	1,930	95.00	83.00	2,110
3390	20"	"	1.714	2,610	110	95.00	2,810
33 - 11004	**DUCTILE IRON PIPE**						**33 - 11004**
0990	Ductile iron pipe, cement lined, slip-on joints						
1000	4"	LF	0.067	19.25	4.22	3.69	27.25
1010	6"	"	0.071	22.25	4.47	3.91	30.75
1020	8"	"	0.075	29.00	4.75	4.15	38.00
1030	10"	"	0.080	40.00	5.06	4.43	49.50
1040	12"	"	0.096	49.25	6.08	5.32	61.00
1060	14"	"	0.120	62.00	7.60	6.65	76.00
1080	16"	"	0.133	76.00	8.44	7.38	92.00
1100	18"	"	0.150	86.00	9.50	8.31	100

© 2020 BNi Publications Inc. ALL RIGHTS RESERVED

DISTRIBUTION PIPING

ID Code	Description: Component Descriptions	Output: Unit of Meas.	Output: Manhr / Unit	Unit Costs: Material Cost	Unit Costs: Labor Cost	Unit Costs: Equipment Cost	Unit Costs: Total Cost
33 - 11004	**DUCTILE IRON PIPE, Cont'd...**						**33 - 11004**
1120	20"	LF	0.171	98.00	10.75	9.50	120
1190	Mechanical joint pipe						
1200	4"	LF	0.092	20.75	5.84	5.11	31.75
1210	6"	"	0.100	24.75	6.33	5.54	36.50
1220	8"	"	0.109	32.50	6.91	6.04	45.50
1230	10"	"	0.120	42.75	7.60	6.65	57.00
1240	12"	"	0.160	54.00	10.25	8.86	73.00
1260	14"	"	0.185	68.00	11.75	10.25	90.00
1280	16"	"	0.218	75.00	13.75	12.00	100
1300	18"	"	0.240	84.00	15.25	13.25	110
1320	20"	"	0.267	97.00	17.00	14.75	130
1480	Fittings, mechanical joint						
1500	90 degree elbow						
1520	4"	EA	0.533	240	34.00		270
1540	6"	"	0.615	320	39.25		360
1560	8"	"	0.800	450	51.00		500
1580	10"	"	1.143	660	73.00		730
1600	12"	"	1.600	880	100		980
1620	14"	"	2.000	1,370	130		1,500
1640	16"	"	2.667	1,710	170		1,880
1660	18"	"	3.200	2,570	200		2,770
1680	20"	"	4.000	2,860	260		3,120
1700	45 degree elbow						
1720	4"	EA	0.533	210	34.00		240
1740	6"	"	0.615	280	39.25		320
1760	8"	"	0.800	400	51.00		450
1780	10"	"	1.143	570	73.00		640
1800	12"	"	1.600	690	100		790
1820	14"	"	2.000	1,140	130		1,270
1840	16"	"	2.667	1,370	170		1,540
1860	18"	"	4.000	2,000	260		2,260
1880	20"	"	4.000	2,370	260		2,630
2000	Tee						
2020	4"x3"	EA	1.000	340	64.00		400
2040	4"x4"	"	1.000	370	64.00		430
2060	6"x3"	"	1.143	430	73.00		500
2080	6"x4"	"	1.143	440	73.00		510
2100	6"x6"	"	1.143	480	73.00		550

© 2020 BNi Publications Inc. ALL RIGHTS RESERVED

DISTRIBUTION PIPING

Description		Output		Unit Costs			
ID Code	Component Descriptions	Unit of Meas.	Manhr / Unit	Material Cost	Labor Cost	Equipment Cost	Total Cost
33 - 11004	**DUCTILE IRON PIPE, Cont'd...**						**33 - 11004**
2120	8"x4"	EA	1.333	600	85.00		680
2140	8"x6"	"	1.333	680	85.00		770
2160	8"x8"	"	1.333	640	85.00		730
2180	10"x4"	"	1.600	800	100		900
2200	10"x6"	"	1.600	880	100		980
2240	10"x8"	"	1.600	910	100		1,010
2260	10"x10"	"	1.600	1,020	100		1,120
2280	12"x4"	"	2.000	900	130		1,030
2300	12"x6"	"	2.000	970	130		1,100
2320	12"x8"	"	2.000	1,050	130		1,180
2340	12"x10"	"	2.000	1,200	130		1,330
2360	12"x12"	"	2.133	1,280	140		1,420
2380	14"x4"	"	2.286	1,570	150		1,720
2400	14"x6"	"	2.286	1,670	150		1,820
2420	14"x8"	"	2.286	1,700	150		1,850
2460	14"x10"	"	2.286	1,750	150		1,900
2480	14"x12"	"	2.462	1,800	160		1,960
2500	14"x14"	"	2.462	1,780	160		1,940
2520	16"x4"	"	2.667	2,030	170		2,200
2540	16"x6"	"	2.667	2,070	170		2,240
2560	16"x8"	"	2.667	1,840	170		2,010
2580	16"x10"	"	2.667	1,870	170		2,040
2600	16"x12"	"	2.667	1,830	170		2,000
2620	16"x14"	"	2.667	1,930	170		2,100
2640	16"x16"	"	2.667	1,970	170		2,140
2660	18"x6"	"	2.909	2,360	190		2,550
2680	18"x8"	"	2.909	2,400	190		2,590
2700	18"x10"	"	2.909	2,440	190		2,630
2720	18"x12"	"	2.909	2,480	190		2,670
2740	18"x14"	"	2.909	2,770	190		2,960
2760	18"x16"	"	2.909	2,740	190		2,930
2780	18"x18"	"	2.909	3,110	190		3,300
2800	20"x6"	"	3.200	3,000	200		3,200
2820	20"x8"	"	3.200	3,030	200		3,230
2840	20"x10"	"	3.200	3,090	200		3,290
2860	20"x12"	"	3.200	3,140	200		3,340
2880	20"x14"	"	3.200	3,230	200		3,430
2900	20"x16"	"	3.200	3,770	200		3,970

© 2020 BNi Publications Inc. ALL RIGHTS RESERVED

DISTRIBUTION PIPING

ID Code	Component Descriptions	Unit of Meas.	Manhr / Unit	Material Cost	Labor Cost	Equipment Cost	Total Cost
33 - 11004	DUCTILE IRON PIPE, Cont'd...						33 - 11004
2920	20"x18"	EA	3.200	3,940	200		4,140
2940	20"x20"	"	3.200	4,030	200		4,230
3000	Cross						
3020	4"x3"	EA	1.333	360	85.00		450
3040	4"x4"	"	1.333	390	85.00		480
3060	6"x3"	"	1.600	400	100		500
3080	6"x4"	"	1.600	430	100		530
3100	6"x6"	"	1.600	470	100		570
3120	8"x4"	"	1.778	660	110		770
3140	8"x6"	"	1.778	710	110		820
3160	8"x8"	"	1.778	780	110		890
3180	10"x4"	"	2.000	910	130		1,040
3200	10"x6"	"	2.000	970	130		1,100
3220	10"x8"	"	2.000	1,050	130		1,180
3240	10"x10"	"	2.000	1,250	130		1,380
3260	12"x4"	"	2.286	1,170	150		1,320
3280	12"x6"	"	2.286	1,310	150		1,460
3300	12"x8"	"	2.286	1,280	150		1,430
3320	12"x10"	"	2.462	1,480	160		1,640
3340	12"x12"	"	2.462	1,600	160		1,760
3360	14"x4"	"	2.667	1,510	170		1,680
3380	14"x6"	"	2.667	1,710	170		1,880
3400	14"x8"	"	2.667	1,780	170		1,950
3420	14"x10"	"	2.667	1,910	170		2,080
3440	14"x12"	"	2.909	2,060	190		2,250
3460	14"x14"	"	2.909	2,260	190		2,450
3480	16"x4"	"	3.200	1,970	200		2,170
3500	16"x6"	"	3.200	2,030	200		2,230
3520	16"x8"	"	3.200	2,140	200		2,340
3540	16"x10"	"	3.200	2,280	200		2,480
3560	16"x12"	"	3.200	2,390	200		2,590
3600	16"x14"	"	3.200	2,590	200		2,790
3620	16"x16"	"	3.200	2,740	200		2,940
3640	18"x6"	"	3.556	2,540	230		2,770
3660	18"x8"	"	3.556	2,630	230		2,860
3680	18"x10"	"	3.556	2,740	230		2,970
3700	18"x12"	"	3.556	2,880	230		3,110
3720	18"x14"	"	3.556	3,430	230		3,660

© 2020 BNi Publications Inc. ALL RIGHTS RESERVED

DISTRIBUTION PIPING

	Description	Output		Unit Costs			
ID Code	Component Descriptions	Unit of Meas.	Manhr / Unit	Material Cost	Labor Cost	Equipment Cost	Total Cost
33 - 11004	**DUCTILE IRON PIPE, Cont'd...**						**33 - 11004**
3740	18"x16"	EA	3.556	3,660	230		3,890
3760	18"x18"	"	3.556	3,860	230		4,090
3780	20"x6"	"	3.810	3,060	240		3,300
3800	20"x8"	"	3.810	3,140	240		3,380
3820	20"x10"	"	3.810	3,280	240		3,520
3840	20"x12"	"	3.810	3,430	240		3,670
3860	20"x14"	"	3.810	3,600	240		3,840
3880	20"x16"	"	3.810	4,170	240		4,410
3900	20"x18"	"	4.000	4,460	260		4,720
3920	20"x20"	"	4.000	4,720	260		4,980
33 - 11006	**PLASTIC PIPE**						**33 - 11006**
0110	PVC, class 150 pipe						
0120	4" dia.	LF	0.060	5.31	3.80	3.32	12.50
0130	6" dia.	"	0.065	10.00	4.10	3.59	17.75
0140	8" dia.	"	0.069	16.00	4.34	3.80	24.25
0150	10" dia.	"	0.075	22.75	4.75	4.15	31.75
0160	12" dia.	"	0.080	33.50	5.06	4.43	43.00
0165	Schedule 40 pipe						
0170	1-1/2" dia.	LF	0.047	1.34	3.00		4.34
0180	2" dia.	"	0.050	1.99	3.19		5.18
0185	2-1/2" dia.	"	0.053	3.01	3.40		6.41
0190	3" dia.	"	0.057	4.09	3.64		7.73
0200	4" dia.	"	0.067	5.78	4.25		10.00
0210	6" dia.	"	0.080	11.00	5.10		16.00
0240	90 degree elbows						
0250	1"	EA	0.133	1.12	8.51		9.63
0260	1-1/2"	"	0.133	2.14	8.51		10.75
0270	2"	"	0.145	3.35	9.28		12.75
0280	2-1/2"	"	0.160	10.25	10.25		20.50
0290	3"	"	0.178	12.25	11.25		23.50
0300	4"	"	0.200	19.75	12.75		32.50
0310	6"	"	0.267	62.00	17.00		79.00
0320	45 degree elbows						
0330	1"	EA	0.133	1.72	8.51		10.25
0340	1-1/2"	"	0.133	3.01	8.51		11.50
0350	2"	"	0.145	3.91	9.28		13.25
0360	2-1/2"	"	0.160	10.25	10.25		20.50

© 2020 BNi Publications Inc. ALL RIGHTS RESERVED

DISTRIBUTION PIPING

	Description	Output		Unit Costs			
ID Code	Component Descriptions	Unit of Meas.	Manhr / Unit	Material Cost	Labor Cost	Equipment Cost	Total Cost
33 - 11006	**PLASTIC PIPE, Cont'd...**						**33 - 11006**
0370	3"	EA	0.178	15.75	11.25		27.00
0380	4"	"	0.200	25.50	12.75		38.25
0390	6"	"	0.267	63.00	17.00		80.00
0400	Tees						
0410	1"	EA	0.160	1.48	10.25		11.75
0420	1-1/2"	"	0.160	2.86	10.25		13.00
0430	2"	"	0.178	4.12	11.25		15.25
0440	2-1/2"	"	0.200	13.50	12.75		26.25
0450	3"	"	0.229	18.00	14.50		32.50
0460	4"	"	0.267	29.25	17.00		46.25
0470	6"	"	0.320	98.00	20.50		120
0490	Couplings						
0510	1"	EA	0.133	0.91	8.51		9.42
0520	1-1/2"	"	0.133	1.30	8.51		9.81
0530	2"	"	0.145	2.01	9.28		11.25
0540	2-1/2"	"	0.160	4.42	10.25		14.75
0550	3"	"	0.178	6.91	11.25		18.25
0560	4"	"	0.200	9.02	12.75		21.75
0580	6"	"	0.267	28.50	17.00		45.50
1000	Drainage pipe						
1005	PVC schedule 80						
1010	1" dia.	LF	0.047	2.03	3.00		5.03
1015	1-1/2" dia.	"	0.047	2.46	3.00		5.46
1020	ABS, 2" dia.	"	0.050	3.14	3.19		6.33
1030	2-1/2" dia.	"	0.053	4.47	3.40		7.87
1040	3" dia.	"	0.057	5.26	3.64		8.90
1050	4" dia.	"	0.067	7.17	4.25		11.50
1055	6" dia.	"	0.080	12.00	5.10		17.00
1060	8" dia.	"	0.063	16.00	4.00	3.50	23.50
1080	10" dia.	"	0.075	21.25	4.75	4.15	30.25
1100	12" dia.	"	0.080	34.75	5.06	4.43	44.25
1105	90 degree elbows						
1110	1"	EA	0.133	3.38	8.51		12.00
1125	1-1/2"	"	0.133	4.22	8.51		12.75
1135	2"	"	0.145	5.09	9.28		14.25
1145	2-1/2"	"	0.160	12.25	10.25		22.50
1155	3"	"	0.178	12.50	11.25		23.75
1165	4"	"	0.200	22.25	12.75		35.00

© 2020 BNi Publications Inc. ALL RIGHTS RESERVED

DISTRIBUTION PIPING

ID Code	Description Component Descriptions	Output Unit of Meas.	Manhr / Unit	Unit Costs Material Cost	Labor Cost	Equipment Cost	Total Cost
33 - 11006	**PLASTIC PIPE, Cont'd...**						**33 - 11006**
1175	6"	EA	0.267	48.75	17.00		66.00
1200	45 degree elbows						
1210	1"	EA	0.133	5.50	8.51		14.00
1220	1-1/2"	"	0.133	6.99	8.51		15.50
1230	2"	"	0.145	8.66	9.28		18.00
1240	2-1/2"	"	0.160	16.25	10.25		26.50
1250	3"	"	0.178	17.25	11.25		28.50
1260	4"	"	0.200	32.75	12.75		45.50
1270	6"	"	0.267	76.00	17.00		93.00
1300	Tees						
1310	1"	EA	0.160	3.57	10.25		13.75
1320	1-1/2"	"	0.160	11.50	10.25		21.75
1330	2"	"	0.178	14.00	11.25		25.25
1340	2-1/2"	"	0.200	16.25	12.75		29.00
1350	3"	"	0.229	17.75	14.50		32.25
1360	4"	"	0.267	33.75	17.00		51.00
1370	6"	"	0.320	67.00	20.50		88.00
1400	Couplings						
1410	1"	EA	0.133	2.98	8.51		11.50
1420	1-1/2"	"	0.133	5.09	8.51		13.50
1430	2"	"	0.145	7.50	9.28		16.75
1440	2-1/2"	"	0.160	15.75	10.25		26.00
1450	3"	"	0.178	16.25	11.25		27.50
1460	4"	"	0.200	17.00	12.75		29.75
1470	6"	"	0.267	28.50	17.00		45.50
2000	Pressure pipe						
2020	PVC, class 200 pipe						
2025	3/4"	LF	0.040	0.26	2.55		2.81
2030	1"	"	0.042	0.38	2.68		3.06
2035	1-1/4"	"	0.044	0.64	2.83		3.47
2040	1-1/2"	"	0.047	0.77	3.00		3.77
2050	2"	"	0.050	1.26	3.19		4.45
2060	2-1/2"	"	0.053	1.91	3.40		5.31
2070	3"	"	0.057	2.92	3.64		6.56
2080	4"	"	0.067	5.09	4.25		9.34
2090	6"	"	0.080	10.25	5.10		15.25
2100	8"	"	0.069	19.75	4.34	3.80	28.00
2200	90 degree elbows						

© 2020 BNi Publications Inc. ALL RIGHTS RESERVED

DISTRIBUTION PIPING

	Description	Output		Unit Costs			
ID Code	Component Descriptions	Unit of Meas.	Manhr / Unit	Material Cost	Labor Cost	Equipment Cost	Total Cost
33 - 11006	**PLASTIC PIPE, Cont'd...**						**33 - 11006**
2210	3/4"	EA	0.133	0.84	8.51		9.35
2220	1"	"	0.133	0.93	8.51		9.44
2230	1-1/4"	"	0.133	1.39	8.51		9.90
2240	1-1/2"	"	0.133	1.77	8.51		10.25
2250	2"	"	0.145	2.77	9.28		12.00
2260	2-1/2"	"	0.160	8.44	10.25		18.75
2270	3"	"	0.178	13.00	11.25		24.25
2280	4"	"	0.200	23.50	12.75		36.25
2290	6"	"	0.267	49.50	17.00		67.00
2300	8"	"	0.400	83.00	25.50		110
2320	45 degree elbows						
2330	3/4"	EA	0.133	0.95	8.51		9.46
2340	1"	"	0.133	1.22	8.51		9.73
2350	1-1/4"	"	0.133	1.76	8.51		10.25
2360	1-1/2"	"	0.133	2.13	8.51		10.75
2370	2"	"	0.145	3.00	9.28		12.25
2380	2-1/2"	"	0.160	4.92	10.25		15.25
2390	3"	"	0.178	11.25	11.25		22.50
2400	4"	"	0.200	21.25	12.75		34.00
2410	6"	"	0.267	44.25	17.00		61.00
2420	8"	"	0.400	90.00	25.50		120
2500	Tees						
2520	3/4"	EA	0.160	0.80	10.25		11.00
2530	1"	"	0.160	1.05	10.25		11.25
2540	1-1/4"	"	0.160	1.51	10.25		11.75
2550	1-1/2"	"	0.160	2.10	10.25		12.25
2560	2"	"	0.178	3.09	11.25		14.25
2570	2-1/2"	"	0.200	4.86	12.75		17.50
2580	3"	"	0.229	15.00	14.50		29.50
2590	4"	"	0.267	21.25	17.00		38.25
2600	6"	"	0.320	63.00	20.50		84.00
2610	8"	"	0.444	130	28.25		160
2700	Couplings						
2710	3/4"	EA	0.133	0.49	8.51		9.00
2720	1"	"	0.133	0.73	8.51		9.24
2730	1-1/4"	"	0.133	0.95	8.51		9.46
2740	1-1/2"	"	0.133	1.04	8.51		9.55
2750	2"	"	0.145	1.46	9.28		10.75

© 2020 BNi Publications Inc. ALL RIGHTS RESERVED

DISTRIBUTION PIPING

	Description	Output		Unit Costs			
ID Code	Component Descriptions	Unit of Meas.	Manhr / Unit	Material Cost	Labor Cost	Equipment Cost	Total Cost
33 - 11006	**PLASTIC PIPE, Cont'd...**						**33 - 11006**
2760	2-1/2"	EA	0.160	3.09	10.25		13.25
2770	3"	"	0.178	4.86	11.25		16.00
2780	4"	"	0.178	6.83	11.25		18.00
2790	6"	"	0.200	18.00	12.75		30.75
2800	8"	"	0.267	32.50	17.00		49.50

SURFACE WATER SOURCES

ID Code	Component Descriptions	Unit of Meas.	Manhr / Unit	Material Cost	Labor Cost	Equipment Cost	Total Cost
33 - 12131	**CORPORATION STOPS**						**33 - 12131**
0090	Stop for flared copper service pipe						
0100	3/4"	EA	0.400	51.00	35.75		87.00
0120	1"	"	0.444	69.00	39.75		110
0130	1-1/4"	"	0.533	190	47.75		240
0140	1-1/2"	"	0.667	220	60.00		280
0150	2"	"	0.800	310	72.00		380
33 - 12132	**THRUST BLOCKS**						**33 - 12132**
0080	Thrust block, 3000# concrete						
0100	1/4 c.y.	EA	1.333	120	110		230
0120	1/2 c.y.	"	1.600	180	130		310
0140	3/4 c.y.	"	2.667	220	220		440
0160	1 c.y.	"	5.333	310	430		740
33 - 12133	**TAPPING SADDLES & SLEEVES**						**33 - 12133**
0080	Tapping saddle, tap size to 2"						
0100	4" saddle	EA	0.400	71.00	25.50		97.00
0120	6" saddle	"	0.500	83.00	32.00		110
0130	8" saddle	"	0.667	96.00	42.50		140
0140	10" saddle	"	0.800	110	51.00		160
0150	12" saddle	"	1.143	130	73.00		200
0160	14" saddle	"	1.600	150	100		250
2000	Tapping sleeve						
2010	4x4	EA	0.533	860	34.00		890
2030	6x4	"	0.615	1,110	39.25		1,150
2050	6x6	"	0.615	1,130	39.25		1,170
2060	8x4	"	0.800	1,160	51.00		1,210
2070	8x6	"	0.800	1,180	51.00		1,230
2090	10x4	"	0.960	1,880	61.00	53.00	1,990
2100	10x6	"	0.960	2,720	61.00	53.00	2,830

© 2020 BNi Publications Inc. ALL RIGHTS RESERVED

SURFACE WATER SOURCES

ID Code	Description: Component Descriptions	Output: Unit of Meas.	Output: Manhr / Unit	Unit Costs: Material Cost	Unit Costs: Labor Cost	Unit Costs: Equipment Cost	Unit Costs: Total Cost
33 - 12133	**TAPPING SADDLES & SLEEVES, Cont'd...**						**33 - 12133**
2120	10x8	EA	0.960	2,840	61.00	53.00	2,950
2130	10x10	"	1.000	2,900	63.00	55.00	3,020
2140	12x4	"	1.000	2,930	63.00	55.00	3,050
2150	12x6	"	1.091	2,950	69.00	60.00	3,080
2160	12x8	"	1.200	3,020	76.00	67.00	3,160
2170	12x10	"	1.333	3,180	84.00	74.00	3,340
2180	12x12	"	1.500	3,290	95.00	83.00	3,470
4000	Tapping valve, mechanical joint						
4010	4" valve	EA	3.000	850	190	170	1,210
4020	6" valve	"	4.000	1,020	250	220	1,500
4030	8" valve	"	6.000	1,520	380	330	2,230
4040	10" valve	"	8.000	2,420	510	440	3,370
4050	12" valve	"	12.000	4,280	760	670	5,710
7980	Tap hole in pipe						
8000	4" hole	EA	1.000		64.00		64.00
8010	6" hole	"	1.600		100		100
8020	8" hole	"	2.667		170		170
8030	10" hole	"	3.200		200		200
8040	12" hole	"	4.000		260		260
33 - 12135	**VALVE BOXES**						**33 - 12135**
0080	Valve box, adjustable, for valves up to 20"						
0100	3' deep	EA	0.267	300	17.00		320
0120	4' deep	"	0.320	360	20.50		380
0130	5' deep	"	0.400	420	25.50		450
33 - 12161	**GATE VALVES**						**33 - 12161**
0100	Gate valve (AWWA) mechanical joint, with adjustable						
0110	4" valve	EA	0.800	1,340	51.00	44.25	1,430
0120	6" valve	"	0.960	1,520	61.00	53.00	1,630
0130	8" valve	"	1.200	2,020	76.00	67.00	2,160
0140	10" valve	"	1.412	3,030	89.00	78.00	3,200
0150	12" valve	"	1.714	4,040	110	95.00	4,240
0160	14" valve	"	2.000	10,120	130	110	10,360
0170	16" valve	"	2.182	13,490	140	120	13,750
0180	18" valve	"	2.400	16,860	150	130	17,150
3000	Flanged, with box, post indicator (AWWA)						
3010	4" valve	EA	0.960	1,210	61.00	53.00	1,320
3020	6" valve	"	1.091	1,420	69.00	60.00	1,550

© 2020 BNi Publications Inc. ALL RIGHTS RESERVED

SURFACE WATER SOURCES

ID Code	Description: Component Descriptions	Output: Unit of Meas.	Output: Manhr / Unit	Unit Costs: Material Cost	Unit Costs: Labor Cost	Unit Costs: Equipment Cost	Unit Costs: Total Cost
33 - 12161	**GATE VALVES, Cont'd...**						**33 - 12161**
3030	8" valve	EA	1.333	2,020	84.00	74.00	2,180
3040	10" valve	"	1.600	3,030	100	89.00	3,220
3050	12" valve	"	2.000	4,400	130	110	4,640
3060	14" valve	"	2.400	10,110	150	130	10,400
3070	16" valve	"	3.000	13,490	190	170	13,850
33 - 12193	**FIRE HYDRANTS**						**33 - 12193**
0080	Standard, 3 way post, 6" mechanical joint						
0100	2' deep	EA	8.000	2,150	510	440	3,100
0120	4' deep	"	9.600	2,310	610	530	3,450
0140	6' deep	"	12.000	2,570	760	670	4,000
0160	8' deep	"	13.714	2,890	870	760	4,520
33 - 12331	**WATER METERS**						**33 - 12331**
0080	Water meter, displacement type						
0090	1"	EA	0.800	220	72.00		290
0100	1-1/2"	"	0.889	750	79.00		830
0190	2"	"	1.000	1,130	89.00		1,220

UTILITY SERVICES

ID Code	Component Descriptions	Unit of Meas.	Manhr / Unit	Material Cost	Labor Cost	Equipment Cost	Total Cost
33 - 21130	**WELLS**						**33 - 21130**
0980	Domestic water, drilled and cased						
1000	4" dia.	LF	0.480	30.50	39.25	46.00	120
1020	6" dia.	"	0.533	33.50	43.50	51.00	130
1040	8" dia.	"	0.600	39.50	49.00	58.00	150
33 - 31001	**CAST IRON FLANGED PIPE**						**33 - 31001**
0100	Cast iron flanged sections						
0110	4" pipe, with one bolt set						
0120	3' section	EA	0.218	60.00	13.75	12.00	86.00
0130	4' section	"	0.240	83.00	15.25	13.25	110
0140	5' section	"	0.267	110	17.00	14.75	140
0150	6' section	"	0.300	130	19.00	16.50	170
0160	8' section	"	0.343	160	21.75	19.00	200
0170	10' section	"	0.480	230	30.50	26.50	290
0180	12' section	"	0.800	240	51.00	44.25	340
0190	15' section	"	1.200	300	76.00	67.00	440
0200	18' section	"	1.600	370	100	89.00	560

© 2020 BNi Publications Inc. ALL RIGHTS RESERVED

UTILITY SERVICES

Description		Output		Unit Costs			
ID Code	Component Descriptions	Unit of Meas.	Manhr / Unit	Material Cost	Labor Cost	Equipment Cost	Total Cost
33 - 31001	**CAST IRON FLANGED PIPE, Cont'd...**						**33 - 31001**
2080	6" pipe, with one bolt set						
2100	3' section	EA	0.240	100	15.25	13.25	130
2102	4' section	"	0.282	150	18.00	15.75	180
2104	5' section	"	0.320	190	20.25	17.75	230
2110	6' section	"	0.369	220	23.50	20.50	260
2120	8' section	"	0.533	270	33.75	29.50	330
2130	10' section	"	0.600	380	38.00	33.25	450
2140	12' section	"	0.800	420	51.00	44.25	520
2150	15' section	"	1.200	520	76.00	67.00	660
2160	18' section	"	1.714	620	110	95.00	820
2165	8" pipe, with one bolt set						
2170	3' section	EA	0.300	160	19.00	16.50	200
2180	4' section	"	0.343	230	21.75	19.00	270
2190	5' section	"	0.400	280	25.25	22.25	330
2200	6' section	"	0.480	340	30.50	26.50	400
2210	8' section	"	0.686	430	43.50	38.00	510
2215	10' section	"	0.800	600	51.00	44.25	700
2230	12' section	"	1.200	650	76.00	67.00	790
2240	15' section	"	1.600	810	100	89.00	1,000
2250	18' section	"	2.000	970	130	110	1,210
3005	10" pipe, with one bolt set						
3010	3' section	EA	0.308	300	19.50	17.00	340
3020	4' section	"	0.353	490	22.25	19.50	530
3030	5' section	"	0.414	570	26.25	23.00	620
3040	6' section	"	0.500	690	31.75	27.75	750
3050	8' section	"	0.727	910	46.00	40.25	1,000
3060	10' section	"	0.857	990	54.00	47.50	1,090
3070	12' section	"	1.333	1,080	84.00	74.00	1,240
3080	15' section	"	1.714	1,350	110	95.00	1,550
3090	18' section	"	2.400	1,620	150	130	1,910
3095	12" pipe, with one bolt set						
3100	3' section	EA	0.333	370	21.00	18.50	410
3120	4' section	"	0.387	550	24.50	21.50	600
3130	5' section	"	0.462	680	29.25	25.50	730
3140	6' section	"	0.545	820	34.50	30.25	880
3150	8' section	"	0.800	1,090	51.00	44.25	1,180
3160	10' section	"	0.923	1,260	58.00	51.00	1,370
3170	12' section	"	1.500	1,600	95.00	83.00	1,780

© 2020 BNi Publications Inc. ALL RIGHTS RESERVED

UTILITY SERVICES

ID Code	Description: Component Descriptions	Output: Unit of Meas.	Output: Manhr / Unit	Unit Costs: Material Cost	Unit Costs: Labor Cost	Unit Costs: Equipment Cost	Unit Costs: Total Cost
33 - 31001	**CAST IRON FLANGED PIPE, Cont'd...**						**33 - 31001**
3180	15' section	EA	2.000	1,810	130	110	2,050
3190	18' section	"	2.667	2,080	170	150	2,400
33 - 31002	**CAST IRON FITTINGS**						**33 - 31002**
0100	Mechanical joint, with 2 bolt kits						
0105	90 deg bend						
0110	4"	EA	0.533	88.00	34.00		120
0120	6"	"	0.615	150	39.25		190
0130	8"	"	0.800	330	51.00		380
0140	10"	"	1.143	500	73.00		570
0150	12"	"	1.600	760	100		860
0155	14"	"	2.000	1,040	130		1,170
0160	16"	"	2.667	1,200	170		1,370
0165	45 deg bend						
0170	4"	EA	0.533	73.00	34.00		110
0180	6"	"	0.615	120	39.25		160
0190	8"	"	0.800	260	51.00		310
0200	10"	"	1.143	390	73.00		460
0210	12"	"	1.600	660	100		760
0220	14"	"	2.000	810	130		940
0230	16"	"	2.667	1,060	170		1,230
2000	Tee, with 3 bolt kits						
2010	4" x 4"	EA	0.800	150	51.00		200
2020	6" x 6"	"	1.000	260	64.00		320
2030	8" x 8"	"	1.333	730	85.00		820
2040	10" x 10"	"	2.000	910	130		1,040
2050	12" x 12"	"	2.667	1,620	170		1,790
3000	Wye, with 3 bolt kits						
3010	6" x 6"	EA	1.000	320	64.00		380
3020	8" x 8"	"	1.333	670	85.00		750
3030	10" x 10"	"	2.000	960	130		1,090
3040	12" x 12"	"	2.667	1,880	170		2,050
3045	Reducer, with 2 bolt kits						
3050	6" x 4"	EA	1.000	140	64.00		200
3060	8" x 6"	"	1.333	230	85.00		310
3070	10" x 8"	"	2.000	640	130		770
3080	12" x 10"	"	2.667	770	170		940
6010	Flanged, 90 deg bend, 125 lb.						

© 2020 BNi Publications Inc. ALL RIGHTS RESERVED

UTILITY SERVICES

	Description	Output		Unit Costs			
ID Code	Component Descriptions	Unit of Meas.	Manhr / Unit	Material Cost	Labor Cost	Equipment Cost	Total Cost
33 - 31002	**CAST IRON FITTINGS, Cont'd...**						**33 - 31002**
6020	4"	EA	0.667	180	42.50		220
6040	6"	"	0.800	220	51.00		270
6060	8"	"	1.000	310	64.00		370
6080	10"	"	1.333	550	85.00		630
6090	12"	"	2.000	770	130		900
6095	14"	"	2.667	1,520	170		1,690
6100	16"	"	2.667	2,270	170		2,440
6115	Tee						
6120	4"	EA	1.000	280	64.00		340
6130	6"	"	1.143	390	73.00		460
6140	8"	"	1.333	600	85.00		680
6150	10"	"	1.600	1,100	100		1,200
6160	12"	"	2.000	1,480	130		1,610
6170	14"	"	2.667	3,300	170		3,470
6180	16"	"	4.000	4,960	260		5,220
33 - 31003	**VITRIFIED CLAY PIPE**						**33 - 31003**
0100	Vitrified clay pipe, extra strength						
1020	6" dia.	LF	0.109	5.73	6.91	6.04	18.75
1040	8" dia.	"	0.114	6.87	7.24	6.33	20.25
1050	10" dia.	"	0.120	10.50	7.60	6.65	24.75
1070	12" dia.	"	0.160	15.00	10.25	8.86	34.00
1090	15" dia.	"	0.240	27.50	15.25	13.25	56.00
1120	18" dia.	"	0.267	41.25	17.00	14.75	73.00
1140	24" dia.	"	0.343	75.00	21.75	19.00	120
1160	30" dia.	"	0.480	130	30.50	26.50	190
1180	36" dia.	"	0.686	180	43.50	38.00	260
33 - 31004	**SANITARY SEWERS**						**33 - 31004**
0980	Clay						
1000	6" pipe	LF	0.080	9.41	5.06	4.43	19.00
1020	8" pipe	"	0.086	12.50	5.43	4.75	22.75
1030	10" pipe	"	0.092	15.75	5.84	5.11	26.75
1040	12" pipe	"	0.100	25.00	6.33	5.54	36.75
2980	PVC						
3000	4" pipe	LF	0.060	4.05	3.80	3.32	11.25
3010	6" pipe	"	0.063	8.11	4.00	3.50	15.50
3020	8" pipe	"	0.067	12.25	4.22	3.69	20.25
3030	10" pipe	"	0.071	16.25	4.47	3.91	24.75

© 2020 BNi Publications Inc. ALL RIGHTS RESERVED

UTILITY SERVICES

ID Code	Description: Component Descriptions	Output: Unit of Meas.	Output: Manhr / Unit	Unit Costs: Material Cost	Unit Costs: Labor Cost	Unit Costs: Equipment Cost	Unit Costs: Total Cost
33 - 31004	**SANITARY SEWERS, Cont'd...**						**33 - 31004**
3040	12" pipe	LF	0.075	24.25	4.75	4.15	33.25
5980	Cleanout						
6000	4" pipe	EA	1.000	18.00	64.00		82.00
6010	6" pipe	"	1.000	39.75	64.00		100
6020	8" pipe	"	1.000	120	64.00		180
7980	Connect new sewer line						
8000	To existing manhole	EA	2.667	100	170		270
8010	To new manhole	"	1.600	75.00	100		180

WASTEWATER UTILITY STORAGE TANKS

ID Code	Component Descriptions	Unit of Meas.	Manhr / Unit	Material Cost	Labor Cost	Equipment Cost	Total Cost
33 - 36001	**DRAINAGE FIELDS**						**33 - 36001**
0080	Perforated PVC pipe, for drain field						
0100	4" pipe	LF	0.053	2.71	3.37	2.95	9.04
0120	6" pipe	"	0.057	5.08	3.62	3.16	11.75
33 - 36005	**SEPTIC TANKS**						**33 - 36005**
0980	Septic tank, precast concrete						
1000	1000 gals	EA	4.000	1,020	250	220	1,500
1200	2000 gals	"	6.000	2,740	380	330	3,450
1280	5000 gals	"	12.000	9,350	760	670	10,780
1290	25,000 gals	"	48.000	53,590	3,040	2,660	59,290
1300	40,000 gals	"	80.000	63,560	5,070	4,430	73,060
1310	Leaching pit, precast concrete, 72" diameter						
1320	3' deep	EA	3.000	780	190	170	1,140
1340	6' deep	"	3.429	1,370	220	190	1,780
1360	8' deep	"	4.000	1,740	250	220	2,220

MANHOLES

ID Code	Component Descriptions	Unit of Meas.	Manhr / Unit	Material Cost	Labor Cost	Equipment Cost	Total Cost
33 - 39133	**MANHOLES**						**33 - 39133**
0100	Precast sections, 48" dia.						
0110	Base section	EA	2.000	360	130	110	600
0120	1'0" riser	"	1.600	100	100	89.00	290
0130	1'4" riser	"	1.714	120	110	95.00	320
0140	2'8" riser	"	1.846	180	120	100	400
0150	4'0" riser	"	2.000	340	130	110	580
0160	2'8" cone top	"	2.400	220	150	130	510
0170	Precast manholes, 48" dia.						

© 2020 BNi Publications Inc. ALL RIGHTS RESERVED

MANHOLES

ID Code	Description: Component Descriptions	Output: Unit of Meas.	Output: Manhr / Unit	Unit Costs: Material Cost	Unit Costs: Labor Cost	Unit Costs: Equipment Cost	Unit Costs: Total Cost
33 - 39133	**MANHOLES, Cont'd...**						**33 - 39133**
0180	4' deep	EA	4.800	700	300	270	1,270
0200	6' deep	"	6.000	1,070	380	330	1,780
0250	7' deep	"	6.857	1,220	430	380	2,030
0260	8' deep	"	8.000	1,380	510	440	2,330
0280	10' deep	"	9.600	1,540	610	530	2,680
1000	Cast-in-place, 48" dia., with frame and cover						
1100	5' deep	EA	12.000	630	760	670	2,060
1120	6' deep	"	13.714	830	870	760	2,460
1140	8' deep	"	16.000	1,210	1,010	890	3,110
1160	10' deep	"	19.200	1,410	1,220	1,060	3,690
1480	Brick manholes, 48" dia. with cover, 8" thick						
1500	4' deep	EA	8.000	670	620		1,290
1501	6' deep	"	8.889	840	690		1,530
1505	8' deep	"	10.000	1,080	780		1,860
1510	10' deep	"	11.429	1,340	890		2,230
1600	12' deep	"	13.333	1,680	1,040		2,720
1620	14' deep	"	16.000	2,040	1,240		3,280
3000	Inverts for manholes						
3010	Single channel	EA	3.200	110	250		360
3020	Triple channel	"	4.000	130	310		440
4200	Frames and covers, 24" diameter						
4210	300 lb	EA	0.800	410	51.00		460
4220	400 lb	"	0.889	430	57.00		490
4230	500 lb	"	1.143	500	73.00		570
4240	Watertight, 350 lb	"	2.667	520	170		690
4250	For heavy equipment, 1200 lb	"	4.000	1,130	260		1,390
4980	Steps for manholes						
5000	7" x 9"	EA	0.160	18.25	10.25		28.50
5020	8" x 9"	"	0.178	23.00	11.25		34.25
6080	Curb inlet, 4' throat, cast-in-place						
6100	12"-30" pipe	EA	12.000	380	760	670	1,810
6150	36"-48" pipe	"	13.714	420	870	760	2,050
8000	Raise existing frame and cover, when repaving	"	4.800		300	270	570

 © 2020 BNi Publications Inc. ALL RIGHTS RESERVED

STORM DRAINAGE

ID Code	Description: Component Descriptions	Output: Unit of Meas.	Output: Manhr / Unit	Unit Costs: Material Cost	Unit Costs: Labor Cost	Unit Costs: Equipment Cost	Unit Costs: Total Cost
33 - 41004	**PIPE**						**33 - 41004**
1000	Concrete pipe						
1080	Plain, bell and spigot joint, Class II						
1100	6" pipe	LF	0.109	7.56	6.91	6.04	20.50
1200	8" pipe	"	0.120	7.89	7.60	6.65	22.25
1300	10" pipe	"	0.126	8.03	8.00	7.00	23.00
1400	12" pipe	"	0.133	10.75	8.44	7.38	26.50
1500	15" pipe	"	0.141	14.50	8.94	7.82	31.25
1600	18" pipe	"	0.150	18.00	9.50	8.31	35.75
1650	21" pipe	"	0.160	21.75	10.25	8.86	40.75
1700	24" pipe	"	0.171	27.50	10.75	9.50	47.75
1730	Reinforced, class III, tongue and groove joint						
1750	12" pipe	LF	0.133	15.50	8.44	7.38	31.25
1800	15" pipe	"	0.141	17.50	8.94	7.82	34.25
1850	18" pipe	"	0.150	19.25	9.50	8.31	37.00
1880	21" pipe	"	0.160	25.00	10.25	8.86	44.00
1900	24" pipe	"	0.171	32.75	10.75	9.50	53.00
1950	27" pipe	"	0.185	38.75	11.75	10.25	61.00
1960	30" pipe	"	0.200	42.50	12.75	11.00	66.00
1970	36" pipe	"	0.218	64.00	13.75	12.00	90.00
1980	42" pipe	"	0.240	87.00	15.25	13.25	120
1990	48" pipe	"	0.267	120	17.00	14.75	150
2000	54" pipe	"	0.300	130	19.00	16.50	170
2050	60" pipe	"	0.343	170	21.75	19.00	210
2070	66" pipe	"	0.400	220	25.25	22.25	270
2080	72" pipe	"	0.480	240	30.50	26.50	300
2090	Flared end-section, concrete						
2100	12" pipe	LF	0.133	70.00	8.44	7.38	86.00
2110	15" pipe	"	0.141	83.00	8.94	7.82	100
2120	18" pipe	"	0.150	98.00	9.50	8.31	120
2130	24" pipe	"	0.171	110	10.75	9.50	130
2140	30" pipe	"	0.200	140	12.75	11.00	160
2150	36" pipe	"	0.218	190	13.75	12.00	220
2160	42" pipe	"	0.240	210	15.25	13.25	240
2170	48" pipe	"	0.267	230	17.00	14.75	260
2180	54" pipe	"	0.300	250	19.00	16.50	290
5090	Corrugated metal pipe, coated, paved invert						
5095	16 ga.						
6000	8" pipe	LF	0.080	11.00	5.06	4.43	20.50

© 2020 BNi Publications Inc. ALL RIGHTS RESERVED

STORM DRAINAGE

ID Code	Description: Component Descriptions	Output: Unit of Meas.	Output: Manhr / Unit	Unit Costs: Material Cost	Unit Costs: Labor Cost	Unit Costs: Equipment Cost	Unit Costs: Total Cost
33 - 41004	**PIPE, Cont'd...**						**33 - 41004**
6010	10" pipe	LF	0.083	14.75	5.24	4.58	24.50
6020	12" pipe	"	0.086	16.50	5.43	4.75	26.75
6030	15" pipe	"	0.092	20.25	5.84	5.11	31.25
6040	18" pipe	"	0.100	24.00	6.33	5.54	35.75
6050	21" pipe	"	0.109	29.50	6.91	6.04	42.50
6060	24" pipe	"	0.120	35.00	7.60	6.65	49.25
6070	30" pipe	"	0.133	46.00	8.44	7.38	62.00
6080	36" pipe	"	0.150	63.00	9.50	8.31	81.00
6090	12 ga., 48" pipe	"	0.171	110	10.75	9.50	130
6095	10 ga.						
6100	60" pipe	LF	0.200	140	12.75	11.00	160
6110	72" pipe	"	0.240	180	15.25	13.25	210
6200	Galvanized or aluminum, plain						
6205	16 ga.						
6210	8" pipe	LF	0.080	9.24	5.06	4.43	18.75
6220	10" pipe	"	0.083	13.00	5.24	4.58	22.75
6230	12" pipe	"	0.086	14.75	5.43	4.75	25.00
6240	15" pipe	"	0.092	18.50	5.84	5.11	29.50
6250	18" pipe	"	0.100	22.00	6.33	5.54	33.75
6260	24" pipe	"	0.120	33.25	7.60	6.65	47.50
6270	30" pipe	"	0.133	44.25	8.44	7.38	60.00
6280	36" pipe	"	0.150	55.00	9.50	8.31	73.00
6290	12 ga., 48" pipe	"	0.171	100	10.75	9.50	120
6300	10 ga., 60" pipe	"	0.200	140	12.75	11.00	160
6400	Galvanized or aluminum, coated oval arch						
6405	16 ga.						
6410	17" x 13"	LF	0.109	34.00	6.91	6.04	47.00
6420	21" x 15"	"	0.120	45.75	7.60	6.65	60.00
6425	14 ga.						
6430	28" x 20"	LF	0.133	64.00	8.44	7.38	80.00
6440	35" x 24"	"	0.171	96.00	10.75	9.50	120
6445	12 ga.						
6450	42" x 29"	LF	0.200	110	12.75	11.00	130
6460	57" x 38"	"	0.240	170	15.25	13.25	200
6470	64" x 43"	"	0.253	200	16.00	14.00	230
6500	Oval arch culverts, plain						
6505	16 ga.						
6510	17" x 13"	LF	0.109	16.00	6.91	6.04	29.00

© 2020 BNi Publications Inc. ALL RIGHTS RESERVED

STORM DRAINAGE

ID Code	Description: Component Descriptions	Output: Unit of Meas.	Output: Manhr / Unit	Unit Costs: Material Cost	Unit Costs: Labor Cost	Unit Costs: Equipment Cost	Unit Costs: Total Cost
33 - 41004	PIPE, Cont'd...						33 - 41004
6520	21" x 15"	LF	0.120	23.00	7.60	6.65	37.25
6525	14 ga.						
6530	28" x 20"	LF	0.133	42.75	8.44	7.38	59.00
6540	35" x 24"	"	0.171	54.00	10.75	9.50	74.00
6545	12 ga.						
6550	57" x 38"	LF	0.200	87.00	12.75	11.00	110
6560	64" x 43"	"	0.240	110	15.25	13.25	140
6570	71" x 47"	"	0.253	150	16.00	14.00	180
6600	Nestable corrugated metal pipe						
6615	16 ga.						
6620	10" pipe	LF	0.083	12.75	5.24	4.58	22.50
6630	12" pipe	"	0.086	16.00	5.43	4.75	26.25
6640	15" pipe	"	0.092	20.75	5.84	5.11	31.75
6650	18" pipe	"	0.100	24.00	6.33	5.54	35.75
6660	24" pipe	"	0.120	33.75	7.60	6.65	48.00
6670	30" pipe	"	0.133	41.75	8.44	7.38	58.00
6680	14 ga., 36" pipe	"	0.150	48.00	9.50	8.31	66.00
9680	Headwalls, cast-in-place, 30 deg wingwall						
9700	12" pipe	EA	2.000	420	160		580
9740	15" pipe	"	2.000	510	160		670
9750	18" pipe	"	2.286	620	190		810
9760	24" pipe	"	2.286	920	190		1,110
9770	30" pipe	"	2.667	1,110	220		1,330
9780	36" pipe	"	4.000	1,200	330		1,530
9790	42" pipe	"	4.000	1,500	330		1,830
9800	48" pipe	"	5.333	1,590	430		2,020
9810	54" pipe	"	6.667	1,800	540		2,340
9820	60" pipe	"	8.000	2,160	650		2,810
9880	4" cleanout for storm drain						
9900	4" pipe	EA	1.000	670	64.00		730
9910	6" pipe	"	1.000	810	64.00		870
9920	8" pipe	"	1.000	1,120	64.00		1,180
9925	Connect new drain line						
9930	To existing manhole	EA	2.667	140	170		310
9940	To new manhole	"	1.600	120	100		220

© 2020 BNi Publications Inc. ALL RIGHTS RESERVED

DRAINAGE AND CONTAINMENT

ID Code	Description Component Descriptions	Output Unit of Meas.	Manhr / Unit	Unit Costs Material Cost	Labor Cost	Equipment Cost	Total Cost
33 - 44131	**CATCH BASINS**						**33 - 44131**
0100	Standard concrete catch basin						
1021	Cast-in-place, 3'8" x 3'8", 6" thick wall						
1030	2' deep	EA	6.000	540	380	330	1,250
1040	3' deep	"	6.000	730	380	330	1,440
1050	4' deep	"	8.000	950	510	440	1,900
1060	5' deep	"	8.000	1,120	510	440	2,070
1070	6' deep	"	9.600	1,250	610	530	2,390
1201	4'x4', 8" thick wall, cast-in-place						
1210	2' deep	EA	6.000	580	380	330	1,290
1220	3' deep	"	6.000	810	380	330	1,520
1230	4' deep	"	8.000	1,070	510	440	2,020
1240	5' deep	"	8.000	1,240	510	440	2,190
1250	6' deep	"	9.600	1,370	610	530	2,510
5000	Frames and covers, cast iron						
5010	Round						
5020	24" dia.	EA	2.000	430	130		560
5030	26" dia.	"	2.000	480	130		610
5040	28" dia.	"	2.000	570	130		700
5080	Rectangular						
5100	23"x23"	EA	2.000	380	130		510
5120	27"x20"	"	2.000	460	130		590
5130	24"x24"	"	2.000	450	130		580
5140	26"x26"	"	2.000	490	130		620
5200	Curb inlet frames and covers						
5210	27"x27"	EA	2.000	770	130		900
5220	24"x36"	"	2.000	560	130		690
5230	24"x25"	"	2.000	520	130		650
5240	24"x22"	"	2.000	450	130		580
5250	20"x22"	"	2.000	590	130		720
9120	Airfield catch basin frame and grating, galvanized						
9140	2'x4'	EA	2.000	790	130		920
9160	2'x2'	"	2.000	550	130		680

© 2020 BNi Publications Inc. ALL RIGHTS RESERVED

STORMWATER MANAGEMENT

	Description	Output		Unit Costs			
ID Code	Component Descriptions	Unit of Meas.	Manhr / Unit	Material Cost	Labor Cost	Equipment Cost	Total Cost
33 - 46190	**UNDERDRAIN**						**33 - 46190**
1480	Drain tile, clay						
1500	6" pipe	LF	0.053	4.52	3.37	2.95	10.75
1520	8" pipe	"	0.056	7.21	3.53	3.09	13.75
1530	12" pipe	"	0.060	14.50	3.80	3.32	21.50
1580	Porous concrete, standard strength						
1600	6" pipe	LF	0.053	5.14	3.37	2.95	11.50
1620	8" pipe	"	0.056	5.56	3.53	3.09	12.25
1630	12" pipe	"	0.060	7.35	3.80	3.32	14.50
1640	15" pipe	"	0.067	13.25	4.22	3.69	21.25
1650	18" pipe	"	0.080	17.75	5.06	4.43	27.25
1800	Corrugated metal pipe, perforated type						
1810	6" pipe	LF	0.060	7.49	3.80	3.32	14.50
1820	8" pipe	"	0.063	8.85	4.00	3.50	16.25
1830	10" pipe	"	0.067	10.75	4.22	3.69	18.75
1840	12" pipe	"	0.071	15.25	4.47	3.91	23.75
1860	18" pipe	"	0.075	18.75	4.75	4.15	27.75
1980	Perforated clay pipe						
2000	6" pipe	LF	0.069	5.98	4.34	3.80	14.00
2020	8" pipe	"	0.071	8.02	4.47	3.91	16.50
2030	12" pipe	"	0.073	14.00	4.60	4.03	22.75
2480	Drain tile, concrete						
2500	6" pipe	LF	0.053	4.08	3.37	2.95	10.50
2520	8" pipe	"	0.056	6.35	3.53	3.09	13.00
2530	12" pipe	"	0.060	12.75	3.80	3.32	19.75
4980	Perforated rigid PVC underdrain pipe						
5000	4" pipe	LF	0.040	2.10	2.53	2.21	6.85
5100	6" pipe	"	0.048	4.04	3.04	2.66	9.74
5150	8" pipe	"	0.053	6.17	3.37	2.95	12.50
5200	10" pipe	"	0.060	9.43	3.80	3.32	16.50
5210	12" pipe	"	0.069	14.50	4.34	3.80	22.75
6980	Underslab drainage, crushed stone						
7000	3" thick	SF	0.008	0.33	0.50	0.44	1.28
7120	4" thick	"	0.009	0.45	0.58	0.51	1.54
7140	6" thick	"	0.010	0.68	0.63	0.55	1.86
7160	8" thick	"	0.010	0.90	0.66	0.57	2.13
7180	Plastic filter fabric for drain lines	"	0.008	0.50	0.51		1.01
9000	Gravel fill in trench, crushed or bank run, 1/2" to 3/4"	CY	0.600	36.25	38.00	33.25	110

© 2020 BNi Publications Inc. ALL RIGHTS RESERVED

ENERGY DISTRIBUTION

	Description	Output		Unit Costs			
ID Code	Component Descriptions	Unit of Meas.	Manhr / Unit	Material Cost	Labor Cost	Equipment Cost	Total Cost
33 - 51001	**GAS DISTRIBUTION**						**33 - 51001**
0100	Gas distribution lines						
1000	Polyethylene, 60 psi coils						
1010	1-1/4" dia.	LF	0.053	1.92	4.76		6.68
1020	1-1/2" dia.	"	0.057	2.61	5.10		7.71
1030	2" dia.	"	0.067	3.30	5.96		9.26
1040	3" dia.	"	0.080	6.99	7.15		14.25
1045	30' pipe lengths						
1050	3" dia.	LF	0.089	5.73	7.94		13.75
1060	4" dia.	"	0.100	8.95	8.94		18.00
1070	6" dia.	"	0.133	14.25	12.00		26.25
1080	8" dia.	"	0.160	26.25	14.25		40.50
2000	Steel, schedule 40, plain end						
2010	1" dia.	LF	0.067	5.42	5.96		11.50
2020	2" dia.	"	0.073	9.02	6.50		15.50
2030	3" dia.	"	0.080	13.00	7.15		20.25
2040	4" dia.	"	0.160	15.75	10.25	8.86	34.75
2050	5" dia.	"	0.171	29.75	10.75	9.50	50.00
2060	6" dia.	"	0.200	39.75	12.75	11.00	64.00
2070	8" dia.	"	0.218	49.75	13.75	12.00	76.00
5000	Natural gas meters, direct digital reading, threaded						
5050	250 cfh @ 5 lbs	EA	1.600	130	140		270
5060	425 cfh @ 10 lbs	"	1.600	320	140		460
5080	800 cfh @ 20 lbs	"	2.000	450	180		630
5090	1,000 cfh @ 25 lbs	"	2.000	1,330	180		1,510
5130	1,400 cfh @ 100 lbs	"	2.667	3,090	240		3,330
5140	2,300 cfh @ 100 lbs	"	4.000	4,300	360		4,660
5150	5,000 cfh @ 100 lbs	"	8.000	6,310	720		7,030
5500	Gas pressure regulators						
5505	Threaded						
5510	3/4"	EA	1.000	61.00	89.00		150
5520	1"	"	1.333	64.00	120		180
5530	1-1/4"	"	1.333	67.00	120		190
5540	1-1/2"	"	1.333	440	120		560
5560	2"	"	1.600	450	140		590
5565	Flanged						
5570	3"	EA	2.000	1,590	180		1,770
5580	4"	"	2.667	2,370	240		2,610

© 2020 BNi Publications Inc. ALL RIGHTS RESERVED

HYDROCARBON STORAGE

ID Code	Description: Component Descriptions	Output: Unit of Meas.	Output: Manhr / Unit	Unit Costs: Material Cost	Unit Costs: Labor Cost	Unit Costs: Equipment Cost	Unit Costs: Total Cost
33 - 56001	**STORAGE TANKS**						**33 - 56001**
0080	Oil storage tank, underground, single wall, no excv.						
0090	Steel						
1000	500 gals	EA	3.000	4,040	190	170	4,400
1020	1,000 gals	"	4.000	5,470	250	220	5,950
1040	4,000 gals	"	8.000	8,380	510	440	9,330
1060	5,000 gals	"	12.000	9,830	760	670	11,260
1080	10,000 gals	"	24.000	17,500	1,520	1,330	20,350
1980	Fiberglass, double wall						
2000	550 gals	EA	4.000	11,370	250	220	11,850
2020	1,000 gals	"	4.000	14,620	250	220	15,100
2030	2,000 gals	"	6.000	15,880	380	330	16,590
2060	4,000 gals	"	12.000	18,440	760	670	19,870
2080	6,000 gals	"	16.000	20,120	1,010	890	22,020
2090	8,000 gals	"	24.000	22,580	1,520	1,330	25,430
2095	10,000 gals	"	30.000	23,150	1,900	1,660	26,710
2100	12,000 gals	"	40.000	25,100	2,530	2,220	29,850
2120	15,000 gals	"	53.333	26,920	3,380	2,960	33,250
2140	20,000 gals	"	60.000	32,980	3,800	3,330	40,110
2520	Above ground						
2530	Steel, single wall						
2540	275 gals	EA	2.400	2,290	150	130	2,580
2560	500 gals	"	4.000	5,720	250	220	6,200
2570	1,000 gals	"	4.800	7,810	300	270	8,380
2580	1,500 gals	"	6.000	9,960	380	330	10,670
2590	2,000 gals	"	8.000	12,320	510	440	13,270
2620	5,000 gals	"	12.000	14,470	760	670	15,900
3020	Fill cap	"	0.800	140	72.00		210
3040	Vent cap	"	0.800	140	72.00		210
3100	Level indicator	"	0.800	220	72.00		290

© 2020 BNi Publications Inc. ALL RIGHTS RESERVED

STEAM ENERGY DISTRIBUTION

ID Code	Description: Component Descriptions	Output: Unit of Meas.	Output: Manhr / Unit	Unit Costs: Material Cost	Unit Costs: Labor Cost	Unit Costs: Equipment Cost	Unit Costs: Total Cost
33 - 63330	**STEAM METERS**						**33 - 63330**
0100	In-line turbine, direct reading, 300 lb, flanged						
0120	2"	EA	1.000	3,880	89.00		3,970
0130	3"	"	1.333	4,160	120		4,280
0140	4"	"	1.600	4,570	140		4,710
0145	Threaded, 2"						
0150	5" line	EA	8.000	7,350	720		8,070
0170	6" line	"	8.000	7,490	720		8,210
0180	8" line	"	8.000	7,620	720		8,340
0190	10" line	"	8.000	8,040	720		8,760
0200	12" line	"	8.000	8,180	720		8,900
0210	14" line	"	8.000	8,460	720		9,180
0220	16" line	"	8.000	9,010	720		9,730

POWER & COMMUNICATIONS

ID Code	Description: Component Descriptions	Output: Unit of Meas.	Output: Manhr / Unit	Unit Costs: Material Cost	Unit Costs: Labor Cost	Unit Costs: Equipment Cost	Unit Costs: Total Cost
33 - 71160	**UTILITY POLES & FITTINGS**						**33 - 71160**
0980	Wood pole, creosoted						
1000	25'	EA	2.353	450	200		650
1030	40'	"	3.791	860	310		1,170
1060	55'	"	7.547	1,340	630		1,970
1065	Treated, wood preservative, 6"x6"						
1070	8'	EA	0.500	100	41.50		140
1120	16'	"	1.600	250	130		380
1150	20'	"	2.000	360	170		530
1155	Aluminum, brushed, no base						
1160	8'	EA	2.000	650	170		820
1190	20'	"	3.200	1,020	270		1,290
1230	40'	"	6.250	3,060	520		3,580
1235	Steel, no base						
1240	10'	EA	2.500	760	210		970
1250	20'	"	3.810	1,110	320		1,430
1300	35'	"	6.250	1,880	520		2,400
2000	Concrete, no base						
2020	13'	EA	5.517	980	460		1,440
2100	30'	"	12.121	2,670	1,010		3,680
2180	50'	"	18.182	5,920	1,510		7,430
2220	60'	"	20.000	7,550	1,660		9,210

© 2020 BNi Publications Inc. ALL RIGHTS RESERVED

POWER & COMMUNICATIONS

ID Code	Description: Component Descriptions	Output: Unit of Meas.	Output: Manhr / Unit	Unit Costs: Material Cost	Unit Costs: Labor Cost	Unit Costs: Equipment Cost	Unit Costs: Total Cost
33 - 71190	**ELECTRIC MANHOLES**						**33 - 71190**
0980	Precast, handhole, 4' deep						
1000	2'x2'	EA	3.478	480	290		770
1020	3'x3'	"	5.556	640	460		1,100
1040	4'x4'	"	10.256	1,380	850		2,230
1060	Power manhole, complete, precast, 8' deep						
1080	4'x4'	EA	14.035	1,980	1,160		3,140
1100	6'x6'	"	20.000	2,640	1,660		4,300
1140	8'x8'	"	21.053	3,130	1,750		4,880
1180	6' deep, 9' x 12'	"	25.000	3,460	2,080		5,540
1980	Cast-in-place, power manhole, 8' deep						
2000	4'x4'	EA	14.035	2,350	1,160		3,510
2020	6'x6'	"	20.000	3,030	1,660		4,690
2040	8'x8'	"	21.053	3,360	1,750		5,110

© 2020 BNi Publications Inc. ALL RIGHTS RESERVED

DIVISION 34
TRANSPORTATION

BASE COURSES AND BALLASTS

ID Code	Description: Component Descriptions	Output: Unit of Meas.	Output: Manhr / Unit	Unit Costs: Material Cost	Unit Costs: Labor Cost	Unit Costs: Equipment Cost	Unit Costs: Total Cost
34 - 11130	**RAILROAD BALLAST, RAIL, APPURTENANCES**						**34 - 11130**
0080	Rail						
1010	90 lb	LF	0.010	29.00	0.60	0.53	30.25
1020	100 lb	"	0.010	33.25	0.60	0.53	34.50
1030	115 lb	"	0.010	37.25	0.60	0.53	38.50
1080	132 lb	"	0.010	41.50	0.60	0.53	42.75
1090	Rail relay						
1100	90 lb	LF	0.010	13.50	0.60	0.53	14.75
1120	100 lb	"	0.010	14.75	0.60	0.53	16.00
1140	115 lb	"	0.010	18.00	0.60	0.53	19.25
1160	132 lb	"	0.010	21.75	0.60	0.53	23.00
1170	New angle bars, per pair						
1540	90 lb	EA	0.012	110	0.76	0.66	110
1560	100 lb	"	0.012	120	0.76	0.66	120
1600	115 lb	"	0.012	150	0.76	0.66	150
1620	132 lb	"	0.012	170	0.76	0.66	170
1630	Angle bar relay						
1640	90 lb	EA	0.012	44.50	0.76	0.66	46.00
1660	100 lb	"	0.012	45.50	0.76	0.66	47.00
1680	115 lb	"	0.012	47.75	0.76	0.66	49.25
1700	132 lb	"	0.012	51.00	0.76	0.66	52.00
1800	New tie plates						
2020	90 lb	EA	0.009	14.50	0.54	0.47	15.50
2040	100 lb	"	0.009	15.25	0.54	0.47	16.25
2060	115 lb	"	0.009	16.50	0.54	0.47	17.50
2080	132 lb	"	0.009	17.50	0.54	0.47	18.50
2090	Tie plate relay						
2100	90 lb	EA	0.009	4.56	0.54	0.47	5.57
2120	100 lb	"	0.009	6.41	0.54	0.47	7.42
2140	115 lb	"	0.009	6.41	0.54	0.47	7.42
2180	132 lb	"	0.009	7.90	0.54	0.47	8.91
2190	Track accessories						
2250	Wooden cross ties, 8'	EA	0.060	56.00	3.80	3.32	63.00
2260	Concrete cross ties, 8'	"	0.120	140	7.60	6.65	150
2280	Tie plugs, 5"	"	0.006	6.85	0.38	0.33	7.56
2300	Track bolts and nuts, 1"	"	0.006	5.59	0.38	0.33	6.30
2320	Lockwashers, 1"	"	0.004	1.46	0.25	0.22	1.93
2340	Track spikes, 6"	"	0.024	1.38	1.52	1.33	4.23
2360	Wooden switch ties	BF	0.006	2.37	0.38	0.33	3.08

© 2020 BNi Publications Inc. ALL RIGHTS RESERVED

BASE COURSES AND BALLASTS

ID Code	Description: Component Descriptions	Output: Unit of Meas.	Output: Manhr / Unit	Unit Costs: Material Cost	Unit Costs: Labor Cost	Unit Costs: Equipment Cost	Unit Costs: Total Cost
34 - 11130	**RAILROAD BALLAST, RAIL, APPURTENANCES, Cont'd...34 - 11130**						
2380	Rail anchors	EA	0.022	5.68	1.38	1.20	8.27
2400	Ballast	TON	0.120	17.75	7.60	6.65	32.00
2420	Gauge rods	EA	0.096	42.75	6.08	5.32	54.00
2460	Compromise splice bars	"	0.160	540	10.25	8.86	560
2470	Turnout						
3020	90 lb	EA	24.000	16,940	1,520	1,330	19,790
3060	100 lb	"	24.000	17,850	1,520	1,330	20,700
3080	110 lb	"	24.000	19,200	1,520	1,330	22,050
3100	115 lb	"	24.000	19,650	1,520	1,330	22,500
3120	132 lb	"	24.000	21,460	1,520	1,330	24,310
3130	Turnout relay						
3160	90 lb	EA	24.000	10,840	1,520	1,330	13,690
3180	100 lb	"	24.000	11,970	1,520	1,330	14,820
3200	110 lb	"	24.000	12,430	1,520	1,330	15,280
3220	115 lb	"	24.000	13,100	1,520	1,330	15,950
3240	132 lb	"	24.000	14,230	1,520	1,330	17,080
3250	Railroad track in place, complete						
3260	New rail						
3320	90 lb	LF	0.240	190	15.25	13.25	220
3340	100 lb	"	0.240	190	15.25	13.25	220
3360	110 lb	"	0.240	200	15.25	13.25	230
3380	115 lb	"	0.240	200	15.25	13.25	230
3400	132 lb	"	0.240	210	15.25	13.25	240
3410	Rail relay						
3420	90 lb	LF	0.240	110	15.25	13.25	140
3440	100 lb	"	0.240	120	15.25	13.25	150
3450	110 lb	"	0.240	120	15.25	13.25	150
3460	115 lb	"	0.240	130	15.25	13.25	160
3500	132 lb	"	0.240	130	15.25	13.25	160
3510	No. 8 turnout						
3520	90 lb	EA	32.000	38,610	2,030	1,770	42,410
3540	100 lb	"	32.000	42,930	2,030	1,770	46,730
3560	110 lb	"	32.000	47,030	2,030	1,770	50,830
3580	115 lb	"	32.000	48,060	2,030	1,770	51,860
3600	132 lb	"	32.000	49,090	2,030	1,770	52,890
3610	No. 8 turnout relay						
3620	90 lb	EA	32.000	28,080	2,030	1,770	31,880
3640	100 lb	"	32.000	29,520	2,030	1,770	33,320

© 2020 BNi Publications Inc. ALL RIGHTS RESERVED

BASE COURSES AND BALLASTS

	Description	Output		Unit Costs			
ID Code	Component Descriptions	Unit of Meas.	Manhr / Unit	Material Cost	Labor Cost	Equipment Cost	Total Cost
34 - 11130	**RAILROAD BALLAST, RAIL, APPURTENANCES, Cont'd...34 - 11130**						
3650	110 lb	EA	32.000	31,020	2,030	1,770	34,820
3660	115 lb	"	32.000	34,520	2,030	1,770	38,320
3700	132 lb	"	32.000	36,710	2,030	1,770	40,510
3800	Railroad crossings, asphalt, based on 8" thick x 20'						
3900	Including track and approach						
4020	12' roadway	EA	6.000	910	380	330	1,620
4040	15' roadway	"	6.857	1,070	430	380	1,880
4060	18' roadway	"	8.000	1,250	510	440	2,200
4080	21' roadway	"	9.600	1,400	610	530	2,540
4100	24' roadway	"	12.000	1,580	760	670	3,010
4200	Precast concrete inserts						
4420	12' roadway	EA	2.400	1,450	150	130	1,740
4440	15' roadway	"	3.000	1,740	190	170	2,100
4460	18' roadway	"	4.000	2,080	250	220	2,560
4480	21' roadway	"	4.800	2,680	300	270	3,250
4500	24' roadway	"	5.333	3,240	340	300	3,870
4600	Molded rubber, with headers						
4820	12' roadway	EA	2.400	8,340	150	130	8,630
4840	15' roadway	"	3.000	10,440	190	170	10,800
4860	18' roadway	"	4.000	12,360	250	220	12,840
4880	21' roadway	"	4.800	13,630	300	270	14,200
4900	24' roadway	"	5.333	16,610	340	300	17,240

© 2020 BNi Publications Inc. ALL RIGHTS RESERVED

DIVISION 41
HANDLING EQUIPMENT

HOISTS AND CRANES

	Description	Output		Unit Costs			
ID Code	Component Descriptions	Unit of Meas.	Manhr / Unit	Material Cost	Labor Cost	Equipment Cost	Total Cost
41 - 22133	**INDUSTRIAL HOISTS**						**41 - 22133**
1000	Industrial hoists, electric, light to medium duty						
1010	500 lb	EA	4.000	9,870	330		10,200
1020	1000 lb	"	4.211	10,400	350		10,750
1030	2000 lb	"	4.444	10,860	370		11,230
1040	3000 lb	"	4.706	11,240	390		11,630
1050	4000 lb	"	5.000	11,850	420		12,270
1060	5000 lb	"	5.333	14,200	440		14,640
1070	6000 lb	"	5.517	16,100	460		16,560
1080	7500 lb	"	5.714	18,230	470		18,700
1090	10,000 lb	"	5.926	45,950	490		46,440
1100	15,000 lb	"	6.154	57,640	510		58,150
1110	20,000 lb	"	6.667	68,200	550		68,750
1120	25,000 lb	"	7.273	71,160	600		71,760
1130	30,000 lb	"	8.000	74,270	660		74,930
1200	Heavy duty						
1210	500 lb	EA	4.000	16,010	330		16,340
1220	1000 lb	"	4.211	22,400	350		22,750
1240	2000 lb	"	4.444	24,760	370		25,130
1250	3000 lb	"	4.706	25,670	390		26,060
1260	4000 lb	"	5.000	26,810	420		27,230
1270	5000 lb	"	5.333	27,340	440		27,780
1280	6000 lb	"	5.517	29,620	460		30,080
1290	7500 lb	"	5.714	34,020	470		34,490
1300	10,000 lb	"	5.926	36,300	490		36,790
1310	15,000 lb	"	6.154	44,050	510		44,560
1320	20,000 lb	"	6.667	53,460	550		54,010
1330	25,000 lb	"	7.273	60,000	600		60,600
1340	30,000 lb	"	8.000	66,680	660		67,340
1450	Air powered hoists						
1460	500 lb	EA	4.000	9,270	330		9,600
1470	1000 lb	"	4.000	9,720	330		10,050
1480	2000 lb	"	4.211	9,870	350		10,220
1490	4000 lb	"	4.706	10,780	390		11,170
1500	6000 lb	"	6.154	11,620	510		12,130
2000	Overhead traveling bridge crane						
2010	Single girder, 20' span						
2030	3 ton	EA	12.000	25,300	760	670	26,730
2040	5 ton	"	12.000	29,050	760	670	30,480

© 2020 BNi Publications Inc. ALL RIGHTS RESERVED

HOISTS AND CRANES

	Description	Output		Unit Costs			
ID Code	Component Descriptions	Unit of Meas.	Manhr / Unit	Material Cost	Labor Cost	Equipment Cost	Total Cost
41 - 22133	**INDUSTRIAL HOISTS, Cont'd...**						**41 - 22133**
2045	7.5 ton	EA	12.000	36,490	760	670	37,920
2050	10 ton	"	15.000	36,790	950	830	38,570
2060	15 ton	"	15.000	45,380	950	830	47,160
2080	30' span						
2090	3 ton	EA	12.000	31,280	760	670	32,710
2100	5 ton	"	12.000	37,870	760	670	39,300
2120	10 ton	"	15.000	52,740	950	830	54,520
2130	15 ton	"	15.000	59,640	950	830	61,420
2170	Double girder, 40' span						
2180	3 ton	EA	26.667	49,230	1,690	1,480	52,400
2190	5 ton	"	26.667	51,820	1,690	1,480	54,990
2195	7.5 ton	"	26.667	51,280	1,690	1,480	54,450
2200	10 ton	"	34.286	64,310	2,170	1,900	68,380
2210	15 ton	"	34.286	70,680	2,170	1,900	74,750
2220	25 ton	"	34.286	122,420	2,170	1,900	126,490
2230	50' span						
2250	3 ton	EA	26.667	58,560	1,690	1,480	61,730
2260	5 ton	"	26.667	59,790	1,690	1,480	62,960
2265	7.5 ton	"	26.667	61,480	1,690	1,480	64,650
2270	10 ton	"	34.286	67,300	2,170	1,900	71,370
2280	15 ton	"	34.286	82,790	2,170	1,900	86,860
2290	25 ton	"	34.286	109,230	2,170	1,900	113,300
41 - 22135	**JIB CRANES**						**41 - 22135**
0100	Self supporting, swinging 8' boom, 200 deg rotation						
0120	1000 lb	EA	6.667	4,710	600		5,310
0140	2000 lb	"	6.667	5,160	600		5,760
0160	3000 lb	"	13.333	5,540	1,200		6,740
0180	4000 lb	"	13.333	6,070	1,200		7,270
0200	6000 lb	"	13.333	7,360	1,200		8,560
0220	10,000 lb	"	13.333	10,400	1,200		11,600
0230	Wall mounted, 180 deg rotation						
0240	2000 lb	EA	6.667	2,510	600		3,110
0260	3000 lb	"	6.667	3,040	600		3,640
0280	4000 lb	"	13.333	1,900	1,200		3,100
0300	6000 lb	"	13.333	3,800	1,200		5,000
0320	10,000 lb	"	13.333	7,210	1,200		8,410

© 2020 BNi Publications Inc. ALL RIGHTS RESERVED

Assemblies

The following Assemblies Tables list dozens of commonly specified types of construction systems, each with all of the details that make up its costs. These assemblies are developed by using the unit price data from the front of this manual, provide a bridge between unit prices and square foot costs, and can be used in conjunction with either. The data has been updated to reflect current construction costs.

BUILDING CATEGORY	PAGE

BNi® Building News
A DESIGN COST DATA COMPANY
DATA YOU CAN TRUST

A SUBSTRUCTURE

A1010 FOUNDATIONS

A1010.15 FOOTINGS

A1010.15.10 Strip Footing, 1' thick 2' wide — Unit: LF

		Quan.	Unit	Material	Labor	Equip.	Total
031124201220	Keyway forms (5 uses),2x4	1	LF	$0.31	$3.26	$0.00	$3.57
032100605000	Reinforcing,straight dowels,24" long,1" dia	1	EA	$5.83	$1.66	$0.00	$7.49
032100701020	Reinforcing,foundations,#5-#6	0.005	TON	$8.25	$4.76	$0.00	$13.01
033100502000	Place concrete,footing,3500# or 4000#,by chute	0.07	CY	$9.83	$1.19	$0.00	$11.02
033500101040	Concrete finish,broom,darby	2	SF	$0.00	$1.28	$0.00	$1.28
312316502220	Hand compaction of backfill,with air tamper	0.2	CY	$0.00	$7.30	$0.00	$7.30
312333601140	Trenching,hydraulic excavator,1/2 cy,medium soil	0.75	CY	$0.00	$1.84	$2.40	$4.24
031113602090	Job built,wall footing forms,continuous,5 uses	2	SF	$1.91	$10.87	$0.00	$12.78
				$26.13	**$32.16**	**$2.40**	**$60.69**

A1010.15.15 Strip Footing, 18" thick 3' wide — Unit: LF

		Quan.	Unit	Material	Labor	Equip.	Total
031113602090	Job built,wall footing forms,continuous,5 uses	3	SF	$2.87	$16.31	$0.00	$19.18
312333601140	Trenching,hydraulic excavator,1/2 cy,medium soil	1.125	CY	$0.00	$2.76	$3.60	$6.36
312316502220	Hand compaction of backfill,with air tamper	0.4	CY	$0.00	$14.59	$0.00	$14.59
033500101040	Concrete finish,broom,darby	3	SF	$0.00	$1.92	$0.00	$1.92
033100502000	Place concrete,footing,3500# or 4000#,by chute	0.15	CY	$21.07	$2.55	$0.00	$23.62
032100701020	Reinforcing,foundations,#5-#6	0.01	TON	$16.50	$9.51	$0.00	$26.01
031124201220	Keyway forms (5 uses),2x4	1	LF	$0.31	$3.26	$0.00	$3.57
032100605000	Reinforcing,straight dowels,24" long,1" dia	1	EA	$5.83	$1.66	$0.00	$7.49
				$46.58	**$52.56**	**$3.60**	**$102.74**

A1010.20 FOUNDATION WALLS

A1010.20.10 Foundation Wall 6' high, 1' thick — Unit: SF

		Quan.	Unit	Material	Labor	Equip.	Total
032101101040	Reinforcing,wall,#7-#8	0.003	TON	$4.71	$2.22	$0.00	$6.93
071113101020	Dampproofing,asphalt,troweled,1 coat	2	SF	$1.36	$3.41	$0.00	$4.77
033500104160	Break ties and patch holes	2	SF	$0.00	$2.04	$0.00	$2.04
033500104020	Wall finishes,burlap rub,w/cement paste	2	SF	$0.26	$1.70	$0.00	$1.96
031114103190	Job built,ext wall forms,8' high,5 uses	2	SF	$2.66	$10.87	$0.00	$13.53
031003004030	Snap ties,long-end with washers, 12" long	1	EA	$2.26	$0.00	$0.00	$2.26
033100901040	Place wall conc,2500# or 3000#,to 8',by pump	0.037	CY	$5.17	$1.89	$1.30	$8.36
				$16.42	**$22.13**	**$1.30**	**$39.85**

A1010.20.15 Foundation Wall 10' high, 18" thick — Unit: SF

		Quan.	Unit	Material	Labor	Equip.	Total
033500104020	Wall finishes,burlap rub,w/cement paste	2	SF	$0.26	$1.70	$0.00	$1.96
033500104160	Break ties and patch holes	2	SF	$0.00	$2.04	$0.00	$2.04
033100901040	Place wall conc,2500# or 3000#,to 8',by pump	0.055	CY	$7.69	$2.81	$1.92	$12.42
032101101040	Reinforcing,wall,#7-#8	0.005	TON	$7.86	$3.70	$0.00	$11.56
031114103290	Wall forms,ext,job built,over 8' high wall,5 uses	2	SF	$3.25	$13.05	$0.00	$16.30
031003004030	Snap ties,long-end with washers, 12" long	1	EA	$2.26	$0.00	$0.00	$2.26
071113101020	Dampproofing,asphalt,troweled,1 coat	2	SF	$1.36	$3.41	$0.00	$4.77
				$22.68	**$26.71**	**$1.92**	**$51.31**

A SUBSTRUCTURE

A1010 FOUNDATIONS

A1010.25 SPREAD FOOTINGS

		Quan.	Unit	Material	Labor	Equip.	Total
A1010.25.10	**Spread Footing 6' by 6' by 2' thick**						**EA**
031113903100	Job built,mat foundation forms,5 uses	48	SF	$46.56	$313.15	$0.00	$359.71
032100605000	Reinforcing,straight dowels,24" long,1" dia	4	EA	$23.32	$6.66	$0.00	$29.98
032100901040	Reinforcing,slab,bars,#7-#8	0.147	TON	$230.99	$122.38	$0.00	$353.37
033100805060	Foundation mat,2500#-3000# conc,over 20 cy,pump	2.677	CY	$375.97	$51.28	$35.14	$462.39
033500101040	Concrete finish,broom,darby	36	SF	$0.00	$22.99	$0.00	$22.99
				$676.84	**$516.46**	**$35.14**	**$1,228.44**

		Quan.	Unit	Material	Labor	Equip.	Total
A1010.25.20	**Spread Footing 12' x 12' by 3' thick**						**EA**
031113903100	Job built,mat foundation forms,5 uses	72	SF	$69.84	$469.73	$0.00	$539.57
033500101040	Concrete finish,broom,darby	36	SF	$0.00	$22.99	$0.00	$22.99
033100805060	Foundation mat,2500#-3000# conc,over 20 cy,pump	4	CY	$561.77	$76.62	$52.50	$690.89
032100901040	Reinforcing,slab,bars,#7-#8	0.221	TON	$347.26	$183.98	$0.00	$531.24
032100605000	Reinforcing,straight dowels,24" long,1" dia	4	EA	$23.32	$6.66	$0.00	$29.98
				$1,002.19	**$759.98**	**$52.50**	**$1,814.67**

A1010.30 PILE CAPS

		Quan.	Unit	Material	Labor	Equip.	Total
A1010.30.20	**Pile Cap 4' by 4' by 2' thick**						**EA**
031113801600	Pile cap forms,job built,square/rect,5 uses	32	SF	$41.24	$208.77	$0.00	$250.01
032100605000	Reinforcing,straight dowels,24" long,1" dia	4	EA	$23.32	$6.66	$0.00	$29.98
032101601060	Reinforcing,pile caps,#9-#10	0.09	TON	$141.42	$108.98	$0.00	$250.40
033100702020	Pile cap concrete,3500# or 4000#,by pump	1.18	CY	$175.03	$51.66	$35.40	$262.09
033500101040	Concrete finish,broom,darby	16	SF	$0.00	$10.22	$0.00	$10.22
				$381.01	**$386.29**	**$35.40**	**$802.70**

		Quan.	Unit	Material	Labor	Equip.	Total
A1010.30.25	**Pile Cap 4' by 8' by 18" thick**						**EA**
031113801600	Pile cap forms,job built,square/rect,5 uses	36	SF	$46.39	$234.86	$0.00	$281.25
033500101040	Concrete finish,broom,darby	32	SF	$0.00	$20.43	$0.00	$20.43
033100702020	Pile cap concrete,3500# or 4000#,by pump	1.77	CY	$262.54	$77.50	$53.10	$393.14
032100605000	Reinforcing,straight dowels,24" long,1" dia	4	EA	$23.32	$6.66	$0.00	$29.98
032101601060	Reinforcing,pile caps,#9-#10	0.09	TON	$141.42	$108.98	$0.00	$250.40
				$473.67	**$448.43**	**$53.10**	**$975.20**

A SUBSTRUCTURE

A1020 SPECIAL FOUNDATIONS

A1020.10 PILES, CONCRETE

		Quan.	Unit	Material	Labor	Equip.	Total
A1020.10.10	**Prestressed concrete piling, less than 60', 12" sq .....Unit: LF**						
316313501002	Prestressed concrete piling,less than 60',12" sq	1	LF	$29.08	$3.40	$4.00	$36.48
A1020.10.15	**Prestressed concrete piling, less than 60', 18" sq .....Unit: LF**						
316313501008	Prestressed concrete piling,less than 60',18" sq	1	LF	$51.49	$3.82	$4.49	$59.80
A1020.10.20	**Prestressed concrete piling, less than 60', 24" sq .....Unit: LF**						
316313501012	Prestressed concrete piling,less than 60',24" sq	1	LF	$89.08	$4.01	$4.72	$97.81
A1020.10.25	**Prestressed conc piling, more than 60' long, 12" sq .....Unit: LF**						
316313501120	Prestressed conc piling,more than 60' long,12" sq	1	LF	$29.67	$2.80	$3.29	$35.76
A1020.10.30	**Prestressed conc piling, more than 60' long, 18" sq .....Unit: LF**						
316313501180	Prestressed conc piling,more than 60' long,18" sq	1	LF	$51.69	$2.95	$3.47	$58.11
A1020.10.35	**Prestressed conc piling, more than 60' long, 24" sq .....Unit: LF**						
316313501220	Prestressed conc piling,more than 60' long,24" sq	1	LF	$83.29	$3.07	$3.61	$89.97
A1020.10.40	**Prestressed, strt cyl, less than 60', 12" dia .....Unit: LF**						
316313501500	Prestressed,strt cyl,less than 60',12" dia	1	LF	$27.05	$3.56	$4.18	$34.79
A1020.10.45	**Prestressed, strt cyl, less than 60', 18" dia .....Unit: LF**						
316313501580	Prestressed,strt cyl,less than 60',18" dia	1	LF	$56.42	$3.82	$4.49	$64.73
A1020.10.50	**Prestressed, strt cyl, less than 60', 24" dia .....Unit: LF**						
316313501620	Prestressed,strt cyl,less than 60',24" dia	1	LF	$82.70	$4.01	$4.72	$91.43
A1020.10.55	**Prestressed strt cyl, more than 60' long, 12" dia .....Unit: LF**						
316313501700	Prestressed strt cyl,more than 60' long,12" dia	1	LF	$27.05	$2.85	$3.35	$33.25
A1020.10.60	**Prestressed strt cyl, more than 60' long, 18" dia .....Unit: LF**						
316313501760	Prestressed strt cyl,more than 60' long,18" dia	1	LF	$56.42	$3.01	$3.54	$62.97
A1020.10.65	**Prestressed strt cyl, more than 60' long, 24" dia .....Unit: LF**						
316313501800	Prestressed strt cyl,more than 60' long,24" dia	1	LF	$84.35	$3.13	$3.68	$91.16
A1020.12.10	**Mobilization, equip, pile driving rig, minimum ..... EA**						
015400801360	Mobilization, equip, pile driving rig, minimum	1.000	EA	$11,102.78	$0.00	$0.00	$11,102.78
A1020.12.15	**Mobilization, equip, pile driving rig, average ..... EA**						
015400801380	Mobilization, equip, pile driving rig, average	1.000	EA	$22,207.65	$0.00	$0.00	$22,207.65
A1020.12.20	**Mobilization, equip, pile driving rig, maximum ..... EA**						
015400801400	Mobilization, equip, pile driving rig, maximum	1.000	EA	$39,973.66	$0.00	$0.00	$39,973.66
A1020.14.10	**Pile test, 50 ton to 100 ton ..... EA**						
316200101020	Pile test, 50 ton to 100 ton	1.000	EA	$20,674.50	$0.00	$0.00	$20,674.50
A1020.14.15	**Pile test, 50 ton to 300 ton ..... EA**						
316200101060	Pile test, 50 ton to 300 ton	1.000	EA	$33,726.00	$0.00	$0.00	$33,726.00
A1020.14.20	**Pile test, 50 ton to 600 ton ..... EA**						
316200101100	Pile test, 50 ton to 600 ton	1.000	EA	$49,376.25	$0.00	$0.00	$49,376.25

A SUBSTRUCTURE

A1020 SPECIAL FOUNDATIONS

A1020.15 PILES, STEEL

		Quan.	Unit	Material	Labor	Equip.	Total
A1020.15.10	**Friction piles, stl casing, w/4000 psi conc, 12" dia ....Unit: LF**						
316216505040	Friction piles,stl casing,w/4000 psi conc,12" dia	1	LF	$21.53	$3.91	$4.60	$30.04
A1020.15.15	**Friction piles, stl casing, w/4000 psi conc, 18" dia ....Unit: LF**						
316216505100	Friction piles,stl casing,w/4000 psi conc,18" dia	1	LF	$33.45	$4.61	$5.41	$43.47
A1020.15.20	**Concrete filled, 3000#, up to 40', 12" dia ....Unit: LF**						
316216601140	Concrete filled,3000#,up to 40',12" dia	1	LF	$35.90	$6.02	$7.08	$49.00
A1020.15.25	**Concrete filled, 3000#, up to 40', 18" dia ....Unit: LF**						
316216601200	Concrete filled,3000#,up to 40',18" dia	1	LF	$62.16	$6.81	$8.00	$76.97
A1020.15.30	**Pipe piles, non-filled, 12" dia ....Unit: LF**						
316216602060	Pipe piles,non-filled,12" dia	1	LF	$33.34	$4.61	$5.41	$43.36
A1020.15.35	**Pipe piles, non-filled, 18" dia ....Unit: LF**						
316216602120	Pipe piles,non-filled,18" dia	1	LF	$52.30	$5.22	$6.13	$63.65
A1020.15.40	**Pipe piles, standard point, 12" dia ....Unit: EA**						
316216602760	Pipe piles,standard point,12" dia	1	EA	$183.92	$127.70	$0.00	$311.62
A1020.15.45	**Pipe piles, standard point, 18" dia ....Unit: EA**						
316216602820	Pipe piles,standard point,18" dia	1	EA	$356.34	$170.27	$0.00	$526.61
A1020.15.50	**Pipe piles, heavy duty point, 12" dia ....Unit: EA**						
316216602940	Pipe piles,heavy duty point,12" dia	1	EA	$344.85	$170.27	$0.00	$515.12
A1020.15.55	**Pipe piles, heavy duty point, 18" dia ....Unit: EA**						
316216603000	Pipe piles,heavy duty point,18" dia	1	EA	$517.28	$204.32	$0.00	$721.60
A1020.16.10	**Mobilization, equip, pile driving rig, minimum .... EA**						
015400801360	Mobilization, equip, pile driving rig, minimum	1.000	EA	$11,102.78	$0.00	$0.00	$11,102.78
A1020.16.15	**Mobilization, equip, pile driving rig, average .... EA**						
015400801380	Mobilization, equip, pile driving rig, average	1.000	EA	$22,207.65	$0.00	$0.00	$22,207.65
A1020.16.20	**Mobilization, equip, pile driving rig, maximum .... EA**						
015400801400	Mobilization, equip, pile driving rig, maximum	1.000	EA	$39,973.66	$0.00	$0.00	$39,973.66
A1020.17.10	**Pile test, 50 ton to 100 ton .... EA**						
316200101020	Pile test, 50 ton to 100 ton	1.000	EA	$20,674.50	$0.00	$0.00	$20,674.50
A1020.17.15	**Pile test, 50 ton to 300 ton .... EA**						
316200101060	Pile test, 50 ton to 300 ton	1.000	EA	$33,726.00	$0.00	$0.00	$33,726.00
A1020.17.20	**Pile test, 50 ton to 600 ton .... EA**						
316200101100	Pile test, 50 ton to 600 ton	1.000	EA	$49,376.25	$0.00	$0.00	$49,376.25
A1020.20.10	**Treated wood piles, 12" butt, 8" tip, 25' long .... LF**						
316219000100	Treated wood piles, 12" butt, 8" tip, 25' long	1.000	LF	$18.10	$7.56	$8.72	$34.38

All prices are updated to January 1, 2021 and are national averages.
For a more in-depth information contact Design, Cost and Data at 800-533-5680, or go to www.DCD.com

A SUBSTRUCTURE

A1020 SPECIAL FOUNDATIONS

A1020.20 PILES, WOOD

		Quan.	Unit	Material	Labor	Equip.	Total
A1020.20.15	**Treated wood piles, 12" butt, 8" tip, 40' long .... Unit: LF**						
316219000125	Treated wood piles,12" butt,8" tip,40' long	1	LF	$19.28	$4.89	$5.75	$29.92
A1020.20.20	**Treated wood piles, 12" butt, 7" tip, 40' long .... Unit: LF**						
316219000130	Treated wood piles,12" butt,7" tip,40' long	1	LF	$21.74	$4.89	$5.75	$32.38
A1020.20.25	**Treated wood piles, 12" butt, 7" tip, 60' long .... Unit: LF**						
316219000160	Treated wood piles,12" butt,7" tip,60' long	1	LF	$24.79	$3.26	$3.83	$31.88
A1020.21.10	**Mobilization, equip, pile driving rig, minimum .... EA**						
015400801360	Mobilization, equip, pile driving rig, minimum	1.000	EA	$11,102.78	$0.00	$0.00	$11,102.78
A1020.21.15	**Mobilization, equip, pile driving rig, average .... EA**						
015400801380	Mobilization, equip, pile driving rig, average	1.000	EA	$22,207.65	$0.00	$0.00	$22,207.65
A1020.21.20	**Mobilization, equip, pile driving rig, maximum .... EA**						
015400801400	Mobilization, equip, pile driving rig, maximum	1.000	EA	$39,973.66	$0.00	$0.00	$39,973.66
A1020.22.10	**Pile test, 50 ton to 100 ton .... EA**						
316200101020	Pile test, 50 ton to 100 ton	1.000	EA	$20,674.50	$0.00	$0.00	$20,674.50
A1020.22.15	**Pile test, 50 ton to 300 ton .... EA**						
316200101060	Pile test, 50 ton to 300 ton	1.000	EA	$33,726.00	$0.00	$0.00	$33,726.00
A1020.22.20	**Pile test, 50 ton to 600 ton .... EA**						
316200101100	Pile test, 50 ton to 600 ton	1.000	EA	$49,376.25	$0.00	$0.00	$49,376.25

A1020.25 CAISSONS

		Quan.	Unit	Material	Labor	Equip.	Total
A1020.25.10	**Caisson, incl 3000# conc, in stable ground, 18" dia .... Unit: LF**						
316400101020	Caisson,incl 3000# conc,in stable ground,18" dia	1	LF	$17.77	$15.66	$18.40	$51.83
A1020.25.15	**Caisson, incl 3000# conc, in stable ground, 24" dia .... Unit: LF**						
316400101040	Caisson,incl 3000# conc,in stable ground,24" dia	1	LF	$28.42	$16.31	$19.17	$63.90
A1020.25.20	**Caisson, incl 3000# conc, in stable ground, 36" dia .... Unit: LF**						
316400101080	Caisson,incl 3000# conc,in stable ground,36" dia	1	LF	$60.57	$22.37	$26.29	$109.23
A1020.25.25	**Caisson, incl 3000# conc, in stable ground, 48" dia .... Unit: LF**						
316400101100	Caisson,incl 3000# conc,in stable ground,48" dia	1	LF	$108.20	$26.10	$30.67	$164.97
A1020.27.15	**Test, 50 ton to 300 ton .... EA**						
316200101060	Pile test, 50 ton to 300 ton	1.000	EA	$33,726.00	$0.00	$0.00	$33,726.00
A1020.27.20	**Test, 50 ton to 600 ton .... EA**						
316200101100	Pile test, 50 ton to 600 ton	1.000	EA	$49,376.25	$0.00	$0.00	$49,376.25
A1020.21.15	**Mobilization, equip, rig, average .... EA**						
015400801380	Mobilization, equip, pile driving rig, average	1.000	EA	$22,207.65	$0.00	$0.00	$22,207.65
A1020.21.20	**Mobilization, equip, rig, maximum .... EA**						
015400801400	Mobilization, equip, pile driving rig, maximum	1.000	EA	$39,973.66	$0.00	$0.00	$39,973.66

A SUBSTRUCTURE

A1020 SPECIAL FOUNDATIONS

A1020.30 GRADE BEAM

		Quan.	Unit	Material	Labor	Equip.	Total
A1020.30.10	**GRADE BEAM, 1' Wide x 2' High .... LF**						
032100801040	Reinforcing,grade beams,#7-#8	0.05	TON	$78.57	$39.18	$0.00	$117.75
312333603820	Spread dumped fill or gravel,by hand w/air tamp	6	SY	$0.00	$6.13	$2.04	$8.17
312333603740	Spread dumped fill or gravel,no comp,6" layers	2	SY	$0.00	$1.69	$2.00	$3.69
312333601140	Trenching,hydraulic excavator,1/2 cy,medium soil	0.75	CY	$0.00	$1.84	$2.40	$4.24
033500104160	Break ties and patch holes	2	SF	$0.00	$2.04	$0.00	$2.04
033100601120	Place conc,grade beam,3000# or 4000#,by pump	0.07	CY	$10.38	$2.68	$1.84	$14.90
031113701100	Job built,grade beam forms,5 uses	4	SF	$4.10	$21.75	$0.00	$25.85
031003004030	Snap ties,long-end with washers, 12" long	1	EA	$2.26	$0.00	$0.00	$2.26
033500104020	Wall finishes,burlap rub,w/cement paste	4	SF	$0.53	$3.41	$0.00	$3.94
				$95.84	**$78.72**	**$8.28**	**$182.84**
A1020.30.25	**GRADE BEAM, 18" Wide x 3' High .... Unit: LF**						
033500104160	Break ties and patch holes	2	SF	$0.00	$2.04	$0.00	$2.04
031113701100	Job built,grade beam forms,5 uses	6	SF	$6.15	$32.62	$0.00	$38.77
312333603820	Spread dumped fill or gravel,by hand w/air tamp	6	SY	$0.00	$6.13	$2.04	$8.17
312333603740	Spread dumped fill or gravel,no comp,6" layers	2	SY	$0.00	$1.69	$2.00	$3.69
312333601140	Trenching,hydraulic excavator,1/2 cy,medium soil	1.8	CY	$0.00	$4.42	$5.76	$10.18
032100801040	Reinforcing,grade beams,#7-#8	0.12	TON	$188.56	$94.02	$0.00	$282.58
031003004030	Snap ties,long-end with washers, 12" long	1	EA	$2.26	$0.00	$0.00	$2.26
033100601120	Place conc,grade beam,3000# or 4000#,by pump	0.18	CY	$26.70	$6.90	$4.72	$38.32
033500104020	Wall finishes,burlap rub,w/cement paste	6	SF	$0.79	$5.11	$0.00	$5.90
				$224.46	**$152.93**	**$14.52**	**$391.91**

All prices are updated to January 1, 2021 and are national averages.
For a more in-depth information contact Design, Cost and Data at 800-533-5680, or go to www.DCD.com

A SUBSTRUCTURE

A1030 SLAB ON GRADE

A1030.10 SLAB ON GRADE

		Quan.	Unit	Material	Labor	Equip.	Total
A1030.10.05	**SLAB ON GRADE, 4" thick, 20' by 20' ....Unit: SF**						
033500101020	Concrete finish,broom,screed	1	SF	$0.00	$0.64	$0.00	$0.64
312313104030	Prepare and roll subbase,average	0.111	SY	$0.00	$0.05	$0.06	$0.11
312313103020	Base course,bank run gravel,4" deep	0.111	SY	$0.35	$0.04	$0.05	$0.44
033500101060	Concrete finish,float	1	SF	$0.00	$0.85	$0.00	$0.85
033100802040	Place conc,slab/mat,3500# or 4000#,by pump	0.013	CY	$1.93	$0.28	$0.19	$2.40
032100901000	Reinforcing,slab,bars,#3-#4	0.001	TON	$1.88	$1.11	$0.00	$2.99
031124205040	Metal formwork,straight edge,4" high	0.1	LF	$2.35	$0.41	$0.00	$2.76
072600101010	Vapor barrier,polyethylene,6 mil	1	SF	$0.08	$0.26	$0.00	$0.34
				$6.59	**$3.64**	**$0.30**	**$10.53**

		Quan.	Unit	Material	Labor	Equip.	Total
A1030.10.10	**SLAB ON GRADE, 6" thick, 20' by 20' ....Unit: SF**						
032100901020	Reinforcing,slab,bars,#5-#6	0.002	TON	$3.31	$1.90	$0.00	$5.21
312313103020	Base course,bank run gravel,4" deep	0.111	SY	$0.35	$0.04	$0.05	$0.44
312313104030	Prepare and roll subbase,average	0.111	SY	$0.00	$0.05	$0.06	$0.11
072600101010	Vapor barrier,polyethylene,6 mil	1	SF	$0.08	$0.26	$0.00	$0.34
033100802040	Place conc,slab/mat,3500# or 4000#,by pump	0.019	CY	$2.82	$0.42	$0.28	$3.52
031113904004	Mat foundation form,job built,edge,6" high,5 use	0.1	LF	$0.10	$0.50	$0.00	$0.60
033500101060	Concrete finish,float	1	SF	$0.00	$0.85	$0.00	$0.85
033500101020	Concrete finish,broom,screed	1	SF	$0.00	$0.64	$0.00	$0.64
				$6.66	**$4.66**	**$0.39**	**$11.71**

		Quan.	Unit	Material	Labor	Equip.	Total
A1030.10.20	**SLAB ON GRADE, 8" thick, 20' by 20' ....Unit: SF**						
033500101060	Concrete finish,float	1	SF	$0.00	$0.85	$0.00	$0.85
312313104030	Prepare and roll subbase,average	0.111	SY	$0.00	$0.05	$0.06	$0.11
072600101010	Vapor barrier,polyethylene,6 mil	1	SF	$0.08	$0.26	$0.00	$0.34
033500101020	Concrete finish,broom,screed	1	SF	$0.00	$0.64	$0.00	$0.64
033100802040	Place conc,slab/mat,3500# or 4000#,by pump	0.025	CY	$3.71	$0.55	$0.38	$4.64
032100901020	Reinforcing,slab,bars,#5-#6	0.003	TON	$4.97	$2.85	$0.00	$7.82
031113904014	Mat foundation form,job built,edge,12" high,5 use	0.1	LF	$0.09	$0.54	$0.00	$0.63
312313103020	Base course,bank run gravel,4" deep	0.111	SY	$0.35	$0.04	$0.05	$0.44
				$9.20	**$5.78**	**$0.49**	**$15.47**

B SHELL

B SHELL

B1010 FLOOR CONSTRUCTION

B1010.10 COLUMNS, CONCRETE

		Quan.	Unit	Material	Labor	Equip.	Total
B1010.10.10	**C.I.P. COLUMNS, SQUARE, 12" by 12", 10' high .... EA**						
033500104020	Wall finishes,burlap rub,w/cement paste	40	SF	$5.27	$34.05	$0.00	$39.32
031113201290	Job built,column forms,12"x 12",5 uses	40	SF	$52.89	$414.22	$0.00	$467.11
031124202070	Chamfer strips,wood,1" wide	40	LF	$20.21	$57.99	$0.00	$78.20
032100301020	Reinforcing,columns,#9-#10	0.086	TON	$134.61	$104.14	$0.00	$238.75
033100202020	Place concrete,columns,2500#-3500#,by pump	0.37	CY	$54.89	$18.90	$12.95	$86.74
				$267.87	**$629.30**	**$12.95**	**$910.12**
B1010.10.15	**C.I.P. COLUMNS, ROUND, 1' dia., 10' high .... EA**						
033500104020	Wall finishes,burlap rub,w/cement paste	31.4	SF	$4.14	$26.73	$0.00	$30.87
033100202020	Place concrete,columns,2500#-3500#,by pump	0.291	CY	$43.17	$14.86	$10.18	$68.21
031113202060	Column,round fiber form,1 use,12" dia	10	LF	$71.32	$133.14	$0.00	$204.46
032100301020	Reinforcing,columns,#9-#10	0.086	TON	$134.61	$104.14	$0.00	$238.75
				$253.24	**$278.87**	**$10.18**	**$542.29**

B1010.20 COLUMNS, STEEL

		Quan.	Unit	Material	Labor	Equip.	Total
B1010.20.10	**COLUMNS, W Series (Wide Flange), 8" by 8", 10' high .... EA**						
051200100140	Beams and girders,A-36,bolted	0.15	TON	$461.18	$53.38	$62.73	$577.29
051200101040	Column base plates,over 150 lb each	150	LB	$212.85	$90.00	$0.00	$302.85
078100101040	Fireproofing,sprayed on,1" thick,on columns	3.33	SF	$3.04	$3.40	$0.00	$6.44
				$677.07	**$146.78**	**$62.73**	**$886.58**
B1010.20.15	**COLUMNS, W Series (Wide Flange), 12" by 12", 10' high .... EA**						
078100101040	Fireproofing,sprayed on,1" thick,on columns	5	SF	$4.57	$5.11	$0.00	$9.68
051200100140	Beams and girders,A-36,bolted	0.225	TON	$691.76	$80.07	$94.09	$865.92
051200101040	Column base plates,over 150 lb each	150	LB	$212.85	$90.00	$0.00	$302.85
				$909.18	**$175.18**	**$94.09**	**$1,178.45**

B SHELL

B1010 FLOOR CONSTRUCTION

B1010.25 COLUMNS, WOOD

		Quan.	Unit	Material	Labor	Equip.	Total
B1010.25.10	**COLUMNS, Wood 8" by 8", 10' high....EA**						
061300101440	Mill framing,column,douglas fir,8 x 8	10	LF	$102.08	$76.62	$52.50	$231.20
051200101020	Column base plates,up to 150 lb each	50	LB	$86.90	$24.00	$0.00	$110.90
060523101100	Anchor bolts,threaded,incl nuts,3/4" D,15" L	2	EA	$20.81	$8.15	$0.00	$28.96
				$209.79	**$108.77**	**$52.50**	**$371.06**
B1010.25.15	**COLUMNS, Wood 12" by 12", 10' high....EA**						
051200101020	Column base plates,up to 150 lb each	50	LB	$86.90	$24.00	$0.00	$110.90
060523101100	Anchor bolts,threaded,incl nuts,3/4" D,15" L	2	EA	$20.81	$8.15	$0.00	$28.96
061300101480	Mill framing,column,douglas fir,12 x 12	10	LF	$221.01	$85.13	$58.33	$364.47
				$328.72	**$117.28**	**$58.33**	**$504.33**

B1010.30 COLUMNS, PRECAST

		Quan.	Unit	Material	Labor	Equip.	Total
B1010.30.10	**COLUMNS Precast, 10" by 10", 10' high....EA**						
051200101020	Column base plates,up to 150 lb each	50	LB	$86.90	$24.00	$0.00	$110.90
034100200100	Prestressed concrete columns,10"x 10",10' long	1	EA	$331.23	$78.29	$92.00	$501.52
				$418.13	**$102.29**	**$92.00**	**$612.42**
B1010.30.15	**COLUMNS Precast, 12" by 12", 20' high....EA**						
051200101020	Column base plates,up to 150 lb each	50	LB	$86.90	$24.00	$0.00	$110.90
034100200240	Prestressed concrete columns,12"x 12",20' long	1	EA	$915.78	$97.86	$115.00	$1,128.64
				$1,002.68	**$121.86**	**$115.00**	**$1,239.54**

B SHELL

B1010 FLOOR CONSTRUCTION

B1010.50 FLOOR, CONCRETE

		Quan.	Unit	Material	Labor	Equip.	Total
B1010.50.10	**FLOOR SYSTEM, C.I.P. Beams & Slab 20' by 20' Bay Size.........SF**						
033100302020	Place elev slab conc,3500# or 4000#,by pump	0.031	CY	$4.60	$0.73	$0.50	$5.83
033500101020	Concrete finish,broom,screed	1	SF	$0.00	$0.64	$0.00	$0.64
032100401060	Reinforcing,elevated slab,#9-#10	0.007	TON	$10.96	$4.24	$0.00	$15.20
032100101012	Reinforcing,beams-girders,#9-#10	0.003	TON	$4.70	$2.85	$0.00	$7.55
031500106220	Reinf access,chairs,3" high,plain	2	EA	$3.41	$3.33	$0.00	$6.74
031113401100	Job built,slab forms,drop panel,5 uses	0.81	SF	$1.38	$3.64	$0.00	$5.02
031113002100	Job built,beam side forms,5 uses	0.6	SF	$0.96	$3.56	$0.00	$4.52
031113001120	Job built,beam bottom forms,5 uses	0.2	SF	$0.36	$1.86	$0.00	$2.22
015400101180	Staging/scaffolding,measured by SF surface,ave	1	SF	$1.11	$0.00	$0.00	$1.11
033500101060	Concrete finish,float	1	SF	$0.00	$0.85	$0.00	$0.85
				$27.48	**$21.70**	**$0.50**	**$49.68**

B1010.60 FLOOR, PRECAST

		Quan.	Unit	Material	Labor	Equip.	Total
B1010.60.10	**FLOOR SYSTEM, Precast Beam & Slab with Topping 20' by 20' Bay Size.........SF**						
033500101060	Concrete finish,float	1	SF	$0.00	$0.85	$0.00	$0.85
031113301080	Job built curb forms,6" high,5 uses	0.2	LF	$0.19	$1.09	$0.00	$1.28
034100300110	Prestr flat slab,6" thk,4' w,20' span,110 psf	1	SF	$20.17	$1.63	$1.92	$23.72
034100103340	Precast beams,girder,joists,5000 lb/lf LL,20' span	0.2	LF	$32.13	$1.57	$1.84	$35.54
033100302020	Place elev slab conc,3500# or 4000#,by pump	0.012	CY	$1.78	$0.28	$0.19	$2.25
032100905080	Reinforcing,slab,galv mesh,4x4,w4.0xw4.0	1.1	SF	$1.38	$0.61	$0.00	$1.99
031500106220	Reinf access,chairs,3" high,plain	1	EA	$1.71	$1.66	$0.00	$3.37
033500101040	Concrete finish,broom,darby	1	SF	$0.00	$0.64	$0.00	$0.64
				$57.36	**$8.33**	**$3.95**	**$69.64**

B1010.70 FLOOR, CONCRETE & STEEL ON WALL

		Quan.	Unit	Material	Labor	Equip.	Total
B1010.70.10	**FLOOR SYSTEM, Concrete Deck on Open Web Joists on Wall System 30' Span.........SF**						
031500106220	Reinf access,chairs,3" high,plain	1	EA	$1.71	$1.66	$0.00	$3.37
032100905080	Reinforcing,slab,galv mesh,4x4,w4.0xw4.0	1.05	SF	$1.32	$0.58	$0.00	$1.90
033100302020	Place elev slab conc,3500# or 4000#,by pump	0.012	CY	$1.78	$0.28	$0.19	$2.25
033500101040	Concrete finish,broom,darby	1	SF	$0.00	$0.64	$0.00	$0.64
033500101060	Concrete finish,float	1	SF	$0.00	$0.85	$0.00	$0.85
053100101020	Open type decking,galv,1-1/2" d,18 ga	1.05	SF	$3.88	$0.69	$0.80	$5.37
				$8.69	**$4.70**	**$0.99**	**$14.38**

B SHELL

B1010 FLOOR CONSTRUCTION

B1010.75 FLOOR, CONCRETE & W SERIES STEEL

		Quan.	Unit	Material	Labor	Equip.	Total
B1010.75.10	**FLOOR SYSTEM, Concrete on W (Wide Flange) System 20' by 20' Bay Size .... SF**						
031113301080	Job built curb forms,6" high,5 uses	0.2	LF	$0.19	$1.09	$0.00	$1.28
078100101120	Fireproofing,1-1/2" thick,on beams	1.93	SF	$3.10	$2.82	$0.00	$5.92
053100101020	Open type decking,galv,1-1/2" d,18 ga	1	SF	$3.69	$0.65	$0.77	$5.11
051200100140	Beams and girders,A-36,bolted	0.015	TON	$46.12	$5.34	$6.27	$57.73
033500101060	Concrete finish,float	1	SF	$0.00	$0.85	$0.00	$0.85
033500101040	Concrete finish,broom,darby	1	SF	$0.00	$0.64	$0.00	$0.64
033100302020	Place elev slab conc,3500# or 4000#,by pump	0.012	CY	$1.78	$0.28	$0.19	$2.25
031500106220	Reinf access,chairs,3" high,plain	1	EA	$1.71	$1.66	$0.00	$3.37
032100401020	Reinforcing,elevated slab,#5-#6	0.002	TON	$3.30	$1.48	$0.00	$4.78
				$59.89	**$14.81**	**$7.23**	**$81.93**

B1010.80 FLOOR, LAMINATED WOOD

		Quan.	Unit	Material	Labor	Equip.	Total
B1010.80.10	**FLOOR SYSTEM, Laminated Wood System 20' by 20' Bay Size.................................... SF**						
060523101284	Joist and beam hangers,18 ga,2 x 10	0.2	EA	$0.45	$1.45	$0.00	$1.90
061813103550	Laminated beams, gluelam beam, 9-1/2" wide x 30"	0.1	LF	$9.57	$0.47	$0.32	$10.36
061813103280	Laminated beams, gluelam beam, 7-1/2" wide x 18"	0.15	LF	$10.17	$0.61	$0.42	$11.20
061100301220	Floor joists,16" oc,2x10	1	SF	$1.62	$1.12	$0.00	$2.74
061600102080	Sub-flooring,plywood,CDX,3/4" thick	1	SF	$1.60	$1.09	$0.00	$2.69
				$23.41	**$4.74**	**$0.74**	**$28.89**

B1010.85 FLOOR, WOOD

		Quan.	Unit	Material	Labor	Equip.	Total
B1010.85.10	**FLOOR SYSTEM, Wooden Beams & Joists 16' by 16' Bay Size .... SF**						
060523101284	Joist and beam hangers,18 ga,2 x 10	0.25	EA	$0.57	$1.81	$0.00	$2.38
061100301220	Floor joists,16" oc,2x10	1	SF	$1.62	$1.12	$0.00	$2.74
061300101044	Mill framing,beam,douglas fir,6 x 12	0.188	LF	$2.20	$1.07	$0.73	$4.00
061300101090	Mill framing,beam,douglas fir,8 x 16	0.125	LF	$2.51	$0.77	$0.52	$3.80
061600102080	Sub-flooring,plywood,CDX,3/4" thick	1	SF	$1.60	$1.09	$0.00	$2.69
				$8.50	**$5.86**	**$1.25**	**$15.61**

All prices are updated to January 1, 2021 and are national averages.
For a more in-depth information contact Design, Cost and Data at 800-533-5680, or go to www.DCD.com

B SHELL

B1020 ROOF CONSTRUCTION

B1020.15 ROOF, CONCRETE

		Quan.	Unit	Material	Labor	Equip.	Total
B1020.15.10	**ROOF SYSTEM, W Series Structural System, Concrete Deck, 20' by 20' Bay Size ...... SF**						
031500106220	Reinf access,chairs,3" high,plain	1	EA	$1.71	$1.66	$0.00	$3.37
078100101120	Fireproofing,1-1/2" thick,on beams	1.7	SF	$2.73	$2.48	$0.00	$5.21
053100101020	Open type decking,galv,1-1/2" d,18 ga	1	SF	$3.69	$0.65	$0.77	$5.11
051200100140	Beams and girders,A-36,bolted	0.012	TON	$36.89	$4.27	$5.02	$46.18
033500101040	Concrete finish,broom,darby	1	SF	$0.00	$0.64	$0.00	$0.64
032100905080	Reinforcing,slab,galv mesh,4x4,w4.0xw4.0	1	SF	$1.26	$0.56	$0.00	$1.82
031113301080	Job built curb forms,6" high,5 uses	1	LF	$0.97	$5.44	$0.00	$6.41
033100302020	Place elev slab conc,3500# or 4000#,by pump	0.01	CY	$1.48	$0.24	$0.16	$1.88
				$48.73	**$15.94**	**$5.95**	**$70.62**

B1020.20 ROOF, CONCRETE & STEEL JOISTS ON WALLS

		Quan.	Unit	Material	Labor	Equip.	Total
B1020.20.10	**ROOF SYSTEM, Concrete Deck on Open Web Joists on Wall System 30' Span .... SF**						
032100905080	Reinforcing,slab,galv mesh,4x4,w4.0xw4.0	1.05	SF	$1.32	$0.58	$0.00	$1.90
053100101020	Open type decking,galv,1-1/2" d,18 ga	1.05	SF	$3.88	$0.69	$0.80	$5.37
052100100120	Joist,K series	0.003	TON	$6.10	$0.78	$0.92	$7.80
033500101040	Concrete finish,broom,darby	1	SF	$0.00	$0.64	$0.00	$0.64
031500106220	Reinf access,chairs,3" high,plain	1	EA	$1.71	$1.66	$0.00	$3.37
033100302020	Place elev slab conc,3500# or 4000#,by pump	0.012	CY	$1.78	$0.28	$0.19	$2.25
				$14.79	**$4.63**	**$1.91**	**$21.33**

B1020.25 ROOF, CONCRETE & STEEL JOISTS & GIRDERS

		Quan.	Unit	Material	Labor	Equip.	Total
B1020.25.10	**ROOF SYSTEM, Concrete Deck & Open Web Joists and Girders 20' by 20' Bay Size .. SF**						
052100100120	Joist,K series	0.004	TON	$8.13	$1.04	$1.23	$10.40
053100101020	Open type decking,galv,1-1/2" d,18 ga	1.05	SF	$3.88	$0.69	$0.80	$5.37
033500101040	Concrete finish,broom,darby	1	SF	$0.00	$0.64	$0.00	$0.64
033100302020	Place elev slab conc,3500# or 4000#,by pump	0.012	CY	$1.78	$0.28	$0.19	$2.25
032100905080	Reinforcing,slab,galv mesh,4x4,w4.0xw4.0	1.05	SF	$1.32	$0.58	$0.00	$1.90
031500106220	Reinf access,chairs,3" high,plain	1	EA	$1.71	$1.66	$0.00	$3.37
				$16.82	**$4.89**	**$2.22**	**$23.93**

B SHELL

B2010 EXTERIOR WALLS

B2010.10 WALL

		Quan.	Unit	Material	Labor	Equip.	Total
B2010.10.10	**WALL SYSTEM, Brick Veneer on Concrete Backup Block, 10 ' high**						**SF**
042200100160	CMU,lightweight block,hollow,load bearing,8"	1	SF	$2.74	$5.18	$0.00	$7.92
040523101120	Horizontal joint reinf,truss type,6" wide,8" wall	1	LF	$0.21	$0.26	$0.00	$0.47
040523103440	Brick anchor,corrugated,3-1/2" long,16 ga	1	EA	$0.57	$1.04	$0.00	$1.61
040523501020	Masonry flashing,0.030" elastomeric	0.1	SF	$0.13	$0.41	$0.00	$0.54
042113101020	Standard brick,red (6.4/sf),veneer	1	SF	$5.67	$10.37	$0.00	$16.04
				$9.32	**$17.26**	**$0.00**	**$26.58**
B2010.10.15	**WALL SYSTEM, Brick Veneer on Metal Stud Wall, 10 ' high**						**SF**
071100101140	Silicone dampprofing,spray on,brick,2 coats	1	SF	$1.07	$0.51	$0.00	$1.58
092116100144	Metal stud,20 ga,3-5/8",24" oc	1	SF	$0.47	$1.09	$0.00	$1.56
072116102120	Insul,wall,foil back,3",R11	1	SF	$0.72	$0.57	$0.00	$1.29
061600307000	Wall sheathing,gypsum,1/2" thick	1	SF	$0.59	$1.00	$0.00	$1.59
042113101020	Standard brick,red (6.4/sf),veneer	1	SF	$5.67	$10.37	$0.00	$16.04
040523103440	Brick anchor,corrugated,3-1/2" long,16 ga	1	EA	$0.57	$1.04	$0.00	$1.61
040523501020	Masonry flashing,0.030" elastomeric	0.1	SF	$0.13	$0.41	$0.00	$0.54
				$9.22	**$14.99**	**$0.00**	**$24.21**
B2010.10.20	**WALL SYSTEM, Stucco on Metal Stud Wall, 10 ' high**						**SF**
040523501020	Masonry flashing,0.030" elastomeric	0.1	SF	$0.13	$0.41	$0.00	$0.54
061600307000	Wall sheathing,gypsum,1/2" thick	1	SF	$0.59	$1.00	$0.00	$1.59
071113100120	Building paper,15# felt	1	SF	$0.20	$2.04	$0.00	$2.24
072116102120	Insul,wall,foil back,3",R11	1	SF	$0.72	$0.57	$0.00	$1.29
092116100144	Metal stud,20 ga,3-5/8",24" oc	1	SF	$0.47	$1.09	$0.00	$1.56
092236202400	Stucco lath,paper backed,max	1	SY	$6.22	$9.32	$0.00	$15.54
092400103000	Stucco,portland,3 coats,1" thick,sand finish	1	SY	$8.44	$26.42	$0.00	$34.86
				$16.77	**$40.85**	**$0.00**	**$57.62**
B2010.10.25	**WALL SYSTEM, C.I.P. Concrete, 10 ' high**						**SF**
033100901040	Place wall conc,2500# or 3000#,to 8',by pump	0.037	CY	$5.17	$1.89	$1.30	$8.36
033500104020	Wall finishes,burlap rub,w/cement paste	2	SF	$0.26	$1.70	$0.00	$1.96
033500104160	Break ties and patch holes	2	SF	$0.00	$2.04	$0.00	$2.04
031114103290	Wall forms,ext,job built,over 8' high wall,5 uses	2	SF	$3.25	$13.05	$0.00	$16.30
031003004030	Snap ties,long-end with washers, 12" long	1	EA	$2.26	$0.00	$0.00	$2.26
032101101040	Reinforcing,wall,#7-#8	0.005	TON	$7.86	$3.70	$0.00	$11.56
				$18.80	**$22.38**	**$1.30**	**$42.48**

B SHELL

B2010 EXTERIOR WALLS

B2010.10 WALL (Cont)

		Quan.	Unit	Material	Labor	Equip.	Total
B2010.10.30	**WALL SYSTEM, Precast wall, 8'x 20', gray cem, liner fin, 8" wall ..... SF**						
034500100160	Precast wall,8'x 20',gray cem,liner fin,8" wall	1	SF	$19.31	$1.22	$1.44	$21.97
B2010.10.35	**WALL SYSTEM, Tilt-up Concrete, 20' wide by 20' high, 12" thick ..... SF**						
033500101040	Concrete finish,broom,darby	1	SF	$0.00	$0.64	$0.00	$0.64
033500101060	Concrete finish,float	1	SF	$0.00	$0.85	$0.00	$0.85
033100802060	Place conc,slab/mat,3500# or 4000#,by hand buggy	0.025	CY	$3.71	$0.85	$0.00	$4.56
032100901040	Reinforcing,slab,bars,#7-#8	0.002	TON	$3.14	$1.66	$0.00	$4.80
031113905100	Formwork for openings,5 uses	0.16	SF	$0.23	$1.49	$0.00	$1.72
031113904014	Mat foundation form,job built,edge,12" high,5 use	0.2	LF	$0.18	$1.09	$0.00	$1.27
				$7.26	**$6.58**	**$0.00**	**$13.84**
B2010.10.40	**WALL SYSTEM, Concrete Block, 10 ' high ..... SF**						
040523101020	Standard steel bar reinforcing,horizontal,#3 - #4	2	LB	$1.28	$4.98	$0.00	$6.26
040523501020	Masonry flashing,0.030" elastomeric	0.1	SF	$0.13	$0.41	$0.00	$0.54
042200104100	CMU,hollow,split ground face,12"	1	SF	$5.28	$6.22	$0.00	$11.50
040516103060	Grouting concrete block,12"	1	SF	$2.71	$1.60	$1.40	$5.71
				$9.40	**$13.21**	**$1.40**	**$24.01**

B SHELL

B2020 EXTERIOR WINDOWS

B2020.10 STEEL WINDOWS

		Quan.	Unit	Material	Labor	Equip.	Total
B2020.10.10	**Steel windows, primed, casements, operable, minimum**						**Unit: SF**
085123101020	Steel windows,primed,casements,operable,minimum	1	SF	$54.23	$4.24	$0.00	$58.47
B2020.10.15	**Steel windows, primed, casements, operable, maximum**						**Unit: SF**
085123101040	Steel windows,primed,casements,operable,maximum	1	SF	$81.27	$4.80	$0.00	$86.07
B2020.10.20	**Steel windows, primed, casements, fixed sash**						**Unit: SF**
085123101060	Steel windows,primed,casements,fixed sash	1	SF	$43.34	$3.60	$0.00	$46.94
B2020.10.25	**Steel windows, primed, double hung**						**Unit: SF**
085123101080	Steel windows,primed,double hung	1	SF	$81.27	$4.00	$0.00	$85.27
B2020.10.30	**Industrial windows, horizontally pivoted sash**						**Unit: SF**
085123101120	Industrial windows,horizontally pivoted sash	1	SF	$68.97	$4.80	$0.00	$73.77
B2020.10.35	**Industrial windows, fixed sash**						**Unit: SF**
085123101130	Industrial windows,fixed sash	1	SF	$54.22	$4.00	$0.00	$58.22
B2020.10.40	**Industrial windows, security sash, operable**						**Unit: SF**
085123101140	Industrial windows,security sash,operable	1	SF	$86.14	$4.80	$0.00	$90.94
B2020.10.45	**Industrial windows, security sash, fixed**						**Unit: SF**
085123101150	Industrial windows,security sash,fixed	1	SF	$76.39	$4.00	$0.00	$80.39
B2020.10.50	**Picture window**						**Unit: SF**
085123101155	Picture window	1	SF	$36.98	$4.00	$0.00	$40.98
B2020.10.55	**Projecting sash window, minimum**						**Unit: SF**
085123101170	Projecting sash window,minimum	1	SF	$64.05	$4.50	$0.00	$68.55
B2020.10.60	**Projecting sash window, maximum**						**Unit: SF**
085123101180	Projecting sash window,maximum	1	SF	$78.80	$4.50	$0.00	$83.30
B2020.10.65	**Mullions**						**Unit: SF**
085123101930	Mullions	1	LF	$16.88	$3.60	$0.00	$20.48

B2020.15 ALUMINUM WINDOWS

		Quan.	Unit	Material	Labor	Equip.	Total
B2020.15.10	**Aluminum window, fixed, 6 Sf to 8 Sf**						**Unit: SF**
085113100240	Aluminum window,fixed,6 Sf to 8 Sf	1	SF	$19.21	$10.29	$0.00	$29.50
B2020.15.15	**Aluminum window, fixed 12 Sf to 16 Sf**						**Unit: SF**
085113100250	Aluminum window,fixed 12 Sf to 16 Sf	1	SF	$17.07	$8.00	$0.00	$25.07
B2020.15.20	**Aluminum window, projecting, 6 Sf to 8 Sf**						**Unit: SF**
085113100260	Aluminum window,projecting,6 Sf to 8 Sf	1	SF	$42.56	$18.00	$0.00	$60.56
B2020.15.25	**Aluminum window, projecting, 12 Sf to 16 Sf**						**Unit: SF**
085113100270	Aluminum window,projecting,12 Sf to 16 Sf	1	SF	$38.36	$12.00	$0.00	$50.36
B2020.15.30	**Aluminum window, horiz sliding, 6 Sf to 8 Sf**						**Unit: SF**
085113100280	Aluminum window,horiz sliding,6 Sf to 8 Sf	1	SF	$27.68	$9.00	$0.00	$36.68
B2020.15.35	**Aluminum window, horiz sliding, 12 Sf to 16 Sf**						**Unit: SF**
085113100290	Aluminum window,horiz sliding,12 Sf to 16 Sf	1	SF	$25.54	$7.20	$0.00	$32.74
B2020.15.40	**Aluminum window, double hung, 6 Sf to 8 Sf**						**Unit: SF**
085113101160	Aluminum window,double hung,6 Sf to 8 Sf	1	SF	$38.36	$14.40	$0.00	$52.76
B2020.15.45	**Aluminum window, double hung, 10 Sf to 12 Sf**						**Unit: SF**
085113101180	Aluminum window,double hung,10 Sf to 12 Sf	1	SF	$34.09	$12.00	$0.00	$46.09

B SHELL

B2020 EXTERIOR WINDOWS

B2020.30 STOREFRONT

		Quan.	Unit	Material	Labor	Equip.	Total
B2020.30.10	**Storefront, aluminum & glass, minimum**						**Unit: SF**
084100100140	Storefront,aluminum & glass,minimum	1	SF	$29.34	$9.00	$0.00	$38.34
B2020.30.15	**Storefront, aluminum & glass, average**						**Unit: SF**
084100100150	Storefront,aluminum & glass,average	1	SF	$43.66	$10.29	$0.00	$53.95
B2020.30.20	**Storefront, aluminum & glass, maximum**						**Unit: SF**
084100100160	Storefront,aluminum & glass,maximum	1	SF	$87.30	$12.00	$0.00	$99.30

B2020.35 WINDOW WALL

		Quan.	Unit	Material	Labor	Equip.	Total
B2020.35.10	**Window wall system, complete, minimum**						**Unit: SF**
084400103010	Window wall system,complete,minimum	1	SF	$42.64	$7.20	$0.00	$49.84
B2020.35.15	**Window wall system, complete, average**						**Unit: SF**
084400103030	Window wall system,complete,average	1	SF	$68.20	$8.00	$0.00	$76.20
B2020.35.20	**Window wall system, complete, maximum**						**Unit: SF**
084400103050	Window wall system,complete,maximum	1	SF	$157.76	$10.29	$0.00	$168.05

For bronze, add 20% to material
For stainless steel, add 50% to material

B2030 EXTERIOR DOORS

B2030.15 DOORS, WOOD

		Quan.	Unit	Material	Labor	Equip.	Total
B2030.15.10	**Wood Doors**						**Unit: EA**
087100101260	Hinges,4 x 4 butts,steel,standard	1.5	PAIR	$45.87	$0.00	$0.00	$45.87
099123304060	Int painting, trim, brush, second coat, avg	40	LF	$8.45	$8.38	$0.00	$16.83
099123303980	Int painting, trim, brush, first coat, avg	40	LF	$8.45	$9.90	$0.00	$18.35
099123302680	Int painting, doors, wood, brush, second coat, avg	42	SF	$6.65	$28.58	$0.00	$35.23
087100201300	Latchset,heavy duty,cylindrical	1	EA	$191.40	$40.78	$0.00	$232.18
081400900260	Door frame,pine,interior,3-0 x 7-0	1	EA	$140.66	$93.20	$0.00	$233.86
081400101970	Door,flush,solid core,1-3/4",birch,3-0 x 7-0	1	EA	$269.98	$81.55	$0.00	$351.53
062023101070	Molding,casing,wood,11/16 x 3-1/2	40	LF	$109.39	$124.27	$0.00	$233.66
099123302600	Int painting, doors, wood, brush, first coat, avg	42	SF	$8.87	$41.57	$0.00	$50.44
				$789.72	**$428.23**	**$0.00**	**$1,217.95**

B SHELL

B2030 EXTERIOR DOORS

B2030.20 DOORS, STEEL

		Quan.	Unit	Material	Labor	Equip.	Total
B2030.20.10	**Steel Doors**						**EA**
099113101840	Ext painting, doors, metal, roller, first coat, avg	42	SF	$7.21	$19.05	$0.00	$26.26
087100201300	Latchset,heavy duty,cylindrical	1	EA	$191.40	$40.78	$0.00	$232.18
099113102300	Ext painting, door frames, metal, brush, second coat, avg	20	LF	$4.22	$9.07	$0.00	$13.29
099113102220	Ext painting, door frames, metal, brush, first coat, avg	20	LF	$4.22	$17.01	$0.00	$21.23
087100101260	Hinges,4 x 4 butts,steel,standard	1.5	PAIR	$45.87	$0.00	$0.00	$45.87
081113401550	Door frames,mtl,stk,16 ga,6-3/4"x 1-3/4",3-0 x 7-0	1	EA	$198.96	$90.00	$0.00	$288.96
081113101240	Door,metal,flush hollow,20 ga,1-3/4",3-0 x 7-0	1	EA	$498.04	$72.49	$0.00	$570.53
099113101920	Ext painting, doors, metal, roller, second coat, avg	42	SF	$7.21	$12.70	$0.00	$19.91
				$957.13	**$261.10**	**$0.00**	**$1,218.23**

B2030.25 DOORS, ALUMINUM

		Quan.	Unit	Material	Labor	Equip.	Total
B2030.25.10	**Alum door, narrow stile, single, 2-6 x 7-0**						**Unit: EA**
081116101520	Alum door,narrow stile,single,2-6 x 7-0	1	EA	$939.63	$360.00	$0.00	$1,299.63
B2030.25.15	**Alum door, narrow stile, single, 3-0 x 7-0**						**Unit: EA**
081116101540	Alum door,narrow stile,single,3-0 x 7-0	1	EA	$988.28	$360.00	$0.00	$1,348.28
B2030.25.20	**Alum door, narrow stile, single, 3-6 x 7-0**						**Unit: EA**
081116101560	Alum door,narrow stile,single,3-6 x 7-0	1	EA	$1,012.61	$360.00	$0.00	$1,372.61
B2030.25.25	**Alum door, narrow stile, double, 5-0 x 7-0**						**Unit: EA**
081116101580	Alum door,narrow stile,double,5-0 x 7-0	1	EA	$1,567.56	$720.00	$0.00	$2,287.56
B2030.25.30	**Alum door, narrow stile, double, 6-0 x 7-0**						**Unit: EA**
081116101600	Alum door,narrow stile,double,6-0 x 7-0	1	EA	$1,591.88	$720.00	$0.00	$2,311.88
B2030.25.35	**Alum door, narrow stile, double, 7-0 x 7-0**						**Unit: EA**
081116101620	Alum door,narrow stile,double,7-0 x 7-0	1	EA	$1,662.63	$720.00	$0.00	$2,382.63
B2030.25.40	**Alum door, wide stile, single, 2-6 x 7-0**						**Unit: EA**
081116101720	Alum door,wide stile,single,2-6 x 7-0	1	EA	$1,316.52	$360.00	$0.00	$1,676.52
B2030.25.45	**Alum door, wide stile, single, 3-0 x 7-0**						**Unit: EA**
081116101740	Alum door,wide stile,single,3-0 x 7-0	1	EA	$1,360.73	$360.00	$0.00	$1,720.73
B2030.25.50	**Alum door, wide stile, single, 3-6 x 7-0**						**Unit: EA**
081116101760	Alum door,wide stile,single,3-6 x 7-0	1	EA	$1,402.95	$360.00	$0.00	$1,762.95
B2030.25.55	**Alum door, wide stile, double, 5-0 x 7-0**						**Unit: EA**
081116101780	Alum door,wide stile,double,5-0 x 7-0	1	EA	$2,413.95	$720.00	$0.00	$3,133.95
B2030.25.60	**Alum door, wide stile, double, 6-0 x 7-0**						**Unit: EA**
081116101800	Alum door,wide stile,double,6-0 x 7-0	1	EA	$2,524.50	$720.00	$0.00	$3,244.50
B2030.25.70	**Alum door, wide stile, double, 7-0 x 7-0**						**Unit: EA**
081116101820	Alum door,wide stile,double,7-0 x 7-0	1	EA	$2,568.72	$720.00	$0.00	$3,288.72

All prices are updated to January 1, 2021 and are national averages.
For a more in-depth information contact Design, Cost and Data at 800-533-5680, or go to www.DCD.com

B SHELL

B3010 ROOF COVERINGS

B3010.10 BUILT-UP

		Quan.	Unit	Material	Labor	Equip.	Total
B3010.10.10	**Built-Up Roofing, 20' by 20' Bay size**						**SF**
075113102110	Roofing,walkway,built-up,3'x 3'x 3/4" thick	1	SF	$3.82	$2.08	$0.00	$5.90
075113102280	Cant strip,4"x 4",mineral fiber	0.15	LF	$0.07	$0.23	$0.00	$0.30
072113101080	Insul,rigid,glass board,roof,1.63" thick,R6.67	1	SF	$1.40	$0.54	$0.00	$1.94
076200108640	Gravel stop,aluminum,0.032",8"	0.15	LF	$0.61	$0.36	$0.00	$0.97
075113101500	Roofing,built-up,felt incl gravel,3-ply	0.11	SQ	$13.26	$22.84	$0.00	$36.10
				$19.16	**$26.05**	**$0.00**	**$45.21**

B3010.15 MEMBRANE

		Quan.	Unit	Material	Labor	Equip.	Total
B3010.15.10	**Membrane Roofing, 20' by 20' Bay size**						**SF**
075300108100	Walkway for membrane roofs,1/2" thick	1	SF	$2.70	$2.08	$0.00	$4.78
076200108640	Gravel stop,aluminum,0.032",8"	0.15	LF	$0.61	$0.36	$0.00	$0.97
075300108000	Ballast, 3/4" through 1-1/2" dia. river gravel,	1	SF	$0.52	$0.62	$0.00	$1.14
075300102110	Elastic sheet roofing,epdm rubber,60 mil	1	SF	$2.44	$0.78	$0.00	$3.22
075113102280	Cant strip,4"x 4",mineral fiber	0.15	LF	$0.07	$0.23	$0.00	$0.30
072113101080	Insul,rigid,glass board,roof,1.63" thick,R6.67	1	SF	$1.40	$0.54	$0.00	$1.94
				$7.74	**$4.61**	**$0.00**	**$12.35**

C INTERIORS

C1010 INTERIOR PARTITIONS

C1010.10 MASONRY

		Quan.	Unit	Material	Labor	Equip.	Total
C1010.10.10	**Concrete Block Partitions .... SF**						
042200100160	CMU,lightweight block,hollow,load bearing,8"	1	SF	$2.74	$5.18	$0.00	$7.92
092300101020	Gypsum plaster,trowel finish,2 coats,walls	0.11	SY	$0.47	$1.97	$0.00	$2.44
096516101020	Vinyl cove molding,6" high	0.2	LF	$0.50	$0.26	$0.00	$0.76
099123304180	Int painting, walls, roller, first coat, avg	1	SF	$0.17	$0.20	$0.00	$0.37
099123304260	Int painting, walls, roller, second coat, avg	1	SF	$0.17	$0.18	$0.00	$0.35
040523101120	Horizontal joint reinf,truss type,6" wide,8" wall	1	LF	$0.21	$0.26	$0.00	$0.47
				$4.26	**$8.05**	**$0.00**	**$12.31**

C1010.15 DRYWALL

		Quan.	Unit	Material	Labor	Equip.	Total
C1010.15.10	**Drywall Partitions/Metal Stud Framing .... SF**						
096516101020	Vinyl cove molding,6" high	0.2	LF	$0.50	$0.26	$0.00	$0.76
099123304180	Int painting, walls, roller, first coat, avg	1	SF	$0.17	$0.20	$0.00	$0.37
092900100400	Drywall,1/2" thick,nailed,walls	1	SF	$0.39	$0.59	$0.00	$0.98
092116100142	Metal stud,20 ga,3-5/8",16" oc	1	SF	$0.63	$1.30	$0.00	$1.93
092900101224	Drywall,taping & finishing joints,average	1	SF	$0.07	$0.54	$0.00	$0.61
099123304260	Int painting, walls, roller, second coat, avg	1	SF	$0.17	$0.18	$0.00	$0.35
				$1.93	**$3.07**	**$0.00**	**$5.00**

C1010.20 PLASTER

		Quan.	Unit	Material	Labor	Equip.	Total
C1010.20.10	**Plaster Partitions/Metal Stud Framing .... SF**						
099123304260	Int painting, walls, roller, second coat, avg	1	SF	$0.17	$0.18	$0.00	$0.35
092116100142	Metal stud,20 ga,3-5/8",16" oc	1	SF	$0.63	$1.30	$0.00	$1.93
092236101090	Gypsum lath,plain,1/2" thick,clipped	0.11	SY	$0.54	$0.40	$0.00	$0.94
092300101020	Gypsum plaster,trowel finish,2 coats,walls	0.11	SY	$0.47	$1.97	$0.00	$2.44
096516101020	Vinyl cove molding,6" high	0.2	LF	$0.50	$0.26	$0.00	$0.76
099123304180	Int painting, walls, roller, first coat, avg	1	SF	$0.17	$0.20	$0.00	$0.37
				$2.48	**$4.31**	**$0.00**	**$6.79**

C INTERIORS

C1020 INTERIOR DOORS

C1020.10 METAL DOOR

		Quan.	Unit	Material	Labor	Equip.	Total
C1020.10.10	**Metal Door/Metal Frame**						**SF**
099123302300	Int painting, doors, metal, roller, second coat, avg	42	SF	$8.87	$12.03	$0.00	$20.90
099123302220	Int painting, doors, metal, roller, first coat, avg	42	SF	$8.87	$17.59	$0.00	$26.46
087100201300	Latchset,heavy duty,cylindrical	1	EA	$191.40	$40.78	$0.00	$232.18
087100101260	Hinges,4 x 4 butts,steel,standard	1.5	PAIR	$45.87	$0.00	$0.00	$45.87
081113101240	Door,metal,flush hollow,20 ga,1-3/4",3-0 x 7-0	1	EA	$498.04	$72.49	$0.00	$570.53
081113401550	Door frames,mtl,stk,16 ga,6-3/4"x 1-3/4",3-0 x 7-0	1	EA	$198.96	$90.00	$0.00	$288.96
				$952.01	**$232.89**	**$0.00**	**$1,184.90**

C1020.20 WOOD DOOR

		Quan.	Unit	Material	Labor	Equip.	Total
C1020.20.10	**Wood Door/Wood Frame**						**SF**
099123302680	Int painting, doors, wood, brush, second coat, avg	42	SF	$6.65	$28.58	$0.00	$35.23
062023101070	Molding,casing,wood,11/16 x 3-1/2	40	LF	$109.39	$124.27	$0.00	$233.66
081400101970	Door,flush,solid core,1-3/4",birch,3-0 x 7-0	1	EA	$269.98	$81.55	$0.00	$351.53
081400900260	Door frame,pine,interior,3-0 x 7-0	1	EA	$140.66	$93.20	$0.00	$233.86
087100101260	Hinges,4 x 4 butts,steel,standard	1.5	PAIR	$45.87	$0.00	$0.00	$45.87
087100201300	Latchset,heavy duty,cylindrical	1	EA	$191.40	$40.78	$0.00	$232.18
099123303980	Int painting, trim, brush, first coat, avg	40	LF	$8.45	$9.90	$0.00	$18.35
099123304060	Int painting, trim, brush, second coat, avg	40	LF	$8.45	$8.38	$0.00	$16.83
099123302600	Int painting, doors, wood, brush, first coat, avg	42	SF	$8.87	$41.57	$0.00	$50.44
				$789.72	**$428.23**	**$0.00**	**$1,217.95**

C1020.25 WOOD DOOR, METAL FRAME

		Quan.	Unit	Material	Labor	Equip.	Total
C1020.25.10	**Wood Door/Metal Frame**						**SF**
099123302600	Int painting, doors, wood, brush, first coat, avg	42	SF	$8.87	$41.57	$0.00	$50.44
099123302680	Int painting, doors, wood, brush, second coat, avg	42	SF	$6.65	$28.58	$0.00	$35.23
087100201300	Latchset,heavy duty,cylindrical	1	EA	$191.40	$40.78	$0.00	$232.18
087100101260	Hinges,4 x 4 butts,steel,standard	1.5	PAIR	$45.87	$0.00	$0.00	$45.87
081400101970	Door,flush,solid core,1-3/4",birch,3-0 x 7-0	1	EA	$269.98	$81.55	$0.00	$351.53
081113401550	Door frames,mtl,stk,16 ga,6-3/4"x 1-3/4",3-0 x 7-0	1	EA	$198.96	$90.00	$0.00	$288.96
				$721.73	**$282.48**	**$0.00**	**$1,004.21**

C INTERIORS

C1030 FITTINGS

C1030.10 WASHROOM ACCESSORIES

		Quan.	Unit	Material	Labor	Equip.	Total
C1030.10.10	**Toilet partition, plastic lam, ceiling mount**						**Unit: EA**
102113000120	Toilet partition,plastic lam,ceiling mount	1	EA	$1,204.16	$217.47	$0.00	$1,421.63
C1030.10.12	**Toilet partition, plastic lam, floor mount**						**Unit: EA**
102113000140	Toilet partition,plastic lam,floor mount	1	EA	$793.95	$163.10	$0.00	$957.05
C1030.10.14	**Toilet partition, metal, ceiling mount**						**Unit: EA**
102113000160	Toilet partition,metal,ceiling mount	1	EA	$821.98	$217.47	$0.00	$1,039.45
C1030.10.16	**Toilet partition, metal, floor mount**						**Unit: EA**
102113000180	Toilet partition,metal,floor mount	1	EA	$783.45	$163.10	$0.00	$946.55
C1030.10.20	**Wheelchair partition, plastic lam, ceiling mount**						**Unit: EA**
102113000200	Wheelchair partition,plastic lam,ceiling mount	1	EA	$1,798.07	$217.47	$0.00	$2,015.54
C1030.10.22	**Wheelchair partition, plastic lam, floor mount**						**Unit: EA**
102113000210	Wheelchair partition,plastic lam,floor mount	1	EA	$1,577.17	$163.10	$0.00	$1,740.27
C1030.10.25	**Wheelchair partition, painted metal, ceiling mount**						**Unit: EA**
102113000240	Wheelchair partition,painted metal,ceiling mount	1	EA	$1,284.34	$217.47	$0.00	$1,501.81
C1030.10.27	**Wheelchair partition, painted metal, floor mount**						**Unit: EA**
102113000260	Wheelchair partition,painted metal,floor mount	1	EA	$1,166.18	$163.10	$0.00	$1,329.28
C1030.10.30	**Urinal screen, plastic lam, wall hung**						**Unit: EA**
102113002000	Urinal screen,plastic lam,wall hung	1	EA	$554.84	$81.55	$0.00	$636.39
C1030.10.32	**Urinal screen, plastic lam, floor mount**						**Unit: EA**
102113002100	Urinal screen,plastic lam,floor mount	1	EA	$500.89	$81.55	$0.00	$582.44
C1030.10.34	**Urinal screen, porcelain enamel steel, floor mount**						**Unit: EA**
102113002120	Urinal screen,porcelain enamel st,floor mount	1	EA	$642.17	$81.55	$0.00	$723.72
C1030.10.36	**Urinal screen, painted metal, floor mount**						**Unit: EA**
102113002140	Urinal screen,painted metal,floor mount	1	EA	$423.84	$81.55	$0.00	$505.39
C1030.10.38	**Urinal screen, stainless steel, floor mount**						**Unit: EA**
102113002160	Urinal screen,stainless steel,floor mount	1	EA	$809.14	$81.55	$0.00	$890.69
C1030.10.40	**Toilet partn, front door/side divider, porcelain enamel steel**						**Unit: EA**
102113005040	Toilet partn,front door/side divider,porc enl stl	1	EA	$1,322.87	$163.10	$0.00	$1,485.97
C1030.10.42	**Toilet partn, front door/side divider, painted steel**						**Unit: EA**
102113005060	Toilet partn,front door/side divider,painted steel	1	EA	$777.02	$163.10	$0.00	$940.12
C1030.10.44	**Toilet partn, front door/side divider, stnls stl**						**Unit: EA**
102113005080	Toilet partn,front door/side divider,stnls stl	1	EA	$1,926.51	$163.10	$0.00	$2,089.61
C1030.15.10	**Bath accessory, ash receiver, wall mounted, alum**						**Unit: EA**
102816001040	Bath accessory,ash receiver,wall mounted,alum	1	EA	$167.99	$32.62	$0.00	$200.61
C1030.15.12	**Commercial hand dryer, surface mounted, 110 volt**						**Unit: EA**
102816001300	Commercial hand dryer,surface mounted,110 volt	1	EA	$810.15	$81.55	$0.00	$891.70

C INTERIORS

C1030 FITTINGS

C1030.15 WASHROOM ACCESSORIES (Cont.)

		Quan.	Unit	Material	Labor	Equip.	Total
C1030.15.14	**Mirror, 1/4" plate glass, up to 10 Sf**						**Unit: SF**
102816001420	Mirror,1/4" plate glass,up to 10 Sf	1	SF	$12.40	$6.52	$0.00	$18.92
C1030.15.16	**Mirror, SS frame, 18"x 24"**						**Unit: EA**
102816001440	Mirror,SS frame,18"x 24"	1	EA	$94.83	$21.75	$0.00	$116.58
C1030.15.18	**Mirror, SS frame, 24"x36"**						**Unit: EA**
102816001510	Mirror,SS frame,24"x36"	1	EA	$122.61	$36.24	$0.00	$158.85
C1030.15.20	**Mirror, SS frame, 30"x30"**						**Unit: EA**
102816001560	Mirror,SS frame,30"x30"	1	EA	$376.62	$65.24	$0.00	$441.86
C1030.15.22	**Mirror, SS frame, 48"x72"**						**Unit: EA**
102816001600	Mirror,SS frame,48"x72"	1	EA	$738.35	$108.73	$0.00	$847.08
C1030.15.24	**Mirror, SS frame, 18"x 24", w/shelf**						**Unit: EA**
102816001640	Mirror,SS frame,18"x 24",w/shelf	1	EA	$291.95	$26.10	$0.00	$318.05
C1030.15.26	**Sanitary napkin dispenser, SS, wall mounted**						**Unit: EA**
102816001820	Sanitary napkin dispenser,SS,wall mounted	1	EA	$704.48	$43.49	$0.00	$747.97
C1030.15.28	**Toilet tissue disp, SS, wall mounted, single**						**Unit: EA**
102816001920	Toilet tissue disp,SS,wall mounted,single	1	EA	$79.31	$16.31	$0.00	$95.62
C1030.15.30	**Toilet tissue disp, SS, wall mounted, double**						**Unit: EA**
102816001940	Toilet tissue disp,SS,wall mounted,double	1	EA	$152.21	$18.64	$0.00	$170.85
C1030.15.32	**Towel dispenser, SS, flush mounted**						**Unit: EA**
102816001950	Towel dispenser,SS,flush mounted	1	EA	$304.42	$36.24	$0.00	$340.66
C1030.15.34	**Towel dispenser, SS, surface**						**Unit: EA**
102816001960	Towel dispenser,SS,surface	1	EA	$433.86	$32.62	$0.00	$466.48
C1030.15.36	**Combination towel dispenser & waste receptacle**						**Unit: EA**
102816001970	Combination towel dispenser & waste receptacle	1	EA	$660.04	$43.49	$0.00	$703.53
C1030.15.38	**Waste receptacle, SS, wall mounted**						**Unit: EA**
102816002100	Waste receptacle,SS,wall mounted	1	EA	$498.77	$54.37	$0.00	$553.14

C2010 STAIR CONSTRUCTION

C2010.10 STAIRS, CONCRETE

		Quan.	Unit	Material	Labor	Equip.	Total
C2010.10.10	**Stair, Reinforced Concrete, 4' wide by 12' high**						**EA**
033101102160	Place concrete,stairs,3500# or 4000#,by pump	3	CY	$442.87	$131.35	$90.00	$664.22
055213100140	Railing,pipe,1-1/4" D,welded,3 rail,primed	20	LF	$812.90	$360.00	$0.00	$1,172.90
032101004210	Stair nosing, galv steel, 4' long	20	EA	$495.00	$88.80	$0.00	$583.80
032101001040	Reinforcing,stairs,#7-#8	0.25	TON	$392.83	$237.86	$0.00	$630.69
031114001060	Job built,stairway forms,5 uses	80	SF	$141.50	$745.60	$0.00	$887.10
055213100180	Railing,1-1/4" D,wall mounted,welded,primed	20	LF	$424.16	$221.54	$0.00	$645.70
				$2,709.26	**$1,785.15**	**$90.00**	**$4,584.41**

All prices are updated to January 1, 2021 and are national averages.
For a more in-depth information contact Design, Cost and Data at 800-533-5680, or go to www.DCD.com

C INTERIORS

C3020 FLOOR FINISHES

C3020.20 FLOORING, TILE

		Quan.	Unit	Material	Labor	Equip.	Total
C3020.20.10	**Unglazed tile, cement bed, face mount, 2"x 2" ....Unit: SF**						
093013106150	Unglazed tile,cement bed,face mount,2"x 2"	1	SF	$9.38	$5.18	$0.00	$14.56
C3020.20.12	**Unglazed tile, cement bed, face mount, 6"x 6" ....Unit: SF**						
093013106164	Unglazed tile,cement bed,face mount,6"x 6"	1	SF	$3.12	$4.44	$0.00	$7.56
C3020.20.14	**Unglazed tile, cement bed, face mount, 12"x 12" ....Unit: SF**						
093013106166	Unglazed tile,cement bed,face mount,12"x 12"	1	SF	$2.75	$3.89	$0.00	$6.64
C3020.20.16	**Unglazed tile, adhesive, white grout, 2"x 2" ....Unit: SF**						
093013106230	Unglazed tile,adhesive,white grout,2"x 2"	1	SF	$7.81	$5.18	$0.00	$12.99
C3020.20.18	**Unglazed tile, adhesive, white grout, 6"x 6" ....Unit: SF**						
093013106262	Unglazed tile,adhesive,white grout,6"x 6"	1	SF	$2.61	$4.44	$0.00	$7.05
C3020.20.20	**Unglazed tile, adhesive, white grout, 12"x 12" ....Unit: SF**						
093013106264	Unglazed tile,adhesive,white grout,12"x 12"	1	SF	$2.29	$3.89	$0.00	$6.18

C3020.25 FLOORING, CARPET

		Quan.	Unit	Material	Labor	Equip.	Total
C3020.25.10	**Carpet, acrylic, 28 oz, medium ....Unit: SY**						
096800201000	Carpet,acrylic,28 oz,medium	0.11	SY	$1.94	$0.80	$0.00	$2.74
C3020.25.12	**Carpet, acrylic, 28 oz, heavy ....Unit: SY**						
096800201020	Carpet,acrylic,28 oz,heavy	0.11	SY	$2.33	$0.80	$0.00	$3.13
C3020.25.14	**Commercial carpet, nylon, 28 oz, medium ....Unit: SY**						
096800202120	Commercial carpet,nylon,28 oz, medium	0.11	SY	$3.37	$0.80	$0.00	$4.17
C3020.25.16	**Commercial carpet, nylon, 32 oz, heavy ....Unit: SY**						
096800202140	Commercial carpet,nylon,32 oz,heavy	0.11	SY	$4.10	$0.80	$0.00	$4.90
C3020.25.18	**Commercial carpet, wool, 30 oz, medium ....Unit: SY**						
096800202150	Commercial carpet,wool,30 oz,medium	0.11	SY	$5.56	$0.80	$0.00	$6.36
C3020.25.20	**Commercial carpet, wool, 36 oz, medium ....Unit: SY**						
096800202160	Commercial carpet,wool,36 oz,medium	0.11	SY	$5.86	$0.80	$0.00	$6.66
C3020.25.22	**Commercial carpet, wool, 42 oz, heavy ....Unit: SY**						
096800202180	Commercial carpet,wool,42 oz,heavy	0.11	SY	$7.76	$0.80	$0.00	$8.56

C3020.25 FLOORING,CARPET TILE

		Quan.	Unit	Material	Labor	Equip.	Total
C3020.25.24	**Carpet tile, foam back, needle punch, minimum ....Unit: SF**						
096800203022	Carpet tile,foam back,needle punch,minimum	1	SF	$4.08	$1.30	$0.00	$5.38
C3020.25.26	**Carpet tile, foam back, needle punch, average ....Unit: SF**						
096800203024	Carpet tile,foam back,needle punch,average	1	SF	$4.72	$1.45	$0.00	$6.17
C3020.25.28	**Carpet tile, foam back, needle punch, maximum ....Unit: SF**						
096800203026	Carpet tile,foam back,needle punch,maximum	1	SF	$7.48	$1.63	$0.00	$9.11

C INTERIORS

C3020 FLOOR FINISHES

C3020.30 FLOORING, CARPET PAD

		Quan.	Unit	Material	Labor	Equip.	Total
C3020.30.10	**Carpet pad, jute, minimum**						**Unit: SY**
096800101020	Carpet pad,jute,minimum	0.11	SY	$0.50	$0.33	$0.00	$0.83
C3020.30.12	**Carpet pad, jute, average**						**Unit: SY**
096800101022	Carpet pad,jute,average	0.11	SY	$0.66	$0.36	$0.00	$1.02
C3020.30.14	**Carpet pad, jute, maximum**						**Unit: SY**
096800101024	Carpet pad,jute,maximum	0.11	SY	$0.99	$0.40	$0.00	$1.39
C3020.30.20	**Carpet pad, sponge rubber, minimum**						**Unit: SY**
096800101040	Carpet pad,sponge rubber,minimum	0.11	SY	$0.60	$0.33	$0.00	$0.93
C3020.30.22	**Carpet pad, sponge rubber, average**						**Unit: SY**
096800101042	Carpet pad,sponge rubber,average	0.11	SY	$0.79	$0.36	$0.00	$1.15
C3020.30.24	**Carpet pad, sponge rubber, maximum**						**Unit: SY**
096800101044	Carpet pad,sponge rubber,maximum	0.11	SY	$1.13	$0.40	$0.00	$1.53
C3020.30.30	**Carpet pad, urethane, 3/8", minimum**						**Unit: SY**
096800101060	Carpet pad,urethane,3/8",minimum	0.11	SY	$0.60	$0.33	$0.00	$0.93
C3020.30.32	**Carpet pad, urethane, 3/8", average**						**Unit: SY**
096800101062	Carpet pad,urethane,3/8",average	0.11	SY	$0.70	$0.36	$0.00	$1.06
C3020.30.34	**Carpet pad, urethane, 3/8", maximum**						**Unit: SY**
096800101064	Carpet pad,urethane,3/8",maximum	0.11	SY	$0.91	$0.40	$0.00	$1.31

C3030 CEILING FINISHES

		Quan.	Unit	Material	Labor	Equip.	Total
C3030.10.10	**Plaster Ceilings**						**SF**
099123301920	Int painting, ceilings, roller, second coat, avg	1	SF	$0.17	$0.21	$0.00	$0.38
099123301840	Int painting, ceilings, roller, first coat, avg	1	SF	$0.17	$0.25	$0.00	$0.42
092236201050	Lath,metal,diamond expd,galv,2.5 lb,ceiling,nailed	0.11	SY	$0.46	$1.03	$0.00	$1.49
092116104540	Furring,on ceilings,1-1/2" channel,16" oc	1	SF	$0.47	$2.72	$0.00	$3.19
092300101000	Gypsum plaster,trowel finish,2 coats,ceilings	0.11	SY	$0.47	$2.09	$0.00	$2.56
				$1.74	**$6.30**	**$0.00**	**$8.04**
C3030.15.10	**Drywall Ceilings**						**SF**
092116104540	Furring,on ceilings,1-1/2" channel,16" oc	1	SF	$0.47	$2.72	$0.00	$3.19
092900100240	Drywall,1/2" thick,clipped,ceilings	1	SF	$0.42	$0.72	$0.00	$1.14
092900101226	Drywall,taping & finishing joints,maximum	1	SF	$0.11	$0.65	$0.00	$0.76
099123301840	Int painting, ceilings, roller, first coat, avg	1	SF	$0.17	$0.25	$0.00	$0.42
099123301920	Int painting, ceilings, roller, second coat, avg	1	SF	$0.17	$0.21	$0.00	$0.38
				$1.34	**$4.55**	**$0.00**	**$5.89**
C3030.25.10	**Acoustical Ceilings**						**SF**
095100105520	Ceiling suspension system,T-bar,2'x 2'	1	SF	$1.38	$0.72	$0.00	$2.10
095100103060	Acoustic tile,fiberglass,12"x 12",3/4" thick	1	SF	$2.32	$1.45	$0.00	$3.77
				$3.70	**$2.17**	**$0.00**	**$5.87**

All prices are updated to January 1, 2021 and are national averages.
For a more in-depth information contact Design, Cost and Data at 800-533-5680, or go to www.DCD.com

D SERVICES

D1010 ELEVATORS

D1010.10 ELEVATORS, HYDRAULIC

		Quan.	Unit	Material	Labor	Equip.	Total
D1010.10.10	**Elevator, hydraulic, 3 stops, 3 op, 50 fpm, 2000 lb ..... Unit: EA**						
142100101500	Elevator,hydraulic,3 stops,3 op,50 fpm,2000 lb	1	EA	$101,249	$1,267	$1,108	$103,624
D1010.10.15	**Elevator, hydraulic, 3 stops, 3 op, 50 fpm, 3000 lb ..... Unit: EA**						
142100101520	Elevator,hydraulic,3 stops,3 op,50 fpm,3000 lb	1	EA	$114,261	$1,322	$1,156	$116,739
D1010.10.20	**Elevator, hydraulic, 3 stops, 3 op, 100 fpm, 2000 lb ..... Unit: EA**						
142100101540	Elevator,hydraulic,3 stops,3 op,100 fpm,2000 lb	1	EA	$110,919	$1,267	$1,108	$113,294
D1010.10.25	**Elevator, hydraulic, 3 stops, 3 op, 100 fpm, 3000 lb ..... Unit: EA**						
142100101560	Elevator,hydraulic,3 stops,3 op,100 fpm,3000 lb	1	EA	$125,139	$1,382	$1,209	$127,730
D1010.10.30	**Elevator, hydraulic, 3 stops, 3 op, 150 fpm, 2000 lb ..... Unit: EA**						
142100101580	Elevator,hydraulic,3 stops,3 op,150 fpm,2000 lb	1	EA	$120,162	$1,267	$1,108	$122,537
D1010.10.32	**Elevator, hydraulic, 3 stops, 3 op, 150 fpm, 3000 lb ..... Unit: EA**						
142100101600	Elevator,hydraulic,3 stops,3 op,150 fpm,3000 lb	1	EA	$140,782	$1,448	$1,266	$143,496

For each additional: 50 fpm add per stop, $3500
For each additional: 500 lb, add per stop, $3500
For each additional: Opening, add, $4200
For each additional: Stop, add per stop, $5300
For each additional: Bonderized steel door, add per opening, $400
For each additional: Colored aluminum door, add per opening, $1500
For each additional: Stainless steel door, add per opening, $650

D1010.15 ELEVATORS, ELECTRIC

		Quan.	Unit	Material	Labor	Equip.	Total
D1010.15.10	**Elevator, electric, 8 stop, 8 openings, 300 fpm, 3000 lb..... EA**						
142100101020	Elevator, electric, 8 stop, 8 op, 300 fpm, 3000 lb	1.000	EA	$270,187	$2,937	$2,510	$275,635
D1010.15.15	**Elevator, electric, 8 stop, 8 openings, 300 fpm, 4000 lb..... EA**						
142100101060	Elevator, electric, 8 stop, 8 op, 300 fpm, 4000 lb	1.000	EA	$287,963	$3,264	$2,788	$294,016
D1010.15.20	**Elevator, electric, 8 stop, 8 openings, 300 fpm, 5000 lb..... EA**						
142100101070	Elevator, electric, 8 stop, 8 op, 300 fpm, 5000 lb	1.000	EA	$319,959	$3,497	$2,988	$326,444
D1010.15.25	**Elevator, electric, 8 stop, 8 openings, 400 fpm, 3000 lb..... EA**						
142100101090	Elevator, electric, 8 stop, 8 op, 400 fpm, 3000 lb	1.000	EA	$283,057	$2,937	$2,510	$288,504
D1010.15.30	**Elevator, electric, 8 stop, 8 openings, 400 fpm, 4000 lb..... EA**						
142100101120	Elevator, electric, 8 stop, 8 op, 400 fpm, 4000 lb	1.000	EA	$307,871	$3,264	$2,788	$313,924
D1010.15.35	**Elevator, electric, 8 stop, 8 openings, 400 fpm, 5000 lb..... EA**						
142100101140	Elevator, electric, 8 stop, 8 op, 400 fpm, 5000 lb	1.000	EA	$350,533	$3,497	$2,988	$357,018

D SERVICES

D1010 ELEVATORS

D1010.15 ELEVATORS, ELECTRIC (Cont.)

		Quan.	Unit	Material	Labor	Equip.	Total
D1010.15.40	**Elevator, electric, 8 stop, 8 openings, 600 fpm, 3000 lb**						**EA**
142100101160	Elevator, electric, 8 stop, 8 op, 600 fpm, 3000 lb	1.000	EA	$398,669	$3,264	$2,788	$404,722
D1010.15.45	**Elevator, electric, 8 stop, 8 openings, 600 fpm, 4000 lb**						**EA**
142100101190	Elevator, electric, 8 stop, 8 op, 600 fpm, 4000 lb	1.000	EA	$412,391	$3,582	$3,060	$419,035
D1010.15.50	**Elevator, electric, 8 stop, 8 openings, 600 fpm, 5000 lb**						**EA**
142100101200	Elevator, electric, 8 stop, 8 op, 600 fpm, 5000 lb	1.000	EA	$423,768	$3,672	$3,137	$430,577
D1010.15.60	**Elevator, electric, 8 stop, 8 openings, 800 fpm, 3000 lb**						**EA**
142100101220	Elevator, electric, 8 stop, 8 op, 800 fpm, 3000 lb	1.000	EA	$472,828	$3,264	$2,788	$478,881
D1010.15.62	**Elevator, electric, 8 stop, 8 openings, 800 fpm, 4000 lb**						**EA**
142100101260	Elevator, electric, 8 stop, 8 op, 800 fpm, 4000 lb	1.000	EA	$483,494	$3,582	$3,060	$490,137
D1010.15.64	**Elevator, electric, 8 stop, 8 openings, 800 fpm, 5000 lb**						**EA**
142100101280	Elevator, electric, 8 stop, 8 op, 800 fpm, 5000 lb	1.000	EA	$486,693	$3,672	$3,137	$493,503

For each additional: 100 fpm add per stop, $13,000
For each additional: 500 lb, add per stop, $6500
For each additional: Opening add per stop, $12,000
For each additional: Stop add per stop, $4800
For each additional: Bypass floor, add per each, $2000
For each additional: Bonderized steel door, add per opening, $150
For each additional: Colored aluminum door, add per opening, $900
For each additional: Stainless steel door, add per opening, $600

D1010 ESCALATORS

D1020.30 ESCALATORS

		Quan.	Unit	Material	Labor	Equip.	Total
D1020.30.10	**Escalators, 32" wide, floor to floor, 12' high**						**EA**
143100101040	Escalators, 32" wide, floor to floor, 12' high	1.000	EA	$180,150	$2,448	$2,091	$184,689
D1020.30.15	**Escalators, 32" wide, floor to floor, 18' high**						**EA**
143100101060	Escalators, 32" wide, floor to floor, 18' high	1.000	EA	$211,986	$3,672	$3,137	$218,795
D1020.30.20	**Escalators, 32" wide, floor to floor, 25' high**						**EA**
143100101080	Escalators, 32" wide, floor to floor, 25' high	1.000	EA	$238,378	$5,875	$5,020	$249,273
D1020.30.25	**Escalators, 48" wide, floor to floor, 12' high**						**EA**
143100101090	Escalators, 48" wide, floor to floor, 12' high	1.000	EA	$200,705	$2,532	$2,163	$205,401
D1020.30.30	**Escalators, 48" wide, floor to floor, 18' high**						**EA**
143100101120	Escalators, 48" wide, floor to floor, 18' high	1.000	EA	$235,309	$3,865	$3,302	$242,477
D1020.30.35	**Escalators, 48" wide, floor to floor, 25' high**						**EA**
143100101140	Escalators, 48" wide, floor to floor, 25' high	1.000	EA	$281,264	$5,875	$5,020	$292,159

D SERVICES

D2010 PLUMBING FIXTURES

D2010.10 WATER CLOSETS

		Quan.	Unit	Material	Labor	Equip.	Total
D2010.10.12	**Water closet, flush tank, floor mounted .... EA**						
224200901600	Toilets,floor support,average	1	EA	$181.50	$71.52	$0.00	$253.02
224213601010	Water closet,flush tank,floor mounted,average	1	EA	$785.29	$238.40	$0.00	$1,023.69
224213609040	Water closet,for trim and rough-in,maximum	1	EA	$399.30	$357.60	$0.00	$756.90
				$1,366.09	**$667.52**	**$0.00**	**$2,033.61**
D2010.10.18	**Water closet, flush tank, floor mount, handicap .... EA**						
224200901600	Toilets,floor support,average	1	EA	$181.50	$71.52	$0.00	$253.02
224213601050	Water closet,flush tank,floor mount,handicap,avrg	1	EA	$974.29	$357.60	$0.00	$1,331.89
224213609040	Water closet,for trim and rough-in,maximum	1	EA	$399.30	$357.60	$0.00	$756.90
				$1,555.09	**$786.72**	**$0.00**	**$2,341.81**
D2010.10.24	**Water closet, bowl, flush valve, floor mount .... EA**						
224213601220	Water closet,bowl,w/flush valve,floor mount,avrg	1	EA	$612.26	$238.40	$0.00	$850.66
224213609040	Water closet,for trim and rough-in,maximum	1	EA	$399.30	$357.60	$0.00	$756.90
224200901600	Toilets,floor support,average	1	EA	$181.50	$71.52	$0.00	$253.02
				$1,193.06	**$667.52**	**$0.00**	**$1,860.58**
D2010.10.30	**Water closet, bowl, flush valve, wall mounted .... EA**						
224213601280	Water closet,bowl,w/flush valve,wall mounted,avrg	1	EA	$652.19	$238.40	$0.00	$890.59
224213609040	Water closet,for trim and rough-in,maximum	1	EA	$399.30	$357.60	$0.00	$756.90
224200901540	Toilets,water closets,wall carrier,average	1	EA	$363.00	$89.40	$0.00	$452.40
				$1,414.49	**$685.40**	**$0.00**	**$2,099.89**

D2010.15 URINALS

		Quan.	Unit	Material	Labor	Equip.	Total
D2010.15.12	**Urinal, flush valve, floor mounted .... EA**						
224200901720	Urinals,floor support,average	1	EA	$193.60	$71.52	$0.00	$265.12
224213501010	Urinal,flush valve,floor mounted,average	1	EA	$705.43	$238.40	$0.00	$943.83
224213509040	Urinal,for trim and rough-in,maximum	1	EA	$432.58	$476.80	$0.00	$909.38
				$1,331.61	**$786.72**	**$0.00**	**$2,118.33**
D2010.15.18	**Urinal, flush valve, wall mounted .... EA**						
224213501080	Urinal,flush valve,wall mounted,average	1	EA	$678.81	$238.40	$0.00	$917.21
224213509040	Urinal,for trim and rough-in,maximum	1	EA	$432.58	$476.80	$0.00	$909.38
224200901660	Urinals,wall carrier,average	1	EA	$205.70	$89.40	$0.00	$295.10
				$1,317.09	**$804.60**	**$0.00**	**$2,121.69**

D SERVICES

D2010 PLUMBING FIXTURES

D2010.20 LAVATORIES

		Quan.	Unit	Material	Labor	Equip.	Total
D2010.20.12	**Lavatory, counter top, porcelain on CI**		**EA**				
224216209040	Lavatory,for trim and rough-in,maximum	1	EA	$559.02	$357.60	$0.00	$916.62
224216202010	Lavatory,counter top,porcelain enamel on CI,avrg	1	EA	$352.72	$178.80	$0.00	$531.52
				$911.74	**$536.40**	**$0.00**	**$1,448.14**
D2010.20.17	**Lavatory, wall hung, china**		**EA**				
224200901140	Lavatory,wall carrier,average	1	EA	$229.90	$89.40	$0.00	$319.30
224216202110	Lavatory,wall hung,china,average	1	EA	$372.68	$178.80	$0.00	$551.48
224216209040	Lavatory,for trim and rough-in,maximum	1	EA	$559.02	$357.60	$0.00	$916.62
				$1,161.60	**$625.80**	**$0.00**	**$1,787.40**
D2010.20.22	**Lavatory, wall hung, china, handicapped**		**EA**				
224200901140	Lavatory,wall carrier,average	1	EA	$229.90	$89.40	$0.00	$319.30
224216202310	Lavatory,wall hung,china,handicapped,average	1	EA	$598.95	$238.40	$0.00	$837.35
224216209040	Lavatory,for trim and rough-in,maximum	1	EA	$559.02	$357.60	$0.00	$916.62
				$1,387.87	**$685.40**	**$0.00**	**$2,073.27**

D2010.25 BATHTUBS

		Quan.	Unit	Material	Labor	Equip.	Total
D2010.25.12	**Bathtub, 5' long**		**EA**				
224219001020	Bathtub,5' long,average	1	EA	$1,397.55	$357.60	$0.00	$1,755.15
224219009020	Bathtub,for trim and rough-in,average	1	EA	$332.75	$357.60	$0.00	$690.35
				$1,730.30	**$715.20**	**$0.00**	**$2,445.50**
D2010.25.22	**Bathtub, 6' long**		**EA**				
224219001080	Bathtub,6' long,average	1	EA	$1,464.10	$357.60	$0.00	$1,821.70
224219009020	Bathtub,for trim and rough-in,average	1	EA	$332.75	$357.60	$0.00	$690.35
				$1,796.85	**$715.20**	**$0.00**	**$2,512.05**
D2010.25.32	**Bathtub, square, whirlpool, 4'x 4'**		**EA**				
224219001140	Bathtub,square,whirlpool,4'x 4',average	1	EA	$3,114.54	$715.20	$0.00	$3,829.74
224219009040	Bathtub,for trim and rough-in,maximum	1	EA	$945.01	$715.20	$0.00	$1,660.21
				$4,059.55	**$1,430.40**	**$0.00**	**$5,489.95**

All prices are updated to January 1, 2021 and are national averages.
For a more in-depth information contact Design, Cost and Data at 800-533-5680, or go to www.DCD.com

D SERVICES

D2010 PLUMBING FIXTURES

D2010.30 SINKS

		Quan.	Unit	Material	Labor	Equip.	Total
D2010.30.22	**Kitchen sink, single, stainless, double bowl...EA**						
224216402100	Kitchen sink,single,stnls stl,double bowl,avrg	1	EA	$432.58	$238.40	$0.00	$670.98
224216409020	Sink,for trim and rough-in,average	1	EA	$532.40	$357.60	$0.00	$890.00
				$964.98	**$596.00**	**$0.00**	**$1,560.98**
D2010.30.42	**Kitchen sink, porcelain on CI, double bowl...EA**						
224216402280	Kitchen sink,porcelain enamel,CI,double bowl,avrg	1	EA	$465.85	$238.40	$0.00	$704.25
224216409020	Sink,for trim and rough-in,average	1	EA	$532.40	$357.60	$0.00	$890.00
				$998.25	**$596.00**	**$0.00**	**$1,594.25**
D2010.35.12	**Service sink, 24"x29"...EA**						
224200901420	Sink,industrial,wall carrier,average	1	EA	$242.00	$89.40	$0.00	$331.40
224216401020	Service sink,24"x29",average	1	EA	$958.32	$238.40	$0.00	$1,196.72
224216409040	Sink,for trim and rough-in,maximum	1	EA	$678.81	$476.80	$0.00	$1,155.61
				$1,879.13	**$804.60**	**$0.00**	**$2,683.73**
D2010.35.22	**Mop sink, 24"x36"x10", average...EA**						
224200901420	Sink,industrial,wall carrier,average	1	EA	$242.00	$89.40	$0.00	$331.40
224216403020	Mop sink,24"x36"x10",average	1	EA	$705.43	$178.80	$0.00	$884.23
224216409040	Sink,for trim and rough-in,maximum	1	EA	$678.81	$476.80	$0.00	$1,155.61
				$1,626.24	**$745.00**	**$0.00**	**$2,371.24**

D2010.40 WATER COOLERS

		Quan.	Unit	Material	Labor	Equip.	Total
D2010.40.10	**Water Cooler...EA**						
224700101020	Electric water cooler,wall mounted	1	EA	$1,040.60	$238.40	$0.00	$1,279.00
224200901040	Water fountain,wall carrier,maximum	1	EA	$116.16	$119.20	$0.00	$235.36
224216409020	Sink,for trim and rough-in,average	1	EA	$532.40	$357.60	$0.00	$890.00
				$1,689.16	**$715.20**	**$0.00**	**$2,404.36**

D SERVICES

D2020 DOMESTIC WATER DISTRIBUTION

D2020.05 COPPER

		Quan.	Unit	Material	Labor	Equip.	Total
D2020.05.05	**3/4" L Copper, Including Hangers and Fittings, 100' Runs ... LF**						
221116202374	Copper,90 ells,3/4"	0.02	EA	$0.05	$0.75	$0.00	$0.80
221116200880	Copper fittings,reducing slip coupling,3/4"	0.08	EA	$0.22	$2.86	$0.00	$3.08
220629100150	A band,copper, 3/4"	0.1	EA	$0.19	$0.53	$0.00	$0.72
221116106190	Copper piping,L tube,3/4"	1	LF	$4.33	$2.38	$0.00	$6.71
				$4.79	**$6.52**	**$0.00**	**$11.31**
D2020.05.07	**1" L Copper, Including Hangers and Fittings, 100' Runs ... LF**						
220629100160	A band,copper,1"	0.1	EA	$0.19	$0.53	$0.00	$0.72
221116106240	Copper piping,L tube,1"	1	LF	$6.51	$2.55	$0.00	$9.06
221116200890	Copper fittings,reducing slip coupling,1"	0.08	EA	$0.47	$3.18	$0.00	$3.65
221116202376	Copper,90 ells,1"	0.02	EA	$0.12	$0.79	$0.00	$0.91
				$7.29	**$7.05**	**$0.00**	**$14.34**
D2020.05.09	**1-1/2" L Copper, Including Hangers and Fittings, 100' Runs ... LF**						
221116202379	Copper,90 ells,1-1/2"	0.08	EA	$1.12	$3.58	$0.00	$4.70
220629100180	A band,copper,1-1/2"	0.1	EA	$0.22	$0.60	$0.00	$0.82
221116106360	Copper piping,L tube,1-1/2"	1	LF	$12.00	$2.98	$0.00	$14.98
221116201000	Copper fittings,reducing slip coupling,1-1/2"	0.02	EA	$0.24	$0.95	$0.00	$1.19
				$13.58	**$8.11**	**$0.00**	**$21.69**
D2020.05.15	**2" L Copper, Including Hangers and Fittings, 100' Runs ... LF**						
221116202381	Copper,90 ells,2"	0.08	EA	$2.02	$3.81	$0.00	$5.83
221116201020	Copper fittings,reducing slip coupling,2"	0.02	EA	$0.40	$1.19	$0.00	$1.59
220629100190	A band,copper,2"	0.1	EA	$0.24	$0.65	$0.00	$0.89
221116106400	Copper piping,L tube,2"	1	LF	$18.73	$3.25	$0.00	$21.98
				$21.39	**$8.90**	**$0.00**	**$30.29**
D2020.05.17	**4" L Copper, Including Hangers and Fittings, 100' Runs ... LF**						
221116202390	Copper,90 ells,4"	0.08	EA	$13.89	$5.72	$0.00	$19.61
220629100220	A band,copper,4"	0.1	EA	$0.55	$0.89	$0.00	$1.44
221116106520	Copper piping,L tube,4"	1	LF	$61.76	$3.97	$0.00	$65.73
221116201050	Copper fittings,reducing slip coupling,4"	0.02	EA	$1.86	$1.79	$0.00	$3.65
				$78.06	**$12.37**	**$0.00**	**$90.43**

All prices are updated to January 1, 2021 and are national averages.
For a more in-depth information contact Design, Cost and Data at 800-533-5680, or go to www.DCD.com

D SERVICES

D2020 DOMESTIC WATER DISTRIBUTION

D2020.05 BLACK STEEL

		Quan.	Unit	Material	Labor	Equip.	Total
D2020.05.30	**3/4" Black Steel, Schedule 40, Including Hangers and Fittings, 100' Runs .... LF**						
221116904160	Malleable iron,threaded,3/4",coupling	0.08	EA	$0.36	$1.91	$0.00	$2.27
221116904090	Malleable iron,threaded,3/4",90 deg ell	0.02	EA	$0.08	$0.48	$0.00	$0.56
221116901100	Black steel,extra heavy pipe,threaded,3/4" pipe	1	LF	$3.64	$2.86	$0.00	$6.50
220529100300	Hanger,3/4" pipe,clevis pipe hanger,black steel	0.1	EA	$0.19	$2.38	$0.00	$2.57
				$4.27	**$7.63**	**$0.00**	**$11.90**
D2020.05.35	**1" Black Steel, Schedule 40, Including Hangers and Fittings, 100' Runs .... LF**						
220529100400	Hanger,1" pipe,clevis pipe hanger,black steel	0.1	EA	$0.20	$2.38	$0.00	$2.58
221116901200	Black steel,extra heavy,threaded,1" pipe	1	LF	$4.68	$3.58	$0.00	$8.26
221116904180	Malleable iron,threaded,1",90 deg ell	0.02	EA	$0.12	$0.57	$0.00	$0.69
221116904240	Malleable iron,threaded,1",coupling	0.08	EA	$0.53	$2.29	$0.00	$2.82
				$5.53	**$8.82**	**$0.00**	**$14.35**
D2020.05.40	**1-1/2" Black Steel, Schedule 40, Including Hangers and Fittings, 100' Runs .... LF**						
221116901400	Black steel,extra heavy,threaded,1-1/2" pipe	1	LF	$7.03	$3.97	$0.00	$11.00
221116904270	Malleable iron,threaded,1-1/2",90 deg ell	0.02	EA	$0.25	$0.72	$0.00	$0.97
220529100600	Hanger,1-1/2" pipe,clevis pipe hanger,black steel	0.1	EA	$0.22	$2.38	$0.00	$2.60
221116904340	Malleable iron,threaded,1-1/2",coupling	0.08	EA	$0.94	$2.86	$0.00	$3.80
				$8.44	**$9.93**	**$0.00**	**$18.37**
D2020.05.45	**2-1/2" Black Steel, Schedule 40, Including Hangers and Fittings, 100' Runs .... LF**						
221116904430	Malleable iron,threaded,2-1/2",coupling	0.08	EA	$3.84	$9.54	$0.00	$13.38
220529100800	Hanger,2-1/2" pipe,clevis pipe hanger,black steel	0.1	EA	$0.42	$2.38	$0.00	$2.80
221116901500	Black steel,extra heavy,threaded,2-1/2" pipe	1	LF	$14.06	$8.94	$0.00	$23.00
221116904370	Malleable iron,threaded,2-1/2",90 deg ell	0.02	EA	$0.96	$1.79	$0.00	$2.75
				$19.28	**$22.65**	**$0.00**	**$41.93**
D2020.05.50	**4" Black Steel, Schedule 40, Including Hangers and Fittings, 100' Runs .... LF**						
221116904590	Malleable iron,threaded,4",coupling	0.08	EA	$10.43	$5.72	$0.00	$16.15
221116904540	Malleable iron,threaded,4",90 deg ell	0.02	EA	$3.04	$2.86	$0.00	$5.90
220529101100	Hanger,4" pipe,clevis pipe hanger,black steel	0.1	EA	$0.64	$2.38	$0.00	$3.02
221116901700	Black steel,extra heavy,threaded,4" pipe	1	LF	$28.37	$14.30	$0.00	$42.67
				$42.48	**$25.26**	**$0.00**	**$67.74**

D SERVICES

D2020 DOMESTIC WATER DISTRIBUTION

D2020.10 WATER HEATERS

		Quan.	Unit	Material	Labor	Equip.	Total
D2020.10.10	**Electric Water Heaters...EA**						
223300101120	Water heater,electric,120 gal	1	EA	$1,669.80	$238.40	$0.00	$1,908.20
223300108200	Water heater, rough-in, Maximum	1	EA	$859.10	$715.20	$0.00	$1,574.30
337116005810	Utility pole,perforated strap for conduit,1-1/2"	6	LF	$18.19	$72.44	$0.00	$90.63
				$2,547.09	**$1,026.04**	**$0.00**	**$3,573.13**

D2020.25 SOLAR WATER HEATERS

		Quan.	Unit	Material	Labor	Equip.	Total
D2020.25.11	**Solar Hot Water heater, hydronic, 100-120 Gallons, minimum...EA**						
223300201010	Solar Water heater, hydronic, 100-120 Gal., min.	1.000	EA	$14,356.65	$0.00	$0.00	$14,356.65
D2020.25.12	**Solar Hot Water heater, hydronic, 100-120 Gallons, average...EA**						
223300201020	Solar Water heater, hydronic, 100-120 Gallons, average	1.000	EA	$15,118.95	$0.00	$0.00	$15,118.95
D2020.25.14	**Solar Hot Water heater, hydronic, 100-120 Gallons, maximum...EA**						
223300201030	Solar Water heater, hydronic, 100-120 Gallons, max.	1.000	EA	$15,881.25	$0.00	$0.00	$15,881.25
D2020.25.20	**Solar Hot Water heater, direct-solar, 100-120 Gallons, minimum...EA**						
223300201050	Solar Water heater, direct, 100-120 Gallons, minimum	1.000	EA	$9,528.75	$0.00	$0.00	$9,528.75
D2020.25.25	**Solar Hot Water heater, direct-solar, 100-120 Gallons, average...EA**						
223300201060	Solar Water heater, direct, 100-120 Gallons, average	1.000	EA	$9,973.43	$0.00	$0.00	$9,973.43
D2020.25.28	**Solar Hot Water heater, direct-solar, 100-120 Gallons, maximum...EA**						
223300201070	Solar heater, direct, 100-120 Gallons, maximum	1.000	EA	$10,418.10	$0.00	$0.00	$10,418.10

D SERVICES

D2030 PIPING, SANITARY WASTE

D2030.10 PLASTIC, DWV

		Quan.	Unit	Material	Labor	Equip.	Total
D2030.10.05	**1-1/2" Plastic, PVC, DWV, Schedule 40, Including Hangers and Fittings, 100' Runs..... LF**						
220629100180	A band,copper,1-1/2"	0.1	EA	$0.22	$0.60	$0.00	$0.82
221116801080	PVC pipe,sch 40,1-1/2" pipe	1	LF	$1.69	$4.47	$0.00	$6.16
221116802460	PVC pipe,sch 40,1-1/2",90 deg elbow	0.02	EA	$0.03	$0.41	$0.00	$0.44
221116802530	PVC pipe,sch 40,1-1/2",coupling	0.08	EA	$0.08	$1.91	$0.00	$1.99
				$2.02	**$7.39**	**$0.00**	**$9.41**
D2030.10.07	**2" Plastic, PVC, DWV, Schedule 40, Including Hangers and Fittings, 100' Runs........... LF**						
221116802640	PVC pipe,sch 40,2",coupling	0.08	EA	$0.12	$2.29	$0.00	$2.41
221116802570	PVC pipe,sch 40,2",90 deg elbow	0.02	EA	$0.05	$0.48	$0.00	$0.53
221116801100	PVC pipe,sch 40,2" pipe	1	LF	$2.14	$5.11	$0.00	$7.25
220629100190	A band,copper,2"	0.1	EA	$0.24	$0.65	$0.00	$0.89
				$2.55	**$8.53**	**$0.00**	**$11.08**
D2030.10.10	**3" Plastic, PVC, DWV, Schedule 40, Including Hangers and Fittings, 100' Runs........... LF**						
221116801120	PVC pipe,sch 40,3" pipe	1	LF	$4.41	$7.15	$0.00	$11.56
221116802780	PVC pipe,sch 40,3",90 deg elbow	0.02	EA	$0.16	$1.19	$0.00	$1.35
221116802830	PVC pipe,sch 40,3",coupling	0.08	EA	$0.37	$4.77	$0.00	$5.14
220629100220	A band,copper,4"	0.1	EA	$0.55	$0.89	$0.00	$1.44
				$5.49	**$14.00**	**$0.00**	**$19.49**
D2030.10.13	**4" Plastic, PVC, DWV, Schedule 40, Including Hangers and Fittings, 100' Runs........... LF**						
221116802930	PVC pipe,sch 40,4",coupling	0.08	EA	$0.54	$5.72	$0.00	$6.26
220629100220	A band,copper,4"	0.1	EA	$0.55	$0.89	$0.00	$1.44
221116801130	PVC pipe,sch 40,4" pipe	1	LF	$6.30	$8.94	$0.00	$15.24
221116802870	PVC pipe,sch 40,4",90 deg elbow	0.02	EA	$0.29	$1.43	$0.00	$1.72
				$7.68	**$16.98**	**$0.00**	**$24.66**

D SERVICES

D2030 PIPING, SANITARY WASTE

D2030.10 CAST IRON

		Quan.	Unit	Material	Labor	Equip.	Total
D2030.10.50	**1-1/2" Cast Iron, No-Hub, Including Hangers and Fittings, 100' Runs .... LF**						
221316005180	CI Pipe,Abv Gr,no hub,1-1/2",coupling	0.02	EA	$0.45	$0.00	$0.00	$0.45
221316005000	CI Pipe,Abv Gr,no hub,1-1/2",1/4 bend	0.08	EA	$0.91	$1.91	$0.00	$2.82
220529100600	Hanger,1-1/2" pipe,clevis pipe hanger,black steel	0.1	EA	$0.22	$2.38	$0.00	$2.60
221316001000	CI Pipe,Abv Gr,no hub,1-1/2" pipe	1	LF	$12.27	$5.11	$0.00	$17.38
				$13.85	**$9.40**	**$0.00**	**$23.25**
D2030.10.55	**4" Cast Iron, No-Hub, Including Hangers and Fittings, 100' Runs.................................. LF**						
221316007100	CI Pipe,Abv Gr,no hub,4",coupling	0.02	EA	$0.44	$0.00	$0.00	$0.44
220529101100	Hanger,4" pipe,clevis pipe hanger,black steel	0.1	EA	$0.64	$2.38	$0.00	$3.02
221316001200	CI Pipe,Abv Gr,no hub,4" pipe	1	LF	$19.57	$11.92	$0.00	$31.49
221316006820	CI Pipe,Abv Gr,no hub,4",1/4 bend	0.08	EA	$2.07	$2.86	$0.00	$4.93
				$22.72	**$17.16**	**$0.00**	**$39.88**
D2030.10.57	**6" Cast Iron, No-Hub, Including Hangers and Fittings, 100' Runs.................................. LF**						
221316007700	CI Pipe,Abv Gr,no hub,6",coupling	0.02	EA	$1.12	$0.00	$0.00	$1.12
221316007600	CI Pipe,Abv Gr,no hub,6",1/4 bend	0.08	EA	$5.24	$4.77	$0.00	$10.01
221316001300	CI Pipe,Abv Gr,no hub,6" pipe	1	LF	$34.50	$14.30	$0.00	$48.80
220529101320	Hanger,6" pipe,clevis pipe hanger,black steel	0.1	EA	$1.02	$2.86	$0.00	$3.88
				$41.88	**$21.93**	**$0.00**	**$63.81**
D2030.10.60	**8" Cast Iron, No-Hub, Including Hangers and Fittings, 100' Runs.................................. LF**						
220529101420	Hanger,8" pipe,clevis pipe hanger,black steel	0.1	EA	$1.68	$2.86	$0.00	$4.54
221316001400	CI Pipe,Abv Gr,no hub,8" pipe	1	LF	$55.45	$23.84	$0.00	$79.29
221316007990	CI Pipe,Abv Gr,no hub,8",1/4 bend	0.08	EA	$9.09	$4.77	$0.00	$13.86
221316008090	CI Pipe,Abv Gr,no hub,8",coupling	0.02	EA	$2.11	$0.00	$0.00	$2.11
				$68.33	**$31.47**	**$0.00**	**$99.80**
D2030.10.62	**10" Cast Iron, No-Hub, Including Hangers and Fittings, 100' Runs................................ LF**						
221316001500	CI Pipe,Abv Gr,no hub,10" pipe	1	LF	$87.00	$28.61	$0.00	$115.61
221316008360	CI Pipe,Abv Gr,no hub,10",1/4 bend	0.08	EA	$18.17	$4.77	$0.00	$22.94
220529101500	Hanger,10" clevis pipe hanger,black steel	0.1	EA	$3.03	$2.86	$0.00	$5.89
221316008420	CI Pipe,Abv Gr,no hub,10",coupling	0.02	EA	$2.78	$0.00	$0.00	$2.78
				$110.98	**$36.24**	**$0.00**	**$147.22**

All prices are updated to January 1, 2021 and are national averages.
For a more in-depth information contact Design, Cost and Data at 800-533-5680, or go to www.DCD.com

D SERVICES

D3010 ENERGY SUPPLY

D3010.10 HVAC SQUARE FOOT COSTS

		Quan.	Unit	Material	Labor	Equip.	Total
D3010.10.10	**HVAC, per building, minimum**						**SF**
019980103030	HVAC, per building, minimum	1.000	SF	$19.44	$0.00	$0.00	$19.44
D3010.10.12	**HVAC, per building, average**						**SF**
019980103050	HVAC, per building, average	1.000	SF	$27.19	$0.00	$0.00	$27.19
D3010.10.14	**HVAC, per building, maximum**						**SF**
019980103070	HVAC, per building, maximum	1.000	SF	$29.19	$0.00	$0.00	$29.19
D3010.10.20	**HVAC, controls, minimum**						**SF**
019980103210	HVAC Controls, minimum	1.000	SF	$3.69	$0.00	$0.00	$3.69
D3010.10.21	**HVAC, controls, average**						**SF**
019980103230	HVAC Controls, average	1.000	SF	$4.05	$0.00	$0.00	$4.05
D3010.10.23	**HVAC, controls, maximum**						**SF**
019980103250	HVAC Controls, maximum	1.000	SF	$4.88	$0.00	$0.00	$4.88
D3010.10.30	**HVAC, ductwork, minimum**						**SF**
019980103410	HVAC Ductwork, minimum	1.000	SF	$2.09	$0.00	$0.00	$2.09
D3010.10.31	**HVAC, ductwork, average**						**SF**
019980103430	HVAC Ductwork, average	1.000	SF	$4.18	$0.00	$0.00	$4.18
D3010.10.32	**HVAC, ductwork, maximum**						**SF**
019980103450	HVAC Ductwork, maximum	1.000	SF	$4.26	$0.00	$0.00	$4.26

D4010 SPRINKLERS

D4010.10 SPRINKLER SQUARE FOOT COSTS

		Quan.	Unit	Material	Labor	Equip.	Total
D4010.10.10	**Sprinkler, per building, minimum**						**SF**
019980102030	Fire Suppress., per building, minimum	1.000	SF	$1.46	$0.00	$0.00	$1.46
D4010.10.12	**Sprinkler, per building, average**						**SF**
019980102050	Fire Suppress., per building, average	1.000	SF	$3.36	$0.00	$0.00	$3.36
D4010.10.14	**Sprinkler, per building, maximum**						**SF**
019980102070	Fire Suppress., per building, maximum	1.000	SF	$4.84	$0.00	$0.00	$4.84

D SERVICES

D5010 ELECTRICAL

D5010.10 ELECTRICAL SQUARE FOOT COSTS

		Quan.	Unit	Material	Labor	Equip.	Total
D5010.10.10	**Electrical, per building, minimum**						**SF**
019980104030	Electrical, per building, minimum	1.000	SF	$24.19	$0.00	$0.00	$24.19
D5010.10.12	**Electrical, per building, average**						**SF**
019980104050	Electrical, per building, average	1.000	SF	$33.43	$0.00	$0.00	$33.43
D5010.10.14	**Electrical, per building, maximum**						**SF**
019980104070	Electrical, per building, maximum	1.000	SF	$50.09	$0.00	$0.00	$50.09
D5010.10.30	**Electrical, rough-in, minimum**						**SF**
019980104210	Electrical Rough-in, minimum	1.000	SF	$2.39	$0.00	$0.00	$2.39
D5010.10.32	**Electrical, rough-in, average**						**SF**
019980104230	Electrical Rough-in, average	1.000	SF	$4.27	$0.00	$0.00	$4.27
D5010.10.34	**Electrical, rough-in, maximum**						**SF**
019980104250	Electrical Rough-in, maximum	1.000	SF	$8.29	$0.00	$0.00	$8.29
D5010.10.50	**Electrical, Lighting, minimum**						**SF**
019980104410	Electrical Lighting, minimum	1.000	SF	$3.21	$0.00	$0.00	$3.21
D5010.10.52	**Electrical, Lighting, average**						**SF**
019980104430	Electrical Lighting, average	1.000	SF	$7.67	$0.00	$0.00	$7.67
D5010.10.54	**Electrical, Lighting, maximum**						**SF**
019980104450	Electrical Lighting, maximum	1.000	SF	$8.86	$0.00	$0.00	$8.86
D5010.10.60	**Electrical, switchgear, minimum**						**SF**
019980104510	Electrical Switchgear, minimum	1.000	SF	$1.67	$0.00	$0.00	$1.67
D5010.10.62	**Electrical, switchgear, average**						**SF**
019980104530	Electrical Switchgear, average	1.000	SF	$2.52	$0.00	$0.00	$2.52
D5010.10.64	**Electrical, switchgear, maximum**						**SF**
019980104550	Electrical Switchgear, maximum	1.000	SF	$3.22	$0.00	$0.00	$3.22

Square Foot Tables

The following Square Foot Tables list hundreds of actual projects for ten building types, each with associated building size, total square foot building cost and percentage of project costs for total mechanical and electrical components. This data provides an overview of construction costs by building type. These costs are for actual projects. The variations within similar building types may be due, among other factors, to size, location, quality and specified components, materials and processes. Depending upon all such factors, specific building costs can vary significantly and may not necessarily fall within the range of costs as presented. The data has been updated to reflect current construction costs.

BUILDING CATEGORY — **PAGE**

BNi® Building News
A DESIGN COST DATA COMPANY
DATA YOU CAN TRUST

All prices are updated to January 1, 2021 and are national averages.
For a more in-depth report of any of these buildings or additional case studies, contact Design, Cost and Data at 800-533-5680, or go to www.DCD.com

PROJECT	DESCRIPTION	CITY	STATE	SIZE	$/SF	NOTES
Commercial						
Bank	School Credit Union Administration Building	Katy	TX	30,700	$216.78	New
	FineMark National Bank & Trust	Fort Myers	FL	20,039	$408.96	New
	Florida Shores Bank	Pompano Beach	FL	11,697	$520.52	New
	Mobiloil Credit Union	Vidor	TX	9,252	$365.04	New
	Beaumont Community Credit Union	Beaumont	TX	3,267	$413.83	New
Office	Allendale Town Center	Allendale	NJ	80,226	$38.20	Addition/Renovation
	Roanoke Electric Cooperative	Ahoskie	NC	52,752	$217.92	New
	Regional Aviation & Training Center	Currituck	NC	39,930	$276.44	New
	Transportation/Warehouse Facility	Monroe	GA	32,400	$191.90	New
	Collection System Operations Facility	Walnut Creek	CA	27,179	$379.11	New
	ULTA - (Shell Only)	Pensacola	FL	10,850	$98.09	Renovation
	Daycare Center	Clawson	MI	4,270	$155.32	Adaptive Reuse
	Campgrounds Office & Retail	Cincinnati	OH	2,300	$340.27	New
Parking	Palm Avenue Parking Garage	Sarasota	FL	287,040	$58.13	New
	Reynolds Street Parking Deck	Augusta	GA	214,000	$69.15	New
	Awty Int. School Parking Structure	Houston	TX	174,582	$61.22	New
Retail	SNG Center - Mixed-Use	Fargo	ND	143,860	$102.89	New
	Roof & Lifeway/Steinmart Renovation	Pensacola	FL	88,299	$36.14	Renovation
	No Frills Supermarket	Omaha	NE	61,000	$110.07	New
	West Oaks Mall Redevelopment	Houston	TX	49,800	$195.89	Renovation
	World of Decor	Deerfield Beach	FL	47,500	$149.51	New
	Sarasota Yacht Club	Sarasota	FL	41,332	$439.28	New
	Karschs Village Market	Barnhart	MO	35,384	$85.80	New
	Fresh Thyme Farmers Market	Fishers	IN	28,784	$188.13	New
	Nashville Hangar Inc.	Nashville	TN	28,702	$197.26	New
	Party Time Plus	Billings	MT	26,000	$100.08	New
	Marshalls - (Shell Only)	Pensacola	FL	25,990	$47.89	Renovation
	Ed Hicks Mercedes-Benz USA	Corpus Christi	TX	25,273	$261.77	New
	Montana Honda & Marine	Billings	MT	22,963	$113.75	Addition
	Fresh Market - (Shell Only)	Pensacola	FL	21,000	$82.32	Renovation
	Theatre Exchange Interior Fit Up	Manitoba	CA	19,344	$132.78	Renovation
	DSW Shoes Renovation	Pensacola	FL	18,000	$92.17	Renovation
	Dormans Lighting & Design	Lutherville	MD	15,220	$124.94	Addition
	The Groves Exterior Renovation	Farmington	MI	15,137	$70.63	Renovation
	Don Gibson Theatre	Shelby	NC	13,386	$372.40	Renovation
	Tri Ford Showroom Expansion	Highland	IL	12,881	$121.80	Addition/Renovation
	Fiat of LeHigh Valley	Easton	PA	11,905	$155.97	New
	Sicardi Art Gallery	Houston	TX	6,175	$245.35	New
	Childrens Mercy Hospital Gift Shop	Kansas City	MO	5,010	$309.23	Tenant Build-out
Restaurant	LaMar Cebicheria Peruana Restaurant	San Francisco	CA	11,000	$316.16	Tenant Build-out
	Ulele Restaurant	Tampa	FL	8,905	$766.28	Adaptive Reuse
	Youells Oyster House	Allentown	PA	6,107	$251.55	New
	Mellow Mushroom Highlands Shell	Louisville	KY	5,802	$101.96	New
	Mellow Mushroom Highlands TBO	Louisville	KY	5,802	$169.11	Tenant Build-out
	Mellow Mushroom Pizza	Wilder	KY	5,500	$241.87	New
	Liberty Microbrewery	Plymouth	MI	3,425	$174.02	Addition
	700 South Deli	Linthicum	MD	3,200	$214.42	Tenant Build-out
	New York Pizza Department (NYPD)	Tempe	AZ	2,338	$328.05	Tenant Build-out
	Airport Restaurant Build Out	Eglin Air Force Base	FL	2,320	$282.41	Tenant Build-out

PROJECT	DESCRIPTION	CITY	STATE	SIZE	$/SF	NOTES
Civic/Government						
Civic Center	Lincoln Center	Fort Collins	CO	38,160	$216.29	Addition/Renovation
	Rockport City Services Building	Rockport	TX	20,062	$223.75	New
	Sinclair Park Community Centre	Manitoba	CA	17,007	$316.81	Addition/Renovation
	Mt. Olive City Hall Complex	Mount Olive	IL	14,360	$121.48	New
	Teaneck Municipal Complex	Teaneck	NJ	12,870	$232.81	Addition/Renovation
	Cobb Community Center Additions	Pensacola	FL	5,200	$324.01	Addition/Renovation
	Newtown Municipal Center	Newtown	OH	5,077	$168.59	Adaptive Reuse
	Nederland City Hall	Nederland	TX	4,983	$375.09	New
Correctional	County Sheriffs Office	Morgantown	WV	31,645	$297.28	New
	Nederland Public Safety Complex	Nederland	TX	21,189	$217.30	Adaptive Reuse
	Detention Center & Sheriffs Office	Spencer	IA	16,983	$421.41	New
	Ogle City Sheriff & Coroner Admin	Oregon	IL	15,377	$283.95	New
	Chautauqua City Jail & Sheriff	Sedan	KS	12,257	$299.47	New
Courthouse	Courthouse HVAC System Replacement	Gainesville	FL	101,000	$43.19	Renovation
	Courthouse Renovation & Restoration	Springfield	IL	47,720	$184.63	Renovation
Fire Department	College Station Fire Station #6	College Station	TX	25,133	$350.16	New
	Mt. Orab Fire Station	Village of Mt. Orab	OH	18,170	$180.68	New
	Richardson Fire Station No. 4	Richardson	TX	14,090	$411.12	New
	Willowfork Fire Station No. 2	Katy	TX	13,358	$302.19	New
	Joint Fire & Rescue Station	Newtown	OH	13,125	$194.79	Addition/Renovation
	Little Miami Fire & Rescue	Fairfax	OH	12,316	$231.30	New
	Fire Station No. 11	Fort Smith	AR	12,155	$350.36	New
	Pearisburg Fire Station	Pearisburg	VA	11,818	$225.50	New
	Ponderosa Fire Station No. 62	Spring	TX	11,163	$302.54	New
	El Dorado Hills Fire Station 84	El Dorado Hills	CA	10,869	$455.55	New
	Wayne Fire Department	Goshen	OH	10,000	$99.16	New
	Fire Station No. 40	Jacksonville	FL	9,703	$399.75	New
	Rosenberg Fire Station No. 3	Rosenberg	TX	8,479	$378.94	New
	Little Rock Fire Station No. 23	Little Rock	AR	8,291	$487.53	New
	Cleveland Volunteer Fire Station	Cleveland	MS	6,910	$340.81	New
Government	Council Center For Scouting	Fargo	ND	20,466	$204.40	New
	Camp Crook Ranger Station	Camp Crook	SD	4,880	$504.83	New
	Beaumont Municipal Tennis Center	Beaumont	TX	4,460	$369.39	Addition
	Florence Transit Hub	Florence	KY	3,115	$527.64	New
	Knox Area Rescue Ministries	Knoxville	TN	1,762	$694.90	New
	Entrance Station Lake Mead	Clark County	NV	480	$2,819.31	New
	Vehicle Charging Stations	Denton	TX	6 spaces	$7,233.53	New
Library	Dover Public Library	Dover	DE	46,424	$443.19	New
	Clinton-Macomb Public Library	Clinton Township	MI	24,723	$129.49	Adaptive Reuse
	Crozet Western Albemarle Library	Crozet	VA	23,199	$359.80	New
	Upper Tampa Bay Regional Library	Tampa	FL	13,630	$249.02	Addition/Renovation
	Palmetto Branch Library	Palmetto	GA	11,200	$496.97	New
	Regional Library Expansion	Valrico	FL	10,970	$305.27	Addition/Renovation
Miscellaneous	City of Pampa Animal Welfare	Pampa	TX	13,578	$289.19	New
	Royal Winnipeg Ballet Renovations	Manitoba	CA	13,237	$78.02	Renovation
	Senior Services - Kitchen Facility	Batavia	OH	6,000	$123.50	New
	Historical Site Locomotive Shelter	Bismarck	ND	2,100	$165.06	New
Office	Federal Building & Courthouse Modernization	Denver	CO	41,600	$357.90	Renovation
	Brazos County Tax Office	Bryan	TX	13,143	$307.93	New
	Illinois Water District Office	Lincoln	IL	8,974	$160.26	New
	Arkansas River Resource Center	Little Rock	AR	4,926	$588.40	New

All prices are updated to January 1, 2021 and are national averages.
For a more in-depth report of any of these buildings or additional case studies, contact Design, Cost and Data at 800-533-5680, or go to www.DCD.com

PROJECT	DESCRIPTION	CITY	STATE	SIZE	$/SF	NOTES
Educational						
Athletic Facility	Physical Activity/Sports Science	Morgantown	WV	117,344	$244.33	New
	Indoor Football Practice Facility	Clemson	SC	81,992	$188.54	New
	Jesuit College Locker Room Addition	Dallas	TX	41,673	$225.42	Addition
	Multi-Purpose Gymnasium	Jacksonville	FL	22,844	$299.13	New
College Classroom	New Mexico Tech Geology Building	Socorro	NM	86,813	$325.86	New
	CSU Concourse & Training Room	Fort Collins	CO	53,050	$152.21	Addition/Renovation
	Jack Williamson Liberal Arts Center	Portales	NM	52,480	$244.58	Renovation
	UNLV Literature & Law Building	Las Vegas	NV	44,830	$203.35	Renovation
	Southern State Community College	Mount Orab	OH	43,833	$195.65	New
	SERT Building Iowa Lakes College	Estherville	IA	42,940	$134.09	Adaptive Reuse
Elementary	Cibolo Valley Elementary School	Cibolo	TX	153,130	$264.86	New
	Hill Farm Elementary School	Bryant	AR	88,800	$316.82	New
	CREC International Magnet School	South Windsor	CT	63,923	$435.46	New
	Pineville Elementary School	Pineville	WV	51,650	$231.89	New
	Crownpoint Elementary School	Crownpoint	NM	48,592	$432.07	New
	Janney Elementary School Addition	Washington	DC	10,000	$638.61	Addition
High School	Hmong College Prep Academy	Saint Paul	MN	154,434	$90.68	Addition/Renovation
	HFC High School North Building	Flossmoor	IL	136,555	$221.96	Addition/Renovation
	Somerset High School	Von Ormy	TX	125,800	$192.78	New
	Takoma Education Campus	Washington	DC	119,000	$238.43	Renovation
	High School Addition & Renovation	Decatur	GA	83,816	$276.83	Addition/Renovation
	Palmer Catholic Academy	Ponte Vedra Beach	FL	34,209	$150.00	New
	Elmwood High School Addition	Elmwood Park	IL	31,630	$371.42	Addition/Renovation
	High School Fine Arts Building	Heber Springs	AR	30,505	$433.80	New
	Alamo Heights High School Fine Arts	San Antonio	TX	25,536	$348.60	Addition/Renovation
	St. Patrick Catholic School	Jacksonville	FL	23,227	$319.96	New
	Springdale School Alteration	Corbett	OR	13,680	$136.51	Renovation
	Goddard School Addition Renovation	Anderson Township	OH	12,489	$90.96	Addition/Renovation
	ISD Outdoor Education Center	Sabine Pass	TX	10,193	$345.68	New
	Indian Mountain School Student Center	Lakeville	CT	9,335	$331.86	Addition
	First Impressions Academy	Fayetteville	NC	7,752	$190.94	New
	High School South Campus Field	Cincinnati	OH	5,580	$256.08	New
Middle School	Timberline Middle School	Waukee	IA	187,375	$171.88	New
	Red Bank Middle School	Chattanooga	TN	158,637	$284.58	New
	Jaime Escalante Middle School	Pharr	TX	156,538	$213.09	New
	Conservatory Green ECE-8 School	Denver	CO	113,616	$190.04	New
	Midland School Addition	Floral	AR	30,150	$255.54	Addition
Laboratory/Research	Science & Technology Building	Fayetteville	NC	65,048	$539.12	New
	NSU/US Geological Survey	Davie	FL	24,000	$112.03	Tenant Build-out
	Northeast Technology Center	Pryor	OK	11,909	$336.06	New
	Environmental Education Center	Bushkill Township	PA	9,275	$582.64	New
	Research & Education Center	Homestead	FL	5,760	$735.26	New
Multi-Purpose	BGSU Student Rec Center	Bowling Green	OH	179,549	$70.62	Renovation
	Kennedy Center Theatre/Studio Arts	Clinton	NY	96,100	$356.65	New
	Classroom & Administration Building	Houston	TX	65,234	$230.55	New
	NM State U Pete Domenici Building	Las Cruces	NM	53,341	$290.60	Addition/Renovation
	Widener University Freedom Hall	Chester	PA	36,700	$343.98	New
	Alumni Hall, Lincoln Park	Midland	PA	29,027	$280.51	New
	GSU Piedmont North Dining Hall	Atlanta	GA	12,300	$393.83	Addition
	NAU Dining Hall Expansion Phase II	Flagstaff	AZ	10,096	$502.26	New
	Neighborhood Resource Center	Richmond	TX	6,935	$216.82	New
	Heber Springs Cafeteria Remodel	Heber Springs	AR	5,585	$244.33	Renovation

PROJECT	DESCRIPTION	CITY	STATE	SIZE	$/SF	NOTES
		Hotels				
Hotels	Omni Dallas Hotel	Dallas	TX	1,161,450	$443.81	New
	John Ascuagas Nugget Hotel/Casino	Sparks	NV	449,820	$151.41	Addition
	Le Centre On Fourth Embassy Suites	Louisville	KY	408,229	$119.82	Adaptive Reuse
	Minneapolis Marriott West	Minneapolis	MN	237,362	$172.95	New
	Sheraton Centre Park Hotel	Dallas	TX	231,031	$247.51	New
	AmeriSuites	Chicago	IL	191,600	$156.50	Addition/Renovation
	Hampton Inn & Suites Hotel	Chicago	IL	162,000	$170.91	New
	Sheraton Harbor Island Hotel Tower	San Diego	CA	144,126	$208.37	Addition
	Compri Hotel	Los Angeles	CA	110,150	$184.54	New
	Best Western Columbia Hotel	San Diego	CA	108,040	$142.76	New
	Spooky Nook Warehouse Hotel	Manheim	PA	92,726	$135.95	Adaptive Reuse
	The Atrium Motel	Norfolk	VA	75,889	$145.99	New
	Staybridge Hotel At Preston Ridge	Alpharetta	GA	74,607	$187.19	New
	Hampton Inn & Suites	Allentown	PA	71,686	$156.12	New
	Hampton Inn Hotel	Carol Stream	IL	71,000	$191.72	New
	The Inn On Lake Superior	Duluth	MN	65,345	$154.22	New
	The Lancaster Hotel	Houston	TX	64,310	$322.73	Renovation
	Fairfield Inn	Helena	MT	31,009	$150.71	New
	The Edison Hotel	Miami	FL	28,875	$124.94	Renovation
	Western Executive Inn	Billings	MT	21,984	$111.55	New
	Country Hearth Inn	Preston	MN	21,028	$130.97	New
	Lawrence Welk Resort Hotel	Escondido	CA	19,874	$149.39	Addition
	Hanalei Hotel Conference Center	San Diego	CA	8,587	$237.22	Addition
	Summit At Vail, Multi-Purpose Lodge	Vail	CO	6,000	$283.23	New

PROJECT	DESCRIPTION	CITY	STATE	SIZE	$/SF	NOTES
		Industrial				
Manufacturing	Brentwood Industries Manufacturing	Reading	PA	205,000	$43.98	New
	Lee Steel Corporate Plant	Romulus	MI	200,625	$95.66	New
	Siemens Westinghouse Fuel Cell Facility	Munhall	PA	191,090	$104.16	New
	Manufacturing Plant & Headquarters	Lansing	MI	188,975	$100.99	Addition
	Concepts Direct	Longmont	CO	117,900	$123.18	New
	Nypro Inc.	Clinton	MA	102,475	$126.36	Addition
	SWF Industrial	Wrightsville	PA	76,218	$77.85	New
	Headquarters & Manufacturing Facility	Lower Nazareth Township	PA	62,980	$37.29	Renovation
	Aerzen USA (Office/Manufacturing)	Coatesville	PA	40,000	$190.27	New
	Lee Steel Corporate Expansion	Wyoming	MI	34,821	$80.83	New
	Prescott Aerospace	Prescott	AZ	31,400	$116.45	New
	Battery Innovation Center	Newberry	IN	30,080	$454.35	New
	American Steel	Billings	MT	25,957	$84.46	New
	Phillip S. Luttazi Town Garage	Dover	MA	21,913	$145.33	New
	ITT Flygt - Industrial Facility	Milford	OH	16,991	$137.81	New
	Broadmoor Golf Maintenance	Colorado Springs	CO	16,064	$249.82	New
	Cooper B-Line Expansion	Highland	IL	15,290	$176.51	Addition/Renovation
	Robberson Ford Collision Center	Bend	OR	15,089	$144.89	New
	Brown Industrial Building	Truckee	CA	13,345	$143.13	New
	CTC Vehicle Maintenance Shops	Killeen	TX	11,250	$135.34	New
	Storage & Shop Facility	Billings	MT	8,763	$93.40	New
	Central Plant with Equipment Bay	Mesa	AZ	8,500	$962.99	New
Office	Woodlands Business Center	Richmond	VA	48,000	$83.66	New
	Wiregrass Research Center	Headland	AL	9,740	$258.37	New
Office/Warehouse	American Superconductor	Devens	MA	354,000	$177.79	New
	Castcon Stone Inc.	Saxonburg	PA	47,000	$119.08	New
	Minnesota DNR Headquarters	Tower	MN	37,802	$176.35	New
	Office & Warehouse	Miami	FL	14,815	$144.48	New
	DOT Office & Maintenance Building	Hillsboro	OH	10,876	$254.87	New
Warehouse	Distribution Center	Windsor	CT	303,750	$36.25	New
	Zany Brainy Distribution Center	Bridgeport	NJ	250,000	$43.29	New
	Galderma - Warehouse	Fort Worth	TX	70,000	$93.26	New
	Manzana Products Warehouse	Sebastopol	CA	41,395	$69.46	New
	Tactical Equip Maintenance Facility	Fort Campbell	KY	35,290	$250.61	New
	Sonoma Wine Company Canopy	Graton	CA	26,000	$58.96	New
	Administration/Chemical Storage Building	Killeen	TX	23,837	$356.50	New
	DOT Truck Storage Building	Hillsboro	OH	18,400	$97.26	New
	F.I. Storage Facility	Kentwood	MI	13,125	$71.52	New
	50 Columbia Drive Warehouse	Pooler	GA	10,000	$89.76	New
	DOT Salt Storage Building	Hillsboro	OH	9,100	$98.06	New
	Organizational Storage Facility	Fort Campbell	KY	8,040	$148.79	New
	Dwan Maintenance Building	Bloomington	MN	7,240	$200.94	Addition/Renovation
	Maintenance/Storage Building	Batavia	OH	7,200	$86.43	New
	DOT Cold Storage Building	Hillsboro	OH	5,040	$104.56	New
	Job Corp Warehouse	Hartford	CT	3,800	$495.71	New
	DOT Materials Storage Building	Hillsboro	OH	1,920	$116.90	New
	Aerial Vehicle Storage	Fort Campbell	KY	1,800	$223.41	New
	Petro, Oil, Lubricant Storage	Fort Campbell	KY	640	$312.31	New
	Hazardous Waste Storage Building	Fort Campbell	KY	640	$333.76	New

PROJECT	DESCRIPTION	CITY	STATE	SIZE	$/SF	NOTES
Medical						
Clinic	HealthCare Emergency/Trauma Center	Topeka	KS	115,000	$421.22	Addition
	Sadler Clinic (Shell Only)	Conroe	TX	61,599	$131.13	New
	Pinellas County Health Department	Largo	FL	54,965	$273.25	Retrofit
	Sanford Moorhead Clinic	Moorhead	MN	49,250	$267.79	New
	County Health Department	Port Charlotte	FL	47,564	$303.63	New
	Sadler Clinic	Conroe	TX	41,066	$124.23	Tenant Build-out
	PineMed Medical Plaza	The Woodlands	TX	30,398	$138.54	New
	Outpatient Specialty Clinic	Vancouver	WA	20,139	$344.70	New
	Ambulatory Surgery Center	Stroudsburg	PA	19,929	$369.95	Addition/Renovation
	Thundermist Health Center	West Warwick	RI	18,217	$213.31	Adaptive Reuse
	E Texas Community Health Services	Nacogdoches	TX	12,500	$101.02	Retrofit
	North Mobile Health Center	Mt. Vernon	AL	6,765	$301.35	New
Dental Office	Construct Dental Clinic Roseburg	Roseburg	OR	7,750	$524.39	New
	Kitchens Pediatric Dental Clinic	Little Rock	AR	6,068	$315.67	New
	Dental Office Shell & Parking	Olympia	WA	5,302	$179.11	New
	Evans Family Dental	Austin	TX	2,354	$219.51	Tenant Build-out
Hospital	Union Hospital Addition	Terre Haute	IN	492,348	$329.92	Addition
	Regional Medical Center	Lafayette	LA	410,273	$552.16	New
	Houston Medical Pavilion	Warner Robins	GA	180,000	$65.05	Adaptive Reuse
	Langley AFB Hospital Renovation	Langley Air Force Base	VA	160,000	$551.26	Renovation
	Cass Regional Medical Center	Harrisonville	MO	137,524	$387.62	New
	Texas Spine & Joint Hospital	Tyler	TX	115,789	$256.73	Addition/Renovation
	Oktibbeha County Hospital Expansion	Starkville	MS	87,116	$336.41	New
	Childrens Mercy Hospital	Independence	MO	54,682	$324.12	New
	Oktibbeha County Hospital Renovation	Starkville	MS	30,263	$167.91	Renovation
	El Rio Community Health Center	Tucson	AZ	26,998	$247.56	New
	Pondella Public Health Center	Fort Myers	FL	26,400	$308.35	Renovation
	Topeka Ear Nose & Throat	Topeka	KS	24,073	$319.70	New
	Rapha Primary Care	Fayetteville	NC	19,907	$137.25	Renovation
	UNM Hospitals North Valley Center	Albuquerque	NM	16,500	$323.25	New
	El Rio Community Health Center	Tucson	AZ	14,000	$281.06	New
	VA Medical Center Area G Renovation	Houston	TX	12,000	$341.99	Renovation
	Surgical Suite Expansion	Dobbs Ferry	NY	9,000	$390.34	Renovation
	Legacy Emergency Room	Allen	TX	8,432	$647.41	New
	Oral & Maxillofacial Surgery Center	Fayetteville	NC	2,214	$550.86	Renovation
Nursing Home/Rehab	Senior Living Community	Hoschton	GA	56,251	$143.56	New
	Retirement Community	Carlisle	PA	47,075	$161.66	Addition/Renovation
	Assisted Living & Memory Center	Dacula	GA	38,221	$166.98	New
	Homestead Village Nursing Care	Lancaster	PA	28,149	$109.33	Renovation
	Jewish Services For The Aging	Tucson	AZ	24,993	$203.54	New
	Short-Term Rehabilitation	Olathe	KS	13,800	$273.81	Addition
	St. Katharine Retirement Center	El Reno	OK	12,000	$323.60	New/Addition
Office	Orthopedic Hospital/Medical Office	Allentown	PA	79,807	$205.10	Renovation
	Tomball Medical Office Building	Tomball	TX	54,380	$139.46	New
	Olathe Health Education Center	Olathe	KS	50,258	$311.86	New
	Home & Hospice Care	Providence	RI	47,734	$196.72	Renovation
	NE Georgia Medical Plaza 400	Dawsonville	GA	26,997	$274.47	Adaptive Reuse
	VA Medical Center/Pharmacy	Waco	TX	19,171	$159.81	Renovation
	Cancer Specialists of North Florida	Jacksonville	FL	18,654	$291.51	New
	MJHS Hospice Residence	N.Y.C.	NY	12,500	$150.84	Renovation
	Medical Office Building	Pelham	NH	8,399	$271.46	New
	Podiatry Group	Marietta	GA	6,768	$55.20	Renovation
	Marietta Podiatry Group	Marietta	GA	4,400	$252.11	New

All prices are updated to January 1, 2021 and are national averages.
For a more in-depth report of any of these buildings or additional case studies, contact Design, Cost and Data at 800-533-5680, or go to www.DCD.com

PROJECT	DESCRIPTION	CITY	STATE	SIZE	$/SF	NOTES
Office						
Office	Restaurant Support Center	Lenexa	KS	186,465	$266.92	New
	5000 NASA Boulevard	Fairmont	WY	132,000	$284.61	New
	Rockford Construction Office	Grand Rapids	MI	71,144	$113.72	Adaptive Reuse
	Woodlawn Office Bldg. (Shell Only)	Louisville	KY	60,000	$143.86	New
	Rosecrance Ware Center	Rockford	IL	44,800	$141.98	Adaptive Reuse
	Professional Center (Shell)	White Marsh	MD	43,025	$180.98	New
	Swan Skyline Office Plaza (Shell)	Tucson	AZ	37,200	$122.07	New
	Infinite Energy Phase IV	Gainesville	FL	36,500	$319.97	New
	Freedom Plaza Building	Cookeville	TN	28,488	$250.17	New
	Landmark Professional Building	Clayton	NC	27,231	$237.21	New
	FC Gulf Freeway Building (Shell)	Houston	TX	24,084	$270.05	New
	White Street Building (Shell Only)	Marietta	GA	23,809	$220.38	New
	Columbia Shores Office Condo	Vancouver	WA	22,574	$182.90	New
	Office & Design Studio	Chicago	IL	20,244	$128.41	Tenant Build-out
	Pinnacle III Office Tenant Finish	Leawood	KS	18,409	$61.13	Tenant Build-out
	Commerce Park (Shell)	Suwanee	GA	17,097	$125.37	New
	PCWA Business Center Interior	Auburn	CA	12,085	$63.59	Renovation
	Tenth Avenue Holdings Offices	N.Y.C.	NY	11,000	$89.26	Tenant Build-out
	Office Park - Building A (Shell)	Fort Collins	CO	10,000	$274.11	New
	Longshoremens Welfare Fund Building	Savannah	GA	8,160	$402.45	New
	Garry Street Office Building	Manitoba	CA	7,506	$160.58	Renovation
	Reserve Advisors	Milwaukee	WI	5,300	$51.48	Tenant Build-out
	510 Armory Street Office	Boston	MA	5,100	$98.57	Renovation
	Offices of Bonsall Shafferman	Bethlehem	PA	4,950	$72.91	Tenant Build-out
	FCT Capital Partners	Houston	TX	4,100	$85.29	Tenant Build-out
	Martin Rogers Associates Office	Wilkes-Barre	PA	4,000	$136.48	Addition/Renovation
	Cowan & Kohne Financial	Suwanee	GA	3,713	$128.52	Tenant Build-out
	212 Archer Street Office	Bel Air	MD	3,600	$181.80	New
	Utilities Analyses Inc.	Suwanee	GA	3,513	$119.21	Tenant Build-out
	Visual Lizard Interior Fit-Up	Manitoba	CA	2,430	$95.65	Tenant Build-out
	Richardson State Farm	Houston	TX	2,200	$115.65	Tenant Build-out
Mixed-Use	Office/Retail/Parking Mixed-Use	Jackson	MS	228,407	$262.41	New
	Korte & Luitjohan Office & Shop	Highland	IL	26,000	$126.09	New
Medical Office	Evanston Medical Office Building	Evanston	WY	9,157	$289.02	New
	Advanced Medical Group	Suwanee	GA	4,433	$122.72	Tenant Build-out
	Bothell Dental Office Build Out	Bothell	WA	2,203	$203.79	Tenant Build-out
Headquarters	CONSUL Energy Corporation Headquarters	Southpointe, Canonsburg	PA	317,500	$243.40	New
	Fairmont Supply Corporate Headquarters	Southpointe, Canonsburg	PA	75,255	$160.85	New
	Practice Velocity Corporate Headquarters	Machesney Park	IL	64,318	$101.07	Adaptive Reuse
	Enterprise Integration Headquarters	Jacksonville	FL	57,723	$62.88	Renovation
	Linear Technology	Cary	NC	20,000	$326.65	New
	PIPS Technology Inc.	Knoxville	TN	19,884	$259.65	New
	Lee Steel Corporate Offices	Novi	MI	15,781	$172.57	Renovation
	Weaver Cooke Headquarters	Goldsboro	NC	15,464	$310.53	New
	In Capital Holdings	Boca Raton	FL	13,000	$174.52	Tenant Build-out
	ACCION Regional Headquarters	Albuquerque	NM	7,580	$334.09	New
Civic Office	Miss Department of Environmental Quality	Jackson	MS	121,170	$97.32	Renovation
	County Central Office Complex	Pensacola	FL	74,630	$294.96	New
	State of WV Office Building	Fairmont	WV	70,442	$262.81	New
	JAX Chamber of Commerce Renovation	Jacksonville	FL	20,110	$216.99	Renovation

Recreational

PROJECT	DESCRIPTION	CITY	STATE	SIZE	$/SF	NOTES
Educational	The Pavilion at Ole Miss	Oxford	MS	235,301	$461.96	New
	University Laker Turf Building	Allendale	MI	137,662	$153.17	New
	CSU Recreation Center	Chico	CA	110,245	$476.63	New
	Intramural Recreation Penn State	State College	PA	59,303	$422.80	Addition/Renovation
	Center For Women's Athletics	Fayetteville	AR	39,183	$376.49	New
	Pickens Recreation Center	Pickens	SC	20,400	$187.94	New
	High School Concessions & Press Box	Loganville	GA	1,506	$482.87	New
Health Club	Brooklyn Yard Fitness Club (Shell)	Portland	OR	63,987	$118.32	New
	Title Boxing Club	Cedar Hill	TX	4,330	$73.35	Tenant Build-out
Recreational	Community Recreation Center	Williston	ND	223,787	$459.86	New
	Spirit Lake Casino & Resort	St. Michael	ND	112,277	$62.37	Renovation
	Phipps Tropical Forest	Pittsburgh	PA	80,000	$335.63	New
	Family Recreation Center	Colonie	NY	70,256	$245.30	New
	Youth Activity Center	Joplin	MO	62,056	$103.88	New
	New Holland Recreational Center	New Holland	PA	51,256	$135.45	Renovation
	Community College Recreation Center	Cedar Rapids	IA	43,500	$170.93	New
	Anderson Recreation Center	Anderson	SC	34,282	$383.25	New
	East Park Community Center	Nashville	TN	33,000	$307.43	New
	Church Family Life Center	Clemson	SC	31,509	$331.75	New/Renovation
	C.K. Ray Recreation Center	Conroe	TX	30,380	$156.57	Addition/Renovation
	St. Raphael Athletic & Wellness Center	Pawtucket	RI	30,268	$271.07	New
	The Forge For Families	Houston	TX	29,860	$255.57	New
	Christian Life Center	Birmingham	MI	26,966	$349.81	Addition
	Children's Sports Center	Woodbury	MN	26,219	$125.10	New
	Trinity River Audubon Center	Dallas	TX	20,791	$964.61	New
	Baptist Church Activity Center	Indianapolis	IN	16,636	$154.67	New
	Boys & Girls Club Syracuse	Syracuse	NY	12,107	$189.04	Addition
	Job Corp Recreational Building	Hartford	CT	11,300	$212.11	New
	Presbyterian Family Life Center	Strawberry Plains	TN	11,236	$172.35	New
	McDaniel Yacht Basin	North East	MD	7,620	$194.11	New
	Bicentennial Park	Cincinnati	OH	4,050	$909.31	New
	Bahosky Softball Complex	Bronx	NY	3,800	$491.75	New
Swimming Center	Resort & Indoor Waterpark	Cortland	NY	175,060	$274.72	New
	The Aquatic Center	Tunica	MS	45,008	$327.33	New
	Spirit Lake Phase 4	St. Michael	ND	26,630	$374.25	Addition/Renovation
	Family Aquatic Center	Beachwood	OH	7,500	$1,216.24	New
	Community Aquatic Park & Center	Billings	MT	6,730	$764.93	New
Theater	Cinema & IMAX Theatre	Lansing	MI	13,750	$231.22	New
	Academy Theater	N.Y.C.	NY	4,593	$234.91	Renovation
YMCA	David D. Hunting YMCA	Grand Rapids	MI	162,966	$210.71	New
	YMCA Recreational Center	Ann Arbor	MI	83,377	$286.80	New
	Floyd Co. YMCA & Aquatic Center	New Albany	IN	82,324	$349.20	New
	Wade Walker Park Family YMCA	Stone Mountain	GA	59,134	$364.48	New
	Alexandria YMCA	Alexandria	MN	55,150	$176.35	New
	Greater Nashua YMCA	Nashua	NH	49,980	$217.06	New
	Lancaster YMCA Harrisburg Ave.	Lancaster	PA	42,502	$369.13	New
	Greater Kingsport Family YMCA	Kingsport	TN	40,007	$263.67	New
	Eastside YMCA	Knoxville	TN	39,984	$261.97	New
	Highland County Family YMCA	Hillsboro	OH	33,228	$156.09	New
	Houston Texans YMCA	Houston	TX	31,628	$370.40	New
	Cypress Creek YMCA	Houston	TX	25,699	$156.22	Addition/Renovation

All prices are updated to January 1, 2021 and are national averages.
For a more in-depth report of any of these buildings or additional case studies, contact Design, Cost and Data at 800-533-5680, or go to www.DCD.com

PROJECT	DESCRIPTION	CITY	STATE	SIZE	$/SF	NOTES
			Religious			
Church	First United Methodist Church	Orlando	FL	121,536	$291.57	New
	Beautiful Savior Lutheran Church	Plymouth	MN	69,700	$119.25	New
	Solid Rock Baptist Church	Berlin	NJ	58,359	$107.82	New
	North Side Baptist Church	Greenville	SC	48,087	$275.35	New
	St. Martha Catholic Church	Porter	TX	46,748	$544.05	New
	Gracepoint Gospel Fellowship Church	Ramapo	NY	46,595	$202.36	New
	Good Shepherd Methodist Church	Odessa	TX	41,003	$322.30	New
	Davisville Church Addition	Southampton	PA	36,090	$190.55	Addition
	Immaculate Catholic Church	Columbia	IL	34,000	$266.45	New
	Keystone Community Church	Ada	MI	29,775	$156.89	New
	Grace Church	Des Moines	IA	29,296	$212.86	New
	Good Shepherd Church	Naperville	IL	27,869	$210.85	Addition/Renovation
	River Hills Baptist Church	Corpus Christi	TX	27,404	$319.22	New
	Good Shepherd Catholic Church	Smithville	MO	24,810	$219.30	New
	Prince of Peace Catholic Church	Chesapeake	VA	24,740	$274.77	Addition/Renovation
	Sanctuary Addition Christian Church	Oklahoma City	OK	23,820	$164.70	Addition
	St. Sylvester Catholic Church	Gulf Breeze	FL	22,000	$431.86	New
	St. Peters Catholic Sanctuary	Fallbrook	CA	20,764	$475.86	New
	Notre Dame Catholic Church	Houston	TX	20,280	$438.99	New
	St Eugene Catholic Church	Oklahoma City	OK	20,000	$481.16	New
	Chapin Presbyterian Church	Chapin	SC	19,900	$350.82	New
	St. Michaels Catholic Church	Glen Allen	VA	19,770	$376.38	New
	Shrine of Holy Spirit	Branson	MO	19,200	$358.18	New
	Wildwood United Methodist Church	Magnolia	TX	19,000	$255.93	New
	Episcopal Church of the Nativity	Scottsdale	AZ	18,288	$128.00	Adaptive Reuse
	Hardin Church of Christ	Knoxville	TN	17,149	$128.68	New
	First United Methodist Church	Crossville	TN	15,816	$503.64	New
	Good Shepherd Episcopal Church	Silver Spring	MD	15,200	$300.51	Addition/Renovation
	Ascension Catholic Church	LaPlace	LA	15,057	$396.69	New
	St. Patrick Catholic Church	Jacksonville	FL	14,139	$283.55	New
	Our Lady of Guadalupe Catholic Church	Rosenberg	TX	12,910	$437.35	New
	St. Timothy's Episcopal Church	Creve Coeur	MO	12,682	$218.03	New/Renovation
	St. Paul Lutheran Church	Pomaria	SC	12,072	$356.98	New
	Covenant Baptist Church	Florida City	FL	10,725	$266.36	New
	First United Methodist Church	Katy	TX	10,503	$323.51	Addition/Renovation
	Lake Ann United Methodist Church	Lake Ann	MI	9,975	$223.55	New
	United Methodist Church	Odenton	MD	8,783	$349.29	New
	Kent R. Hance Chapel at Texas Tech	Lubbock	TX	6,530	$634.03	New
	Haven for Hope Chapel	San Antonio	TX	2,232	$476.23	New
Multi-Purpose	Baptist Church Multi-Purpose Bldg	Maryville	TN	41,656	$118.82	New
	Good Shepherd Parish Center	San Diego	CA	28,752	$163.89	New
	Christian Life Center	Kansas City	MO	26,320	$318.31	New
	Baptist Church Outreach Center	Fort Smith	AR	25,000	$279.49	New
	St Rafael Administration Building	San Diego	CA	24,276	$176.52	New
	Catholic Church Social Hall	Chula Vista	CA	23,596	$293.69	New
	United Methodist Church	West Chester	PA	11,935	$256.82	Addition/Renovation
	Student Ministry Center	Knoxville	TN	11,700	$318.18	New
	Catholic Pastoral Ministries Center	Spring	TX	10,135	$408.50	New
	Christian Renewal Center	Dickinson	TX	8,500	$224.58	New
	Holy Family St Lawrence Parish Center	Essex Junction	VT	7,900	$235.52	New
	Presbyterian Church Addition	Gap	PA	7,414	$270.36	Addition/Renovation
	New Hope Church Addition/Alteration	Saint Louis	MO	5,564	$207.01	Renovation

PROJECT	DESCRIPTION	CITY	STATE	SIZE	$/SF	NOTES
Residential						
Apartment	Solace Apartments	Virginia Beach	VA	331,681	$104.79	New
	1221 Broadway Lofts	San Antonio	TX	205,137	$148.00	Adaptive Reuse
	Sustainable Fellwood Phase I	Savannah	GA	124,037	$138.45	New
	Kelly Cullen Community	San Francisco	CA	98,385	$573.66	Adaptive Reuse
	Bachelors Enlisted Quarters	Camp Williams	UT	76,253	$260.84	New
	Mockingbird Terrace Homes	Louisville	KY	71,110	$165.78	New
	Homeless Men's Residential	San Antonio	TX	67,908	$274.80	New
	Homeless Women's/Family Residence	San Antonio	TX	60,182	$264.80	New
	Magnolia Place	Lancaster	PA	39,714	$166.99	New
	Elkins First Ward Apartments	Elkins	WV	27,000	$119.88	Adaptive Reuse
	Young Burlington Apartments	Los Angeles	CA	24,399	$214.84	Now
	The Lofts at 300 Bowman	Dickson City	PA	23,900	$89.56	Adaptive Reuse
	Peaceful Paths Emergency Svc Campus	Gainesville	FL	22,535	$153.73	New
	Wylie House - Ronald McDonald House	Kansas City	MO	21,885	$194.99	New
	Dogwood Manor Apartments	Oak Ridge	TN	19,975	$176.85	New
	Anderson Village Multi-Family	Austin	TX	12,500	$295.24	New
	Salvation Army Sally's House	Houston	TX	7,812	$257.65	Addition
	Stones River Apartment Complex	Murfreesboro	TN	7,548	$238.26	Addition
	Sunshine Park Apartments Renovation	Gainesville	FL	2,252	$126.29	Renovation
Assisted Living	Kenmore Apartments Senior Housing	Chicago	IL	90,528	$220.91	Renovation
	Country Meadows Retirement	Allentown	PA	53,237	$169.03	New
	Creekside Village Assisted Living	Harrisburg	PA	16,150	$146.90	New
	Landis Homes Retirement Community	Lititz	PA	14,255	$69.43	Renovation
Dormitory	NSU Graduate Student Housing	Davie	FL	203,500	$226.96	Renovation
	Rider University Student Housing	Lawrenceville	NJ	50,500	$244.12	New
	JWU Biscayne Commons Dormitory	Miami	FL	40,048	$267.78	New
	College Residence Dorm	Bloomfield	NJ	25,980	$329.72	Renovation
Single-Family Home	Island Residence	Grosse Ile	MI	19,237	$711.73	New
	MG Residence Restoration	Williamston	MI	9,768	$64.01	Renovation
	Concepcion House	Coral Gables	FL	6,067	$367.06	New
	Leal House	Miami	FL	5,935	$253.59	New/Renovation
	Monserrate Street Residence	Coral Gables	FL	5,885	$458.63	New
	Private Residence	Newburgh	IN	5,566	$365.63	New
	Private Residence	Austin	MN	5,489	$150.64	New
	Private Residence	Lake Wallenpaupack	PA	4,845	$329.92	New
	Island in the Grove	Boca Raton	FL	4,701	$377.14	New
	Private Residence	Benson	AZ	3,660	$183.96	New
	Fairhope Green Home	Fairhope	AL	3,610	$209.44	New
	PATH Concept House	Omaha	NE	3,490	$80.23	New
	Private Residence	La Jolla	CA	3,420	$370.20	New
	Solar House - Private Residence	Fly Creek	NY	3,304	$185.27	New
	Elliott Residence	Fort Collins	CO	3,300	$279.38	New
	Renfrew House	Manitoba	CA	3,206	$199.94	New
	Rosado I Hansen Residence	Tucson	AZ	3,175	$139.47	New
	306 W. Waldburg Residence	Savannah	GA	2,588	$174.90	New
	Nutter Green Home	Milford	OH	2,289	$166.78	New
	Guest House Residence	Ahwatuckee	AZ	1,913	$407.10	New
	Kiwi House	Baton Rouge	LA	1,515	$162.58	New
	Private Residence Renovation	Shavertown	PA	810	$168.99	Renovation

All prices are updated to January 1, 2021 and are national averages.
For a more in-depth report of any of these buildings or additional case studies, contact Design, Cost and Data at 800-533-5680, or go to www.DCD.com

Metro Area Multipliers

The costs as presented in this book attempt to represent national averages. Costs, however, vary among regions, states and even between adjacent localities.

In order to more closely approximate the probable costs for specific locations throughout the U.S., this table of Metro Area Multipliers is provided. These adjustment factors can be used to modify costs obtained from this book to help account for regional variations of construction costs and to provide a more accurate estimate for specific areas. The factors are formulated by comparing costs in a specific area to the costs presented in this Costbook. An example of how to use these factors is shown below. Whenever local current costs are known, whether material prices or labor rates, they should be used when more accuracy is required.

Cost from Costbook Pages X Metro Area Multiplier Divided by 100 = Adjusted Cost

For example, a project estimated to cost $1,000,000 using the Costbook pages can be adjusted to more closely approximate the cost in Los Angeles:

$$\$1,000,000 \times \frac{133}{100} = \$1,330,000$$

© 2020 By BNi Publications Inc. ALL RIGHTS RESERVED

State	Metropolitan Area	Multiplier
AK	ANCHORAGE	130
AL	ANNISTON	77
	AUBURN	75
	BIRMINGHAM	78
	DECATUR	75
	DOTHAN	75
	FLORENCE	75
	GADSDEN	77
	HUNTSVILLE	76
	MOBILE	76
	MONTGOMERY	76
	OPELIKA	75
	TUSCALOOSA	79
AR	FAYETTEVILLE	73
	FORT SMITH	73
	JONESBORO	74
	LITTLE ROCK	77
	NORTH LITTLE ROCK	77
	PINE BLUFF	75
	ROGERS	79
	SPRINGDALE	74
	TEXARKANA	75
AZ	FLAGSTAFF	80
	MESA	80
	PHOENIX	82
	TUCSON	76
	YUMA	76
CA	BAKERSFIELD	130
	CHICO	136
	FAIRFIELD	136
	FRESNO	139
	LODI	139
	LONG BEACH	133
	LOS ANGELES	133
	MERCED	139
	MODESTO	139
	NAPA	136
	OAKLAND	139
	ORANGE COUNTY	133
	PARADISE	136
	PORTERVILLE	132
	REDDING	136
	RIVERSIDE	133
	SACRAMENTO	136
	SALINAS	139
	SAN BERNARDINO	133
	SAN DIEGO	130

© 2020 By BNi Publications Inc. ALL RIGHTS RESERVED

State	Metropolitan Area	Multiplier
CA	SAN FRANCISCO	139
	SAN JOSE	139
	SAN LUIS OBISPO	133
	SANTA BARBARA	133
	SANTA CRUZ	139
	SANTA ROSA	136
	STOCKTON	139
	TULARE	132
	VALLEJO	136
	VENTURA	133
	VISALIA	132
	WATSONVILLE	139
	YOLO	136
	YUBA CITY	136
CO	BOULDER	90
	COLORADO SPRINGS	89
	DENVER	90
	FORT COLLINS	86
	GRAND JUNCTION	87
	GREELEY	88
	LONGMONT	88
	LOVELAND	90
	PUEBLO	87
CT	BRIDGEPORT	127
	DANBURY	127
	HARTFORD	122
	MERIDEN	125
	NEW HAVEN	125
	NEW LONDON	122
	NORWALK	127
	NORWICH	122
	STAMFORD	127
	WATERBURY	125
DC	WASHINGTON	102
DE	DOVER	123
	NEWARK	122
	WILMINGTON	114
FL	BOCA RATON	84
	BRADENTON	82
	CAPE CORAL	81
	CLEARWATER	83
	DAYTONA BEACH	81
	FORT LAUDERDALE	85
	FORT MYERS	79
	FORT PIERCE	81
	FORT WALTON BEACH	79
	GAINESVILLE	79

© 2020 By BNi Publications Inc. ALL RIGHTS RESERVED.

State	Metropolitan Area	Multiplier
FL	JACKSONVILLE	79
	LAKELAND	79
	MELBOURNE	81
	MIAMI	85
	NAPLES	82
	OCALA	80
	ORLANDO	81
	PANAMA CITY	79
	PENSACOLA	78
	PORT ST. LUCIE	81
	PUNTA GORDA	81
	SARASOTA	81
	ST. PETERSBURG	85
	TALLAHASSEE	80
	TAMPA	84
	TITUSVILLE	94
	WEST PALM BEACH	85
	WINTER HAVEN	85
GA	ALBANY	80
	ATHENS	80
	ATLANTA	77
	AUGUSTA	79
	COLUMBUS	76
	MACON	78
	SAVANNAH	79
HI	HONOLULU	133
IA	CEDAR FALLS	100
	CEDAR RAPIDS	105
	DAVENPORT	111
	DES MOINES	111
	DUBUQUE	104
	IOWA CITY	107
	SIOUX CITY	100
	WATERLOO	100
ID	BOISE CITY	83
	POCATELLO	88
IL	BLOOMINGTON	126
	CHAMPAIGN	121
	CHICAGO	144
	DECATUR	119
	KANKAKEE	126
	NORMAL	126
	PEKIN	123
	PEORIA	123
	ROCKFORD	126
	SPRINGFIELD	119
	URBANA	121

© 2020 By BNi Publications Inc. ALL RIGHTS RESERVED

State	Metropolitan Area	Multiplier
IN	BLOOMINGTON	105
	ELKHART	105
	EVANSVILLE	104
	FORT WAYNE	105
	GARY	119
	GOSHEN	105
	INDIANAPOLIS	105
	KOKOMO	105
	LAFAYETTE	105
	MUNCIE	105
	SOUTH BEND	119
	TERRE HAUTE	104
KS	KANSAS CITY	111
	LAWRENCE	109
	TOPEKA	108
	WICHITA	97
KY	LEXINGTON	100
	LOUISVILLE	100
	OWENSBORO	99
LA	ALEXANDRIA	80
	BATON ROUGE	82
	BOSSIER CITY	82
	HOUMA	81
	LAFAYETTE	81
	LAKE CHARLES	81
	MONROE	80
	NEW ORLEANS	81
	SHREVEPORT	82
MA	BARNSTABLE	139
	BOSTON	139
	BROCKTON	134
	FITCHBURG	133
	LAWRENCE	139
	LEOMINSTER	133
	LOWELL	139
	NEW BEDFORD	139
	PITTSFIELD	133
	SPRINGFIELD	125
	WORCESTER	133
	YARMOUTH	139
MD	BALTIMORE	89
	CUMBERLAND	91
	HAGERSTOWN	91
ME	AUBURN	84
	BANGOR	83
	LEWISTON	84
	PORTLAND	84

© 2020 By BNi Publications Inc. ALL RIGHTS RESERVED.

State	Metropolitan Area	Multiplier
MI	ANN ARBOR	118
	BATTLE CREEK	103
	BAY CITY	105
	BENTON HARBOR	118
	DETROIT	118
	EAST LANSING	108
	FLINT	110
	GRAND RAPIDS	87
	HOLLAND	93
	JACKSON	110
	KALAMAZOO	103
	LANSING	108
	MIDLAND	101
	MUSKEGON	93
	SAGINAW	103
MN	DULUTH	122
	MINNEAPOLIS	122
	ROCHESTER	122
	ST. CLOUD	118
	ST. PAUL	123
MO	COLUMBIA	114
	JOPLIN	97
	KANSAS CITY	115
	SPRINGFIELD	98
	ST. JOSEPH	114
	ST. LOUIS	115
MS	BILOXI	69
	GULFPORT	69
	HATTIESBURG	69
	JACKSON	69
	PASCAGOULA	69
MT	BILLINGS	98
	GREAT FALLS	96
	MISSOULA	98
NC	ASHEVILLE	74
	CHAPEL HILL	71
	CHARLOTTE	72
	DURHAM	71
	FAYETTEVILLE	71
	GOLDSBORO	71
	GREENSBORO	71
	GREENVILLE	71
	HICKORY	72
	HIGH POINT	72
	JACKSONVILLE	74
	LENOIR	74
	MORGANTON	72

© 2020 By BNi Publications Inc. ALL RIGHTS RESERVED

State	Metropolitan Area	Multiplier
NC	RALEIGH	70
	ROCKY MOUNT	72
	WILMINGTON	74
	WINSTON SALEM	71
ND	BISMARCK	91
	FARGO	94
	GRAND FORKS	92
NE	LINCOLN	92
	OMAHA	92
NH	MANCHESTER	89
	NASHUA	89
	PORTSMOUTH	90
NJ	ATLANTIC CITY	140
	BERGEN	143
	BRIDGETON	138
	CAPE MAY	138
	HUNTERDON	141
	JERSEY CITY	143
	MIDDLESEX	142
	MILLVILLE	138
	MONMOUTH	127
	NEWARK	143
	OCEAN	138
	PASSAIC	144
	SOMERSET	141
	TRENTON	128
	VINELAND	138
NM	ALBUQUERQUE	81
	LAS CRUCES	81
	SANTA FE	80
NV	LAS VEGAS	128
	RENO	125
NY	ALBANY	116
	BINGHAMTON	113
	BUFFALO	111
	DUTCHESS COUNTY	116
	ELMIRA	114
	GLENS FALLS	114
	JAMESTOWN	110
	NASSAU	116
	NEW YORK	164
	NEWBURGH	116
	NIAGARA FALLS	125
	ROCHESTER	114
	ROME	116
	SCHENECTADY	116
	SUFFOLK	154

© 2020 By BNi Publications Inc. ALL RIGHTS RESERVED.

State	Metropolitan Area	Multiplier
NY	SYRACUSE	112
	TROY	116
	UTICA	112
OH	AKRON	107
	CANTON	102
	CINCINNATI	103
	CLEVELAND	111
	COLUMBUS	104
	DAYTON	102
	ELYRIA	107
	HAMILTON	103
	LIMA	104
	LORAIN	107
	MANSFIELD	104
	MASSILLON	102
	MIDDLETOWN	102
	SPRINGFIELD	103
	STEUBENVILLE	104
	TOLEDO	107
	WARREN	107
	YOUNGSTOWN	107
OK	ENID	80
	LAWTON	83
	OKLAHOMA CITY	80
	TULSA	82
OR	ASHLAND	103
	CORVALLIS	110
	EUGENE	108
	MEDFORD	103
	PORTLAND	113
	SALEM	110
	SPRINGFIELD	108
PA	ALLENTOWN	118
	ALTOONA	112
	BETHLEHEM	120
	CARLISLE	111
	EASTON	120
	ERIE	110
	HARRISBURG	111
	HAZLETON	115
	JOHNSTOWN	111
	LANCASTER	107
	LEBANON	107
	PHILADELPHIA	133
	PITTSBURGH	110
	READING	119
	SCRANTON	114

© 2020 By BNi Publications Inc. ALL RIGHTS RESERVED

State	Metropolitan Area	Multiplier
PA	SHARON	110
	STATE COLLEGE	112
	WILKES BARRE	115
	WILLIAMSPORT	115
	YORK	109
PR	MAYAGUEZ	72
	PONCE	72
	SAN JUAN	72
RI	PROVIDENCE	125
SC	AIKEN	71
	ANDERSON	69
	CHARLESTON	71
	COLUMBIA	71
	FLORENCE	72
	GREENVILLE	70
	MYRTLE BEACH	69
	NORTH CHARLESTON	71
	SPARTANBURG	71
	SUMTER	71
SD	RAPID CITY	78
	SIOUX FALLS	84
TN	CHATTANOOGA	78
	CLARKSVILLE	77
	JACKSON	76
	JOHNSON CITY	78
	KNOXVILLE	75
	MEMPHIS	78
	NASHVILLE	77
TX	ABILENE	74
	AMARILLO	74
	ARLINGTON	73
	AUSTIN	76
	BEAUMONT	76
	BRAZORIA	76
	BROWNSVILLE	70
	BRYAN	75
	COLLEGE STATION	75
	CORPUS CHRISTI	74
	DALLAS	74
	DENISON	74
	EDINBURG	70
	EL PASO	73
	FORT WORTH	73
	GALVESTON	75
	HARLINGEN	70
	HOUSTON	71
	KILLEEN	73

© 2020 By BNi Publications Inc. ALL RIGHTS RESERVED.

State	Metropolitan Area	Multiplier
TX	LAREDO	72
	LONGVIEW	73
	LUBBOCK	75
	MARSHALL	69
	MCALLEN	70
	MIDLAND	73
	MISSION	70
	ODESSA	73
	PORT ARTHUR	76
	SAN ANGELO	73
	SAN ANTONIO	77
	SAN BENITO	70
	SAN MARCOS	75
	SHERMAN	70
	TEMPLE	73
	TEXARKANA	73
	TEXAS CITY	70
	TYLER	72
	VICTORIA	74
	WACO	73
	WICHITA FALLS	75
UT	OGDEN	77
	OREM	76
	PROVO	76
	SALT LAKE CITY	77
VA	CHARLOTTESVILLE	80
	LYNCHBURG	81
	NEWPORT NEWS	82
	NORFOLK	82
	PETERSBURG	79
	RICHMOND	80
	ROANOKE	83
	VIRGINIA BEACH	82
VT	BURLINGTON	80
WA	BELLEVUE	116
	BELLINGHAM	109
	BREMERTON	111
	EVERETT	115
	KENNEWICK	112
	OLYMPIA	114
	PASCO	112
	RICHLAND	112
	SEATTLE	116
	SPOKANE	93
	TACOMA	116
	YAKIMA	102

© 2020 By BNi Publications Inc. ALL RIGHTS RESERVED

State	Metropolitan Area	Multiplier
WI	APPLETON	112
	BELOIT	115
	EAU CLAIRE	112
	GREEN BAY	111
WI	JANESVILLE	115
	KENOSHA	116
	LA CROSSE	112
	MADISON	114
	MILWAUKEE	118
	NEENAH	112
	OSHKOSH	112
	RACINE	117
	SHEBOYGAN	111
	WAUKESHA	118
	WAUSAU	111
WV	CHARLESTON	114
	HUNTINGTON	117
	PARKERSBURG	112
	WHEELING	110
WY	CASPER	85
	CHEYENNE	86

© 2020 By BNi Publications Inc. ALL RIGHTS RESERVED.

BNi Building News
A DESIGN COST DATA COMPANY
DATA YOU CAN TRUST

© 2020 By BNi Publications Inc. ALL RIGHTS RESERVED

- A -

- B -

- C -

- D -

- N -

- O -

- P -

- T -

- U -

- V -

- W -

- X -

- Y -

- Z -

Other Estimating References from BNi Building News

BNi Building News

GENERAL CONSTRUCTION

2021 COSTBOOK
CSI MasterFormat

Over 12,000 unit costs provide you with cost coverage for all aspects of construction — from sitework and concrete to doors and painting. The *2021 BNi General Construction Costbook* is broken down into material and labor costs, to allow for maximum flexibility and accuracy in estimating. What's more, you get detailed square foot tables as well as a new section on Assemblies.

$129.95

BNi Building News

PUBLIC WORKS

2021 COSTBOOK
CSI MasterFormat

Now you can quickly and easily estimate the cost of all types of public works projects involving roads, excavation, drainage systems and much more. The *BNi Public Works Costbook 2021* is the first place to turn, whether you're preparing a preliminary estimate, evaluating a contractor's bid, or submitting a formal budget proposal. It provides accurate and up-to-date material and labor costs for thousands of cost items, based on the latest national averages and standard labor productivity rates, square-foot tables based on the cost-per-square-foot of hundreds of actual projects — invaluable data for quick, ballpark estimates.

$135.95

BNi Building News

FACILITIES MANAGER'S

2021 COSTBOOK
CSI MasterFormat

The *BNi Facilities Manager's Costbook 2021* is the first place to turn, whether you're preparing a preliminary estimate, evaluating a contractor's bid, or submitting a formal budget proposal. Labor costs are provided and are based on the prevailing rates for each trade and type of work. PLUS you also get equipment costs — including rental and operating costs, and square-foot tables based on the cost-per-square-foot of hundreds of actual projects.

$165.95

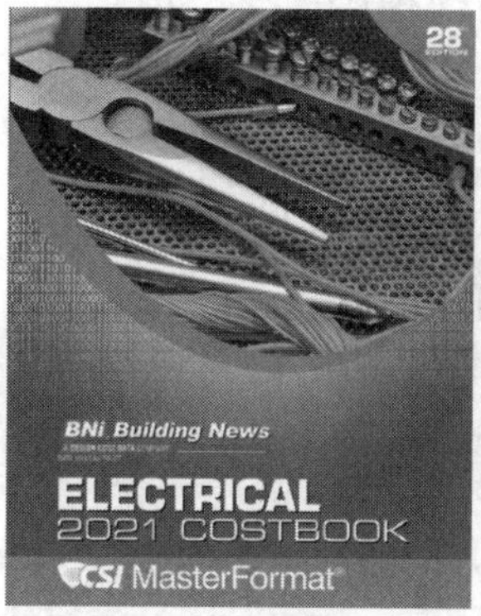

BNi Building News

ELECTRICAL

2021 COSTBOOK
CSI MasterFormat

From meter to duct, conduit to receptacle, The *BNi Electrical Costbook* is the first place to turn, whether you're preparing a preliminary estimate, evaluating a subcontractor's bid, or submitting a formal budget proposal. It puts at your fingertips accurate and up-to-date material and labor costs for thousands of cost items, based on the latest national averages and standard labor productivity rates. What's more, the *2021 BNi Electrical Costbook* includes detailed regional cost modifiers for adjusting your estimate to your local conditions.

$135.95

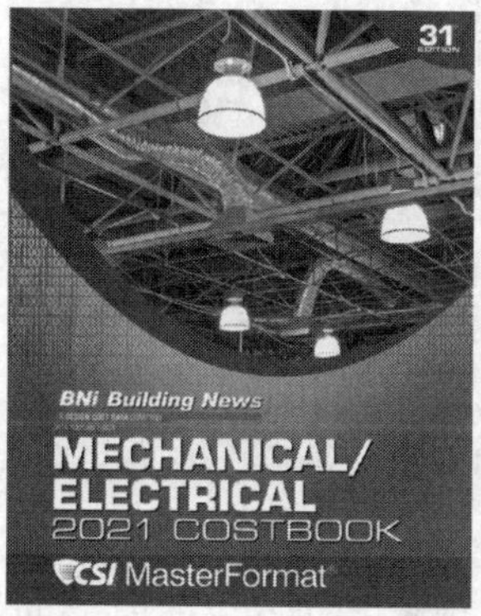

BNi Building News

MECHANICAL/ ELECTRICAL

2021 COSTBOOK
CSI MasterFormat

From pipe to duct to receptacle, this detailed reference book provides extensive coverage of the most technical aspects of building construction. With thousands of current, reliable mechanical and electrical costs at your fingertips, you can estimate quickly and accurately. Geographic Cost Modifiers allow you to tailor your estimates to specific areas of the country.

$135.95

BNi Building News

HOME BUILDER'S

2021 COSTBOOK
CSI MasterFormat

Here's the easy way to estimate the cost of all types of residential construction projects! Accurate and up-to-date material and labor costs for thousands of cost items, based on the latest national averages and standard labor productivity rates. Includes detailed regional cost modifiers for adjusting your estimate to your local conditions. Material costs are included for thousands of items based on current national averages (including allowances for transport, handling and storage).

$115.95

BNi Building News
HOME REMODELER'S
2021 COSTBOOK CSI MasterFormat

The *2021 BNi Home Remodeler's Costbook* lets you quickly and easily estimate the cost of all types of home remodeling projects, including additions, new kitchens and baths, and much more. The *BNi Home Remodeler's Costbook 2021* is the first place to turn, whether you're preparing a preliminary estimate, evaluating a subcontractor's bid, or submitting a formal budget proposal.

This all-new costbook puts at your fingertips accurate and up-to-date material and labor costs for thousands of cost items, based on the latest national averages and standard labor productivity rates.

$115.95

BNi Building News
REMODELING
2021 COSTBOOK CSI MasterFormat

Now you can quickly and easily estimate the cost of all types of home remodeling projects. You'll find yourself turning to the *2021 BNi Remodeling Costbook* again and again, whenever you're preparing a preliminary estimate, evaluating a subcontractor's bid, or submitting a formal budget proposal.

It puts at your fingertips accurate and up-to-date material and labor costs for thousands of cost items. Includes detailed regional cost modifiers for adjusting your estimate to your local conditions.

$135.95

BNi Building News
SQUARE FOOT
2021 COSTBOOK CSI MasterFormat

In this costbook you'll find over 80 detailed square foot cost studies for projects ranging from Civic Government Buildings to Hotels to Industrial and Office Buildings to Residential Buildings and so many more. For each building project you get a detailed narrative with background information on the specific project. In addition, you'll receive unit-in-place costs for nearly 15,000 items and materials used in all types of construction. For each item, you can see labor/equipment and material costs, all clearly broken out.

$115.95

BNi Building News
GREEN BUILDING SQUARE FOOT
2021 COSTBOOK CSI MasterFormat

The new *2021 BNi Green Building Square Foot Costbook* provides you with a comprehensive collection of 57 recent LEED and sustainable building projects along with their actual square-foot costs, broken down by CSI MasterFormat section. For each building, the *2021 BNi Green Building Square Foot Costbook* provides a detailed narrative describing the major features of the actual building, the steps taken to minimize the environmental impact both in its construction and its operation and a square-foot cost breakdown of each building component.

$115.95

DCR
INTERIORS SQUARE FOOT COSTBOOK
2021 EDITION

Unlike other building cost estimating resources, the *2021 DCR Interiors Square Foot Costbook* covers new construction, addition/renovation, adaptive re-use, and tenant build-out.

Each project is broken down by all its interior components presented on a cost-per-square-foot basis. It itemizes the materials used, along with their costs, to assist you in developing a conceptual estimate for interior construction.

$89.95

DCR
MECHANICAL/ ELECTRICAL SQUARE FOOT
2021 COSTBOOK

Unlike other building cost estimating resources, the *2021 DCR Mechanical/Electrical Square Foot Costbook* breaks down the MEP divisions and itemizes the materials used, along with their costs, in actual projects.

In addition, many of the cost studies in this book feature a variety of "green" technologies, such as hydronic pipe, geothermal heating and cooling, solar water heating, and hybrid ventilation air handlers. It lets you instantly see exactly how MEP costs relate to overall building costs, and how much they can vary from one project to another.

$89.95

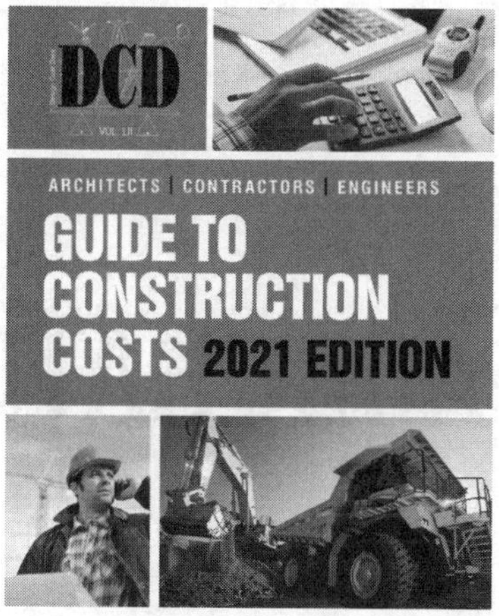

Architects Contractors Engineers

DCD GUIDE TO CONSTRUCTION COSTS

2021 Edition

Find thousands of thoroughly-researched construction material and installation costs you can use when making budgets, checking prices, calculating the impact of change orders or preparing bids. You'll find costs listed for scheduling, testing, temporary facilities, equipment, signage, and more. You'll find costs for every area of construction — from demolition and excavation to material and installation costs for finishes, flooring and much more.

$79.95

$89.95

Engineering News-Record

SQUARE FOOT

COSTBOOK 2021 EDITION

As you know, square-foot costs can vary widely, making them difficult to use for estimating and budgeting. The *2021 ENR Square-Foot Costbook* eliminates this problem by giving you costs that are based on actual projects, not hypothetical models.

This year's projects include:

Restaurants • Offices • Storage buildings • A theatre and studio arts building • A boxing club • A tennis center • A YMCA • A ministries center • And different types of residences

For each building project you get a detailed narrative with background information on the specific project. This lets you put the cost data into context and make appropriate adjustments to your own projections.

Developed in partnership with *Design Cost Data* and *BNi Building News*, this ready-reference costbook also features:

•Illustrations of each building type • A guide to 5-year cost trends for key building materials

• Detailed unit-in-place costs for thousands of items, from asphalt and anchor bolts to vents and wall louvers

STANDARD ESTIMATING PRACTICE

10th Edition

An invaluable "how to" reference manual on the practice of estimating construction projects.

Standard Estimating Practice presents a standard set of practices and procedures proven to create consistent estimates. From the order of magnitude, to conceptual design, design development, construction documents, to the bid and the various types of contracts you'll run up against: Every step is covered in detail — from specs and plan review to what to expect on bidding day.

Standard Estimating Practice also provides:

- Practical advice for using historical data in determining future production rates
- 14 key elements that will influence production rates on every project
- 10 important considerations when including construction equipment in an estimate
- 7 key costs that need to be included in a direct labor burden

$99.95

DCR

ARCHITECT'S SQUARE FOOT COSTBOOK

2021 EDITION

This manual presents detailed square-foot costs for 65 buildings tailored specifically to meet the needs of today's architect. For each project you get a complete cost breakdown of the included systems, so you can easily calculate the impact of modifications and enhancements on your own projects.

This *2021 Architect's Square Foot Costbook* gives you square-foot costs for a wide range of actual projects, from a senior living facility and low-income housing unit to a theater, a restaurant, and a corporate headquarters building. The themes of this year's Architect's Square Foot Costbook are Commercial, Educational, and Residential.

$89.95

Simpl•Est™

Square Foot

THE SIMPLE WAY TO CREATE ACCURATE SQUARE FOOT ESTIMATES!

Register (FREE!) on dcd.com and get:

- A full month of unlimited **FREE** access to Simpl•Est Square Foot
- A FREE copy of the 2021 DCD Guide to Construction Costs *(print and digital!)* as our gift to you!

Just go to www.dcd.com/simpl_est

With Simpl•Est Square Foot, you get a complete estimating solution

Simpl•Est Square Foot is amazingly powerful and amazingly SIMPLE!

You get unlimited access to a cost database featuring nearly 2,000 actual projects!

You get detailed square-foot cost breakdowns for schools, hospitals, municipal buildings, offices, muti-family units, restaurants, retail stores and more!

Powerful search tools let you quickly drill down to find projects that closely match what you or your client is planning ...

You can drill down on projects that match foundation types, roof types, interior and exterior wall types ... you can also drill down on new projects, renovations, adaptive re-use projects and more!

You can run multiple "what if" scenarios and answer the question "How much will it cost to build?"

Check out the costs of different building types. Get a reality check on your estimate. Demonstrate the impact of a change order. Estimate with greater confidence than ever before!

Simpl•Est®

Unit Price

Create a detailed estimate just by clicking on the items from the *BNi General Construction Costbook!* Set the quantities, change the labor rates, and a complete estimate will be printed instantly!

Simpl•Est®

Davis-Bacon

Fine tune your estimate with the most up-to-date prevailing wage rates for all commercial construction trades in hundreds of regions throughout the country.

Construction Project Log Book

The 365 Daily Work Log pages let you keep a detailed record for each day of the year. Additional forms such as Accident Reports and numerous checklists help make sure that you're covered.

With *Construction Project Log Book* those tasks are made as simple as possible.

Document every shipment, machinery rental, delivery, delay, and weather condition — all items that can affect productivity. With the interactive forms you can keep this information on your computer and enter new data daily into the interactive PDF forms. There are forms that actually do the math for you — eliminating typical mistakes.

$39.95

Maintenance Manager's Standard Manual 6th Edition

Since it was first published in 1993, the *Maintenance Manager's Standard Manual* has indeed become the STANDARD reference in the field.

This brand-new Sixth Edition brings it completely up to date, incorporating the latest technology and best practices in all aspects of maintenance management.

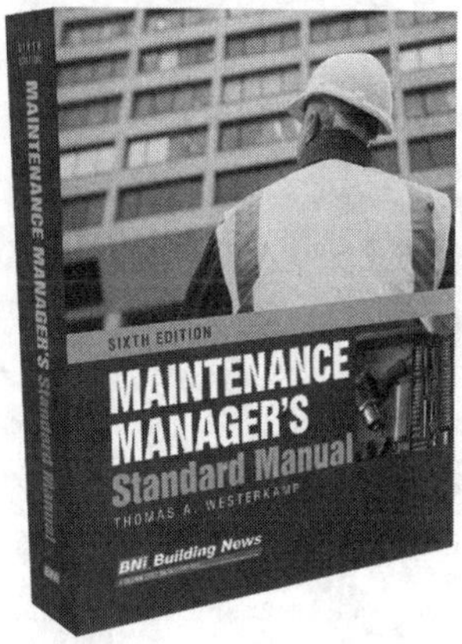

Whether you are a facilities manager, engineer, property owner, developer, or anyone else responsible for maintenance operations; not only does it give you all of the essential ingredients for understanding and carrying out successful day-to-day management of maintenance activities, it provides you with an integrated plan for continuous improvement of the maintenance function.

$99.95

100% Satisfaction Guaranteed!

If not completely satisfied, return the item within 30 days for a complete refund of the purchase price.

This has been the BNi policy for over 74 years.

Is there a title you need and can't find?

Give us a call at 1-888-264-2665; we'll be glad to help.

Order online: www.bnibooks.com

Find hundreds of construction references, forms and contracts to help you with your construction business.

NOTES